PULSE AND SWITCHING CIRCUIT ACTION

Other Books by the Author
PULSE AND SWITCHING CIRCUIT MEASUREMENTS
TRANSISTOR CIRCUIT ACTION

PULSE AND SWITCHING CIRCUIT ACTION

HENRY C. VEATCH

Instructor and Electronics Coordinator
San Leandro Adult School

Technical Writing Supervisor
International Video Corp.

McGRAW-HILL BOOK COMPANY
NEW YORK   ST. LOUIS   SAN FRANCISCO
DÜSSELDORF   JOHANNESBURG   KUALA LUMPUR
LONDON   MEXICO   MONTREAL
NEW DELHI   PANAMA   RIO DE JANEIRO
SINGAPORE   SYDNEY   TORONTO

TO MY FAMILY,
WHOSE HELP HAS MADE THIS
BOOK POSSIBLE.

**PULSE AND
SWITCHING CIRCUIT ACTION**

Copyright © 1971 by McGraw-Hill, Inc. All rights reserved. Printed in the United States of America. No part of this publication may be reproduced, stored in a retrieval system, or transmitted, in any form or by any means, electronic, mechanical, photocopying, recording, or otherwise, without the prior written permission of the publisher.

Library of Congress Catalog Card Number 70-147166

07-067386-1

456789 KPKP 7987654

CONTENTS

PREFACE ix

1 PULSE WAVEFORMS 1

    1-1 Pulse nomenclature 1
    1-2 Pulse aberrations 4
    1-3 The square wave 6
        Questions and problems 9

2 CIRCUIT THEOREM REVIEW 11

    2-1 Thevenin's theorem 11
    2-2 Norton's theorem 14
    2-3 The superposition theorem 16
    2-4 Millman's theorem 20
        Questions and problems 22

3 *RC* NETWORKS 24

    3-1 Capacitors 24
    3-2 Capacitor charge and discharge characteristics 25
    3-3 *RC* time constant 33
    3-4 *RC* coupling in pulse circuits 37
    3-5 *DC* restoration 48
    3-6 Typical *RC* circuits 51
        Questions and problems 64

4 RL NETWORKS — 66

- 4-1 Inductor energy storage — 66
- 4-2 Inductive time constant — 69
- 4-3 Pulse transformers — 70
- Questions and problems — 73

5 THE pn-JUNCTION DIODE — 77

- 5-1 The pn junction — 77
- 5-2 Reverse bias — 80
- 5-3 Forward bias — 82
- 5-4 Diode symbol — 84
- 5-5 Diode characteristics — 84
- Questions and problems — 85

6 DIODE CHARACTERISTICS — 87

- 6-1 General description — 87
- 6-2 Diode characteristics — 89
- 6-3 Diode applications — 93
- Questions and problems — 98

7 THE JUNCTION TRANSISTOR — 100

- 7-1 Fundamentals of transistor characteristics — 100
- 7-2 Transistor equivalent circuits — 103
- 7-3 Typical switching characteristics — 107
- Questions and problems — 108

8 TRANSISTOR CHARACTERISTICS — 110

- 8-1 The transistor as a switch — 110
- 8-2 The dc load line — 117
- 8-3 The basic switching circuit — 119
- 8-4 Circuit limits — 122
- Questions and problems — 125

9 TRANSISTOR SWITCHING CHARACTERISTICS 127

9-1 Transistor switches 127
9-2 Switching modes 149
9-3 Nonresistive loads 152
    Questions and problems 153

10 AMPLIFIER SWITCHING CIRCUITS 156

10-1 The CE (common-emitter) switching circuit 156
10-2 The CB (common-base) switching circuit 182
10-3 The CC (common-collector) switching circuit 184
    Questions and problems 189

11 LOGIC AND LOGIC CIRCUITS 192

11-1 Number systems 192
11-2 Logic symbology 202
11-3 Boolean algebra 214
11-4 Logic circuit design 238
11-5 Logic circuits 243
    Questions and problems 256

12 MULTIVIBRATORS 259

12-1 Introduction to regeneration 260
12-2 Relaxation oscillators 261
12-3 Bistable multivibrators 274
12-4 Monostable multivibrators 289
12-5 The Schmitt-trigger circuit 296
12-6 The blocking oscillator 304
    Questions and problems 306

13 COUNTERS 308

13-1 Binary up-counters 309
13-2 Permuted counters 317
13-3 The shift register 319
13-4 Ring counters 327
13-5 Binary up-down counters 332
    Questions and problems 338

14 MATRICES 340

    14-1 Decoding matrices   340
    14-2 Encoding matrices   349
           Questions and problems   350

15 SPECIAL CIRCUITS AND DEVICES 352

    15-1 Single-ended small-signal linear pulse amplifiers   352
    15-2 Push-pull linear pulse amplifiers   357
    15-3 Biased-up differential amplifiers   361
    15-4 CRT deflection amplifiers   364
    15-5 Nonlinear write amplifiers   369
    15-6 Sawtooth generators   372
    15-7 Staircase generators   377
    15-8 Inductively coupled circuits   383
    15-9 Special devices   388
           Questions and problems   407

16 INTEGRATED CIRCUITS 409

    16-1 Introduction   409
    16-2 Manufacturing processes   411
    16-3 Switching and gating IC's   425
    16-4 IC multivibrators   431
    16-5 LSI and MSI   441
    16-6 MOS integrated circuits   445
           Questions and problems   454

APPENDIX: BOOLEAN ALGEBRA THEOREMS   455

GLOSSARY   460

BIBLIOGRAPHY   462

ANSWERS TO ODD-NUMBERED QUESTIONS AND PROBLEMS   463

INDEX   468

PREFACE

The purpose of this book is to remove much of the mystery surrounding transistors used in pulse and switching circuits and to provide the reader with a clear, easy-to-understand method of working with them. Transistor circuits often seem difficult to understand, but many of their attributes can be described in very simple terms if properly approached. The underlying thought in the development of this book is to immediately eliminate all impractical and complicated methods and replace them with down-to-earth, simplified, practical ones.

All circuit descriptions given herein can be verified in the laboratory, if desired. Any analytical material that cannot be checked on a volt-ohm-milliammeter or an oscilloscope does not appear in this book. As these two instruments are the mainstay of the science of electronics, it is felt that all circuit descriptions of any consequence can be handled in terms of the measurement techniques of these instruments. To assist in the development of this practical approach, a laboratory manual is offered that parallels this text and that will be of great assistance in providing a "hands-on" experience for the student.

This book, by providing simple but factual material, will enable the reader to comprehend basic transistor circuits as used in typical pulse and switching applications. For the most part the math used is no more difficult than Ohm's law. Only a basic knowledge of direct- and alternating-current circuits is presumed, together with an introductory-level knowledge of linear amplifier circuits.

The book is designed to be used in any course where a clear, detailed, and completely practical explanation of basic semiconductor switching circuits is required. Junior colleges, technical institutes, industrial in-plant and home-study programs, adult schools, and individual home-study plans will benefit from this approach. The correlated lab manual, "Pulse and Switching Circuit Measurements," is available from the publisher for use with this text.

The circuits used as examples are taken wherever possible, from actual commercial equipment. Accordingly, they can be built in the laboratory, which will greatly facilitate learning the principles of semiconductor circuits. One complete chapter is devoted to a rather detailed exposition of the integrated circuit. I consider the integrated circuit of such importance that all the necessary material, covering several kinds of circuits, has been gathered together in a single chapter to make reference to it easier. I hope that the somewhat unusual approach taken in this chapter will make the understanding of integrated circuits much easier.

This book is based upon the assumption that the student has a reasonable knowledge of transistorized linear circuits, such as is found in the author's earlier book, "Transistor Circuit Action," also published by McGraw-Hill Book Company. Thus, there is little herein that relates to linear circuits, with the exception of certain circuit attributes common to both types of circuits.

Every effort has been made to keep the material updated. The latest information available has been used, consistent with normal publishing deadlines. Much of this information would not have been available were it not for the cooperation of several firms. Fairchild Semiconductors, Texas Instruments, Incorporated, Motorola Semiconductors and Component Division, and Singer, Friden Division were especially helpful in terms of both material and time. To them goes the author's sincere appreciation. I am also indebted to E. E. Pollock. His penetrating comments, useful advice, and timely corrections were much valued. Many of his suggestions have been incorporated into the book. Particular thanks must be given to Miss Rose Ann Saenz for helping to type the manuscript.

Henry C. Veatch

1

PULSE WAVEFORMS

1-1 PULSE NOMENCLATURE

It has been estimated that perhaps 50% of all electronic equipment operates as it does because of its ability to use a pulse as a signal. Since pulse circuitry is so prominent in the field of electronics, one needs to spend as much time studying pulse circuits as he does studying linear circuits.

A *pulse* has been defined as an *abrupt discontinuity in a voltage or current of relatively short duration*. A pulse may be repetitive, or it may be a one-shot affair. A typical pulse is shown in Fig. 1-1, and each of the various parts is labeled with an appropriate designation. This is a negative-going pulse operating between the levels of 0 (ground) and −6 volts. Each of the designations in this figure will now be discussed, along with a few others.

The *rise time* is measured between the 10 and 90% points on the curve, as indicated. The reason for this is that between 0 and 10% and between 90 and 100% the voltage change occurs very slowly; it is difficult to measure these points with respect to time with any degree of accuracy. Owing to the small stray capacitance and inductance in any circuit, it is impossible to experience an instantaneous change in voltage or current.

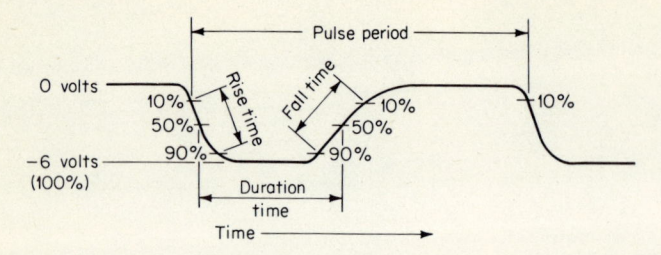

Fig. 1-1. Pulse nomenclature.

Some period of time must elapse before the new level is attained. The rise time is the measure of this change and is that change which occurs first in time (as opposed to the fall time).

The *duration time* is a measure of the length of time the pulse height stays above the 50% level. (There are also other ways of specifying this.) Usually, the duration time can be measured in a straightforward manner, but if the pulse has a peculiar shape, this might become difficult. In a case such as this, one would have to specify the exact meaning of the measurement in order to avoid confusion.

The *fall time* is analogous to the rise time in that it, too, is measured between the 10 and 90% points on the curve. It is the length of time required to decrease from the maximum to the minimum level.

The *pulse period* is the measure of the time required for one complete cycle of a repetitive waveform. It is measured from one point on the waveform to the next corresponding point, say, from the 50% level on one pulse to the 50% level on the next pulse. It includes the rise time, the duration time, the fall time, and the rest period between pulses.

The *repetition rate* of a series of recurrent pulses is a measure of the number of pulses per unit time, usually per second. Since frequency is defined as cycles per second and since the period corresponds to a cycle, it can be said that the frequency of a square wave is so many cycles per second. Thus, the two expressions, frequency and repetition rate, are synonymous in this case.

The *amplitude* of a pulse is the maximum value to which it rises, or the 100% value. The direction of the pulse may be either positive- or negative-going, and it may originate from either a positive value, a negative value, or ground. Some of the possibilities are shown in Fig. 1-2, with typical values given. The shape of a pulse may be nearly anything *except* that of a sine wave. Figure 1-3 shows a few possibilities.

The *duty period* of a pulse is a measure of the "on" time related to the total pulse time, or period, and is the ratio of the first by the second. For

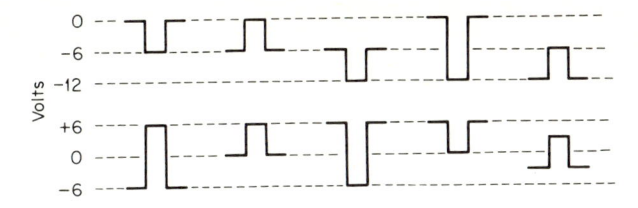

Fig. 1-2. Various pulse amplitudes.

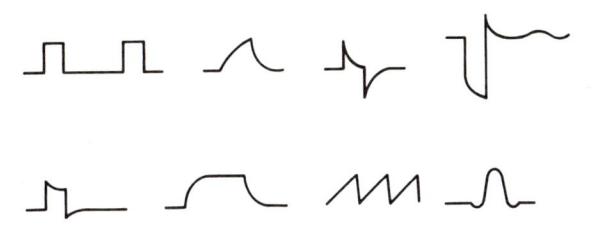

Fig. 1-3. Various pulse shapes.

instance, the pulse shown in Fig. 1-4 has a duty period of

$$P_d = \frac{T_{on}}{\text{period}} = \frac{10}{100} = 0.1$$

That is, the pulse is on for $\frac{1}{10}$ of the total period and would be said to have a duty period of 10%.

The *peak amplitude* of a pulse is a peak measurement from the most negative excursion to the most positive excursion or vice versa. The pulses shown in Fig. 1-4 have an amplitude of -6 volts.

In a practical case, a pulse is not usually as perfect as shown up to this point. Figure 1-5 illustrates a positive nanosecond pulse as it might actually appear on an oscilloscope. The anomalies are clearly indicated.

Very often, a pulse train rides on top of a fixed dc level, as indicated in Fig. 1-6. This figure shows that the pulse, during the rest period, is not at ground potential. Here the voltage at the base line is called the *offset* voltage. An offset may be either a positive or negative voltage.

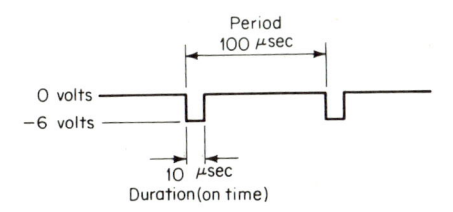

Fig. 1-4. Pulse duration versus pulse period.

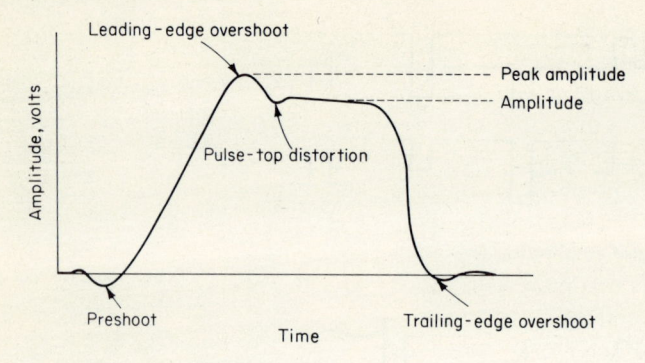

Fig. 1-5. Pulse distortion.

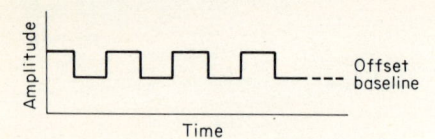

Fig. 1-6. Pulse offset.

1-2 PULSE ABERRATIONS

Any deviation from an ideal pulse is caused by some deviation from an ideal circuit. It is useful to be aware of at least some of the circuit effects that can degrade a theoretically ideal pulse. Because some of these aberrations are discussed elsewhere in this book, only a brief listing and discussion will be given here.

Probably the most often encountered departure from the ideal pulse is that of *droop* or *sag*. A drooping pulse is shown in Fig. 1-7, where the

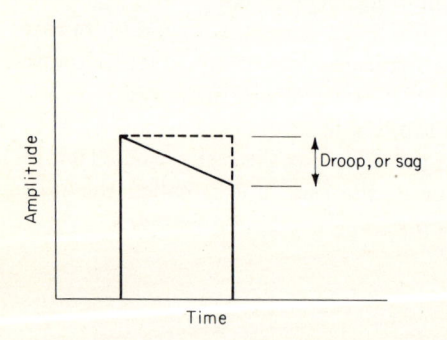

Fig. 1-7. Pulse droop, or sag.

dotted line represents the ideal. This waveform is produced by passing an ideal waveform through a circuit that has poor low-frequency response. The leading and trailing edges represent the high-frequency components in the square wave, while the flat top represents the low-frequency components of the square wave.

Another departure from the ideal pulse train is the *jitter*, which causes the individual pulses to be spaced at slightly different intervals or to have slightly different amplitudes. Any train of pulses suffers to some extent from jitter since at present it is impossible to achieve absolute perfection. Jitter is defined as <u>*any random variation in the repetition rate, width, amplitude, or waveform of a train of pulses.*</u> One of the most common sources of jitter is the ripple voltage remaining on the power-supply busses. Because of this ripple, the voltages are not absolutely constant and so affect the transmission of the pulses differently at different times. Figure 1-8 illustrates both time and amplitude jitter.

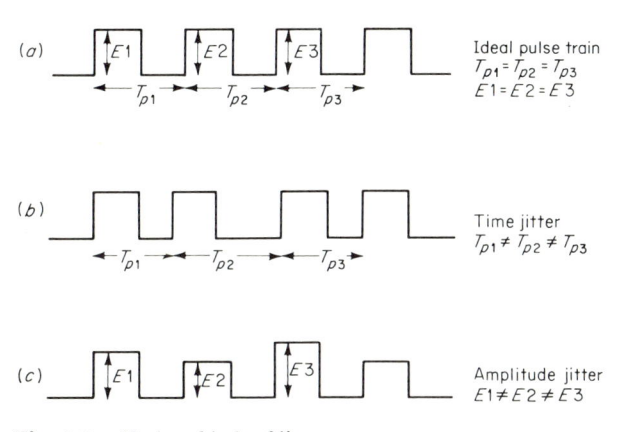

(a) Ideal pulse train
$T_{p1} = T_{p2} = T_{p3}$
$E1 = E2 = E3$

(b) Time jitter
$T_{p1} \neq T_{p2} \neq T_{p3}$

(c) Amplitude jitter
$E1 \neq E2 \neq E3$

Fig. 1-8. Various kinds of jitter.

Many other abnormalities exist. Such conditions as preshoot, overshoot, top distortion, base-line distortion, rise- and fall-time distortion and degradation, as well as others, all detract from the ideal pulse. High-quality equipment is specifically designed to minimize these effects, the degree of success depending upon the current state of the art.

A final example of pulse aberrations is the response of a linear pulse amplifier to a square-wave input. A linear amplifier designed to work with pulses is often, but not always, called a *video amplifier*. Video frequencies are a combination of low and high frequencies and include both sine and square waves. Any amplifier that must handle such a wide range of signals is much more complex than a simple single-frequency amplifier. Such an amplifier is also called a *wideband amplifier*, signifying that low, medium and high frequencies are to be amplified equally.

6 PULSE AND SWITCHING CIRCUIT ACTION

One way of testing such an amplifier is to apply a square wave to the input. The resultant output reveals such defects as poor low-frequency response, poor high-frequency response, phase lead or lag, or departures from equal amplification at all frequencies within the passband. Figure 1-9 shows the results of some of these anomalies.

Applied square wave; also ideal output. (Disregard phase inversion.)

Phase lead at low frequencies; poor low-frequency response.

Phase lag at low frequencies.

Amplitude rises at low frequencies. (No phase shift.)

Amplitude falls at low frequencies. (No phase shift.)

Poor high-frequency response. (No phase shift.)

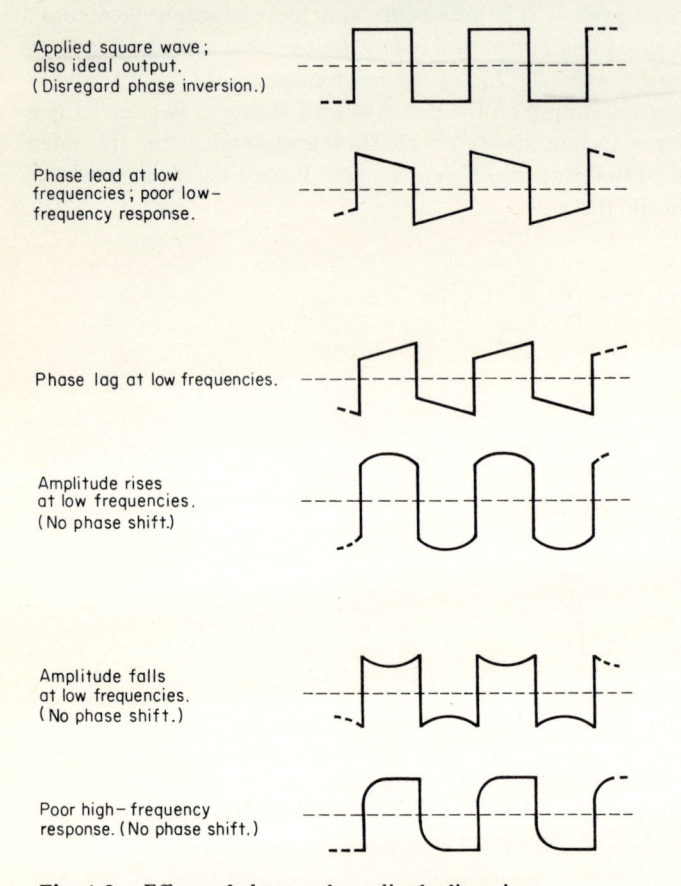

Fig. 1-9. Effects of phase and amplitude distortion.

1-3 THE SQUARE WAVE

Strictly speaking, a square wave is one whose sides are all equal in length. However, the geometry of such a wave depends on the scale used to graph the waveform, hence any rectangular pulse is usually referred to as a square wave.

One way of producing square waves is to use a switching circuit to *chop* the direct current from the power supply. This is a very practical and often-used method. If, however, a square wave is resolved into its constituent parts, it will be noted that these parts are really sinusoidal by nature. In other words, if several sine waves are combined in the correct way, the result is a very presentable square wave. If an infinite number of sine waves of the proper size and time relation are properly combined, the result is a perfect square wave, which is of theoretical interest only.

Figure 1-10 illustrates the manner in which various sine waves can be combined to form a square wave. To keep the example as simple as possible, only three waveforms are combined, and therefore the result only begins to appear as a reasonable square wave. If several additional steps were to be shown, the result would have a flatter top and the vertical sides would become steeper. Figure 1-10a shows a half-cycle of a sine wave at some frequency. The waveform in Fig. 1-10b shows a frequency that is three times greater than that in Fig. 1-10a. Finally, in Fig. 1-10c a frequency that is five times greater is shown.

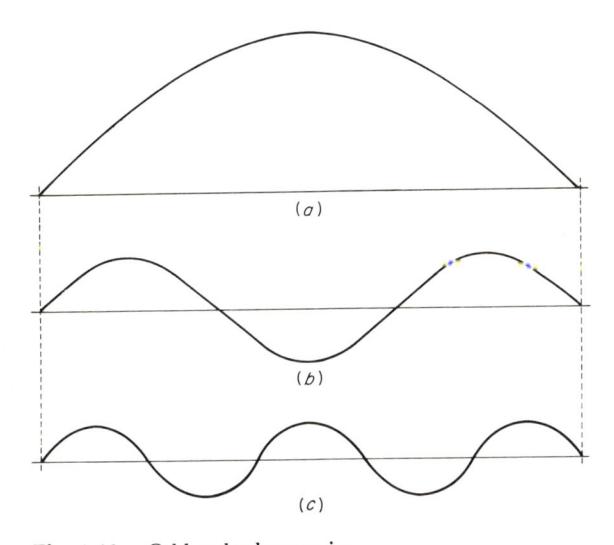

Fig. 1-10. Odd-order harmonics.

If the first two waveforms are combined, the result is as shown in Fig. 1-11. Combining all three waveforms results in the waveform in Fig. 1-12. Note that this closely approximates a square wave. By adding more odd-order multiples of the lowest frequency (seven times, nine times, etc.), properly phased and of the correct amplitude, the end result would be a nearly perfect square wave.

The lowest frequency mentioned above is called the *fundamental* frequency, and the others are called the *harmonics* or *overtones*. It can now be

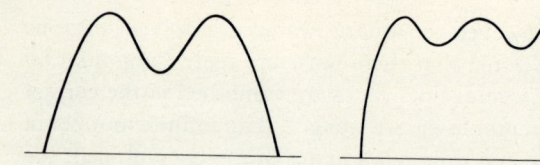

Fig. 1-11. Summation of 1st- and 3d-order harmonics.

Fig. 1-12. Summation of 1st-, 3d-, and 5th-order harmonics.

appreciated that the fundamental frequency determines the lowest frequency involved while the highest harmonic determines the highest frequency inherent in the square wave. Hence, to pass a nearly perfect square wave through any kind of circuit, the circuit must be able to pass the lowest as well as the highest frequencies inherent in the square wave, with no loss or distortion. If any loss or distortion occurs, the output will not be an exact replica of the input.

The circuit of Fig. 1-13 is helpful in visualizing the loss in a circuit. A

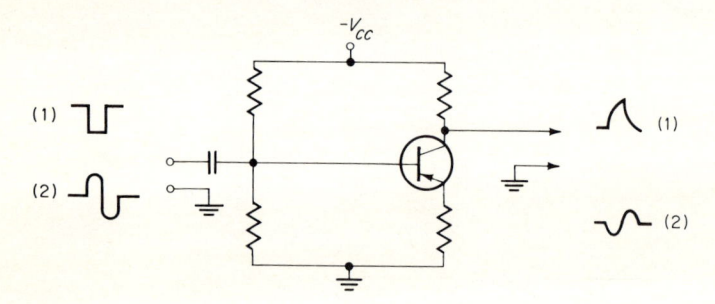

Fig. 1-13. Amplifier distortion—square waves and sine waves.

transistor amplifier is shown with two kinds of inputs. Input 1 is a square wave, while input 2 is a sine wave. For purposes of explanation, the circuit will be explained only in very simple terms at this point. Without regard for the amplifier itself, our attention is directed to the waveforms shown. The sine wave is unique among all other waveforms. When passed through a circuit with poor frequency response, *the waveform is not altered*. The only change that occurs is that the amplitude becomes smaller. This is not true of the square wave (or any other waveshape). Also note that the shape of the square wave is much different at the output than at the input.

Each of these effects is related to the other since each is related to the frequency response of the circuit itself. The measure of the highest frequency inherent in a square wave is intimately related to the rise time of the wave. By the same token, the highest frequency that a circuit can pass without undue loss of amplitude is called the *upper 3-db point* or f_2.

An example of the interrelation between the rise time t_r and the upper 3-db point f_2 will serve to clarify this point. The manufacturer of a widely used oscilloscope advertises a rise time for his instrument of 90 nsec, or 0.09 μsec. That is, under no circumstances can the instrument show a pulse with a rise time faster than 0.09 μsec. If a sinusoidal waveform is impressed, one might inquire what maximum frequency the instrument is capable of indicating with no more than a 3-db loss. Use of either of the following relationships will allow one to be found if the other is known.

$$t_r = \frac{0.35}{f_2} \quad \text{or} \quad f_2 = \frac{0.35}{t_r}$$

In the case of the oscilloscope,

$$f_2 = \frac{0.35}{t_r} = \frac{0.35}{0.09 \times 10^{-6}} \cong 3.9 \text{ MHz}$$

QUESTIONS AND PROBLEMS

1-1 Given the accompanying waveform, find the rise time of this pulse.

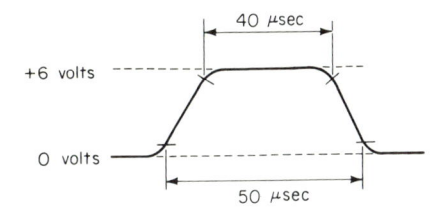

1-2 Given the accompanying waveform, find the rise time exhibited by this pulse.

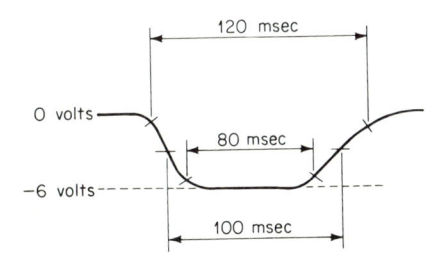

1-3 Refer to Fig. 1-4. Change the pulse period from 100 to 500 μsec. Determine the repetition rate.

1-4 Refer to Fig. 1-4. Change the pulse period from 100 to 1500 μsec and the duration time to 150 μsec. Determine the repetition rate.

1-5 Refer to Fig. 1-4. Change the pulse period from 100 to 10 μsec and the duration time to 1 μsec. Determine the duty period.

1-6 Refer to Fig. 1-4. Change the pulse period from 100 to 1.5 μsec and the duration time to 0.010 μsec. Find (a) the repetition rate and (b) the duty period.

1-7 A certain amplifier is said to have a high-frequency cutoff of 1 MHz (1,000,000 Hz). Determine the fastest rise time it is capable of reproducing.

1-8 Refer to Question 1-7. The amplifier is modified so that its upper 3-db point is 250 kHz. Determine the maximum rise time.

1-9 An amplifier exhibits a maximum rise time of 0.035 μsec (35 nsec). Determine the upper 3-db point.

1-10 An amplifier exhibits a maximum rise time of 0.015 msec. Determine the upper 3-db point.

2
CIRCUIT THEOREM REVIEW

If the reader feels quite proficient in such areas as Thevenin's theorem, Norton's theorem, the principle of superposition, etc., he may bypass this chapter. However, so much of the circuit description that occurs later in the text is based upon these powerful circuit-analysis methods, that a brief coverage of these is given for the convenience of those who may wish to refresh their memories. It is assumed that the reader has a good foundation in such subjects as Ohm's law and Kirchhoff's law, for these are the very basis of all electronic circuit analysis.

2-1 THEVENIN'S THEOREM

This useful theorem is important in the simplification of circuits containing several elements. Its use allows an equivalent circuit to be constructed consisting of a constant-voltage generator in series with a single impedance that drives the load. An ideal constant-voltage generator is one that has zero internal impedance. Its output voltage is therefore constant with changing current demands.

Probably the most widely known equivalent circuit yielded by Thevenin's theorem is the equivalent circuit of a triode vacuum tube, which is illus-

12 PULSE AND SWITCHING CIRCUIT ACTION

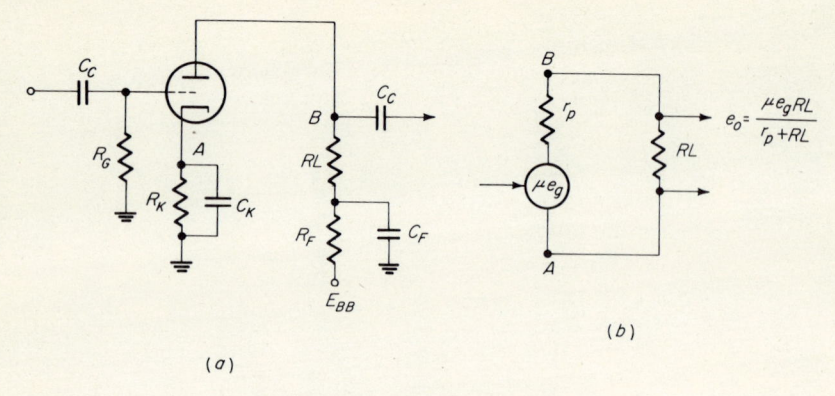

Fig. 2-1. Illustration of Thevenin's theorem. (*a*) The original circuit; (*b*) the Thevenin-equivalent circuit.

trated in Fig. 2-1. For signal (ac) conditions Fig. 2-1*b* is the Thevenin-equivalent circuit of that shown in Fig. 2-1*a*. Because the internal impedance of the tube is quite appreciable, it must be taken into account.

Although it is not our purpose to delve into vacuum-tube theory to any extent, the circuit shown is a perfect example of the application of Thevenin's theorem. Briefly, the equivalent circuit is derived as follows. In a triode amplifier the maximum theoretical voltage gain is equal to the mu (μ) of the tube. However, in practice, the measured amount of gain is always less than this, and the reason is explicit in the Thevenin-equivalent circuit. The tube is considered to represent a perfect generator, whose output is equal to μ (amplification factor) times e_g (input voltage), or μe_g. This perfect generator is considered to have zero internal impedance.

Now, because the true internal impedance must be considered, it is shown as an external resistance in series with the generator. It can now be appreciated that internal impedance must be considered as a separate part of the circuit. Note that it still is a part of the generator circuit since it is shown between points A and B. Only the portions of the complete circuit through which signal current flows and across which a voltage drop occurs are shown. Hence, only r_p, RL, and the generator are given in the equivalent circuit. All other components are either bypassed (R_K and R_F) or their reactance is essentially zero (C_K and C_F). From this equivalent circuit, then, the output-voltage equation can be derived. A voltage is developed by signal current across RL that is the result of total current. Because r_p is usually appreciable, a significant portion of the total available voltage is dropped across it, leaving something less to be dropped across RL.

RL and r_p are treated as a simple voltage-divider circuit, where the size of the two resistors determines the percentage of total voltage dropped across each resistor. In a dc circuit, the voltage division would be expressed as follows:

$$E_{RL} = \frac{RL}{r_p + RL} E_{\text{bat}}$$

In the tube circuit, the following signal values are used:

$$e_o = \frac{RL}{r_p + RL} \mu e_g = \frac{\mu e_g RL}{r_p + RL}$$

This equation states that the maximum possible output voltage is μ times e_g, and this is reduced by the relation between RL and r_p.

As a numerical example, assume that $\mu = 20$, $r_p = 7$ kilohms, and $RL = 14$ kilohms. The voltage output for a 1-mv (rms) input is then easily found:

$$e_o = \frac{\mu e_g RL}{r_p + RL} = \frac{20 \times 0.001 \times 14 \text{ kilohms}}{7 \text{ kilohms} + 14 \text{ kilohms}} = \frac{280}{21 \text{ kilohms}} = 13 \text{ mv}$$

Hence, the voltage gain of this tube in this circuit is 13. Without the Thevenin-equivalent circuit it would be very difficult to derive a mathematical expression for voltage gain.

As a further example of this important principle of circuit analysis, consider the simple example shown in Fig. 2-2. The source, or battery, is shown under several conditions of load or no-load. E_{oc} is the open-

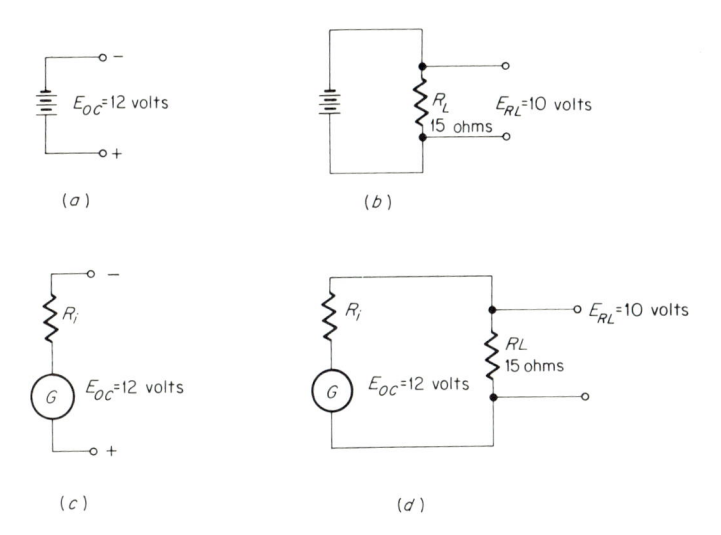

Fig. 2-2. A Thevenin-equivalent circuit using batteries.

circuit voltage (no-load) while E_L is the loaded output voltage. To represent the two conditions simultaneously (load and no-load), the Thevenin-equivalent circuit must be developed; in effect, it allows the conditions in Fig. 2-2a and b to be specified in a single drawing.

Given only the above information, the value of the internal resistance of the battery must be found. It is this internal resistance, of course, which causes the drop in terminal voltage as the load is applied. The physical relation of the internal resistance to the battery is shown in Fig. 2-2c, but the entire Thevenin-equivalent circuit in Fig. 2-2d must be used to find the value of it. The drop across R_i must be

$$E_{R_i} = E_{oc} - E_{RL} = 12 - 10 = 2 \text{ volts}$$

Now, the total circuit current can be found since the values of RL and E_{RL} are known:

$$I_T = I_{RL} = \frac{E_{RL}}{R_L} = \frac{10}{15} = 0.667 \text{ amp}$$

At this point, the *total* circuit resistance can be determined from the total current and total voltage:

$$R_T = \frac{E_T}{I_T} = \frac{12}{0.667} \cong 18 \text{ ohms}$$

Finally, R_i can be given a value:

$$R_i = R_T - R_L = 18 - 15 = 3 \text{ ohms}$$

This, then, is a much more exact representation of the actual circuit action than, for example, Fig. 2-2b.

2-2 NORTON'S THEOREM

This theorem is useful in analyzing circuits in which there is a constant-current generator. It allows a complex network to be reduced to an equivalent circuit consisting of an ideal constant-current generator shunted by a single impedance. The ideal source, or generator, is considered to have an infinitely high internal impedance. In the practical case, however, this ideal cannot be attained. Hence, the Norton-equivalent circuit assumes an ideal generator shunted by an equivalent impedance, or resistance, that exactly represents the circuit action.

Consider Fig. 2-3, which illustrates a constant-current generator with an internal resistance of 5 megohms and a load of 1 megohm. The load current I is delivered to the load in some amount determined by both R_G

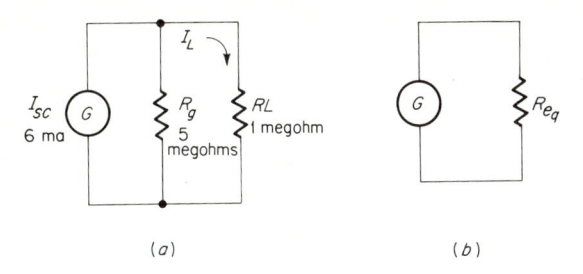

Fig. 2-3. (*a*) A constant-current generator; (*b*) the Norton-equivalent circuit.

(the generator resistance) and R_L. The short-circuit current I_{sc}, which is shown as 6 ma, is the value that would flow if $R_L = 0$ ohms. To evaluate for the unknown value I_L, the following relation is used:

$$I_L = \frac{I_{sc} \times R_G}{R_G + R_L} = \frac{0.006 \times 5{,}000{,}000}{6{,}000{,}000} = 5 \text{ ma}$$

Reducing the circuit to its Norton equivalent yields the circuit shown in Fig. 2-3*b*. Here, the equivalent resistance R_{eq} is the effective combination of R_G and R_L:

$$R_{eq} = \frac{R_G \times R_L}{R_G + R_L} = \frac{5{,}000{,}000 \times 1{,}000{,}000}{5{,}000{,}000 + 1{,}000{,}000} = 833 \text{ kilohms}$$

An interesting and illuminating point regarding such a circuit is the essentially constant-current output with a changing load. For example, if the load is changed to 500 kilohms, the output current for this new condition can be easily found:

$$I_L = \frac{I_{sc} \times R_G}{R_G + R_L} = \frac{0.006 \times 5{,}000{,}000}{5{,}000{,}000 + 500{,}000} \cong 5.45 \text{ ma}$$

Thus, by halving the load resistance, the current changed by only 450 µa. This, of course, is the reason for the name *constant-current generator*. Only when the load resistance approaches the value of the internal resistance will the current change by any appreciable amount.

It is important to realize that, under some conditions, even a relatively low-impedance source can be considered a constant-current generator if the load impedance is one-fifth of the generator impedance or smaller. Hence, a generator having an internal impedance of 100 ohms will appear to be a constant-current generator if its load is 20 ohms or less. Then, a small change in load will yield virtually no change in load current. The same generator, on the other hand, will act as a constant-voltage generator if its load resistance is 500 ohms or more. In this case, a reasonable change in load resistance will cause a negligible change in load voltage.

2-3 THE SUPERPOSITION THEOREM

This important theorem is probably one of the most useful in simplifying more or less complex networks. Basically, the theorem states that the current flowing at any point in a complex circuit can be determined by considering one source at a time, with all other generators replaced with their internal impedances. The actual current, then, is the sum of the individual currents that would flow considering one generator at a time.

The first example is given in Fig. 2-4. If it is necessary to find the drop across $R3$, the principles of superposition can be applied as follows. First, replace one of the sources ($B2$, for example) with its internal resistance, which for simplicity we shall assume to be zero. This is now a simple series-parallel circuit, as shown in Fig. 2-4b. Using this simple equivalent, the total current is found:

$$I_T = \frac{E}{R1 + R2R3/(R2 + R3)} = \frac{5}{100 + 50} \cong 33.33 \text{ ma}$$

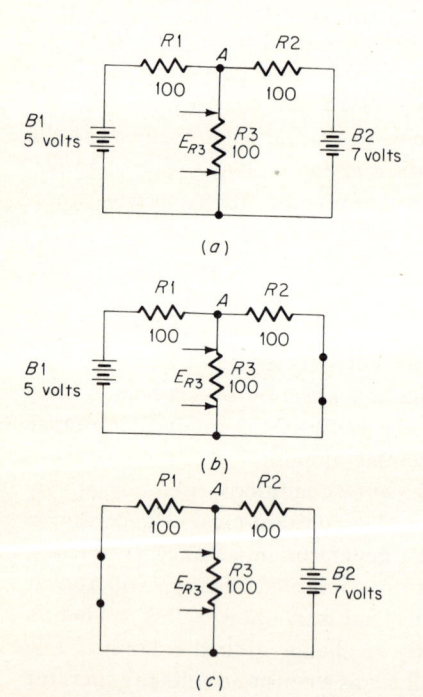

Fig. 2-4. A circuit illustrating the principle of superposition.

Next, the branch current through R3 is found:

$$I_{R3} = I_T \frac{R2}{R2+R3} = 33.33 \text{ ma} \frac{100}{200} = 16.67 \text{ ma}$$

This, then, is the current which would flow through R3 if B2 were to be removed and replaced with its internal resistance.

Now, the other source is removed and replaced with its internal resistance, which again shall be considered zero. The equivalent circuit shown in Fig. 2-4c can be used to determine the current through R3 for this condition:

$$I_T = \frac{E}{R2 + R1R3/(R1+R3)} = \frac{7}{150} = 46.66 \text{ ma}$$

$$I_{R3} = I_T \frac{R1}{R1+R3} = 0.04666 \frac{100}{200} = 23.33 \text{ ma}$$

Noting that in both cases current will flow upward, causing point A to become more positive, the two interim values of current are therefore additive. Hence, the true value of current through R3 is the sum of the interim values:

$$I_{R3} = 16.67 + 23.33 = 40.0 \text{ ma}$$

The drop across R3, then, is easily found:

$$E_{R3} = I_{R3} \times R3 = (40.0 \times 10^{-3})100 = 4.0 \text{ volts}$$

Noting the direction of current flow, it is obvious that the upper end of R3 is more positive in relation to the lower end.

The circuit just described can be drawn in a different way, as shown in Fig. 2-5. This is the manner of presentation that is usually used on schematic diagrams, and the correlation between the two methods should be understood.

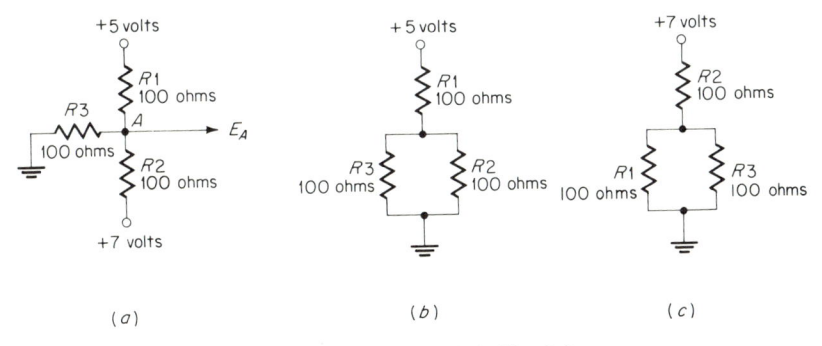

Fig. 2-5. Alternate way of drawing the circuit in Fig. 2-4.

18 PULSE AND SWITCHING CIRCUIT ACTION

A shortcut method is often used for multiple-source circuits. The circuit given in Fig. 2-6 will be analyzed for the voltage at *A*, relative to ground, by this method. To reduce the number of steps required to complete the solution, the equivalent circuit is altered slightly from the preceding example. Note Fig. 2-6a; a break is first made in a convenient

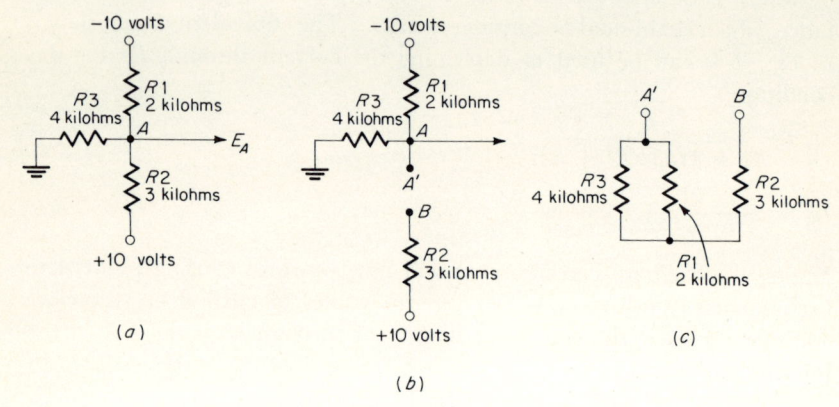

Fig. 2-6. An alternate approach to superposition.

branch that will allow the voltage at the required point to be determined. Because now there is no current flowing in this leg, the potential difference across the break can easily be determined:

$$E_{A'} = -10 \frac{R3}{R3 + R1} = -6.67 \text{ volts}$$

$$E_B = +10 \text{ volts (by inspection)}$$

$$E_B - E_{A'} = 10 - (-6.67) = 16.67 \text{ volts}$$

The current that actually flows in this break, when there is no break, is of the same value as that produced by the above voltage (16.67 volts) applied across the equivalent resistance *as viewed from the break*, looking back into the circuit. Hence, the resistance at terminals *A'* and *B* must be found. The equivalent circuit at *c* is useful to determine this value.

$$R_T = R2 + \frac{R1 R3}{R1 + R3} = 3 \text{ kilohms} + \frac{4 \text{ kilohms} \times 2 \text{ kilohms}}{4 \text{ kilohms} + 2 \text{ kilohms}}$$

$$= 4.33 \text{ kilohms}$$

The actual current flowing in *R2* is now found by the following expression:

$$I_{A'-B} = I_{R2} = \frac{E_{A'-B}}{R_T} = \frac{16.67}{4300} \cong 3.88 \text{ ma}$$

Once the current through R3 is found, the voltage drop can be calculated:

$E_{R2} = I_{R2} \times R2 = 0.00388 \times 3000 \cong 11.6$ volts

Finally, the voltage at A, relative to ground, can be found:

$E_A = 10 - 11.6 \cong -1.6$ volts

By using this method, all generators are replaced with their internal resistance in one step, thus often reducing and simplifying the work.

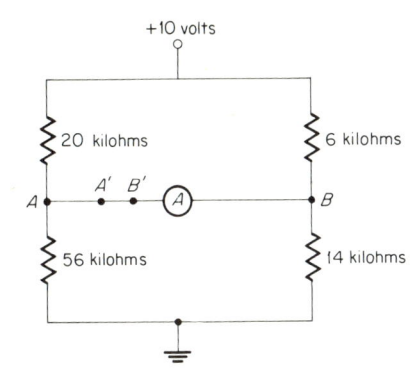

Fig. 2-7. A bridge circuit to be analyzed by superposition.

The bridge circuit shown in Fig. 2-7 will be used as the next example. This circuit is solved in much the same manner as the previous one. First, a break is made in the circuit at the points labeled A' and B' (Fig. 2-8). The voltage difference is then found between these two points. It

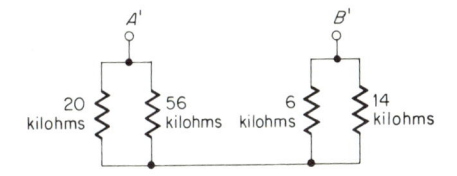

Fig. 2-8. The bridge circuit viewed from A' and B'.

will be instructive for the reader to verify the following results. Find the branch current flowing through the meter.

$E_{A'} = 7.368$ volts
$E_{B'} = 7.0$ volts
$E_{A-B} = 0.368$ volt
$R_{eq} = 18.94$ kilohms
$I_{branch} = 19.4$ μa

2-4 MILLMAN'S THEOREM

Millman's theorem is found to be of greatest use in solving for unknowns in circuits that have a large number of sources. When more than two sources are evident in a circuit, other methods, although satisfactory, are often tedious. Replacing each generator, or source, with its internal resistance and then computing a set of figures for each condition can consume much time. In many cases, this is not necessary, and Millman's theorem can then be used to good advantage.

As a first example, a simple circuit is shown in Fig. 2-9. While other

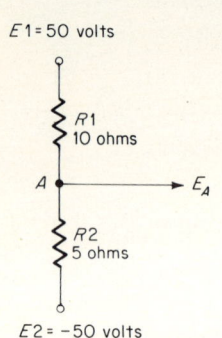

Fig. 2-9. A circuit to be solved by Millman's theorem.

methods could be employed in such an example, Millman's method will be used to illustrate the principles. Then, more sophisticated examples will be given to show that even very complex circuits can easily be solved by this system.

Millman's theorem uses *apparent* voltages and *conductances* to determine the unknown values. The basic formula is given below, and the value of E_A is then determined.

$$E_A = \frac{E1/R1 + E2/R2}{1/R1 + 1/R2} = \frac{50/10 + (-50/5)}{1/10 + 1/5} = \frac{5 + (-10)}{0.1 + 0.2}$$
$$= \frac{-5}{0.3} = -16.67 \text{ volts}$$

Note that the numerator of the equation uses the apparent voltage drop across $R1$, not the true drop; that is, point A is assumed to be at ground potential even though in reality it is not. The expressions $E1/R1$ and $E2/R2$ yield the Thevenin-equivalent current, which would flow through the equivalent resistance to produce the required voltage at point A. Also, note that the signs of the sources must be used. The denominator is simply the sum of the conductances, which a little thought will reveal to be the same as the reciprocal of the Thevenin-equivalent resistance at the junction of the two resistors. Hence, the following statements specify the real meaning of Millman's theorem:

$$E_A = I_{th} \times R_{th} = \frac{E1}{R1} + \frac{-E2}{R2} \times \frac{1}{1/R1 + 1/R2}$$

$$= [5 + (-10)] \frac{1}{\frac{1}{10} + \frac{1}{5}} = -5 \times 3.33 = -16.67 \text{ volts}$$

The foregoing problem was one of an unloaded voltage divider. These principles are just as useful for solving a loaded voltage divider. Such a circuit is shown in Fig. 2-10. This circuit is the same as the one previously

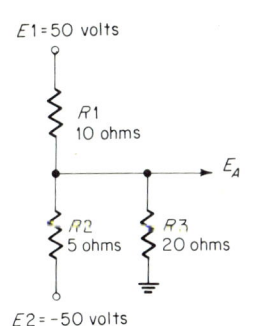

Fig. 2-10. A loaded voltage divider solved by the use of Millman's theorem.

given, but with a load resistor connected to the output line. In this instance, the fact that $R3$ is returned to ground must be considered:

$$E_A = \frac{E1/R1 + (-E2/R2) + E3/R3}{1/R1 + 1/R2 + 1/R3}$$

$$= \frac{{}^{50}\!/_{10} + (-{}^{50}\!/_{5}) + {}^{0}\!/_{20}}{\frac{1}{10} + \frac{1}{5} + \frac{1}{20}} = \frac{5 + (-10) + 0}{0.1 + 0.2 + 0.05} = \frac{-5}{0.35} = -14.3 \text{ volts}$$

Note that the 0/20 expression in the numerator is simply dropped from the equation since $R3$ does not contribute to a source voltage at A but does contribute to the Thevenin resistance.

Much more complex circuits can be readily evaluated by this method. Consider Fig. 2-11, which shows a somewhat more intricate circuit.

22 PULSE AND SWITCHING CIRCUIT ACTION

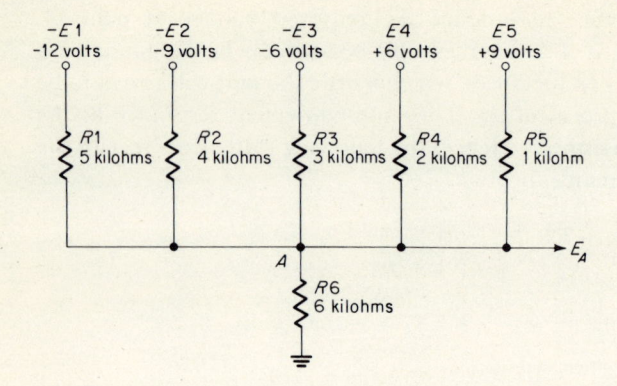

Fig. 2-11. A complex circuit solved by use of Millman's theorem.

Regardless of its complexity, the voltage at $A(E_A)$ is relatively easy to determine.

$$\begin{aligned}
E_A &= \frac{(-E1/R1) + (-E2/R2) + (-E3/R3) + E4/R4 + E5/R5 + E6/R6}{1/R1 + 1/R2 + 1/R3 + 1/R4 + 1/R5 + 1/R6} \\
&= \frac{(-12/5000) + (-9/4000) + (-6/3000) + 6/2000 + 9/1000 + 0/6000}{1/5000 + 1/4000 + 1/3000 + 1/2000 + 1/1000 + 1/6000} \\
&= \frac{(-0.0024) + (-0.00225) + (-0.002) + 0.003 + 0.009}{0.0002 + 0.00025 + 0.000333 + 0.0005 + 0.001 + 0.000167} \\
&= \frac{0.00535}{0.00245} = 2.1837 \cong 2.18 \text{ volts}
\end{aligned}$$

Hence, the voltage at point A is $+2.18$ volts with respect to ground, and the result is arrived at with a minimum of effort, considering the complexity of the circuit.

QUESTIONS AND PROBLEMS

2-1 The open-circuit voltage of a battery is 15 volts. When a 12-ohm resistor is connected across it, its terminal voltage drops to 12 volts.
(a) Determine the total internal resistance.
(b) Determine the total current.
(c) Determine the drop across R_i.

2-2 The battery in Question 2-1 has been in use for some time and is therefore somewhat discharged. When the 12-ohm resistor is again connected as before, the terminal voltage drops to 10 volts.

(a) Determine the total internal resistance.
(b) Determine the total current.
(c) Determine the drop across R_i.

2-3 Refer to Fig. 2-2a and b. E_{oc} is to be changed to 18 volts. Determine the value of R_i if E_{RL} is 15 volts.

2-4 Refer to Fig. 2-2a and b. E_{oc} is to be changed to 25 volts. Determine the value of R_i if E_{RL} is 15 volts.

2-5 Refer to Fig. 2-2a and b. E_{oc} is to be changed to 10 volts, and R_L will be 15 kilohms. Determine R_i if E_{RL} is 9 volts.

2-6 Refer to Fig. 2-2. E_{oc} is to be changed to 10 volts, and R_L will be 13 kilohms. Determine R_i if E_{RL} is 8 volts.

2-7 Refer to Fig. 2-5. $R1$, $R2$, and $R3$ are to be changed to 200 ohms. Determine E_A.

2-8 Refer to Fig. 2-5. $R1$, $R2$, and $R3$ are to be changed to 1000 ohms. Determine E_A.

2-9 Refer to Fig. 2-11. $R5$ is to be completely removed from the circuit. Determine E_A.

2-10 Refer to Fig. 2-11. $R3$ is to be changed to a 1-kilohm value. Determine E_A.

3
RC NETWORKS

3-1 CAPACITORS

As we know, a capacitor is a device used for its energy-storage capabilities. In the usual application of capacitors in typical communications circuits, the items usually studied are capacitive reactance, phase angle, impedance, resonance, etc. However, in the application of capacitors to pulse circuits, these are of lesser importance and we usually are more concerned with such things as time constants, pulse widths, differentiation, etc.

First, the general characteristics of capacitors will be described briefly, with particular emphasis upon their ability to store electric energy. Then, a few fundamental ideas of how a capacitor reacts in several simple circuit applications will be given. Finally, several of the more complex ideas regarding capacitors used in pulse circuits will be considered.

The amount of electric energy stored in a capacitor that has attained a charge can be expressed in several ways. In terms of quantity of charge, or coulombs (actually, the number of electrons), and the resultant potential from plate to plate, the charge can be expressed by the following relationship:

$$Q = CV$$

where Q = coul (1 coul is 6.28×10^{18} electrons)
C = capacitance, farads
V = plate-to-plate voltage

As an example, a 100-μf capacitor is charged to 100 volts. What is the energy storage in coulombs?

$$Q = CV = (100 \times 10^{-6})(1 \times 10^2) = 100 \times 10^{-4} = 0.01 \text{ coul}$$

A more useful formula is one that states the energy storage in terms of work that the charge could do, expressed in joules, or watt-seconds. For example, if the energy stored in a capacitor were to be discharged into a load, some amount of work would be accomplished; i.e., power would be expended over some period of time. In the foregoing example, the energy stored in terms of joules (J) is easily determined:

$$J = \tfrac{1}{2}CV^2 = \tfrac{1}{2}(100 \times 10^{-6})(1 \times 10^4) = \frac{100 \times 10^{-2}}{2} = 50 \times 10^{-2} \text{ joules}$$

This is the product of watts and seconds, and the capacitor in the example is capable of expending 0.5 joules, or 0.5 watts for a period of 1 second. Or, if the energy stored in the capacitor is dissipated over a period of 1 msec, the power expended would be 500 watts, which represents the same amount of work.

In a practical circuit application, one does not usually need to know the actual amount of energy storage. Instead, the major concern is with the voltage present at some point in the circuit, as well as the point in time that this voltage exists relative to some reference time.

3-2 CAPACITOR CHARGE AND DISCHARGE CHARACTERISTICS

The mechanism of capacitor charge and discharge is a most important aspect in the study of capacitor function in typical pulse circuits. These basic ideas must be thoroughly understood before studying the actual circuitry. The following discussion assumes some familiarity with capacitors on the part of the reader. It is not intended, nor should it be construed, as an adequate treatment for the beginner. For most readers, this discussion will be only a review, and very little new material will be found in the first few pages.

One of the fundamental rules of capacitor action is that the potential from plate to plate cannot be changed instantly. The amount of time necessary to either charge or discharge a capacitor is usually of great

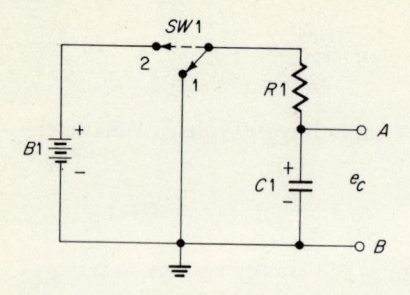

Fig. 3-1. A simple circuit illustrating capacitor charge and discharge.

importance in pulse circuits. A simple example that will help to visualize how a capacitor charges and discharges is shown in Fig. 3-1. With the switch in position 1, the battery is disconnected from the circuit. Hence, there is initially neither voltage at terminals A and B nor current through $R1$. When the switch is thrown to position 2, the battery is connected into the circuit and current will begin to flow, even though the capacitor represents an open circuit to direct current.

Now, current will flow until the capacitor becomes fully charged, which is the condition in which the plate-to-plate voltage across the capacitor just equals the applied voltage. The current that flows during this time is called the *charging current*. It is identified as charging current by the fact that as it flows the voltage across the capacitor becomes greater. The maximum value of this current is determined by the series resistor and the applied voltage.

Electrons flow out the negative post of the battery and into the lower plate of the capacitor. At the same time, electrons are pulled away from the other plate, returning to the positive post of the battery. When a large number of negative and positive charges have accumulated on the capacitor plates, the voltage from plate to plate will exactly equal the battery voltage. Because this voltage opposes the battery voltage, current flow ceases until the switch is returned to position 1.

With the switch back in position 1, the accumulated charge now discharges, with electrons from the negative plate being forced around the circuit toward and onto the positive plate. This is called *discharge current*. Finally, the charge on the upper plate exactly equals the charge on the lower plate. The voltage across the capacitor has fallen to zero, and current is again zero. (The foregoing description ignores the fact that the charge on a capacitor actually resides in the dielectric. As far as circuit action is concerned, it makes little difference where we consider the charge to be.)

Very generally, this describes the capacitor action when voltage is applied to the circuit and then taken away. Much more detail must be added to this description to facilitate an understanding of capacitor circuits. In the following material we shall use the same basic circuit and try to find ways of easily describing the various aspects of capacitor charge and discharge.

The Voltage Charge Curve

The most convenient method of describing the voltage rise across a charging capacitor is to use a graph that plots the rising voltage versus time. The voltage across a capacitor cannot change instantly, and while the capacitor charges, the voltage from plate to plate is continuously changing. Thus, a graph of the changing voltage is useful to describe this action. (The purpose of the resistor in this circuit, Fig. 3-1, will be discussed later.)

Figure 3-2 illustrates the circuit shown previously in Fig. 3-1, where the curved line represents the way the voltage builds up to the battery voltage during the charge time. Note that the voltage e_c does not increase linearly. The shape of this curve is exponential and follows a precise mathematical relationship. The point of origin represents both time zero (t_0) and voltage zero ($e_c = 0$ volts). This is the time that the switch makes contact with position 2. At this instant the capacitor begins to charge. (In this graph the assumption is made that the capacitor is initially discharged completely.)

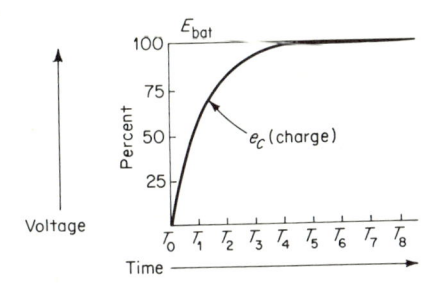

Fig. 3-2. Capacitor charge curve (voltage).

As time progresses, the voltage e_c from plate to plate rises rapidly at first and at time 1 (T_1) it has increased from zero to better than 60% of the final value. During the next equal interval from T_1 to T_2, it rises only to about 86%, an increase of perhaps 23%. During the next interval the capacitor voltage increases from about 86 to 95%, an increase of only 9%.

Thus, the *rate* of rise of e_c decreases as time goes by, each interval to the right representing a smaller increase than the one just to its left. Beyond T_5, the voltage across the capacitor is so nearly equal to the applied voltage

28   PULSE AND SWITCHING CIRCUIT ACTION

that for all practical purposes it is considered to be equal. Thus, by using such a curve the voltage rise across a capacitor can be described in terms of time.

The interval between T_0 and T_1 is called the *time constant* of the circuit, which will be discussed in detail later in this chapter. The curve of Fig. 3-1 is generalized so as to be applicable to any capacitor circuit. In any certain case, with values for voltage, current, resistance, and capacitance known, appropriate values along both the ordinate (in terms of volts) and the abscissa (in terms of time constants) will allow the curve to describe a particular circuit rather than the general case. Examples using actual circuit values will be given subsequently.

The Voltage Discharge Curve

If the capacitor in Fig. 3-1 has been allowed time to reach full charge ($e_c = E_{\text{bat}}$), then when the switch is thrown to position 1, the capacitor will discharge. As discharge current flows from the negative plate to the positive plate, the charge differential decreases and so the plate-to-plate voltage decreases. This, too, can be shown graphically, as in Fig. 3-3. Again, voltage is plotted against time, and the decrease of voltage from plate to plate is evident as time progresses.

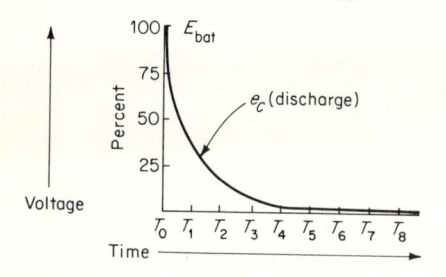

Fig. 3-3. Capacitor discharge curve (voltage).

Initially, the capacitor is fully charged to 100% of the battery voltage. As the switch in Fig. 3-1 is placed in position 1, a complete path is evident from plate to plate and current will begin to flow. The capacitor now acts as a *source*. As soon as current begins to flow, the voltage across the capacitor begins to decrease, following in every case the curve of Fig. 3-3. At time 1, only about 37% of the initial voltage is left; at time 2 about 15% is left, and so on. At time 5, the capacitor is, for all practical purposes, discharged, only a tiny amount of charge being left.

If the curve were to be applied to an actual circuit, the ordinate would be labeled in volts while the abscissa would be labeled in time constants, as before. The discharge curve is also exponential in form, following the same general mathematical formula mentioned earlier.

The Current Charge and Discharge Curves

Thus far, the capacitor charge and discharge has been discussed in terms of voltage across the capacitor. Equally important is the current flowing during both charge and discharge. Measuring the current in such a circuit poses certain problems. If the current were a smooth direct current, the circuit would be broken and a milliammeter inserted to complete the circuit. The current could then be read directly. However, in this case the current, when it is flowing, is constantly changing. At other times it is zero (when the capacitor is completely charged or discharged). Therefore, an indirect method is resorted to so that this important measurement can be made.

Ohm's law tells us that the voltage across a resistor is exactly proportional to the current through it. Hence, the resistor in series with the capacitor will have a voltage developed across it that is a replica of the current flowing through it. Since it is much easier to measure a changing voltage than a changing current, the voltage across the resistor will now be considered.

The circuit shown in Fig. 3-4 is identical to that shown in Fig. 3-1, with

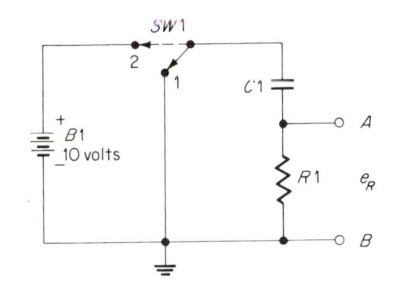

Fig. 3-4. A circuit illustrating capacitor charge and discharge current.

one exception. The position of $R1$ relative to $C1$ is reversed, and this simply allows the voltage developed across $R1$ to be referred to ground. With the switch in position 1 for some time, the capacitor is discharged; hence, no current or voltage is apparent in the RC circuit. As the switch is transferred to position 2, the capacitor begins to charge and charging current flows through $R1$. Point A will become more positive since electrons are flowing from B to A ($-$ to $+$). When the capacitor becomes fully charged, current ceases and there is no voltage between A and B since no current flows through the resistor.

30 PULSE AND SWITCHING CIRCUIT ACTION

If the switch is thrown back to position 1, the capacitor must discharge since a complete path is available. The lower plate of the capacitor is negative (excess of electrons), while the upper plate is positive (deficient in electrons). When the discharge current flows, electrons emerge from the lower plate through $R1$, $SW1$, and back to the upper plate. Current now flows through $R1$ from A to B, making point A more negative than B. When the capacitor loses its charge, current again ceases and there is no further activity until the switch is again transferred.

Figure 3-5 shows the waveform evident at point A, referred to ground.

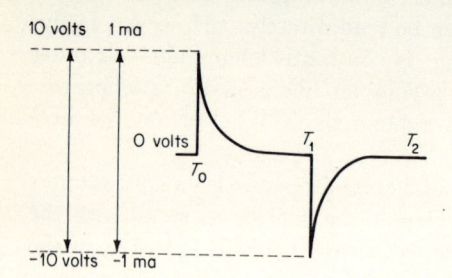

Fig. 3-5. The waveform for the circuit of Fig. 3-4.

At time zero T_0, the voltage at point A jumps to $+10$ volts. This is the first instant that $SW1$ is transferred from 1 to 2. Because point A is at $+10$ volts at this first instant, several interesting observations can be made. For one thing, since the battery voltage is 10 volts, this indicates that the voltage across the capacitor is still zero. While the switch was in position 1, $C1$ was discharged.

Because the capacitor was initially discharged and cannot change its state of charge instantly, it must still have 0 volts across its plates. As the switch is transferred to 2, the upper plate becomes connected to $+10$ volts. To maintain zero charge, plate to plate, the lower plate must rise to $+10$ volts also. Hence, the full 10 volts is dropped across the resistor at the first instant!

Now, because the full battery voltage appears across $R1$ at the first instant, the initial value of current flowing in $R1$ (which is, of course, capacitor-charging current) is determined solely by Ohm's law:

$$I_{R1(T_0)} = \frac{E_B}{R1}$$

That is, at the first instant, current reaches a maximum value that is the same as if the resistor itself were across the 10-volt battery. Of course, as time goes by, current decreases exponentially and so does the voltage across $R1$.

Because the electron current is flowing up through the resistor, point A is positive with respect to B (ground). This direction of current flow is arbitrarily assigned a positive value to agree with the voltage producing it. Eventually the current decreases to zero as the capacitor becomes fully charged. If, as soon as current becomes zero, the switch is thrown to position 1 again, the waveform produced would appear as shown. Note particularly that point A instantly goes to -10 volts with respect to ground. This requires a detailed explanation.

Remember that, just prior to time 1, the capacitor is fully charged, with the upper plate at $+10$ volts and the lower plate (point A) at ground. Thus, a full 10-volt difference exists from plate to plate at this time. As the switch contact is transferred, the upper plate is directly placed at ground potential (0 volts). The capacitor still has a full charge of 10 volts, with the upper plate positive with respect to the lower plate. Therefore, the lower plate is 10 volts more negative than ground, or -10 volts, as shown. Only after some time elapses (T_1 to T_2) can the capacitor discharge and point A gradually drops toward ground from -10 volts. When the capacitor loses all charge, current again ceases and the voltage across $R1$ becomes zero once again.

Note that in this last action, electron current flows out the lower plate of the capacitor through $R1$ in a *downward* direction, thus verifying that point A must be more negative than ground during discharge. This current is given a negative sign, again to agree with the voltage associated with it.

The waveform of Fig. 3-5, then, could equally well describe either the voltage drop across $R1$ or the current flowing into or out of the capacitor. By assigning a value to the resistor, the maximum current values can be easily determined. Assume $R1$ is a 10-kilohm resistor. Maximum current is then

$$I_{max} = \frac{E_B}{R1} = \frac{10}{10 \text{ kilohms}} = 1 \text{ ma}$$

The curve is also labeled in terms of milliamperes, as shown. Thus the same curve describes both e_c and I_c if appropriate values are provided on the ordinate.

The RC circuit shown in Fig. 3-1 (less the battery and switch) is variously called a *low-pass filter* or an *integrator*. When used with sinusoidal signals, the capacitor bypasses the high-frequency signals to ground, allowing only low-frequency signals to appear at the output (point A). When used with square-wave pulses, it is termed an *integrator*, and the waveform of Fig. 3-2 is an example of an integrated waveform.

By the same token, the RC circuit of Fig. 3-4 is called either a *high-pass filter* or a *differentiator*, again depending upon whether sinusoidal or pulse signals are used. The waveform of Fig. 3-5 is an example of a differentiated waveform.

Capacitor Charge and Discharge using Input Pulses

The circuits just shown are not practical circuits, obviously, except when used as a medium to explain basic ideas. The battery and switch were used to simulate a pulse, and in reality the movable contact of the switch (swinger) has a square wave evident upon it in relation to ground. With the movable contact at position 1, 0 volts, or ground, is evident. If the switch were alternately moved from one position to the other, it would generate a waveform similar to that shown in Fig. 3-6.

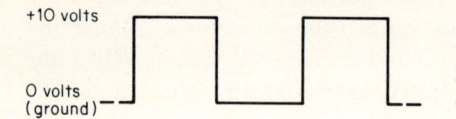

Fig. 3-6. A square wave used in lieu of a switch.

To assist the reader in applying what has just been learned to an actual circuit, the same circuits with a square-wave input are shown in Figs. 3-7 and 3-8. The input to each circuit is as shown in Fig. 3-6, while the output is shown with each circuit. With the proper values of R and C the waveforms shown would be produced. Choosing the values is, however, critical and will be discussed in detail later.

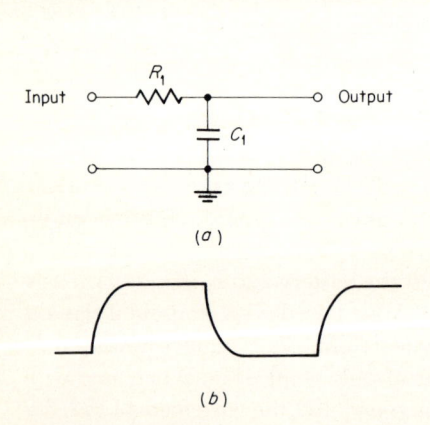

Fig. 3-7. The integrator. (*a*) Circuit; (*b*) output.

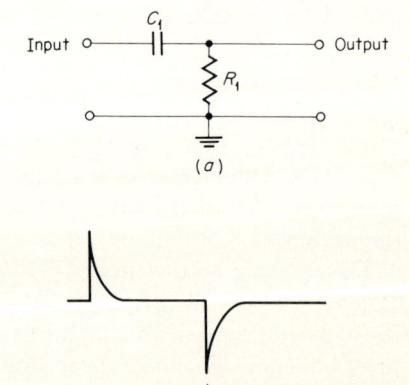

Fig. 3-8. The differentiator. (*a*) Circuit; (*b*) output.

3-3 RC TIME CONSTANT

If a capacitor were connected to a dc power source through perfect conductors (zero resistance) and if the supply had zero internal resistance, the capacitor would instantaneously charge to the supply voltage. This would theoretically require an infinitely large current, and since this is impossible, it is likewise impossible to charge a capacitor in zero time. In all practical cases the capacitor can reach full charge ($e_c = E_{bat}$) only after a certain time has elapsed. Although a capacitor can be charged very quickly at times, perhaps 0.1 to 0.01 μsec, this is still a finite time and as such is both measurable and calculable.

The amount of time required to reach full charge depends upon the relative amounts of resistance and capacitance in the circuit. In most circuits the amount of time the capacitor takes to become fully charged is very important. So that this can be conveniently measured, the time constant is used to describe the element of time in connection with the charge and discharge of a capacitor.

Generally, the larger the value of either the resistance R or the capacitance C, the longer the time constant; that is, the longer it takes the capacitor to reach some state of charge. Conversely, the smaller the value of R and C, the shorter the time constant. Mathematically, the time constant is the simple product of R and C and is expressed as

$$t_c = R \times C$$

where t_c = time for the capacitor voltage to rise from 0 to 63.2% of full charge, seconds
C = capacitance, farads
R = resistance, ohms

The time constant, then, represents an interval of time. In Fig. 3-2 the interval from the point of origin to T_1 represents one time constant; the interval from T_3 to T_4 represents one time constant; the interval from T_0 to T_5 represents five time constants, etc. In the practical case, a capacitor never quite reaches full charge but can be considered to have reached full charge in five times the amount of one time constant (actually attaining 99.3% of full charge).

A typical capacitor charge and discharge curve is shown in Fig. 3-9. This is often called the *universal charge curve*. The shape of these curves is that of an exponential, as previously mentioned, rising or falling very rapidly at first, gradually sloping off to a nearly horizontal line as it progresses to the right. The various percentages of full charge are shown for different values of the time constant. This curve has been shown before but was not labeled in terms of time constants as such.

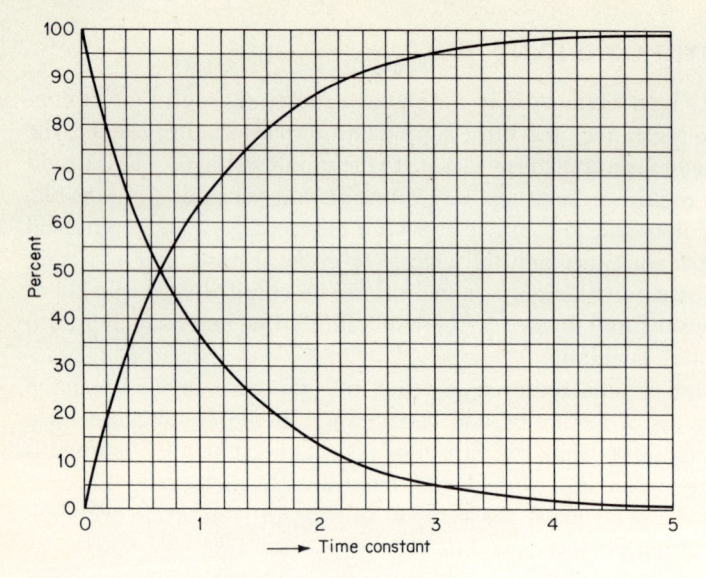

Fig. 3-9. The universal time-constant curve.

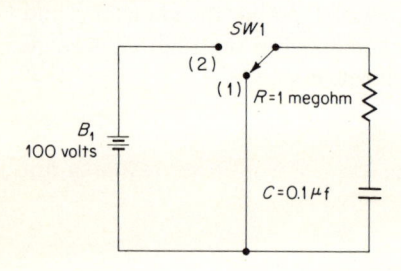

Fig. 3-10. A circuit used to illustrate the time constant. Simple capacitor, charge-discharge circuit.

To illustrate the action of capacitor charge, the circuit of Fig. 3-10 is offered. If the capacitor is initially discharged (0 volts from plate to plate), it will begin to accept a charge at the instant the switch transfers to position 2. The length of time necessary for the plate-to-plate voltage to increase to 63.2 volts (63.2% of 100 volts) is given by one time constant. Using the values given, the time constant of these components can be easily computed:

$$t_c = R \times C = (1 \times 10^6)(0.1 \times 10^{-6}) = 0.1 \text{ second}$$

Thus, in the circuit shown, the voltage across $C1$ will rise to 63.2% of full charge (0.632 × 100 volts) in 0.1 second. In two time constants (0.2 second) the capacitor voltage has risen to about 86.5% of full charge, or

86.5 volts. After five time constants have elapsed, the capacitor is considered to be fully charged and would have 99.3 volts from plate to plate. This is so near to 100 volts that it is considered to be 100 volts. After 10 time constants have elapsed, e_c will be equal to 99.995275 volts in the circuit shown. As previously mentioned, the capacitor never quite reaches full charge but so nearly approaches it after five or more time constants that the difference can be disregarded.

The discharge curve shown in Fig. 3-11 (also shown in Fig. 3-9) can be

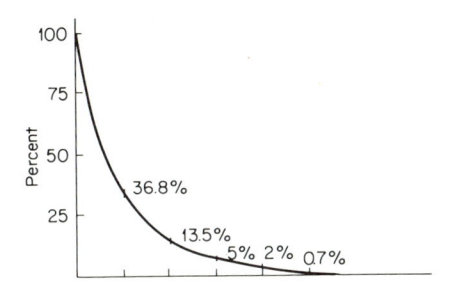

Fig. 3-11. Universal time-constant curve (discharge).

used to show how much charge remains on the capacitor during discharge after some number of time constants has elapsed. For example, after one time constant, the capacitor has lost 63.2% of the original voltage and so $100 - 63.2 = 36.8\%$ remains. When two time constants have gone by, the capacitor has lost 86.5% of its original charge and so 13.5% remains. When five or more time constants have elapsed, the capacitor is considered to be completely discharged, although some small charge still remains (0.7%).

When the case arises where the charge on a capacitor must be known for a period of time that is not an integral time constant, then the universal-time-constant curve is not as useful. Also, often the curve itself is not available, and then it becomes necessary to be able to calculate the charge or discharge characteristics of the *RC* circuit. We shall consider first the charge equation for a typical *RC* circuit. The charge equation applicable to a circuit such as that shown in Fig. 3-10 is

$$e_c = E(1 - \epsilon^{-t/RC})$$

where e_c = instantaneous plate-to-plate capacitor voltage
 E = maximum applied voltage
 ϵ = epsilon, the base of the natural log system: 2.718
 t = time of interest after charge has begun
 RC = time constant, $R \times C$, ohms and farads

Using the circuit of Fig. 3-10 as an example, the capacitor voltage will be determined at the point in time 0.06 second after charge has begun:

$$e_c = E(1 - \epsilon^{-t/RC}) = 100(1 - 2.718^{-0.06/0.1}) = 100\left(1 - \frac{1}{2.718^{0.6}}\right)$$

$$= 100\left(1 - \frac{1}{1.82}\right) = 100(1 - 0.549) \cong 100 \times 0.451 = 45.1 \text{ volts}$$

Thus, the capacitor will have accumulated 45 volts in only 0.6 time constant. Because this formula is used to construct the universal curve, we should be able to verify one or more points on the curve by using the formula. It has been arbitrarily stated that at the end of one time constant the capacitor's voltage will rise to 63.2% of full charge. This can now be verified, as shown in the following example:

$$e_c = E(1 - \epsilon^{-t/RC}) = 100(1 - 2.718^{-0.1/0.1}) = 100\left(1 - \frac{1}{2.718}\right)$$

$$= 100(1 - 0.368) = 100 \times 0.632 = 63.2 \text{ volts}$$

Obviously, 63.2 volts is 63.2% of 100 volts. The universal-time-constant curve, then, can be constructed by calculating several points and connecting them with a smooth line.

The discharge curve of Fig. 3-11 (or Fig. 3-9) can also be constructed by using a slight variation of the charge-curve formula. Additionally, any point on the curve can be determined, as shown in the following example. Assume the capacitor voltage is to be found after 0.18 second has elapsed (for the same circuit):

$$e_c = E(\epsilon^{-t/RC}) = 100(2.718^{-0.18/0.1}) = 100\,\frac{1}{2.718^{1.8}} = 100\,\frac{1}{6.05}$$

$$= 100 \times 0.1653 \cong 16.5 \text{ volts}$$

Now, it is often useful to know the amount of current flowing during charge or discharge. The following formula is useful in determining this:

$$i = I(\epsilon^{-t/RC})$$

where i = instantaneous current at specific time

$I = \dfrac{E}{R}$, maximum current possible

ϵ = epsilon = 2.718

t = time of interest after charge or discharge has begun

RC = time constant

Again using Fig. 3-10 as an example, assume it is desired to know the current in the resistor at the instant 0.04 second after charge has begun:

$$I_R = i = I(\epsilon^{-t/RC}) = \frac{E}{R}\epsilon^{t/RC} = \frac{100}{1 \times 10^6} 2.718^{-0.04/0.1}$$

$$= 100 \times 10^{-6} \frac{1}{2.718^{0.4}} = 100 \times 10^{-6} \frac{1}{1.492} = (100 \times 10^{-6})0.6702$$

$$= 67.0 \; \mu a$$

At the instant charge begins, current flow is 100 μa, and after 0.04 second elapses, the current at this instant is 67 μa.

3-4 *RC* COUPLING IN PULSE CIRCUITS

Coupling a pulse by way of an *RC* circuit involves essentially the same basic mechanism as in the case of a sinusoidal signal. However, there are differences. The basic reason an *RC* network is used in the first place is the fact that a capacitor will not allow a dc voltage to pass, while an ac voltage will easily pass. That is, as far as direct current is concerned, the capacitor appears as an open circuit, but with an alternating current impressed, it appears as a short circuit for suitable frequencies.

The reason for some of the differences between the pulse circuit and the circuit used with sinusoidal signals lies in the fact that a train of square-wave pulses exhibits at the same time some of the characteristics of both alternating and direct current. Hence, an *RC* coupler used in pulse work must be investigated much more thoroughly than if it were to be used with sinusoidals. At the very least, the components must be viewed in a different light.

The Coupler Defined

A typical *RC* coupling circuit for use with pulses is shown in Fig. 3-12. The input consists of a continuous train of square waves between the levels of +6 and +12 volts. The output is seen to be a replica of the input except for the dc levels involved. The dc level has been eliminated since the waveform is centered around the 0-volt line. That is, the waveform is positive as much as it is negative, and the dc level averaged over a period of time encompassing an integral number of time intervals is zero. On the other hand, the average dc level at the input is something between 6 and 12 volts. This is, of course, blocked by the capacitor and so cannot appear at the output.

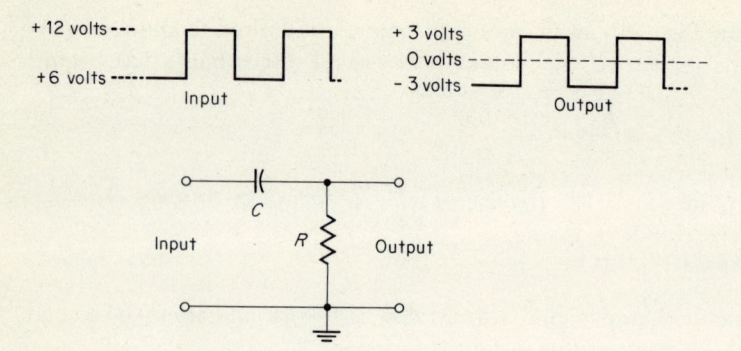

Fig. 3-12. A typical *RC* coupling circuit showing waveforms.

The first and most important requirement concerning the values of *R* and *C* is that the time constant must be long compared with the pulse width. A typical square-wave pulse is shown in Fig. 3-13. The pulse period is the time elapsed during one complete cycle. The pulse width is shown as one-half of this, or 50 μsec.

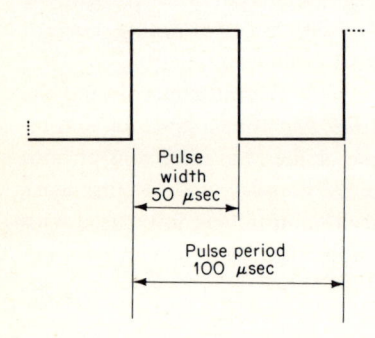

Fig. 3-13. Pulse width versus pulse period.

Pulse Width Versus Time Constant

A rule of thumb tells us that if the time constant of *C* and *R* is equal to or greater than 100 times the pulse width, the circuit will pass the square wave with no more than a minimum of distortion. Figure 3-14 illustrates a square wave that has been passed through a coupler with inadequate time constant.

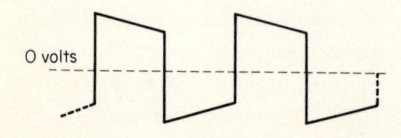

Fig. 3-14. Square-wave droop, or sag, caused by inadequate time constant.

Note the droop, or sag, at the top and bottom of the pulse. As will be discussed later in detail, the droop is caused by the capacitor having charged somewhat toward the applied voltage. When the capacitor accepts a large percentage of charge, the voltage across the resistor droops. Because the output is taken across the resistor, the voltage does not remain essentially constant.

By making the time constant very long relative to the pulse width, the capacitor cannot charge very much in the very short time allowed by the pulse. Since by the time a pulse terminates the capacitor has virtually no change in its state of charge, nearly the full voltage still exists across the resistor. Hence, no readily visible droop occurs, and the original pulse shape is retained.

To determine the minimum value of R and C for a specific case, assume that the pulse shown in Fig. 3-13 is to be coupled through the circuit of Fig. 3-12. The time constant is to be 100 times the pulse width.

$t_c = 100 \times 50 \ \mu sec = 5000 \ \mu sec = 5 \ msec$

The formula for the time constant is shown:

$t_c = R \times C$

With t_c known, a value for R can be assumed and the proper size capacitor can be determined by solving for C. Assume a resistance of 100 kilohms:

$C = \dfrac{t_c}{R} = \dfrac{5 \ msec}{100 \ kilohms} = \dfrac{5 \times 10^{-3}}{1 \times 10^{5}} = 5 \times 10^{-8} = 0.05 \ \mu f$

With these values used in the circuit, then, the droop would be negligible. To determine how much droop occurs, it is only necessary to calculate the voltage across the capacitor at the end of a 50-μsec period.

$e_c = E(1 - \epsilon^{-t/RC}) = 6(1 - 2.718^{(-50 \times 10^{-6})/(5 \times 10^{3})}) = 6\left(1 - \dfrac{1}{2.718^{0.01}}\right)$

$= 6\left(1 - \dfrac{1}{1.01}\right) = 6(1 - 0.99) = 6 \times 0.01 = 0.06 \ volt$

At the end of one pulse, the capacitor has charged to 0.06 volt, leaving 5.94 volts across the resistor. For all practical purposes, this is nearly a full 6 volts. A longer time constant would result in less droop and an even smaller charge on the capacitor.

Inadequate Time Constant

If the *RC* time constant is too short or if the pulse is longer in duration, a very severe droop occurs and the pulse is badly distorted. In some instances this is desirable, but in a coupling circuit it is to be avoided at all costs. An *RC* coupler can be defined as an *RC* circuit arranged to block dc that transmits a square-wave pulse with no more than 1% droop. That is, if the leading edge is 10 volts in amplitude, the top of the pulse should droop by no more than $0.01 \times 10 = 0.1$ volt.

To illustrate excessive droop, assume that an *RC* coupler has a time constant of 1 second. With an output pulse of 10 msec, as shown in Fig. 3-15,

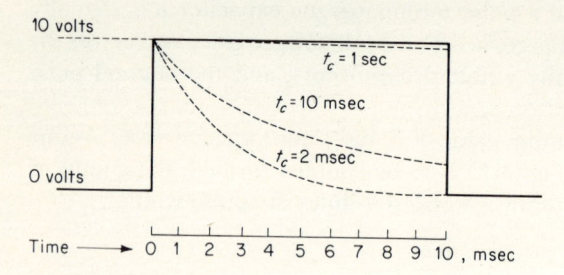

Fig. 3-15. Various degrees of droop for different time constants (fixed pulse width).

the droop is barely noticeable, as indicated by the dotted line labeled "1 sec." If the same 10-msec pulse were to be applied to a circuit having only a 10-msec time constant, the output would follow the curved line labeled "10 msec." Note that the pulse top has nearly fallen back to the base line.

If the circuit time constant is made to be 2 msec, the 10-msec pulse is exactly five times the time constant and the capacitor just reaches full charge at the pulse termination. Thus, the output, viewed across the resistor, just drops to zero as the pulse terminates. This, of course, is a very severe droop, and in fact the original waveform is so severely altered that it is given the special name *differentiated waveform*, as previously mentioned. A discussion of this particular waveshape will be given later in some detail.

Assume for a moment in the case of the circuit described above, with $t_c = 10$ msec, that it is necessary to calculate the amount of droop in terms of voltage at the output. The voltage that exists across the capacitor after 10 msec has elapsed is easily determined:

$$e_c = E(1 - \epsilon^{-t/RC}) = 10(1 - 2.718^{-0.01/0.01}) = 10\left(1 - \frac{1}{2.718}\right) = 10(1 - 0.368)$$
$$= 10 \times 0.632 = 6.32 \text{ volts}$$

If 6.32 volts is across the capacitor at the end of one time constant, then the output voltage is

$e_o = E - e_c = 10 - 6.32 = 3.68$ volts

(When the input pulse terminates, at time 10, the output waveform extends below the 0-volt base line. This is not shown.)

Step Transmission

In our previous discussions of *RC* circuits we have deliberately ignored an important point which now must be considered. It has perhaps occurred to the reader that no extended explanation has been offered giving the reason why the initial step in a square wave is passed unchanged by the capacitor in a coupler. To one who has worked around pulse circuits extensively, the fact that this does occur is self-evident, but the reason why it happens is perhaps not so self-evident. Also, the resulting implications are often obscure.

A circuit is shown in Fig. 3-16 that will be used to explain this occurrence.

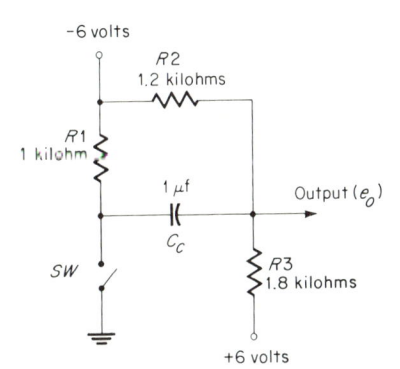

Fig. 3-16. A circuit used to show step transmission.

The switch *SW* is a single-pole, single-throw type. It is either open or closed. Because the capacitor cannot pass direct current, the switch must be alternately opened and closed if a varying output is to be produced. Because the switch actually represents a transistor, the overall circuit action is quite important. Many pulse and digital circuits operate upon the basic principles inherent in this simple configuration.

The output of this circuit is determined by the action of the switch. It will first be assumed that the switch is open, and appropriate circuit conditions will be noted. Then the switch will be closed and the new circuit conditions determined. The voltages evident at the output wire, relative to ground, are shown in Fig. 3-17.

First, it is very important to realize that there are three areas of interest in connection with the capacitor action: the voltage at the left plate of C_c, the voltage at the right plate of C_c (both referred to ground), and the voltage from plate to plate. Figure 3-17 suggests this in view of the three waveforms given.

The static condition of the capacitor must first be known. While the switch is open, the left plate of C_c is returned through resistor R_1 to -6

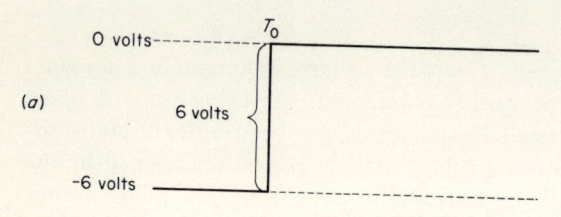

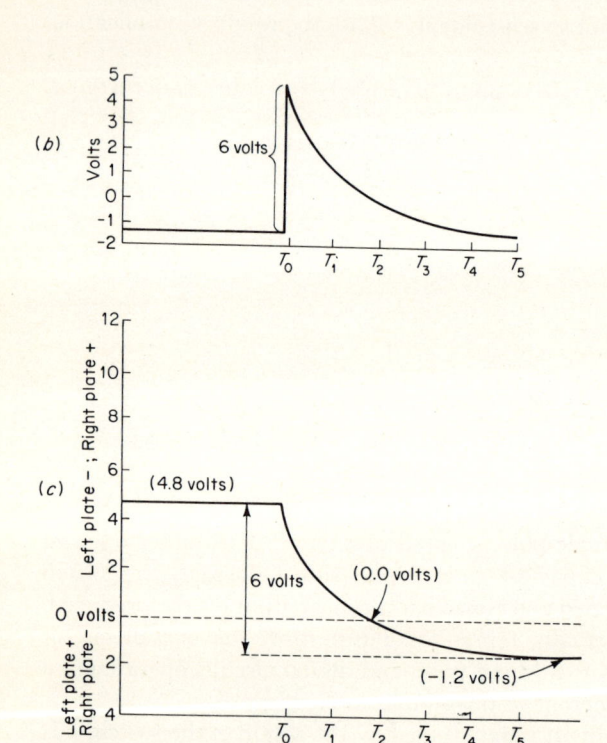

Fig. 3-17. The waveforms necessary to explain step transmission. (a) The left plate of C_c referred to ground (e_i); (b) the right plate of C_c referred to ground; (c) plate-to-plate voltage.

volts. On the assumption that the switch has been open for some time, the left plate of C_c is at a potential of -6 volts relative to ground. The right plate is returned to a voltage divider, and so its potential is somewhere between -6 and $+6$ volts.

$$e_o = -V_{cc} + \frac{R2}{R2 + R3} V_{total} = -6 + \frac{1.2 \text{ kilohms}}{1.2 \text{ kilohms} + 1.8 \text{ kilohms}} 12$$
$$= -6 + 4.8 = -1.2 \text{ volts}$$

Hence, the right plate is 1.2 volts more negative than ground. The total plate-to-plate voltage $e_{p\text{-}p}$ is the difference between -6 and -1.2 volts.

$$e_{p\text{-}p} = (-6) - (-1.2) = 4.8 \text{ volts}$$

Across the capacitor, then, there is a difference of potential that is equal to 4.8 volts. Note that this is in no way directly referred to ground and that the left plate is the more negative of the two.

Now, at time zero (T_0), the switch is suddenly closed. (The assumption is made that the switch closes instantly, with no chatter or bounce.) The time constant of the RC network is very long, so that at this instant the capacitor cannot change its state of charge. That is, the *plate-to-plate* voltage cannot noticeably change, although the voltages at the plates may change *relative to ground*.

The left plate of the capacitor, then, drops to ground almost instantly, making a full 6-volt excursion in the positive direction (-6 volts to ground). Because the capacitor must retain its plate-to-plate voltage for some time, the right plate *must also make a 6-volt positive excursion*. Since it starts from -1.2 volts relative to ground, it goes to a potential 6 volts more positive than this. Thus, $-1.2 + 6 = +4.8$ volts. This is shown in Fig. 3-17b.

Note that just prior to T_0 the difference of potential from plate to plate, as indicated on the waveforms in Fig. 3-17a and b, is -6 to $-1.2 = 4.8$ volts. An instant after T_0 the difference of potential from plate to plate is 0 to $4.8 = 4.8$ volts. Even though the potentials relative to ground are now different, the plate-to-plate voltage is still the same. Of course, after some time has elapsed (say, at T_1), the capacitor has had time to begin to charge, and it does so.

Figure 3-17a shows the voltage at the left plate of C_c plotted against time. As expected, once the switch is closed, the left plate of the capacitor is held firmly at ground, as shown. The waveform in Fig. 3-17b, however, shows an exponential curve that indicates that the capacitor is changing its state of charge over the five time constants shown. At the instant of the transition, the right plate of C_c has risen to $+4.8$ volts relative to ground. The voltage divider, however, still tries to place −1.2 volts at this point, and so the capacitor must eventually return to this voltage at its right plate. It does so following the exponential curve, as shown, and after five time constants returns to −1.2 volts on the right plate, thus generating the waveform shown.

The waveform in Fig. 3-17c shows the plate-to-plate voltage during the five time constants. The area above 0 volts represents the situation where the right plate is more positive than the left, while in the area below 0 volts the left plate is the more positive. Initially, the right plate is 4.8 volts more positive than the left. At the instant of transition, even though the voltages have changed relative to ground, the plate-to-plate voltage is unchanged. Then, as time progresses, the right plate begins to go toward minus. After the full five time constants have elapsed, the plate-to-plate voltage is 1.2 volts, with the left plate at ground and the right plate at −1.2 volts, relative to ground.

These three waveforms, then, completely specify the action of C_c for the conditions set forth. It is easily seen that the sharp step is caused by the fact that C_c cannot change its state of charge, plate to plate, as rapidly as the input waveform changes. Thus, the output side of the capacitor is in essence a duplicate of the input side for a total voltage change.

Base-line Shift

A point concerning *RC* coupling circuits that is often obscure is the so-called *base-line shift* that occurs when a signal is passed through a coupler. This effect is quite noticeable when an oscilloscope probe is first touched to a circuit with a waveform such as shown in Fig. 3-18a. If the scope is ac coupled, the waveform appears somewhat as shown in Fig. 3-19b, slowly drifting upward toward a final stationary position near the center of the screen. This effect is caused by the coupling capacitor(s) in the scope and is directly related to the fact that a capacitor will not pass direct current. In order that the concepts of this phenomenon can be readily grasped, the idea of the average dc level of a waveform must first be understood.

Considering the waveform of Fig. 3-18a, it is seen that the pulse string operates between the levels of 0 (ground) and −6 volts with a symmetrical waveshape. If a dc meter were used to measure this waveform, it would yield a reading of −3 volts. The meter movement will average the total values with respect to time; hence the average dc value of this particular waveform is −3 volts. This can be easily calculated by using the following relationship:

$$E_{av} = E_{max} \times \frac{T_{p2}}{T_{p1} + T_{p2}}$$

where E_{av} = average dc voltage
E_{max} = peak value of the pulse
T_{p1} = time the pulse remains at one extremity (½ cycle)
T_{p2} = time the pulse remains at the other extremity (½ cycle)
$T_{p1} + T_{p2}$ = total pulse period (1 cycle)
Using the previously given pulses as an example, the average dc level is easily found (assume T_{p1} = 5 msec = T_{p2}):

$$E_{av} = E_{max} \times \frac{T_{p2}}{T_{p1} + T_{p2}} = -6 \left(\frac{5 \text{ msec}}{10 \text{ msec}}\right) = -6 \times 0.5 = -3 \text{ volts}$$

If the waveform were nonsymmetrical, the average dc level would be either greater or less, depending upon the length of time the waveform stays at zero. For instance, assume T_{p1} = 2 msec and T_{p2} = 8 msec. Determine the average dc level.

$$E_{av} = E_{max} \times \frac{T_{p2}}{T_{p1} + T_{p2}} = -6 \times \frac{8 \times 10^{-3}}{2 \times 10^{-3} + 8 \times 10^{-3}} = -6 \times 0.8$$
$$= -4.8 \text{ volts}$$

The foregoing relationship is valid for all rectangular pulses. If the pulse has a different shape, an account must be made of this and the above formula suitably modified.

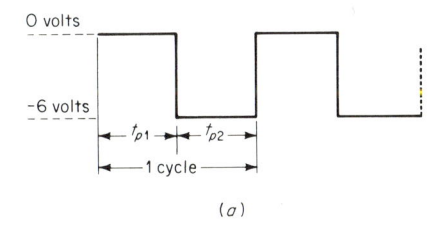

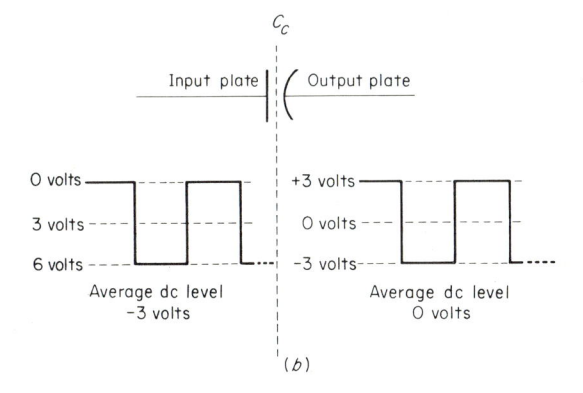

Fig. 3-18. Waveforms explaining base-line shift.

When a pulse such as that in Fig. 3-18a is passed through a coupling capacitor, *the average dc level at the input side must become the 0-volt level at the output side.* This important fact is illustrated in Fig. 3-18b. The average dc level on the input side of C_c is -3 volts relative to ground. This becomes the 0-volt level on the output side, where it is seen that the pulse here operates between $+3$ and -3 volts. There must be no average dc level at the output since by definition the capacitor cannot pass direct current. Hence, any dc level, or component, of the input waveform is blocked and cannot appear at the output.

If the input does not have a symmetrical shape, the output 0-volt line will not exactly bisect the peak-to-peak value. An example of this kind of waveform is given subsequently.

When a train of pulses is passed through a coupling network the first several pulses undergo a rather unusual change. Initially, the input is zero and the output, of course, is also zero. In Fig. 3-19 we see the result

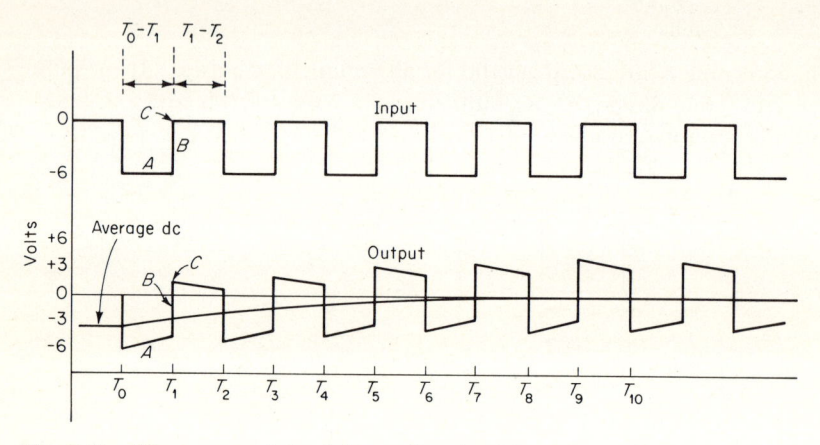

Fig. 3-19. The average dc level shown changing from -3 to 0 volts as a train of pulses is applied to an *RC* coupler.

of passing the pulse train through a coupling circuit. The output cannot immediately center around the 0-volt line, and several pulses must go by before the waveform settles around zero. (During this period of time the capacitor is actually passing a dc component.)

The slope (or droop) at A is exaggerated in order to more clearly show what happens. The degree of slope is a function of how much charge the capacitor attains, and so the longer the time constant relative to the pulse length, the smaller the charge and the smaller the slope. But some slope will exist no matter how long the time constant. At the end of the first pulse, the output voltage has dropped toward zero slightly, and when the input falls toward zero (B), it makes a positive-going excursion of 6 volts. The output must also make a positive-going excursion of 6 volts, but in this case it is starting from perhaps −5.0 volts. Six volts in a positive direction from −5.0 volts must be +1 volt, and this is shown at C.

The small charge on the capacitor, obtained during the interval marked "T_0 to T_1," now begins to discharge, but remember the circuit time constant is very long and the capacitor discharges only a small amount of its charge when the input again changes, going to −6 volts at T_2. Since there is still some charge upon the capacitor, the output waveform cannot start at zero again but rather at whatever value is still left across the capacitor, say, +0.5 volt. Six volts more negative than +0.5 volt is −5.5 volts, and at T_2 the output is seen to be at −5.5 volts. This action continues, each successive pulse shifting away from the 0-volt line until the pulses are bisected by the base (or 0-volt) line, after which no further change occurs and the pulse levels are equally distributed around the base line. Note that what actually happened is that the average dc level becomes 0 volts, as, of course, it must.

If the input pulse is nonsymmetrical, i.e., the period T_0-T_1 is not the same as the period T_1-T_2, the base line will not exactly bisect the peak-to-peak values. Such an example is shown in Fig. 3-20a. Now, the average dc level of this waveform is

$$E_{av} = \frac{T_{p1}}{T_{p1} + T_{p2}} \times -8 \text{ volts} = \frac{1}{1+3} \times -8 = \frac{1}{4} \times -8 = -2 \text{ volts}$$

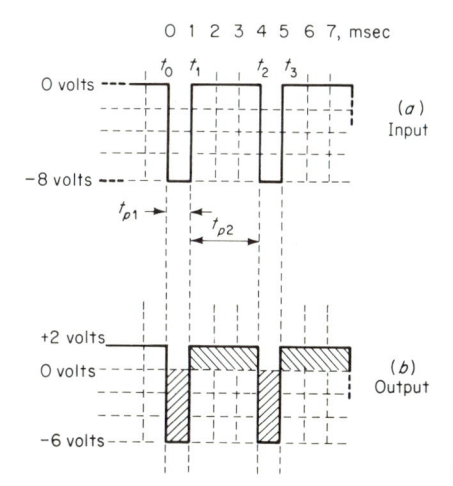

Fig. 3-20. Base-line shift using nonsymmetrical pulses.

48 PULSE AND SWITCHING CIRCUIT ACTION

(Note that whether T_{p1} or T_{p2} is placed in the numerator depends upon which represents the value *away* from ground.) When this waveform is passed through an *RC* coupler, the average dc level becomes the new base line, as shown in Fig. 3-20*b*. Note that the average voltage above the base line is the same as the average voltage below the base line for one complete cycle. That is, as drawn, there are three squares above and three squares below the 0-volt line. The excursion above the base line is only 2 volts but remains there for three units of time. When the waveform falls below the base line, it remains there for only one unit of time but goes to -6 volts. Hence, the product of volts times time is the same either side of 0 volts. This, of course, must be true since again the capacitor cannot allow the dc component to pass.

3-5 DC RESTORATION

At times, when using an *RC* coupler, the loss of the dc level is not desirable. In these instances a restoration of the level is necessary. Dc restoration consists of arranging a circuit to reestablish a fixed dc potential in relation to the original waveform. The fixed dc potential is often ground, although it does not have to be.

The simpler forms of dc restorers are used with repetitive waveforms, usually. A circuit to provide dc restoration is shown in Fig. 3-21. The

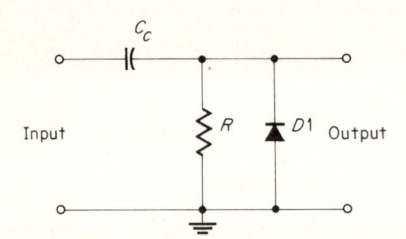

Fig. 3-21. A circuit providing positive dc restoration.

waveforms for this circuit are given in Fig. 3-22*a*, *b*, and *c*. Figure 3-22*a* shows the input voltage, which is seen to be a square wave operating between the limits of 10 and 20 volts.

In a coupling circuit without dc restoration (diode *D*1 removed, Fig. 3-21), the output would center around the 0-volt base line, as shown in Fig. 3-22*b*. With the output direct current restored (Fig. 3-22*c*), the waveform operates between ground and $+10$ volts. With the circuit as shown, the most negative portion of the waveform is always clamped to ground and hence is said to be restored to this fixed dc level.

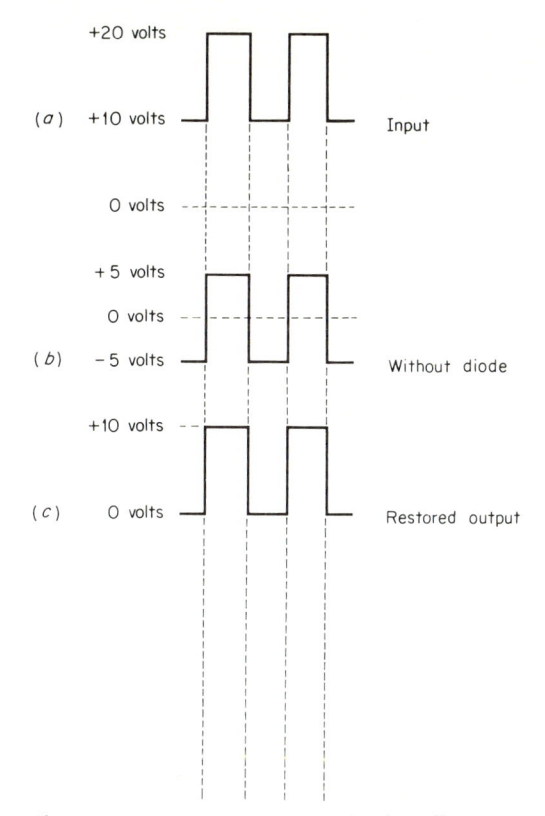

Fig. 3-22. Waveforms for the circuit of Fig. 3-21.

A dc restorer arranged to provide positive output pulses is called a *positive dc restorer*. It is just as possible to provide negative output pulses, as indicated in Fig. 3-23. The only significant difference between the two circuits is that the diode is reversed in each case. The output is shown as a 10-volt negative pulse referred to ground. Thus, an output that is either

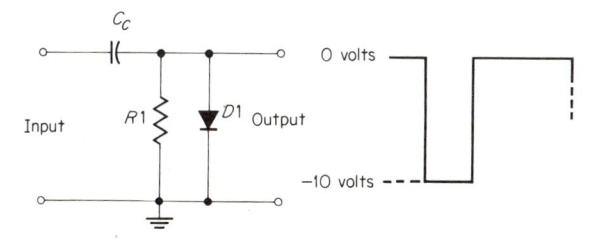

Fig. 3-23. A circuit providing negative dc restoration.

50 PULSE AND SWITCHING CIRCUIT ACTION

more positive or more negative than ground is possible. By returning the diode to a voltage other than ground, any other reference is possible.

How the circuit works to restore a dc level is best understood by using the circuit and waveform shown in Fig. 3-24. Assume that the input is a square-wave series of pulses with an amplitude of 8 volts. The actual levels between which the input operates are immaterial since the output is clamped to ground.

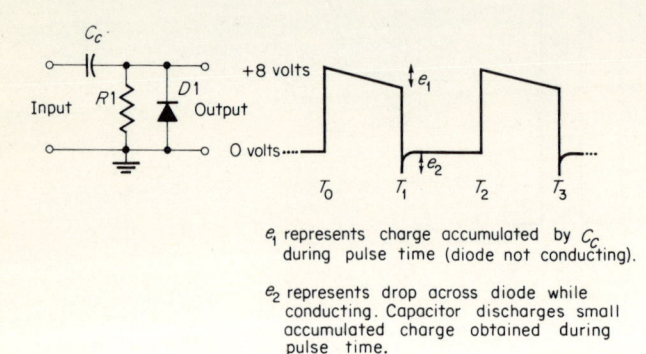

e_1 represents charge accumulated by C_c during pulse time (diode not conducting).

e_2 represents drop across diode while conducting. Capacitor discharges small accumulated charge obtained during pulse time.

Fig. 3-24. A circuit and waveforms used to explain dc restoration.

The waveform shown in Fig. 3-24 is exaggerated, particularly with respect to e_1, which represents the droop caused by the finite time constant of C_c and $R1$. Normally $C_c \times R1$ is made as long as practical to preserve the shape of the square-wave input. In Fig. 3-24, however, the droop is accentuated to be more clear.

Recall from a previous discussion that the amount of droop, or sag, is determined by the amount of charge accumulated during the pulse. Hence, the longer the time constant, the smaller the charge attained. Some charge, however, is accepted by the capacitor no matter how long the time constant is made.

At time 1 (T_1), the input waveform makes an 8-volt excursion in the negative direction. This drives the cathode of $D1$ very negative and causes it to become forward-biased. Thus, $D1$ conducts with a characteristically low forward resistance, perhaps on the order of 10 ohms.

Because now the circuit time constant is determined by C_c and $D1$, the capacitor very quickly rids itself of the accumulated charge. Hence, except for the small drop across the diode indicated as e_2 (again exaggerated), the output waveform is essentially at ground. With no capacitor current through either *R*1 or *D*1, after the first instant there is no voltage drop and the output can be nothing but 0 volts.

Actually, the only purpose of the diode is to remove the small accumulated charge on the capacitor between pulses. It is this accumulated charge that, without the diode, builds up, pulse after pulse, and that causes the base-line shift. By eliminating the buildup of capacitor charge after each pulse, the centering of the waveform around the midpoint is prevented.

3-6 TYPICAL *RC* CIRCUITS

In this section various *RC* circuits will be discussed. Limited space precludes a discussion of all possible combinations currently in use. It is hoped, however, that a sufficiently wide range of examples is given to provide the reader with a good fundamental knowledge of the subject.

The Differentiator

The first circuit to be covered is that of the differentiator, shown in Fig. 3-25. Note that it appears identical to that of an *RC* coupler and, in fact,

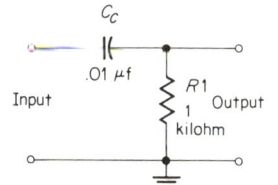

Fig. 3-25. A differentiating circuit.

is identical. Any *RC* coupler may be used as a differentiator if the relationship between the pulse length T_p and the circuit time constant t_c is adjusted accordingly.

If the circuit is to be used as a coupler, t_c is made to be on the order of 100 times the pulse length. On the other hand, if the time constant is made equal to or less than $0.2 \times T_p$, the circuit will differentiate the input waveform.

Some typical waveforms are shown in Fig. 3-26 to acquaint the reader with a differentiated waveshape. If $t_c = 100 T_p$, the output of the *RC* circuit is as shown in Fig. 3-26*b* and the circuit acts as a coupler. If $t_c = 0.2 T_p$ or less, the waveform is said to be differentiated with the characteristic shape shown. The peak-to-peak voltage of the output is very nearly twice that of the input.

52 PULSE AND SWITCHING CIRCUIT ACTION

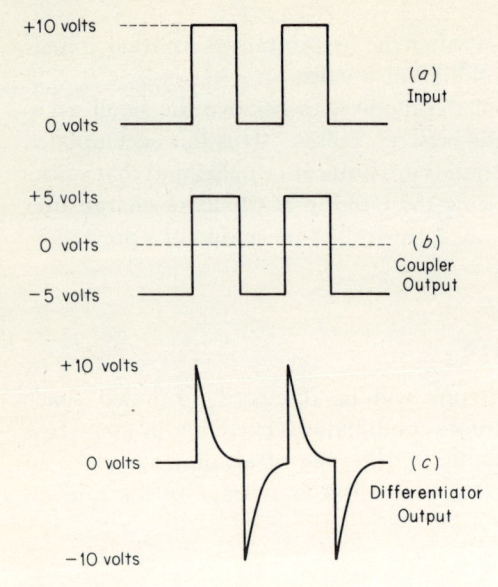

```
+10 volts  ------┐   ┌──┐   ┌──┐
                 │   │  │   │  │        (a)
                 │   │  │   │  │        Input
   0 volts ──────┘   └──┘   └──
```

```
 +5 volts         ┌──┐   ┌──┐
                  │  │   │  │           (b)
   0 volts ───────┤  ├───┤  ├──────     Coupler
                  │  │   │  │           Output
 -5 volts ────────┘  └───┘  └──
```

```
+10 volts        ╱╲      ╱╲
                 │ │     │ │            (c)
   0 volts ──────┘ │  ╱──┘ │  ╱───      Differentiator
                   │ │     │ │          Output
                   ╲╱      ╲╱
-10 volts
```

Fig. 3-26. Input-output waveforms for the differentiator.

The reason for this distinctive waveform is related to the required charge time of the capacitor. With the values shown in the schematic drawing of Fig. 3-25, the time constant of C_c is 10 μsec.

$$t_c = R \times C = (1 \times 10^3)(0.01 \times 10^{-6}) = 0.01 \times 10^{-3} = 10 \text{ μsec}$$

Because a capacitor reaches full charge in five time constants, if the input pulse is exactly 50 μsec in length, the capacitor will reach full charge (99.3%) just as the pulse is ready to terminate. If the capacitor reaches nearly full charge, there is virtually no current through the resistor and so no voltage is developed across it. This is why the output waveform just gets back to zero as each pulse terminates.

Now, just prior to the termination of the first pulse, the left plate of C_c is still at +10 volts but the right plate has fallen to zero. Thus, the capacitor has a full 10-volt charge across its plates, with the left plate the more positive of the two. When the input suddenly goes to ground, the left plate must also go to ground. If the left plate is at ground at this instant, the right (output) plate must be 10 volts more negative than 0 volts, or −10 volts. This is true, of course, because the input voltage changes much more rapidly than the capacitor can change its state of charge. With the input back to 0 volts, the capacitor must now discharge and, with a period of 50 μsec in which to accomplish this, will just be discharged when the new pulse begins.

Often, the relationship between t_c and T_p is not exactly a 5:1 ratio; that is, until now we have been considering the differentiator with $t_c = 0.2 T_p$. When this is not the case (when t_c is less than $0.2 T_p$), the resultant waveform is slightly different. Figure 3-27 illustrates such a case.

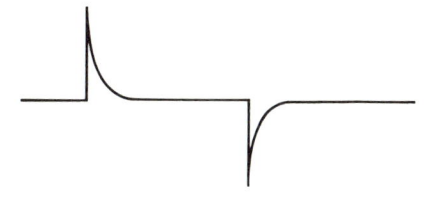

Fig. 3-27. Time constant less than $0.2 t_p$.

Here, the time constant is so much shorter than the pulse width that the capacitor becomes fully charged early in the pulse time. The output therefore falls to zero and remains there until the input pulse terminates.

At times it is necessary to differentiate the *leading edge* of a pulse. In this case the time constant has a great bearing on the amplitude of the output as well as the output width. Figure 3-28 shows a pulse with a rise

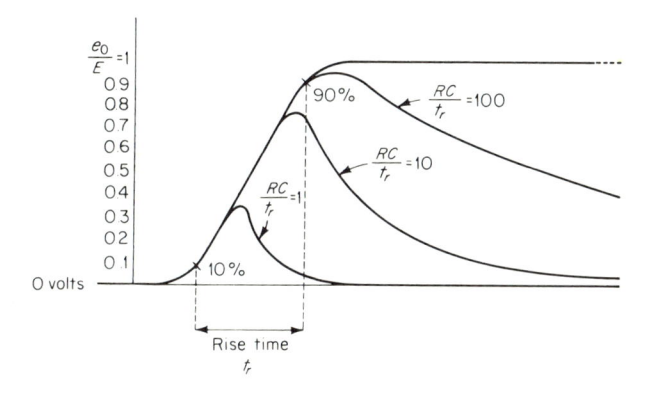

Fig. 3-28. Differentiating the leading edge of a pulse.

time of t_r. If the time constant is equal to the rise time, the output is only 37% of the input amplitude but the pulse width is about equal to the rise time. If the time constant is 10 times the rise time, the output is about 75% of the input amplitude and the pulse width is about 10 times the rise time. Finally, if RC is 100 times the rise time, the output amplitude is nearly equal to the input and the pulse length is very long relative to the input rise time.

The Integrator

An integrator circuit is shown in Fig. 3-29. Although used less frequently than the differentiator, one does encounter it on occasion. Loosely speaking, any circuit similar to this is called an integrator. Actually, however, only when the time constant of $R1$ and $C1$ is long relative to the input pulse length does the circuit actually integrate the input signal.

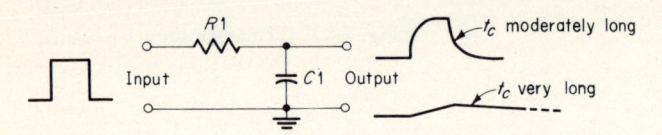

Fig. 3-29. Integrator circuit and waveforms.

The name is derived from the fact that if the time constant is very long, the voltage across the capacitor will be very small in comparison to the drop across the resistor. Therefore, the capacitor current is a function of only the resistor and the output is proportional to the integral of the input.

An example of the use of an integrator is shown in Fig. 3-30. If the

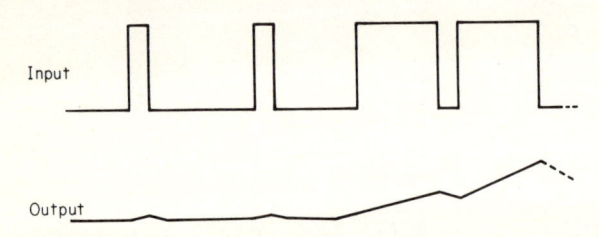

Fig. 3-30. Waveforms of TV sync-separator.

input of the integrator consists of very short pulses ($t_c = 100 T_p$), the integrator does not produce any appreciable output. But if the input suddenly changes to a series of very long pulses, then the output rises to signal the arrival of the broad pulses. Such a circuit is used in the sync-separator stages of a TV receiver. The horizontal sync pulses are very narrow, while the vertical sync pulses are much wider. By using such a circuit the receiver can be controlled by the transmitted pulses, the vertical field being synchronized by the integrator output.

Compensated Voltage Dividers

A voltage divider, or attenuator, used in a dc circuit is a relatively simple device. There are many occasions where a voltage divider must be used with pulses, some of which have sharp leading and trailing edges. In most cases the voltage divider causes badly degraded leading and trailing edges of the square wave. This is illustrated in Fig. 3-31a which shows a

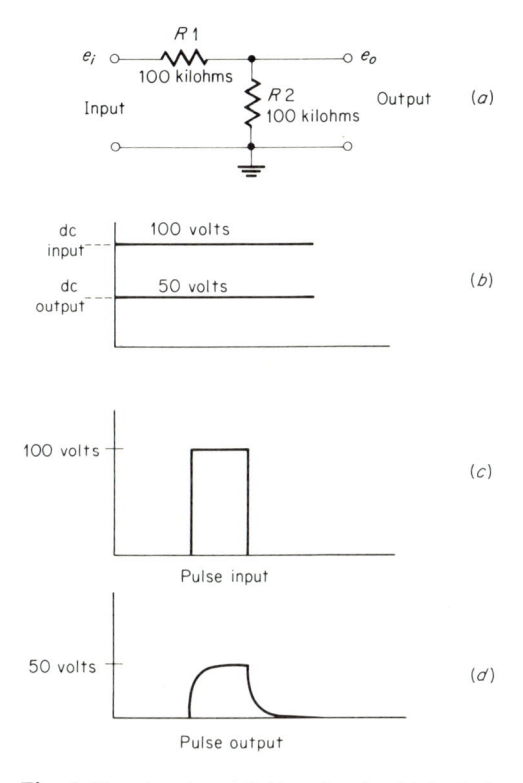

Fig. 3-31. A voltage-divider circuit with both dc and ac waveforms.

voltage divider where the division is 50%. That is, the output is 50% of the input since $R1$ equals $R2$ in value.

If a dc voltage is applied between the input terminals, say, 100 volts, the voltage at the output terminals is

$$e_o = \frac{R2}{R1 + R2} \times E = \frac{100 \text{ kilohms}}{200 \text{ kilohms}} \times 100 = 50 \text{ volts}$$

As long as the output is not loaded too severely, the circuit will produce the expected 50 volts. However, when a short pulse is applied, say, a 6-volt pulse, the output may have a much longer rise time than the input, as illustrated. In most cases, the increase in the rise time is so severe that it cannot be tolerated. The voltage divider, then, requires *compensation* and will be spoken of as a *compensated voltage divider*. Before trying to correct for this defect, one should know the cause. Then, a reasonable solution can be arrived at.

This circuit is usually described in terms of frequency, and although the frequency and time domains are closely related, it is far more meaningful for our purpose to describe such a circuit in terms of pulses and time.

First, the reason for the long slope on the leading and trailing edges is the inevitable capacitance which appears in shunt with $R2$. This capacitance is associated with a printed circuit board, the input capacitance of a vacuum tube, or input capacitance of a transistor or diode. In any event, it always exists and is the chief cause of the poor rise and fall times.

Quite often, $R1$ is a very large value resistor. In the example shown in Fig. 3-32, $R1$ is 9 megohms. Because $C2$ must charge essentially through

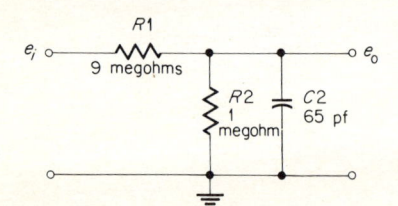

Fig. 3-32. $C2$ is the reason for the rounding of the pulse edges.

$R1$, the time required is very long. Hence, the voltage at the output terminal can only change very slowly. The rise time of this circuit can be calculated quite easily:

$$t_r = 2.2 \times R_t C_t$$

where t_r = rise time
2.2 = constant
R_t = total effective R
C_t = total effective C

With the circuit shown, C_t is, of course, equal to $C2$. The effective resistance, however, is the *parallel* resistance (Thevenized) of $R1$ and $R2$.

$$R_t = \frac{R1 \times R2}{R1 + R2} = \frac{(9 \times 10^6)(1 \times 10^6)}{(9 \times 10^6) + (1 \times 10^6)} = 0.9 \text{ megohms}$$

$t_r = 2.2 \times 0.9 \text{ megohms} \times 65 \text{ pf}$
  $= 2.2(0.9 \times 10^6 \times 65 \times 10^{-12}) = 2.2 \times 58.5 \times 10^{-6}$
  $= 128.7 \times 10^{-6} \cong 129 \text{ } \mu\text{sec}$

Obviously, if this circuit were to be used with microsecond pulses, the rise time of 129 μsec would be much too long. By applying the principles of compensation, this can be greatly improved.

Note in Fig. 3-33 that the rounded portions of the waveform represent

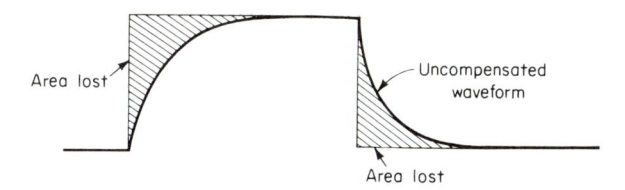

Fig. 3-33. A degraded square wave.

one boundary of the area that is lost relative to a perfect waveshape. If this part of the waveform could be restored, the output voltage would be returned to its original waveshape.

Now, note that the actual rounded waveshape resembles a partially integrated waveform. That is, if the input pulse width were shorter, true integration would be effected by the circuit. To restore the waveform to its original shape, the *inverse of the area lost* must be *added back* to the circuit, as suggested by Fig. 3-34. Since the inverse of integration is differentia-

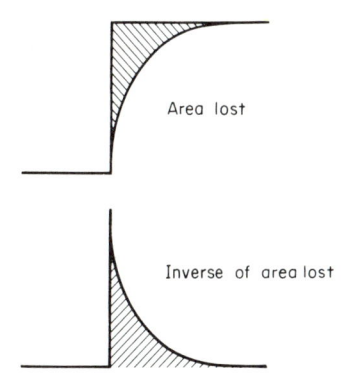

Fig. 3-34. The loss in the pulse edge and the inverse of this loss.

tion, it seems likely that a differentiated waveform is needed, to be added to the original output, thus restoring the waveform to that of a square wave.

Such a circuit has been drawn in Fig. 3-35. $R1$ and $C2$ represent the integrator part of the circuit, while $R2$ and $C1$ represent the differentiator part. If the added capacitor $C1$ is chosen properly, the output will have exactly the same rise time as the input waveform.

58 PULSE AND SWITCHING CIRCUIT ACTION

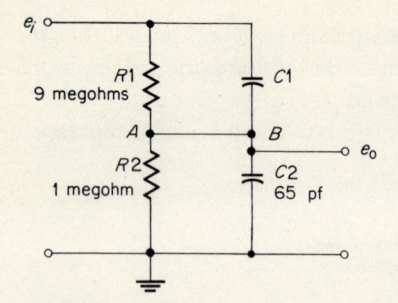

Fig. 3-35. A compensated voltage divider.

By adding $C1$ to the circuit, a differentiated component is added to the waveform. The algebraic sum of the uncompensated output and the differentiated portion added by $C1$ will exactly equal a square wave. This is illustrated in Fig. 3-36. The waveform labeled e_1 is the uncom-

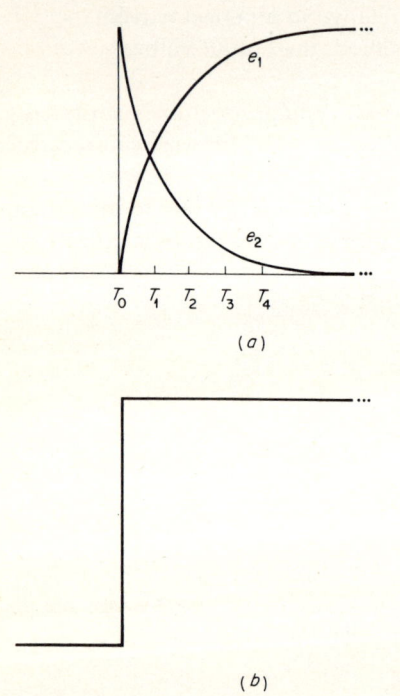

Fig. 3-36. Illustrating waveform summation. (*a*) Waveforms to be summed; (*b*) summation result.

pensated voltage at the output, while e_2 represents the voltage at the output caused by $C1$.

At the first instant (T_0) the output due to e_1 is still zero but e_2 is equal to the full value expected (100%). At T_1 the voltage e_1 has risen to perhaps 50%, while that of e_2 has fallen to 50%. Thus, the sum is

50 + 50 = 100%, and the output is still 100%. As time progresses, say, to T_4, e_1 has risen to about 90%, while e_2 has fallen to 10%. Their sum is still 100%. Beyond T_4, e_2 has fallen to zero, but by now e_1 has risen to 100% where it will stay as long as the input requires. Thus, the waveform has been restored to its original shape, shown in Fig. 3-36b as the summation result.

It remains only to determine the proper value of $C1$ to exactly compensate for the rise time losses. Observing Fig. 3-35 carefully, note that it resembles a bridge circuit. If the voltage drop across the capacitors can be made to exactly equal the drop across the resistors, there will be no current flow in the center arm of the bridge. Hence, no resistive current will flow into the capacitors to affect the capacitors' charge time.

Recall from elementary capacitor theory that the voltage drop across a series string of capacitors is inversely proportional to the value of the capacitors. Noting that $R1$ is nine times larger than $R2$, if $C1$ is to have the same voltage drop across itself that exists across $R1$, it must be nine times smaller than $C2$!

Now the circuit shown in Fig. 3-37 applies. $C1$ must be in the same

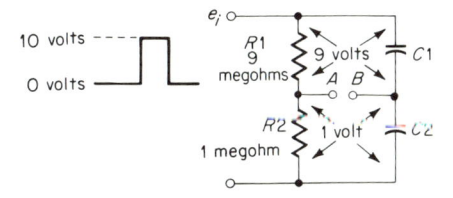

Fig. 3-37. The voltage drops in a balanced RC bridge.

ratio to $C2$ that $R2$ is to $R1$ because of the inverse voltage division across the capacitor. Setting this in a proportion yields the following:

$R1:C2::R2:C1$ and $R1C1 = R2C2$

Solving this for $C1$,

$$C1 = \frac{R2C2}{R1} = \frac{(1 \times 10^6)(65 \times 10^{-12})}{9 \times 10^6}$$
$$= 7.22 \times 10^{-12} = 7.22 \text{ pf}$$

Thus, if $C1$ is 7.22 pf exactly, the voltage division across the capacitor branch is the same as the division across the resistor branch and points A and B will always be at the same potential. Hence, when they are connected together to form the complete circuit of Fig. 3-35, there will be no

60 PULSE AND SWITCHING CIRCUIT ACTION

current through the center leg and the circuit will be completely compensated. The output waveform will be a faithfully attenuated replica of the input.

In practice, the capacitor $C1$ is often a small variable capacitor in instances where the ratio must be satisfied exactly. Component tolerances are normally too wide to allow exact compensation. Hence by using a small trimmer the compensation can be made precise.

There are cases that demand either undercompensation or overcompensation rather than exact compensation. In these cases, the capacitor $C1$ can be made smaller than the calculated value for undercompensation or larger for overcompensation. The waveforms for these instances are shown in Fig. 3-38. It should be noted that rather than make $C1$ smaller, $C2$ can be made larger for the same effect.

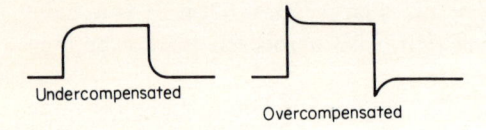

Fig. 3-38. Overcompensation and undercompensation.

The AC Gate

A very simple RC (or AC) gate circuit is shown in Fig. 3-39. A *gate* is a conditional circuit that is considered to have produced an output only with a specified input condition. All other input possibilities will not produce an output. (Many other gate circuits will be discussed subsequently.)

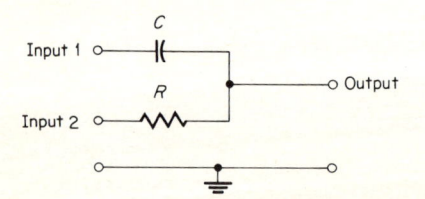

Fig. 3-39. An RC gate.

The purpose of this circuit is to provide an output only under special conditions. The circuit is considered to have produced an output only when a voltage is developed that is more positive than ground. The signal voltages are shown in Fig. 3-40 and are seen to consist of pulses that operate between the levels of ground and -8 volts. The output consists of differentiated pulses derived from the waveform at input 1. Although

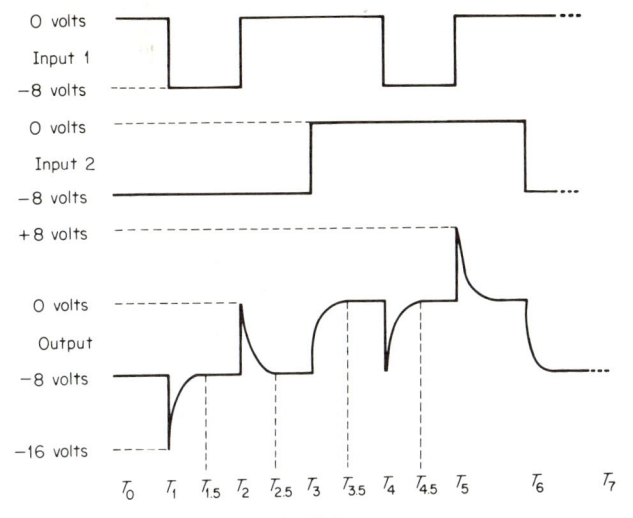

Fig. 3-40. Waveforms for the *RC* gate.

several spikes are evident, only the one that makes an excursion above the 0-volt base line is a true output since the output circuitry will respond only to voltages that are positive with respect to ground.

Figure 3-41 will help to visualize the occurrences in the circuit as the pulses shown are applied. Initially, the input plate of the capacitor is at ground potential, while the output plate is returned through the resistor

	Input, volts	Output, volts
T_0	0	−8
T_1	−8	−16
$T_{1.5}$	−8	−8
T_2	0	0
$T_{2.5}$	0	−8
T_3	0	−8
$T_{3.5}$	0	0
T_4	−8	−8
$T_{4.5}$	−8	0
T_5	0	+8
T_6	0	0
T_7	0	−8

Fig. 3-41. Illustrating capacitor action in the *RC* gate.

to -8 volts. Hence, the capacitor is at first charged to 8 volts. Then, input 1 makes its first excursion from ground to -8 volts. Since the change occurs very rapidly, the capacitor cannot change the state of charge and at the end of the transition must still have 8 volts across its plates. If the left (input) plate is at a potential of -8 volts with respect to ground, the right (output) plate must be at -16 volts, as shown.

There is nothing in the circuit that can hold the output plate at -16 volts, and so the capacitor now recharges to the new state specified by the voltages at inputs 1 and 2. With -8 volts at both inputs, the capacitor loses the 8-volt charge and the output plate goes from -16 to -8 volts. This occurs between the time T_1 to $T_{1.5}$. The time of recovery is a function of $R \times C$. Until another transition occurs, nothing further happens.

At time 2 (T_2) input 1 goes from -8 volts to ground, an 8-volt positive-going excursion. Just prior to the change, the capacitor had zero charge on its plates (-8 to -8 volts). During the transition the capacitor cannot change its state of charge, as before, and it must retain the zero charge from plate to plate. Since the input plate goes from -8 volts to ground, the output plate must also make a positive excursion from its starting point (-8 volts) to a value 8 volts more positive than this. Hence, it too goes to ground. However, it returns to -8 volts through the resistor, and so again the capacitor must recharge to this new condition, with 0 volts at the input and -8 volts at the output.

Now, at time 3, input 2 goes from -8 volts to ground and the capacitor will recharge to *this* new condition. After sufficient time has elapsed, the left plate is at ground while the right plate is at ground also. At time 4 the input-1 waveform again goes from ground to -8 volts. Just prior to the transition the capacitor had zero charge, and so it must still have zero charge when the transition has just terminated. Thus, with the input-1 signal driving the left plate to -8 volts, the right plate must also go to a value 8 volts more negative than it was. It was at 0 volts, and so it is driven to -8 volts but cannot remain there for very long. Since the resistor is returned to ground, the right plate must eventually return to ground. The capacitor is now charged so that the left plate is at -8 volts while the right plate is at ground. This, of course, is an 8-volt charge from plate to plate.

At time 5 the transition at input 1 occurs, and this is the transition that will produce the output. The signal at input 1 goes from -8 volts to ground, an 8-volt excursion in the positive direction. Again, the output plate must also make an excursion that is 8 volts in the positive direction. But now, for the first time, the right plate of the capacitor is at ground just prior to the positive-going transition and so must go 8 volts in the positive direction from ground. Thus, the output is lifted from ground to

+8 volts, and this is defined as the output. Again the right plate cannot remain at this value, and so the waveform decays exponentially toward ground. Finally, the input at the resistor goes back to −8 volts, and the circuit returns to the original condition. In a later chapter this circuit will be used along with many kinds of transistor circuits.

The same circuit could be used to provide an output that is more negative than ground by using voltages that extend to values more positive than ground. A circuit to accomplish this is shown in Fig. 3-42, along with the appropriate waveform.

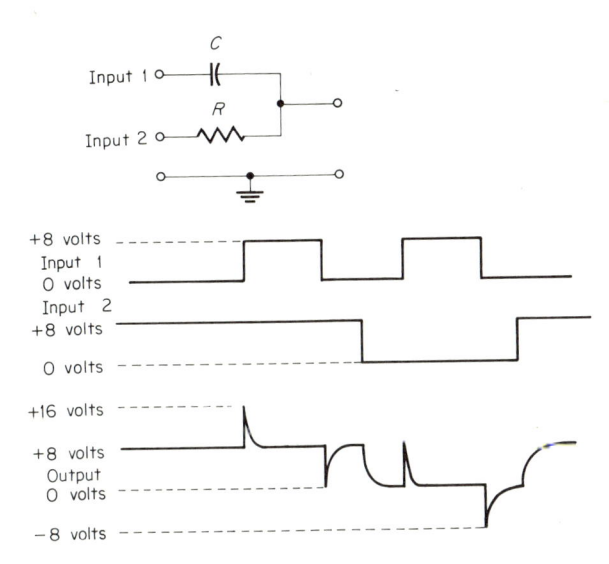

Fig. 3-42. The circuit and output of the RC gate.

In order to function properly the time constant of R and C must be carefully chosen to agree with the pulse widths used. If the pulse is too short (or the time constant too long), the capacitor will not have time to completely recover between transitions and the circuit will fail. As an example of choosing the proper value for R and C, assume that the pulse width at input 1 is 20 μsec. Assume a 10-kilohm resistor value is to be used. Determine the maximum value for the capacitor. Because the capacitor must become fully charged in 20 μsec or less, the time constant must be at least one-fifth this or less. Hence, t_c must be 20 μsec/5 = 4 μsec.

$$t_c = RC \quad \text{and} \quad C = \frac{t_c}{R} = \frac{4 \text{ μsec}}{10 \times 10^3} = 400 \times 10^{-12} \text{ farads}$$

Thus, a 400-pf or smaller capacitor is necessary to assure that the capacitor will reach the new state of charge in time.

QUESTIONS AND PROBLEMS

3-1 A capacitor is charged, and the electric energy stored is known to be 0.35 joule. This is to be discharged into a resistor. What is the total power dissipated by the resistor during discharge if the discharge time is 1 second?

3-2 A 4-μf capacitor is charged to 100 volts. What is the amount of energy stored in joules?

3-3 Refer to Fig. 3-1. $B1 = 20$ volts, $R1 = 1$ kilohm, and $C1 = 1.5$ μf. The maximum current that can flow in this circuit is what value?

3-4 Refer to Fig. 3-1. $B1 = 80$ volts, $R1 = 4$ kilohms, and $C1 = 3$ μf. What is the maximum current that can flow in this circuit?

3-5 Refer to Fig. 3-1. Assuming initial discharge, what is the time required to charge $C1$ completely, for all practical purposes, if $B1 = 12$ volts, $R1 = 18$ kilohms, and $C1 = 1$ μf?

3-6 Refer to Fig. 3-1. On the assumption of initial discharge, what time is required to fully charge $C1$ if $B1 = 24$ volts, $R1 = 810$ kilohms, and $C1 = 0.0001$ μf?

3-7 Refer to Fig. 3-10. The resistor value is to be changed from 1 to 0.5 megohm, but all else remains the same. To what voltage will the capacitor charge after 50 msec? (Assume initial discharge.)

3-8 Refer to Question 3-7. Using the same circuit values, to what voltage will the capacitor charge after 125 msec, assuming initial discharge?

3-9 Refer to Fig. 3-12. The circuit is to be used as a coupler with minimum droop. The pulse width is 200 μsec, and the resistor has a 47-kilohm value. Determine the minimum size of the capacitor.

3-10 Refer to Fig. 3-12. The circuit is to be used as a coupler with minimum droop. The pulse width is 1 msec, and the capacitor value is 10 μf. Determine the minimum size of the capacitor.

3-11 Given the following circuit and input (a continuous waveform), draw the resultant waveform on the graph provided. Each vertical division is 2 volts, and each horizontal division is 1 μsec.

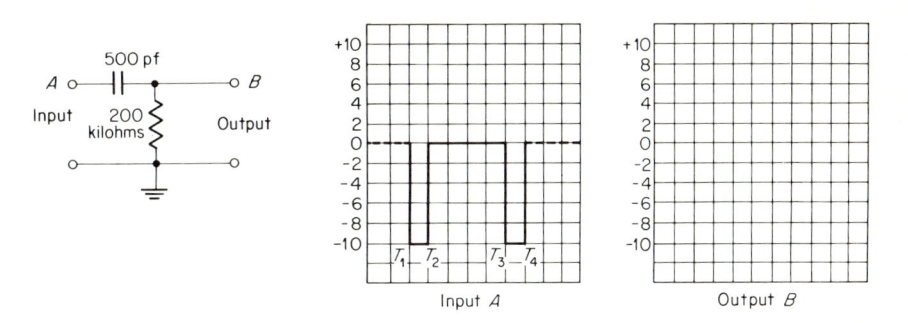

3-12 Given the following circuit and input (a continuous waveform), draw the resultant waveform on the graph provided. Each vertical division is 2 volts, and each horizontal division is 1 μsec.

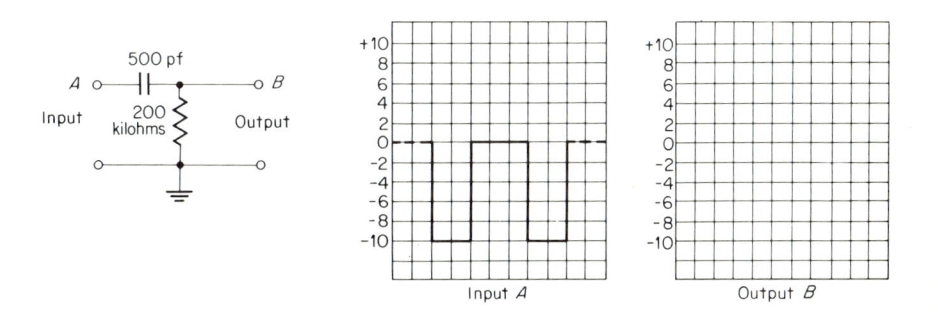

3-13 Given the following circuits, find the value of $C1$ to exactly compensate the divider in each case.

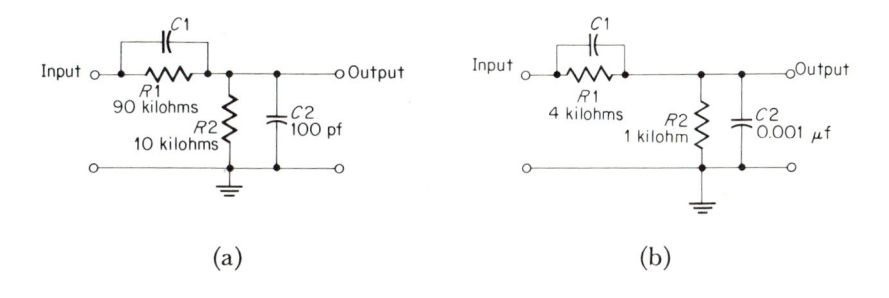

(a) (b)

4
RL NETWORKS

From basic studies of electricity, it is known that a magnetic field surrounds a wire that is carrying an electric current. Also, it is recalled that if a wire is caused to move in a magnetic field so that the lines of force are cut by the wire, an emf is induced in the wire. By the same token, if a magnetic field moves near a stationary wire such that the lines of force cut the wire, an emf is similarly induced in the wire. The following material briefly reviews some effects in typical magnetic circuits and goes on to apply these principles to some typical circuits.

4-1 INDUCTOR ENERGY STORAGE

Consider first Fig. 4-1, which illustrates some of the fundamental relationships of a typical inductive circuit. In Fig. 4-1*a* the switch is open, and hence the current is zero; no magnetic effects are apparent in the circuit then. In Fig. 4-1*b*, the switch has just been closed, and therefore current begins to flow. Current increases more or less gradually, and as it does the magnetic lines of force "grow" outward, becoming more and more intense as current increases.

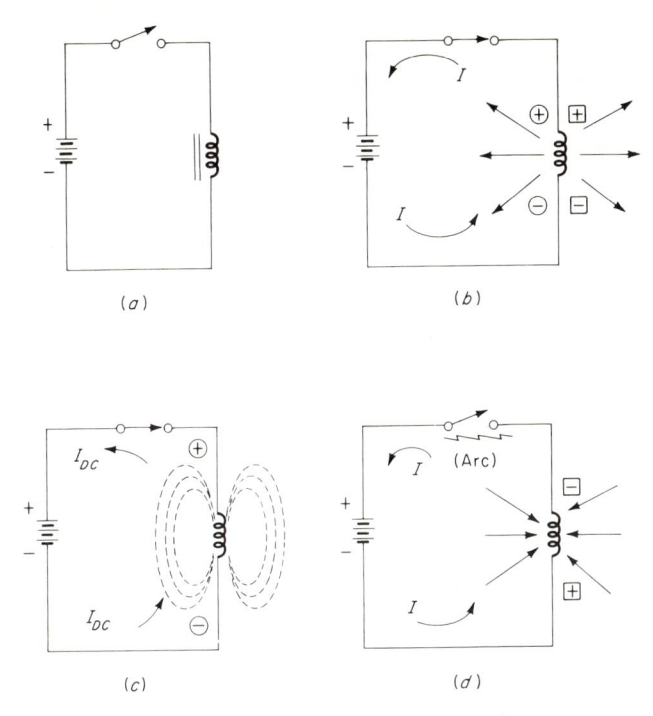

Fig. 4-1. A simple circuit illustrating induced emf.

As the lines of force move outward, they cut across the coil itself generating, or inducing, a voltage across the coil that opposes the applied voltage. Hence, there are two voltages across the coil during the time current is building up toward maximum.

The two voltages are shown on the drawing with different symbols. The normal drop is shown encircled, while the induced-voltage polarity is shown with a square. These values appear to be the same since the polarities are identical, but such is not the case. One of these is a voltage *drop* and as such can only *limit* current flow. The other, however, is a true *source* of voltage and as such must be treated as a source. This voltage, while it exists, tries to cause current flow in a direction opposite to that produced by the battery. Thus, it does more than merely *limit* current flow; it *actively opposes it*.

As the current reaches a maximum value, limited by any series resistance (not shown), the magnetic lines of force no longer are in motion; they surround the coil, as suggested by Fig. 4-1c. Since no motion exists between the lines and the coil, no voltage is induced and the voltage across the coil is simply the drop across the dc resistance of the coil.

When the switch is opened, as shown in Fig. 4-1d, the dc current falls almost immediately to zero. The magnetic lines of force must now fall back toward the coil, and again they cut across the coil, inducing a voltage in the coil. The direction of this induced voltage is such as to try to maintain current flow in the same direction as before the switch was opened. Again, this is now a *source*, and since current (electrons) flows from positive to negative within a source, the lower end of the coil becomes positive while the upper end becomes negative. Note that this polarity is in a direction to aid, rather than oppose, the battery voltage.

If the inductance is large and if the dc current falls to zero very rapidly, the induced voltage becomes very large, often hundreds of times larger than the battery voltage. This is often large enough to ionize the air between the switch contacts, and so an arc occurs. The energy that had been stored in the magnetic field is thus dissipated in overcoming the resistance of the open switch.

Many devices operate on this principle. The ignition coil of an automobile produces about 30,000 volts partly by means of this principle. The high voltage of a TV set is also produced in this same general manner, yielding up to 24,000 volts from a dc supply of, perhaps, 400 volts!

At other times this effect is undesirable and causes no small amount of trouble. This is particularly true of hybrid circuits using such components as relays and transistors. The relay circuits can produce voltage spikes up to perhaps 2000 volts while operating on, for instance, a 50-volt dc supply. Obviously, some method of arc suppression must be employed to eliminate these dangerous voltages, for a transistor can be instantly destroyed by them.

As an example of the voltage-generating capability of such a circuit, consider Fig. 4-2. The induced voltage produced across the coil is a

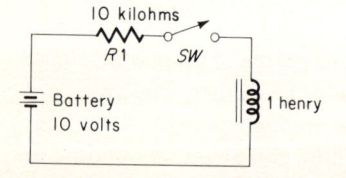

Fig. 4-2. A circuit to demonstrate that $e_{ind} = L(di/dt)$.

function of the value of the inductance (in henrys) and the *rate of change* of current. This relationship is shown in the following equation:

$$e_{ind} = L \frac{di}{dt}$$

where L = inductance, henrys
d_i = change in current, maximum to zero
d_t = time for current to decrease to zero

If the current flowing in the 1-henry coil when the switch is depressed is 1 ma and if it decreases to zero in 1 μsec, the induced voltage can be calculated as follows:

$$e_{ind} = L\frac{d_i}{d_t} = 1 \times \frac{0.001}{0.000001} = 1000 \text{ volts}$$

Obviously, great care must be exercised when using inductors in conjunction with semiconductors, which typically are rated at little more than 30 volts.

The amount of energy stored in the field of an inductor is often necessary to know. This energy, expressed in joules (watt-seconds), is simply expressed in the following relationship:

$$J = \tfrac{1}{2}LI^2$$

As an example, the circuit in Fig. 4-2 has an energy storage that is easily determined.

$$J = \tfrac{1}{2}LI^2 = \tfrac{1}{2}(1 \times 0.001^2) = 0.5 \times 0.000001 = 0.0000005 \text{ joules}$$

Thus, 0.5 μwatt can be dissipated over a period of 1 second, or 0.5 watt can be dissipated in 1 μsec.

4-2 INDUCTIVE TIME CONSTANT

The expression *time constant* has essentially the same meaning for inductive circuits that it does for capacitive circuits. The time constant of an inductance and resistance in series is a measure of the period of time required for current to rise to 63.2% of its maximum value or for the induced voltage to fall to 63.2% from its maximum value. This is illustrated in Fig. 4-3. The formula for the time constant of a circuit containing inductance is expressed as a ratio between L and R:

$$t_c = \frac{L}{R}$$

For example, in the circuit of Fig. 4-2 if R equals 10,000 ohms, the time constant is calculated as follows:

$$t_c = \frac{L}{R} = \frac{1}{10,000} = 0.1 \text{ msec}$$

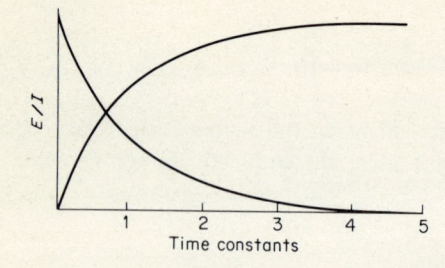

Fig. 4-3. The exponential curves of an inductive circuit.

Hence, in such a circuit the current will rise to 63.2% of its maximum value in 0.1 msec or 100 μsec.

Note that in such a circuit increasing the inductance results in increasing the time constant but increasing the resistance *decreases* the time constant. Thus, time constant and resistance are inversely proportional in the inductive circuit.

To determine the value of current at any point on the rising curve of Fig. 4-3, the following expression is used:

$$i = I_{max}(1 - \epsilon^{-R/L})$$

where I_{max} = ultimate value of dc current
ϵ = epsilon (2.718)
R = series value of resistance, ohms
L = inductance, henrys
i = instantaneous value of current at any desired time

To determine the value of current on the falling curve of Fig. 4-3, the following expression is used, where the definitions are the same as before:

$$i = I_{max}(\epsilon^{-R/L})$$

Because these expressions are evaluated in the same manner as for capacitive circuits, further discussion is deemed unnecessary.

4-3 PULSE TRANSFORMERS

Because inductively coupled circuits are used relatively infrequently in typical pulse circuits, one tends to overlook their importance in the scheme of things. By coupling two circuits together by means of a pulse transformer, several advantages accrue. For one thing, voltages, currents, and impedances may be stepped up or down as needed. Pulse lengths may be made a function of the energy-storage capability of the transformer, impedances may be matched, and so forth.

This general subject can be made to be quite lengthy and detailed, and hence the discussion herein will be necessarily brief. A simple circuit using a pulse transformer is shown in Fig. 4-4. The transformer may be

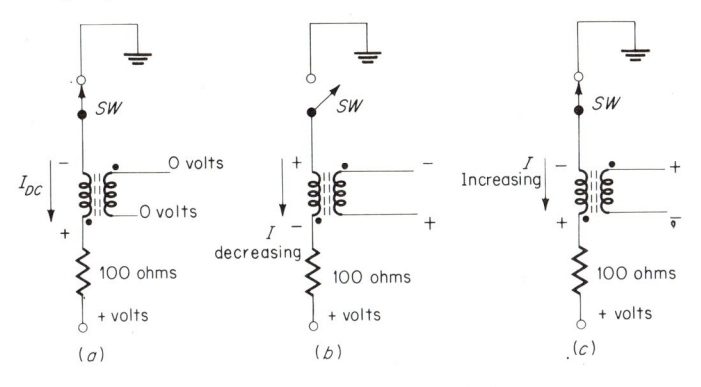

Fig. 4-4. The action of a pulse transformer during a switching sequence.

either a *pot-core* or a *ribbon-core* type. In the former case, the core consists of ferrite, which is a form of powdered iron that is compressed into the proper shape. Figure 4-5 illustrates a typical pot-core transformer. The

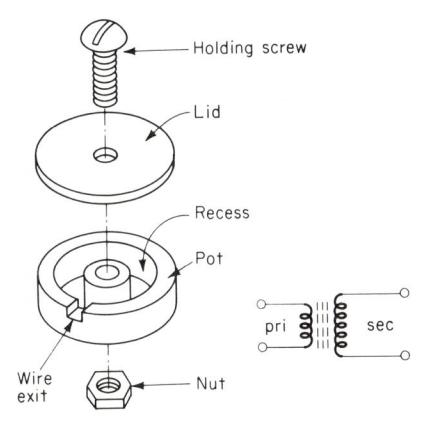

Fig. 4-5. Ferrite pot-core transformer.

windings are usually separate coils of relatively few turns, the size of which is made to fit easily into the recess, with the four wires extending out the wire exit some distance. In the ribbon-core type, the core is in the form of a long ribbon which is wound on a rectangular form. The windings are then wound around the core in the usual fashion. In either case, the inductance of the transformer is usually small, often being on the order of 500 µh or less. Because of this, they are used only with very short pulses, perhaps of a few microseconds or less.

Figure 4-4 shows the circuit action for three different conditions. In Fig. 4-4a the switch is closed and current is flowing. In Fig. 4-4b the switch is shown opened and the assumption is made that the switch will open for about 5 µsec, after which it again closes. Obviously, no mechanical switch can operate so fast, and it is used at this point to illustrate the circuit action only. In reality, the switch is a transistor, which can easily operate at these speeds. Finally, in Fig. 4-4c the switch again closes, allowing current to again flow. To describe the circuit action, the length of time that the switch is open will arbitrarily be set at 5 µsec, while the transformer inductance will be assumed to be 140 µh.

Initially, with the switch closed, electron current is flowing as indicated, limited in value by the applied voltage and the 100-ohm resistor, with the dc resistance of the transformer assumed to be zero. If, however, any appreciable resistance is offered by the primary of the transformer, the resultant voltage drop will be as shown. Since the current is not changing, there is no induced secondary voltage.

At the instant the switch opens (Fig. 4-4b), current begins to decrease. Because the inductance of the transformer will not allow current to stop instantly, some time must elapse before the value of current drops to zero. During this time, while the lines of force are collapsing toward the coil, an emf is induced in both the primary and the secondary, as indicated. The effects of inductance try to keep current flowing as before, and since the primary is now a *source*, electron flow must occur from positive to negative within the primary coil, as shown.

Because the collapsing lines of force must also cut across the secondary coil, a voltage is induced here too. Since the dotted end of the primary is negative at this instant, the dotted end of the secondary is likewise negative. For as long as current continues to flow, the voltage continues to exist at the secondary, decreasing exponentially as the current rate of change decreases.

At this point, one of two things will terminate the voltage, as shown in Fig. 4-4b. If the current flow drops to zero, the voltage across the secondary drops to zero. Alternatively, if the switch is closed again before the current has dropped to zero, the secondary voltage becomes as shown in Fig. 4-4c. Now, the current is again increasing in the primary, and the inductance of the coil opposes the change, again inducing a voltage such as to oppose the applied voltage. This prevents current from immediately increasing to its maximum value, and while it increases, the induced voltage is as shown. Thus, the dotted end of the secondary becomes positive while the lower end becomes negative. This condition will continue until primary current increases to its final value.

If the switch is kept open for a longer time, the voltage existing at the secondary, as shown in Fig. 4-4b, will, as mentioned, terminate of its own accord when the primary current has dropped to zero. If it becomes necessary to compute the time that this voltage exists, the time constant is first determined. Given: $L = 140$ μh; $R = 100$ ohms.

$$t_c = \frac{L}{R} = \frac{140 \times 10^{-6}}{1 \times 10^2} = 140 \times 10^{-8} = 1.4 \text{ μsec}$$

One time constant is therefore 1.4 μsec, and therefore an elapsed time of five time constants will be necessary before the secondary voltage decays to essentially zero:

$$5t_c = 5 \times 1.4 \text{ μsec} = 7 \text{ μsec}$$

Hence, in this instance, if the switch remains open for a time longer than 7 μsec, the voltage at the secondary will decay of its own accord. On the other hand, if the switch is closed for a shorter time, the secondary voltage shown in Fig. 4-4b reverses (Fig. 4-4c) and will decay to zero in approximately 7 μsec.

One further point regarding this circuit action. On the assumption that the switch opens instantly, it would at first appear that current must cease altogether, just as quickly. However, there are several possibilities, consisting of more than one path for current flow for a short period of time. One of these is the stray capacitance of the wiring. Another is the primary-to-secondary capacitance of the transformer, as well as leakage inductance. If these alternate paths are not sufficient for the required amount of current flow, then the switch itself will provide a path across its own contacts, now open. The induced voltage will rise to a very large value, and the switch contacts will arc across as the air between them ionizes and becomes a conducting path. With a value of inductance as small as that given in the previous example, this would not be expected to occur.

QUESTIONS AND PROBLEMS

4-1 In the accompanying figure, what is the direction of current flow (A to B or B to A) if the switch has been in position 1 for some time?

74 PULSE AND SWITCHING CIRCUIT ACTION

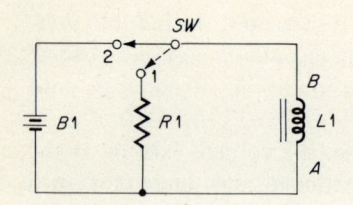

4-2 Refer to the circuit of Question 4-1. The switch has just been transferred from position 1 to 2. At the instant current begins to flow, does it flow from A to B or B to A?

4-3 Refer to the circuit of Question 4-1. The switch has just been transferred from 1 to 2. At the instant current begins to flow, what polarities exist at A and B?

4-4 Refer to the circuit of Question 4-1. The switch has just been transferred from 1 to 2. Which of the following statements most accurately describe the circuit action?
(a) Current begins at a large value and gradually tapers off to zero.
(b) Current rises to a maximum value instantly and remains at maximum.
(c) Current increases gradually toward maximum.

4-5 Refer to the circuit of Question 4-1. The switch has just been transferred from position 2 to 1. Which of the following statements reflect the true circuit action?
(a) Current goes instantly from maximum to zero.
(b) The voltage across the coil disappears instantly.
(c) A large induced voltage appears across the coil.

4-6 Refer to the circuit of Question 4-1. The switch has just been transferred from position 2 to 1. The voltage at A and B is:
(a) Zero
(b) Very small, $A = -$, $B = +$
(c) Very small, $A = +$, $B = -$
(d) Very large, $A = +$, $B = -$
(e) Very large, $A = -$, $B = +$

4-7 Refer to the accompanying circuit. The input pulse is negative-going, relatively short, and R is relatively small in value.

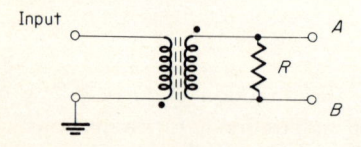

Which of the following waveforms might be viewed at A, referred to B?

(a)
(b)
(c)
(d)
(e)

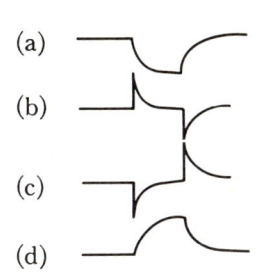

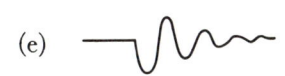

4-8 Refer to the circuit in Question 4-7. The input pulse is to be positive-going rather than negative-going, but all else is to remain the same. Which of the following waveforms might be viewed at A referred to B?

(a)
(b)
(c)
(d)
(e)

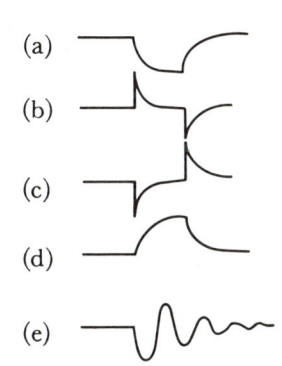

4-9 Refer to the circuit in Question 4-7. Which of the following statements regarding this circuit is true?
(a) The output pulse width is limited by the input pulse amplitude.
(b) The output pulse width is limited by either the input pulse width or the inductance of the transformer, whichever is the greater.
(c) The output pulse width is limited by either the input pulse width or the inductance of the transformer, whichever is the smaller.
(d) The output pulse is independent of any of the above.

4-10 Refer to Fig. 4-2. Determine the induced voltage across the coil when the switch opens (instantly) if $R = 100$ ohms, $L = 1$ henry, and the current decreases at the rate of 1 ma per μsec.

4-11 Refer to Fig. 4-2. Determine the induced voltage across the coil when the switch opens (instantly) if $R = 100$ ohms, $L = 0.001$ henry, and the current decreases at the rate of 1 amp per msec.

4-12 Refer to Fig. 4-2. Determine the induced voltage across the coil when the switch opens (instantly) if $R = 100$ ohms, $L = 0.015$ henry, and the current decreases at the rate of 1 amp per μsec.

5
THE *pn*-JUNCTION DIODE

At this point, the study of semiconductors will begin. It will be assumed that the reader has some familiarity with basic semiconductor physics through the development of the *n*- and *p*-type material. If the reader feels the need to review this area, he is referred to any of the books in the bibliography covering this subject, such as the author's earlier book, "Transistor Circuit Action."[1]

5-1 THE *pn* - JUNCTION

The electrical action of silicon or germanium can be modified by the addition of so-called impurities. This gives rise to the *n*- and *p*-type semiconductors. It will be recalled that the majority carriers in these materials are the electron and the hole, respectively. By themselves, these doped materials are of little practical use to us. If, however, a junction is made, consisting of a piece of *p*-type material joined to a piece of *n*-type material so that the crystal structure is unbroken, a device is produced which *is*

[1] McGraw-Hill Book Company, New York, 1968.

78 PULSE AND SWITCHING CIRCUIT ACTION

extremely useful. We call such a device a *diode*, and its usefulness stems from the fact that it will allow current to flow through it *in one direction only*. The unidirectional properties of a diode allow the "steering" of electric current into a certain path by allowing the passage of current under certain conditions and disallowing it under the reverse conditions. This is called the process of *rectification*, and the device itself is called a *rectifier*, or diode.

How is it possible that by properly joining two nearly identical pieces of material, each of which, by itself, will freely conduct current in any direction, it now refuses to allow conduction in one direction? The answer to this is without doubt one of the more interesting occurrences to be studied in the field of electronics.

Consider, first, the condition of the *p*- and *n*-type germanium just prior to joining. (Actually, we cannot just push two pieces together; the germanium atoms at the junction must offer an unbroken path from the *n* type through to the *p* type. That is to say, the whole crystal is one complete piece, and during manufacture the crystal is doped in alternate layers; when it is cut at the proper place, this will separate the layers so that the junction lies between the cut ends. The description is much simpler, however, if we assume, for the present purpose, that we can join together a small quantity of *p*- and *n*-type germanium.) In Fig. 5-1, a

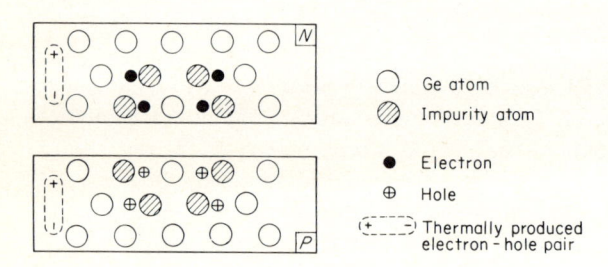

○ Ge atom
⊘ Impurity atom
● Electron
⊕ Hole
⌐‾‾¬ Thermally produced
⌊___⌋ electron-hole pair

Fig. 5-1. Semiconductor materials before forming the junction.

small section of *n* and *p* material is shown just prior to the formation of the junction. Keep in mind that the majority carriers are in constant motion, as are the minority carriers. The minority carriers are thermally produced, and they exist only a short time, after which they recombine. In the meantime, others have been produced; and this process goes on and on, the number that exist at any one time depending upon the temperature of the material. The number of majority carriers is, however, fixed, depending on the number of impurity atoms available. While the above-mentioned particles are in motion, it is important to realize that the germanium and impurity atoms themselves are fixed in place within the structure of the solid material.

We now turn to the condition of the materials at the instant the junction is formed. (Again, we shall take the liberty of assuming that one can merely push the two pieces together to form the junction.) A completely different set of conditions will now exist. As soon as the junction comes into being, two things occur. The free electrons in the *n* material "look" across the junction and see a region that has very few free electrons. At the same time the holes in the *p* material look across the junction and see a region that has very few holes. The electrons and holes (majority carriers) begin to diffuse across the junction, i.e., wander across. Then a few electrons appear in the vicinity of the *p* material, and a few holes appear in the *n* material. The chances are excellent that the hole and the electron will collide, the negative charge on the electron canceling the positive charge of the hole, and both will cease to exist as charge carriers. After several collisions occur, an electrostatic field exists between the *p*- and *n*-type material. The source of this field, or potential is depicted in Fig. 5-2, where only the impurity atoms are shown plus a few Ge atoms.

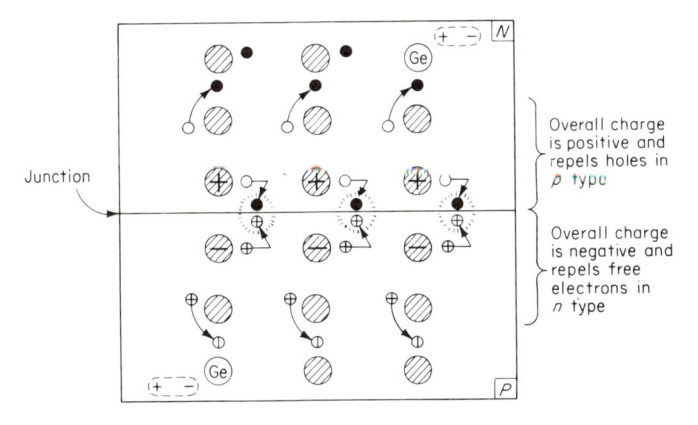

Fig. 5-2. Forming the junction.

Keeping in mind the actual preponderance of Ge atoms compared with impurity atoms, let us see what is happening.

The impurity atoms are, of course, fixed in their individual places. The atom itself is part of the solid structure of the crystal and so cannot move about. When the electron and hole meet, their individual charge is canceled, and this leaves the originating impurity atoms with a net charge; the atom that produced the electron now lacks an electron and so becomes charged positively, whereas the atom that produced the hole now

80   PULSE AND SWITCHING CIRCUIT ACTION

lacks a positive charge and so becomes negative. The electrically charged atoms are called *ions* since they are no longer neutral.

After several collisions occur, the field produced by the sum of the individual impurity-atom charges is great enough to repel the rest of the majority carriers away from the junction. Thus, after a time, a condition of equilibrium exists, and the crystal then remains static; nothing further happens. The net result of this field is that it has produced a region, immediately surrounding the junction, that *has no majority carriers*. The majority carriers have been repelled away from the junction and so are not available as carriers of current; we find that they have been caused to be concentrated nearer the ends of the material, leaving the junction depleted of carriers. The junction is known as the *barrier region*, or depletion region.

If Fig. 5-2 is simplified, we can visualize this more clearly. Figure 5-3

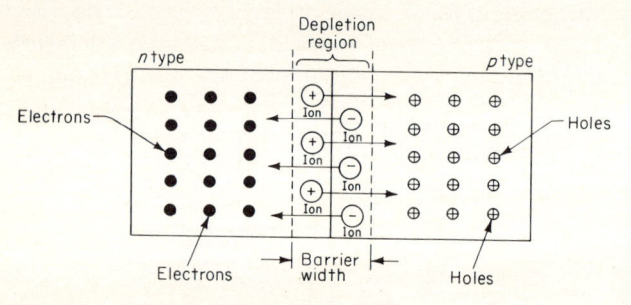

Fig. 5-3. The depletion region.

clearly shows the lack of carriers in the vicinity of the junction and the large number concentrated away from the junction.

5-2 REVERSE BIAS

The diode that has just been described is now capable of exhibiting the property of rectification. Of course, up to this point, we can only assume that this is true, for no mention has been made of current flow through the device. If the semiconductor is connected to a source of voltage, called *bias* when applied to the diode, we can determine how the device reacts.

Figure 5-4 assumes a certain polarity of applied voltage, or bias, that will put a positive voltage on the n side of the diode and a negative voltage on the p side. Notice that the majority carriers are attracted by the battery, pulling them *farther away* from the junction. The barrier width

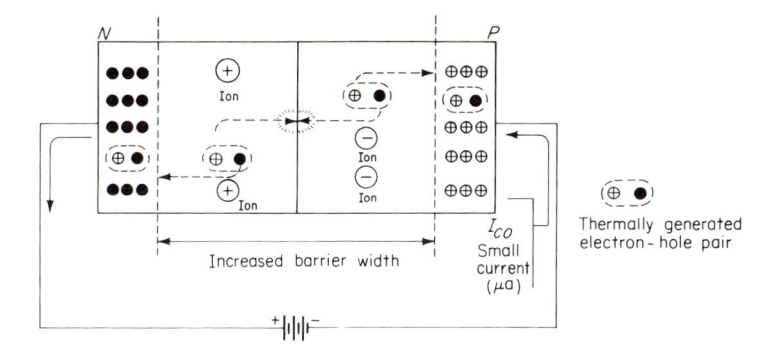

Fig. 5-4. Reverse bias.

has been increased, and when the restraining force of the resultant field within the confines of the barrier region just equals the applied bias, an equilibrium condition is again established. Note that *there is not, nor can be, current flow* that could be attributed to the majority carriers. It is useful to think of the barrier region as an insulator that will not allow current flow.

The above statement that no current can flow is true if we consider only the majority carriers. At some given temperature, some number of electron-hole pairs is generated throughout the volume of the material. We must consider these current carriers if they are generated in the vicinity of the junction and if the applied voltage is as shown in Fig. 5-4. The electron-hole pairs are shown by the symbol ⊕ ● and the majority carriers by ● and ⊕.

The negative voltage applied to the diode will tend to attract the hole thus generated and repel the electron. At the same time, the positive applied voltage will attract the electron toward the battery and repel the hole. *The electron in the p material and the hole in the n material are being forced to move toward each other and will probably combine.* An electron has been removed from the p side and a hole from the n side. When an external voltage is applied, any hole-electron combination in the area of the junction *will cause current to flow in the entire circuit.* An electron from the battery will enter the p material to replace the one lost in the combination, and an electron in the n material will flow out toward the battery. The current that flows, as described above, is due to minority carriers and is usually very small, on the order of a few microamperes. But if the temperature is increased appreciably, the number of minority carriers increases and the current must increase also.

The bias shown in Fig. 5-4 is called *reverse bias* since practically no current exists at normal temperatures. Therefore the small current that does flow is called *reverse current*, labeled I_{co}.

Even with a reasonably large applied voltage, the current would be very small at room temperatures. This suggests that the resistance of the reverse-biased diode is very high. A typical value would lie in the range of 100,000 ohms to well over a megohm.

5-3 FORWARD BIAS

We have seen that, except for a very small reverse current, there is no current flow through the diode if the *n* side is made positive and the *p* side is made negative by an applied voltage, notwithstanding the fact that each section by itself is a reasonably good conductor. We propose now to reverse the battery connections and investigate the difference, if any, between the two polarities.

It is seen that the majority carriers would now be thrown *toward* the junction, rather than drawn away; this is depicted in Fig. 5-5. (Keep in

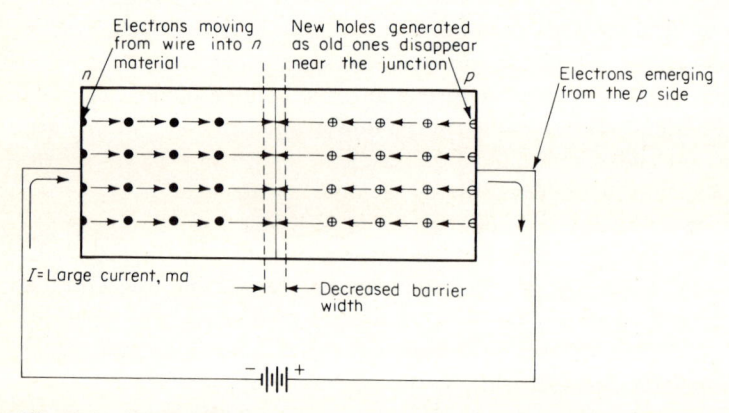

Fig. 5-5. Forward bias.

mind the fact that the hole really does not exist as a separate particle. We say the holes are moving toward the junction; it is just as true that in order for a hole to move in one direction an electron must move in the opposite direction.)

As the holes and the electrons are moved toward each other, a large number of them will collide. As each hole is eliminated at the junction, a new one is formed somewhere in the volume of the *p*-type material.

This must be true because the number of holes is directly dependent upon the number of impurity atoms, and these, of course, cannot be changed or moved or destroyed. At all times there is a statistically constant number of holes present. As each hole is lost at the junction, an electron is just emerging at the wire connected to the p type. The removal of an electron from the p side must result in the generation of a new hole.

The same description could be applied to the n-type material. We cannot destroy the electron itself, but only its effectiveness. Thus for every electron lost in the process of recombination, a new one appears, supplied by the battery.

Now a large number of collisions are occurring at the junction, and it is seen that there is a large continuous current flow throughout the entire circuit. In view of this large current, we can deduce that the resistance of the diode is now *very low*. A typical value might be in the range of a fraction of an ohm to a few hundred ohms.

It might be instructive to trace the flow of current throughout the circuit from start to finish. We shall follow a single electron around the complete circuit, starting at the negative post of the battery. Eventually, this electron will arrive at the terminal connected to the n-type material and will be injected into the body of the crystal. It joins with the existing free electrons and drifts to the right, impelled by the battery voltage toward the junction. As it approaches the junction, a hole from the p side is moving toward it, and right at the junction, the two combine and are lost as separate entities. However, the hole, in order to have moved left toward the electron, must have been produced by a different electron that had moved to the right. Now *this* electron is the one that interests us, and we can see that we have simply traded electrons. The new electron, then, begins a migration to the right, jumping from hole to hole, eventually emerging from the semiconductor to the right, then traveling along the wire to the positive post of the battery, through the battery, and back to the starting point.

We can make a few observations regarding this journey of an electron that may or may not be obvious.

1. The current at any point in the circuit is equal to the current at any other point.

2. If a given number of electrons are moving toward the junction in the n side, the same number of holes are moving toward the junction in the p side.

3. If a given number of electron-hole combinations are occurring per unit time at or near the junction, this number is equal to the number

of electrons flowing per unit time past a point in the external circuit (wire).

In the discussion of diode forward bias, up to this point, we have neglected to consider the reverse current. With the diode forward-biased, we cannot properly apply the same name to this current if it exists. In the n material the electron-hole pairs are still produced, and the electrons thus generated join with the existing free electrons to become a part of the total available current carriers. In the p material, the thermally generated holes join with the existing holes to become a part of the existing current carriers. The normal forward current of a diode is, at normal temperatures, many thousands of times greater than this minute thermal current, and so when a diode is forward-biased, the thermal current can usually be ignored.

5-4 DIODE SYMBOL

The symbol used in schematic drawings is shown in Fig. 5-6, along with the symbol we have used thus far. In either case, electron current flows

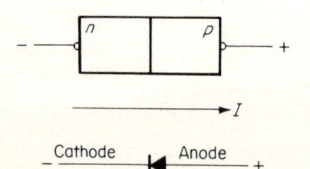

Fig. 5-6. The diode symbol.

from left to right as shown. Also shown is the polarity of forward voltage necessary to cause this current flow.

5-5 DIODE CHARACTERISTICS

Diode characteristics are often shown in graphical fashion as in Fig. 5-7. The voltage labeled $+E$ is forward bias, and the voltage labeled $-E$ is reverse bias. By the same token, $+I$ is the current that flows when forward bias is applied, while $-I$ is the current that flows when reverse bias is applied.

In the forward direction current increases almost linearly for small increases in diode voltage $(+E)$. In the reverse direction almost no current flows until a certain reverse voltage is reached beyond the point

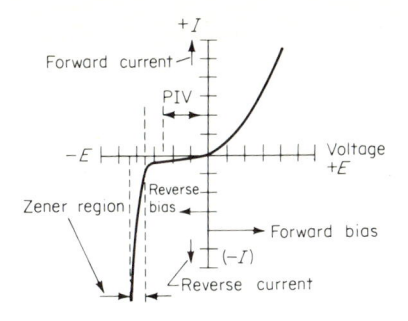

Fig. 5-7. Diode characteristic curves.

labeled "PIV." The peak-inverse voltage (PIV) is the maximum voltage that can be safely applied to a diode. Beyond this, the current again increases rapidly, and this region of operation is called the *zener region*. Operation in the zener region is destructive for the ordinary diode.

Certain diodes, however, are made especially to operate in this region, and they are called *zener diodes*. The symbol for such a device is shown in Fig. 5-8. The zener diode is widely used as a voltage reference, for the

Fig. 5-8. The zener symbol.

voltage drop across the diode remains essentially constant over a wide range of current values.

QUESTIONS AND PROBLEMS

True or false:

5-1 A *pn* junction is produced by the proper joining of *p*- and *n*-type silicon.

5-2 Pure germanium, as a semiconductor, exhibits a few free electrons and holes.

5-3 Forward bias reduces the area encompassed by the depletion region.

5-4 Reverse bias forces the majority carriers toward each other.

5-5 In a *pn* junction, when the *p* material is made more negative than the *n* material, this is called *forward bias*.

5-6 When a diode is reverse-biased, no current flows at all, under any circumstances.

5-7 Typical reverse-biased junction resistance is on the order of several hundred thousand ohms.

5-8 In a forward-biased diode, if the p-type material has twice the volume of the n-type material, there will be a heavier current flow on the p side of the junction than on the n side.

5-9 In a reverse-biased diode, the reverse current I_{co} is small at elevated temperatures.

5-10 A diode in forward bias has very low resistance.

5-11 When a diode is used as a rectifier, it is usually safe to exceed the PIV rating.

5-12 The proper use of diodes allows us to steer current in a predetermined direction.

6
DIODE CHARACTERISTICS

The junction diode, which has been briefly introduced in the last chapter, will now be investigated more thoroughly. First, the general action of a diode will be covered, with particular emphasis upon the circuit action as correlated with the characteristic curve of a typical diode. Then, the static and dynamic characteristics will be discussed, following which will be a section on general applications, with emphasis on the action in a pulse or switching circuit.

6-1 GENERAL DESCRIPTION

A typical characteristic curve of a diode is given in Fig. 6-1. This is quite similar to the one shown previously in Fig. 5-7, but with values for both current and voltage. At the point of origin, there is no current flow, of course, since there is no applied voltage. Hence, the curve passes through zero at this point. As the curve progresses to the right, it goes in the direction of increasing current and voltage. This suggests that if the current is increased in the diode, the voltage drop across it must also increase. At any point on the curve, the values of current and voltage

87

88 PULSE AND SWITCHING CIRCUIT ACTION

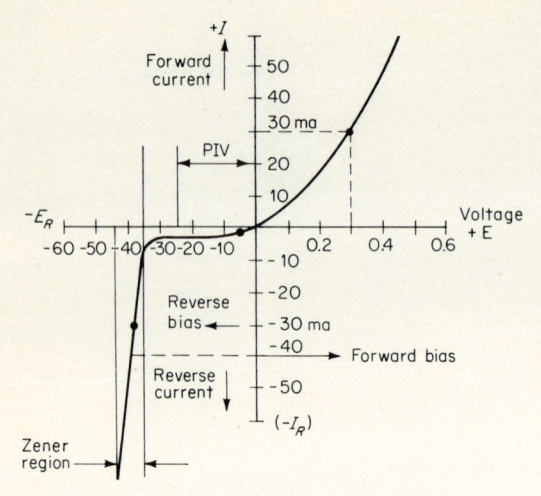

Fig. 6-1. Forward and reverse diode characteristics.

that exist are represented. This is the region of operation that is referred to as the *forward-biased region*.

Note especially that the diode is a current-dependent device; that is, the voltage drop across it is a result of allowing some value of current to flow through it. The current is limited to some particular value, and this causes the voltage drop to be some corresponding value consistant with the curve. Hence, to say that the voltage across a diode is going to be made equal to some value is not correct. Some amount of current is allowed to flow, and this will result in a value of voltage drop across the diode.

When the diode is reverse-biased, it operates in the region of the curve that is located to the left of the ordinate. As the voltage is increased, virtually no current flows, as shown by the curve. Even this small amount is exaggerated and is not drawn to scale. A typical value for this current is, perhaps, 1 to 10 μa. However, as the voltage is increased beyond some critical value, a large current again flows. As the curve begins to slope downward very rapidly, this indicates a rapid increase in current. The diode is now operating in the avalanche region, where, even though the diode is reverse-biased, current is again limited only by some external resistance in series with it. As mentioned before, a normal diode operating in the avalanche region is soon destroyed. To prevent this, the diode must be operated in the region below its peak-inverse voltage, which will prevent its destruction. The general description of the diode characteristic curve is seen to be rather simple.

6-2 DIODE CHARACTERISTICS

To further explain the diode characteristic curve, assume that the simple circuit of Fig. 6-2 is to be used. If battery $A(B_A)$ is connected to the terminals, the diode is reverse-biased and the diode point of operation is indicated by a dot at -6 volts, just to the left of the point of origin in Fig. 6-1. This indicates that diode current is very small, consisting of only leakage current. At room temperature this can be assumed to be on the order of a few microamperes.

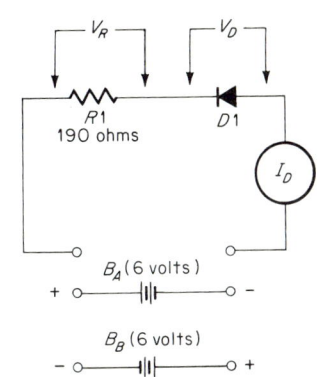

Fig. 6-2. A circuit to determine forward and reverse characteristics of a diode.

A typical manufacturer's data sheet specifies the maximum reverse current at rated voltage to be 1 μa for a particular silicon diode at 25°C (room temperature) and 10 μa at 100°C. Depending upon the diode in question, leakage current can range from a few nanoamperes to several hundred microamperes. Hence, the *relative amount* of reverse current as shown on the graph is grossly exaggerated.

Diodes that are used in certain applications, such as rectifiers or logic circuits, are always operated at reverse voltages less than their peak-inverse-voltage (PIV) rating to avoid the region where appreciable current again flows. As the reverse voltage reaches some critical value, the diode current begins to increase, and this can be attributed to one of two causes.

At low voltages, usually less than 10 volts, the breakdown of the diode is due to *zener* action. Because the volume of material is very small, the voltage gradient within the diode material itself can reach values of several thousand volts per inch. The electrons are literally pulled from their covalent bonds and hence can contribute to current flow. The diode is then said to be operating in the zener region.

If the diode is designed to break down at higher voltages, the mechanism of current generation is slightly different. In this instance, the diode conducts because of *avalanche* breakdown. At higher voltages the first few carriers that break free attain much greater velocity than is the case in zener breakdown. These carriers, then, because of their greater velocity, actually collide with other bonded carriers, thus liberating them. These, in turn, liberate still others, and this cumulative action results in many carriers, which, of course, contributes to a heavy flow of current through the diode.

Such a device is properly called an *avalanche diode*, but it has become customary to identify all such diodes as *zener diodes*. While zener diodes function normally in this region of operation, normal diodes are quickly destroyed by such circuit action.

Now, when battery $B(B_B)$ is applied to the circuit, the diode becomes forward-biased. Note on the graph (Fig. 6-1) that the forward scale ($+E$) is different than the reverse-voltage scale. The forward drop across the diode seldom exceeds 1 volt, and so the scale is incremented in tenths of a volt. The diode-operating point is again indicated with a dot, and, as shown, the drop across the diode is about 0.3 volt, with perhaps 30 ma flowing. This can be proved by the simple Ohm's law. The drop across the resistor is the supply voltage E_{BAT} less the diode drop V_D:

$$V_R = E_{BAT} - V_D = 6 - 0.3 = 5.7 \text{ volts}$$
$$I_{R1} = \frac{V_R}{R} = \frac{5.7}{190} = 30 \text{ ma}$$

The diode current, then, is indeed 30 ma.

Several further points concerning the circuit action of a diode should now be considered. First, the power dissipated by the diode is the product of V_D and I_D:

$$P_D = V_D I_D = 0.3 \times 0.03 = 0.009 \text{ watt}$$

Note that the power dissipated by the diode is very small when forward-biased for moderate values of current.

To illustrate the damaging effect of operation in the reverse direction, assume that 30 ma of *reverse* current is to be allowed to flow. Now, the diode drop is about 40 volts, and the power dissipation can be determined for this new condition:

$$P_D = V_D I_D = 40 \times 0.03 = 1.2 \text{ watts}$$

Hence, operation in this region for any extended period of time will damage the diode unless it can easily dissipate, in this instance, 1.2 watts. Note that while the current through the diode is the same in each instance, the dissipated power is much greater when reverse-biased. This is, of course, caused by the much greater voltage across the diode under these conditions.

Another factor that is often of interest is the static resistance offered to current flow. In the forward direction this resistance R_F is easily found:

$$R_F = \frac{V_D}{I_D} = \frac{0.3}{0.03} = 10 \text{ ohms}$$

Such a low value is, of course, to be expected.

The resistance offered to the circuit when the diode is reverse-biased but still within the PIV rating will be designated R_R and is a function of the applied voltage and the leakage current.

$$R_R = \frac{E_{\text{BAT}}}{I_{co}} = \frac{6}{1 \ \mu a} = 6 \text{ megohms}$$

If leakage current is assumed to be 1 μa, the diode exhibits 6 megohms of resistance. Compared with the usual circuit resistances, this is nearly an open circuit and is normally considered to be such.

Another value that is useful is the dynamic resistance of the diode when forward-biased. That is, the resistance offered by the diode to an alternating current while the diode is continually forward-biased is the ac, or dynamic, resistance. The symbol for this quantity as used herein is r_{df}. The value of r_{df} is determined by the physical and electrical conditions at the junction, as evidenced by the following equation:

$$r_{df} = \frac{KT}{QI_d m}$$

where T = absolute temperature, °K
K = Boltzmann's constant (1.38 × 10^{-23} watt-sec/°C)
Q = electron charge (1.6 × 10^{-19} coul)
m = constant (1 for Ge; between 1 and 2 for Si)
I_d = diode current, dc

Fortunately, at room temperature this reduces to

$$r_{df} = \frac{0.026}{I_d}$$

As an example, a germanium diode having 1 ma of dc current has a dynamic resistance of $0.026/0.001 = 26$ ohms to an ac current superimposed upon the dc current. If the dc current is increased to 10 ma, the ac resistance is reduced to $0.026/0.01 = 2.6$ ohms.

Another facet of diode operation is the static resistance offered by operation in the avalanche, or zener, region. Using the curve of Fig. 6-1, the resistance R_Z will be determined at 30 ma:

$$R_Z = \frac{V_D}{I_R} = \frac{40}{30 \text{ ma}} = 1333 \text{ ohms}$$

To illustrate the advantage of the zener diode used in regulating a voltage source, assume an increase in diode current to 40 ma. As read from the graph the new voltage across the diode is on the order of 41 volts, so a new value must be found:

$$R'_Z = \frac{V'_D}{I'_D} = \frac{41}{40 \text{ ma}} = 1025 \text{ ohms}$$

Note how drastically the diode resistance changes and how little the diode voltage changes. This is typical of zener action, and much larger variations of resistance are encountered in practice.

Because zener diodes are usually used in applications where the current is constantly changing, the dynamic resistance r_z is of greater importance than the static resistance. Using the two sets of numbers above will allow this value to be found:

$$r_z = \frac{\Delta V_D}{\Delta I_D} = \frac{V'_D - V_D}{I'_D - I_D} = \frac{1}{10 \text{ ma}} = 100 \text{ ohms}$$

Hence, the dynamic resistance of such a diode is much lower than the static value.

On the curve of Fig. 6-1 the slope of the diode characteristic in the zener region is exaggerated to clearly show that a slope exists. In an actual case, the slope would be much steeper (more nearly vertical), and thus the dynamic resistance would be much less. Typical values in the range of 0.1 to 100 ohms are found in current data sheets for low-power zener diodes.

Another interesting point concerning the general diode is that under certain conditions it acts as though it were a capacitor, the value of which is a function of the applied voltage. The barrier region of a diode acts very much like the dielectric of a capacitor since this region has essentially no current carriers available. Because the barrier width varies with the applied reverse voltage, the amount of capacitance exhibited also varies. Typical variation of capacitance is from less than 5 to perhaps 100 pf. When diodes are designed to be used in this manner they are called *varactors*, and they are often used in applications requiring variable capacitance that can be adjusted by changing the applied voltage. Many automatically operating circuits are based on this principle.

A simplified equivalent circuit, such as given in Fig. 6-3, is helpful in

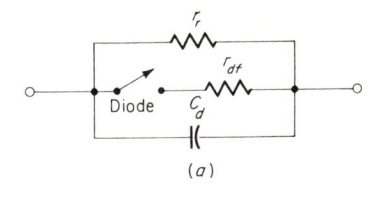

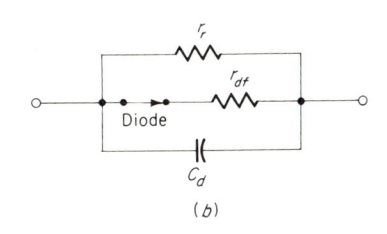

Fig. 6-3. Equivalent circuit of a diode for (a) the reverse-bias condition and (b) the forward-bias condition.

understanding the basic operation of a diode. In Fig. 6-3a the switch is open, which represents the reverse-biased junction. Note that the forward resistance r_{df} is effectively removed from the circuit. However, in parallel with both the junction and r_{df} is the reverse resistance r_r. Thus, some current will flow, but since r_r is typically very large in value the reverse current is quite small. Should the frequency of the impressed voltage be reasonably high, then an alternate path exists through C_d. This, of course, represents the junction capacitance.

Figure 6-3b shows the forward-biased condition, and because the forward resistance is typically quite low (perhaps a few ohms to a few hundred ohms), neither r_r nor C_d have any appreciable influence on diode action.

6-3 DIODE APPLICATIONS

Diodes are used to produce many different results. In this section, the usages will be illustrated and briefly described. It must be remembered that these examples are only representative of the dozens of ways in which diodes can be used.

Figure 6-4 illustrates four circuits that are commonly encountered. Figure 6-4a illustrates a negative-diode ground clamp. A circuit such as this is used to prevent an output excursion below the ground level. Regardless of the amplitude or polarity of the input voltage, the output can go no more negative than the normal forward-biased voltage drop across the diode. This is suggested by the waveforms, also shown. The circuit action, of course is quite simple. When the input voltage is more positive than ground, the diode is reverse-biased and hence can in no

94 PULSE AND SWITCHING CIRCUIT ACTION

way influence the output. When the input goes more negative than ground, the diode becomes forward-biased and in effect shorts-out the incoming signal. The output is said to be clamped to ground during the negative half-cycle. The resistor is shown to emphasize the fact that there must be some current limitation to avoid damaging the diode.

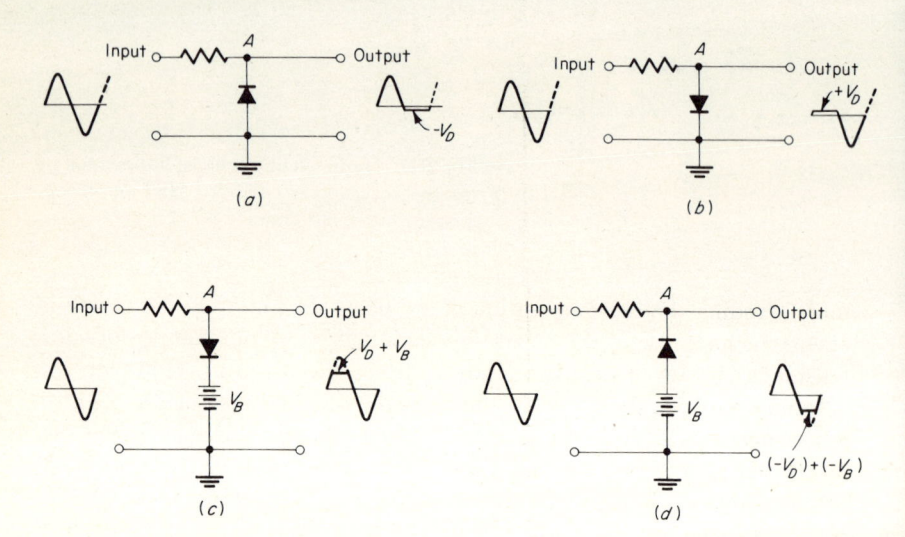

Fig. 6-4. Four circuits illustrating diode applications.

Figure 6-4b shows a very similar circuit that differs only in the direction of the diode. Now, when the input goes more positive than ground, the diode becomes forward-biased and therefore clamps the output to ground during this half-cycle. Such a circuit is called a *positive-diode ground clamp*. The output voltage can go no more positive than the drop across the diode.

A similar circuit is shown in Fig. 6-4c, but it is different in that it employs a battery, which prevents the diode going into forward bias until a certain threshold voltage is exceeded. This circuit is called a *floating positive-diode clamp*, and its output can never exceed $V_B + V_D$. If the battery and diode are turned around, as shown in Fig. 6-4d, the output can go no more negative than $(-V_D) + (-V_B)$.

Diodes are also used as clippers, which are very similar to the clamp circuits. Figure 6-5a and b illustrates first a negative clipper and then a positive clipper. At the output in Fig. 6-5a, only when the input goes positive will there be an output since then the diode is reverse-biased. The negative spike is removed since the diode then becomes forward-biased,

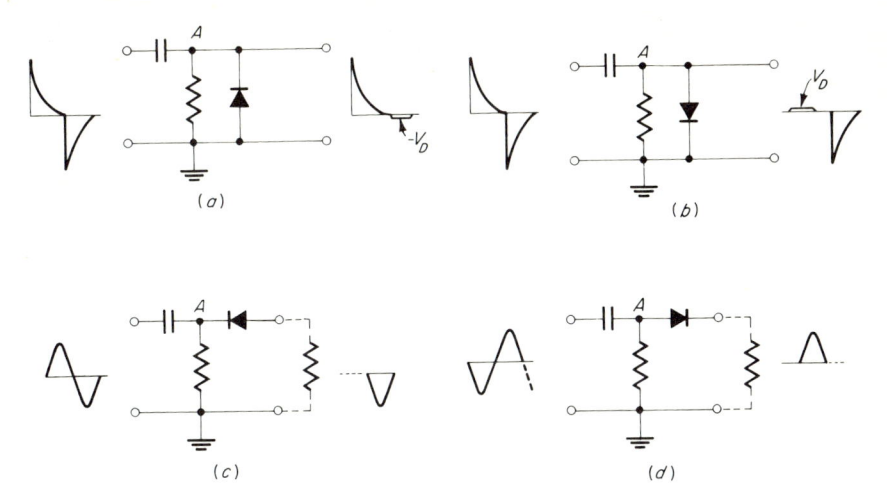

Fig. 6-5. Four circuits illustrating coupling and clipping circuits.

effectively shorting-out the signal. With the diode reversed, Fig. 6-5b, the output spike is negative-going, with the positive-going part eliminated.

The circuits given in Fig. 6-5c and d are examples of diodes used as couplers. Again, the unidirectional characteristics of the diode are used to advantage. In the first case, the output can follow the input only when the input goes negative, while in the second case there is an output only when the input goes positive.

The circuit shown in Fig. 6-6a is known as a *diode gate*, which is one of many so-called "conditional" circuits. A conditional circuit is one that produces an output only when the condition of the inputs agrees with a particular set of values. Such circuits will be dealt with later in great detail, for they are used extensively in digital applications. The accompanying truth table indicates one way in which the circuit can be used. With both inputs firmly connected to ground (0 volt), the output is also at a 0-volt level. If either input A or B goes to a more positive voltage, the output *stays* at ground. Only when *both* inputs go to a more positive voltage will the output also go more positive. Such a circuit is often called an *AND* gate since both A AND B must be presented with a positive signal to produce an output. Chapter 11 deals with this kind of circuit in detail.

Figure 6-6b shows a diode rectifier. When an ac voltage is applied to the input, the output becomes pulsating direct current since the diode does not conduct during one half-cycle. Reversing the diode results in half-

96 PULSE AND SWITCHING CIRCUIT ACTION

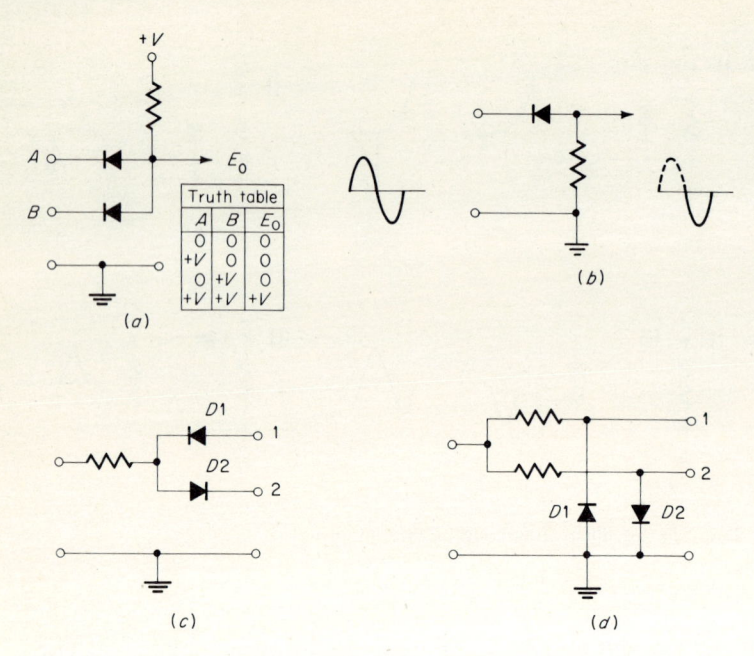

Fig. 6-6. Gating, rectifying, and steering applications of diodes.

cycle output-voltage swings in the positive direction rather than in the negative direction.

The circuits shown in Fig. 6-6c and d are widely used in digital applications. They serve to "steer" a pulse, or signal, toward a certain path, depending on the signal polarity. In Fig. 6-6c, if the input is more positive than ground, $D2$ conducts and output 2 is positive. If, on the other hand, the input is more negative than ground, $D1$ conducts and output 1 is more negative. Hence, positive and negative pulses are, in effect, separated and steered into their respective channels.

Figure 6-6d shows a slightly different arrangement, with the diodes in parallel with the signal. With an input more positive than ground, $D2$ conducts and clamps line 2 to ground. However, with $D1$ reverse-biased, line 1 is allowed to follow the input and thus goes positive. If the input goes in the negative direction, $D1$ conducts and line 2 is allowed to go negative, again channeling the pulses into their respective paths.

Diodes are often used for protection purposes, and an example of this is shown in Fig. 6-7a. With the switch closed, the inductor draws its normal current. The diode is at this time reverse-biased. When the switch opens, the action of the inductor is to produce a large induced voltage

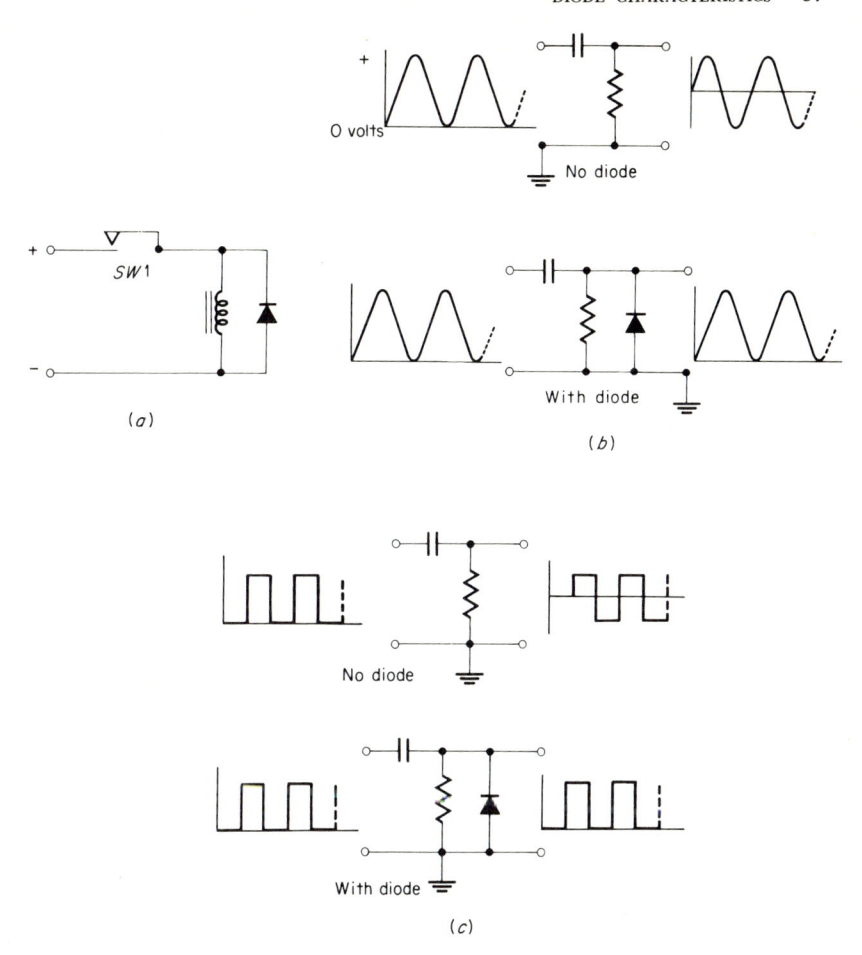

Fig. 6-7. Further diode applications showing a protection diode and several circuits illustrating dc restoration.

(negative at the top, positive at the bottom) and the diode then becomes forward-biased, absorbing the excess energy in its very low internal resistance. The induced energy, then, is, in effect, removed from the surrounding circuitry, and the voltage across the coil will not exceed the normal forward drop across the diode.

A dc restorer circuit is shown in Fig. 6-7b. The circuit is first shown with no diode (and therefore no dc restoration) with attendant waveforms. However, with the diode connected in the circuit, the output waveform can never go more negative than ground. If the diode were reversed, the output waveform would exist completely below

98 PULSE AND SWITCHING CIRCUIT ACTION

0 volts, with the most positive value being ground. Figure 6-7c illustrates the dc restorer used with square waves. (For a complete discussion of dc restoration, see Chap. 3.)

The final example of diode circuitry is given in Fig. 6-8. With properly

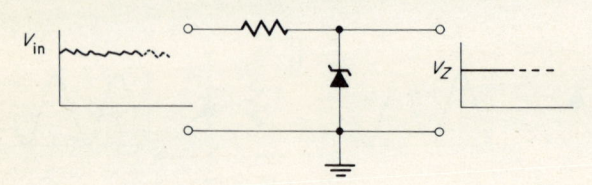

Fig. 6-8. Zener-diode application.

chosen values, the output voltage is constant in spite of input variations. Because the voltage across the zener diode is constant even though the current through it is varying, the voltage V_Z is held constant. This circuit is called a *regulator* because of the smoothing or regulating effect of the diode on circuit voltage variations.

QUESTIONS AND PROBLEMS

6-1 In the accompanying diagram, the source voltage is 12 volts dc. Determine the total current ($V_D \cong 0$ volts).

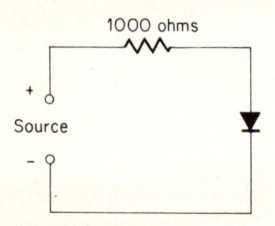

6-2 Refer to the diagram of Question 6-1. The source voltage is 12 volts, and the drop across the series resistor is 12 volts. Determine (a) the current in the circuit and (b) the dc resistance of the circuit.

6-3 Refer to Question 6-2. A new diode is to be placed in the circuit, and V_D when measured is 0.3 volt. Determine (a) the circuit current and (b) the diode dc resistance.

6-4 Refer to Question 6-2. A new diode is to be installed, and its V_D is known to be 0.7 volt. Determine (a) the circuit current and (b) the dc diode resistance.

6-5 A diode is placed in a circuit, and the diode voltage drop is found to be equal to the supply voltage of 10 volts. The diode current is measured as 10 μa. Find the value of R_R.

6-6 A diode is placed in a circuit, and the diode voltage drop is found to be equal to the supply voltage of 10 volts. The diode current is measured as 25 μa. Find the value of R_R.

6-7 A germanium diode is operated at a dc current of 3 ma. Determine the dynamic resistance offered to an ac signal current.

6-8 A germanium diode is operated at a dc current of 0.5 ma. Determine the dynamic resistance offered to an ac signal current.

6-9 A zener diode has a nominal zener voltage of 20 volts. The dc current is 30 ma. Find R_Z.

6-10 A zener diode has a nominal zener voltage of 20 volts. The dc current is 40 ma. Find R_Z.

6-11 Refer to Questions 6-9 and 6-10. The two separate conditions refer to the same diode. Find r_z if the voltage across the diode increases 0.1 volt as the current increases by 10 ma.

6-12 Refer to Questions 6-9 and 6-10. The two separate conditions refer to the same diode. Find r_z if the voltage across the diode increases 0.01 volt as the current increases by 10 ma.

7
THE JUNCTION TRANSISTOR

In this chapter a brief review of basic transistor parameters will be given. If the reader has previously studied transistor fundamentals, little that is new will be found in the first section. Many basic transistor parameters are common to both linear and switching circuits, and these are briefly developed herein. In later sections, parameters that specifically relate to switching circuitry are given. These are further expanded upon in Chaps. 8, 9, and 10 and are ultimately related to actual switching circuits.

7-1 FUNDAMENTALS OF TRANSISTOR CHARACTERISTICS

Underlying the circuit action of all transistors are the various leg currents and the simple relationships between them. To provide the conditions necessary to allow these currents to flow in the correct amount, the transistor must be properly biased. Figure 7-1 illustrates the correct polarity of applied dc voltage to cause the transistor to conduct.

The *npn* unit requires that the collector-to-emitter voltage be applied as shown: positive to the collector and negative to the emitter. The base must be supplied a voltage that is also more positive than the emitter, but by a smaller amount. Furthermore, base current must be limited in value to a specific amount, hence the resistor RB. This resistor simulates a current-limited (constant-current) source.

The collector-base junction of the transistor is reverse-biased by the collector supply voltage V_{CC}. This condition must be met if the transistor is to be used as a linear amplifier. At the same time, the base-emitter junction must be forward-biased; this can be seen to be the case in Fig. 7-1, where the base supply voltage applies a positive potential to the base (p) and a negative potential to the emitter (n).

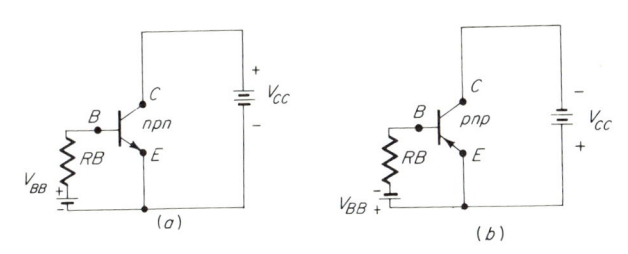

Fig. 7-1. Basic transistor biasing.

The *pnp* transistor is seen to be in a nearly identical circuit. The only significant difference is in the polarity of the applied voltage. This implies, of course, that the currents flowing will be in the opposite direction. Other than this, the two transistors have identical requirements.

The currents flowing in the transistor are identified by the leg of the transistor in which it flows. Hence, emitter current I_E flows in the emitter circuit, collector current I_C flows in the collector circuit, and base current I_B flows in the base circuit. This is illustrated in Fig. 7-2 for an *npn*

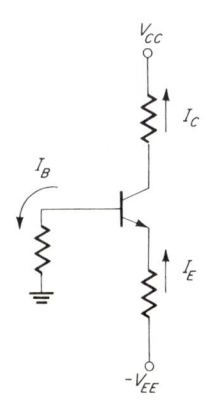

Fig. 7-2. Transistor leg currents.

transistor. The currents shown are given the direction of electron flow; thus I_E flows *into* the emitter, I_C flows *out* of the collector, and I_B flows *out* of the base. If this were a *pnp* unit, the directions would be reversed.

There is a fundamental relationship between these currents that is implied by Fig. 7-2. Because the current flowing up into the emitter divides, with a portion flowing out the base and the remainder flowing out the collector lead, the emitter current must be the *sum* of the other two currents. This is simply an example of Kirchhoff's current law. Stated mathematically, this relationship yields an equation of fundamental importance:

$$I_E = I_C + I_B$$

Many problems in circuit analysis or design can be solved by this simple relationship. For any given circuit, once two of the current values are known, the third can be found. By inserting typical values in this equation for several magnitudes of current, two other relationships can be stated.

In a given transistor, collector current is found to be consistently greater than the base current by a relatively fixed amount, giving rise to the following equation:

$$I_C = \text{constant} \times I_B$$

This constant is called *beta* (h_{FE}) and is an important characteristic of transistors. Over some part of the total operating range of a transistor, its beta remains sufficiently constant to be used as though it were unvarying. Given some value of base current, then, the collector current can be found by the simple product of I_B and the transistor's beta (β).

$$I_C = \beta I_B$$

The beta of a transistor allows any leg current to be found if only one other is known. The following relationships are valuable in circuit-analysis problems:

$$I_C = \beta I_B = \frac{\beta}{\beta + 1} I_E$$

$$I_B = \frac{I_C}{\beta} = \frac{I_E}{\beta + 1}$$

$$\beta = \frac{I_C}{I_B} = \frac{I_C}{I_E - I_C}$$

Note in the foregoing expressions the quantity $\beta/(\beta + 1)$. This is another constant used in transistor work called *alpha* (α). Alpha and beta are interrelated as follows:

$$\alpha = \frac{\beta}{\beta + 1}$$

$$\beta = \frac{\alpha}{1 - \alpha}$$

Beta indicates the degree that I_C is *greater* than I_B; alpha indicates the degree that I_C is *smaller* than I_E. Typical values of alpha range from perhaps 0.95 to 0.995 but never reach unity. Beta, on the other hand, might range from 20 to over 200, depending upon the particular transistor.

As an example of the use of these parameters, assume the transistor in Fig. 7-2 has a published value of alpha equal to 0.98. Its beta can be determined as shown below:

$$\beta = \frac{\alpha}{1 - \alpha} = \frac{0.98}{1 - 0.98} = \frac{0.98}{0.02} = 49$$

These values indicate the relative magnitudes of the three currents. If 1 ma of emitter current is assumed, then I_C and I_B are easily determined:

$$I_C = I_E \times \alpha = 0.001 \times 0.98 = 0.98 \text{ ma}$$

$$I_B = \frac{I_C}{\beta} = \frac{0.98}{49} = 0.02 \text{ ma}$$

Recall from basic transistor theory that neither alpha nor beta is precisely constant. Either, however, can be taken as a constant if values of current are used that are close to the values used to measure alpha or beta. If one uses the value of beta found when measured at 0.1 ma, for instance, in a circuit having a current of 100 ma, beta will be grossly incorrect. But in the range of perhaps 0.05 to 0.2 ma, the measured value will deviate only slightly and hence can be considered to be unvarying.

7-2 TRANSISTOR EQUIVALENT CIRCUITS

To easily analyze transistor circuits and to have a means of visualizing circuit activity, an equivalent circuit of the transistor itself is a useful device. In this section, then, we shall develop an equivalent circuit for a transistor that, while considerably simplified, nevertheless works very well for switching applications. A similar and even simpler equivalent circuit was developed in the author's earlier book, "Transistor Circuit Action,"[1] that was sufficient for the static analysis of linear circuits. For switching circuits, however, the circuits to be used are shown in Fig. 7-3.

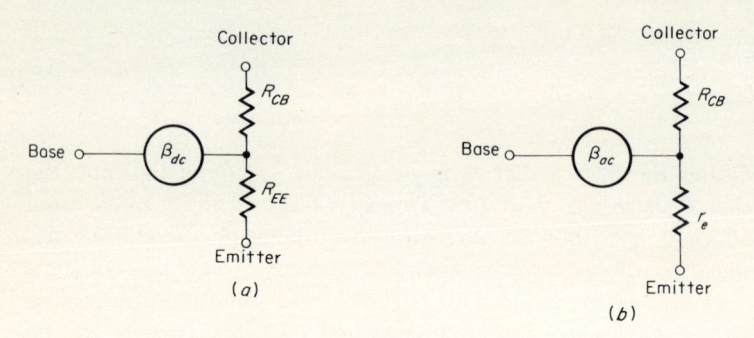

Fig. 7-3. Transistor equivalent circuits for (a) dc and (b) ac conditions for the CE configuration.

The important aspects of this equivalent circuit are, considering the static condition first (Fig. 7-3a), the current generator, the collector-base dc resistance R_{CB}, and the static emitter resistance R_{EE}. From this equivalent circuit can be developed the constants and parameters used for switching circuits.

The current generator simply represents the fact that the collector current is beta times greater than base current. If $\beta = h_{FE} = 100$, then collector current is 100 times greater than base current. The collector-base resistance R_{CB} is defined as V_{CB}/I_C and will assume varying values, depending upon external circuit conditions. Finally, R_{EE} is the ohmic, or bulk, resistance of the emitter material and is defined by V_{BE}/I_E.

The equivalent circuit in Fig. 7-3b is used for the ac (signal) conditions in a linear amplifier. The differences in the two equivalent circuits are slight. The value of β must be the small-signal value h_{fe}, and r_e is the dynamic value of emitter resistance defined as $0.026/I_E$.

Now, recall from basic transistor theory that a junction transistor has three basic modes of operation: active, cutoff, and saturation. The values of the various parts of the equivalent circuit will vary widely, depending upon whether the transistor is in cutoff, in saturation, or in the active region.

Figure 7-4 illustrates these basic modes of operation. Figure 7-4a represents the cutoff condition, where it is seen that since no current flows in the transistor, R_{CB} and R_{EE} are essentially open-circuited. This

[1] McGraw-Hill Book Company, New York, 1968.

assumption, of course, is based on the leakage current I_{cbo} being negligible in value. Under cutoff conditions, the transistor appears to be very nearly an open circuit. The parameters of greatest concern are R_{CE}, the collector-to-emitter resistance ($R_{CB} + R_{EE}$), and R_{BE}, the base-to-emitter resistance. These are quite large in value since both junctions are reverse-biased. The external circuit will therefore determine the value of V_{CE}, V_{BE}, and V_{CB}.

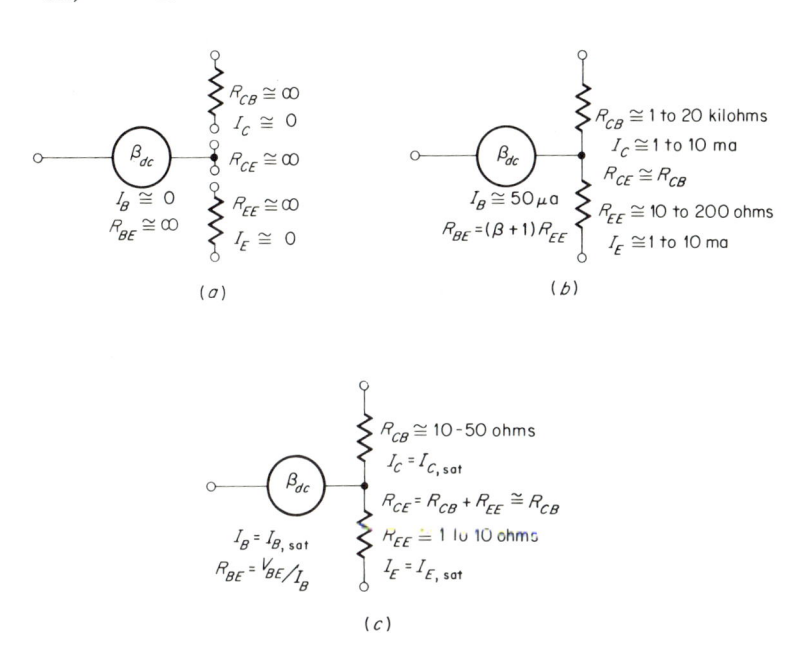

Fig. 7-4. Equivalent circuit for (a) the cutoff condition, (b) the active region, and (c) the saturation region.

Figure 7-4b illustrates a transistor operating in the active region. The different values possible are quite large in number, but typical values are given. The major characteristics of this type of operation are that the leg currents are approximately midway between minimum and maximum and that V_{CE} is on the order of $\frac{1}{2}V_{CC}$. A transistor biased in this manner is in all probability performing as a linear class A amplifier.

Figure 7-4c represents a transistor operating in the saturation region. R_{CB} is a very small value, typically 10 to 50 ohms, while R_{EE} is still smaller, perhaps 1 to 10 ohms. Since R_{CE} is the sum of R_{CB} and R_{EE}, it too can be correspondingly small. R_{BE} is the drop across the base-emitter junction divided by the base current. Because the base current is often large when operating in saturation, this parameter is often small in value.

To indicate the values existing when in one or the other mode of operation, subscripts are used to point out the mode. For example, $V_{CE,\text{sat}}$ is the value of collector-to-emitter voltage that exists when the transistor is in saturation.

To correlate the three modes of operation to actual circuit conditions, Fig. 7-5 is offered. Figure 7-5a illustrates the cutoff condition, where

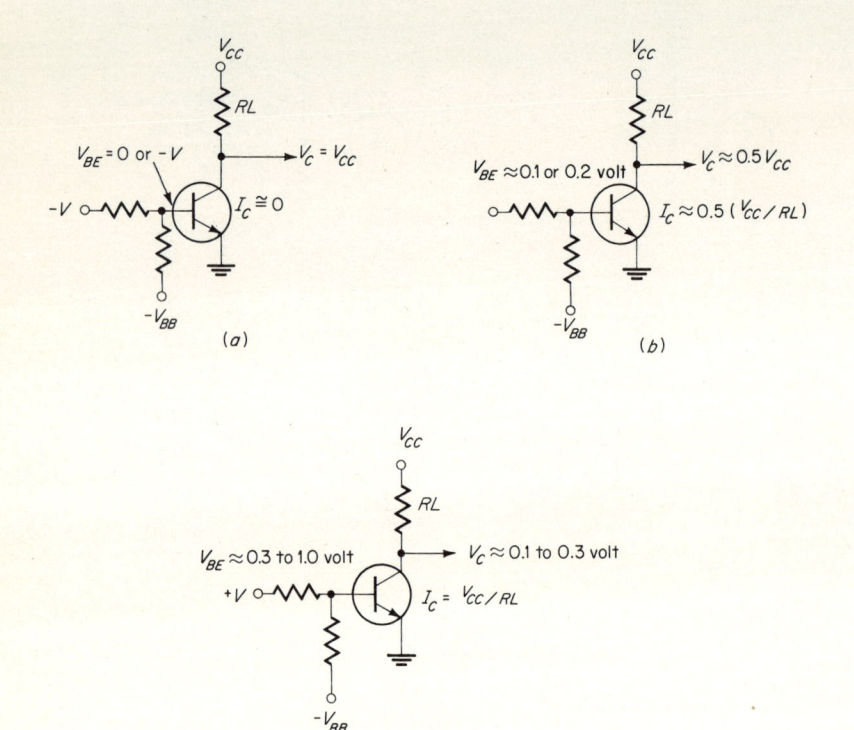

Fig. 7-5. The three operating modes related to circuit action. (a) Cutoff region; (b) active region; and (c) saturation region.

$V_{CE} = V_C = V_{CC}$. This occurs because there is no current flowing through the load resistor and hence no voltage drop appears across it. The transistor is off because the base is sufficiently negative to reverse-bias the base-emitter junction. Next, active-region operation is shown, where the collector voltage is on the order of one-half the collector-supply voltage. The base is slightly positive, and base current is limited to some intermediate value.

Finally, the saturated condition is shown, where both junctions are forward-biased and collector-to-emitter voltage is equal to $V_{CE,\text{sat}}$ which is typically 0.1 or 0.2 volt. On the other hand, $V_{BE,\text{sat}}$ is usually about 0.3 to 0.8 volt. Thus, it is clear that both junctions are forward-biased, a necessary condition for saturation to occur. Saturation-collector current is that value limited by RL and V_{CC}, and the transistor is therefore not able to influence the circuit action to any great extent. The value of $R_{CE,\text{sat}}$ is nearly zero, and for this reason the transistor acts much like a short circuit. The value of $R_{CE,\text{sat}}$ will be determined later from the collector family of curves, along with other parameters.

It is possible to connect a transistor in a circuit in one of three possible ways, as illustrated is Fig. 7-6. The basic characteristics of these configurations will be recalled from earlier studies and are listed briefly in Table 7-1.

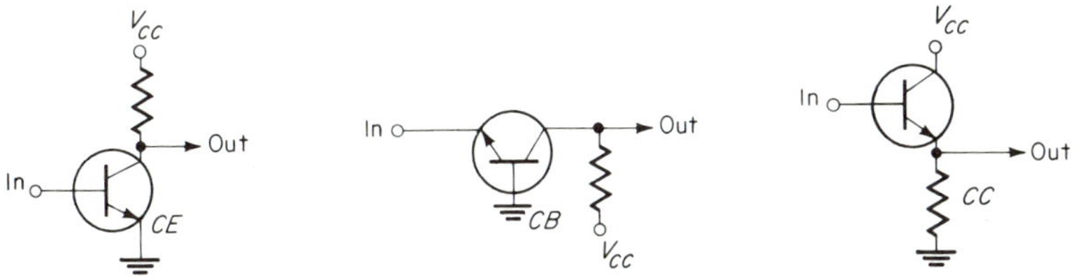

Fig. 7-6. Basic circuit configurations.

Table 7-1
General Characteristics of the Common-emitter, Common-base, and Common-collector Configurations

	Voltage Gain	Current Gain	Input Resistance	Output Resistance	High-frequency Responce	Phase Inversion	Power Gain
CE	High	High	Moderate	Moderate	Poor	Yes	Very High
CB	High	<1	Low	Moderate	Good	No	High
CC	<1	High	High	Low	Good	No	High

7-3 TYPICAL SWITCHING CHARACTERISTICS

In this section a brief description of the parameters appearing in typical manufacturers' data sheets will be given. A widely used germanium transistor (*npn*) will be used as an example, with the data that follows extracted from the actual data sheet.

108   PULSE AND SWITCHING CIRCUIT ACTION

The 2N1304 is an alloy-junction germanium *npn* switching transistor that has been widely used in commercial equipment. The first parameter of interest is BV_{CBO}, the breakdown voltage, collector-to-base, with the emitter open. Measured at $I_C = 100$ μa and $I_E = 0$, this has a value of 25 volts and indicates that the absolute maximum collector-base voltage must be less than this value at all times.

The emitter-base reverse-biased breakdown voltage BV_{EBO}, measured at 100 μa of emitter current and zero collector current, is also 25 volts. In other transistor types, BV_{CBO} and BV_{EBO} may not be equal, with the latter typically the smaller of the two.

Another important parameter is the collector-cutoff current I_{cbo}, which for the 2N1304 is 6 μa maximum, measured at $V_{CB} = 25$ volts and $I_E = 0$. Also of interest is I_{EBO}, the emitter-cutoff current, which is also 6 μa maximum.

Probably the most important characteristic is the forward-transfer ratio $h_{FE}(\beta)$. For this transistor, measured at $V_{CE} = 1$ volt and $I_C = 10$ ma, $h_{FE,\min} = 40$ while $h_{FE,\max} = 200$. Also, the small-signal current gain h_{fe} is given as a typical value of 120 at $V_{CE} = 1$ volt and $F = 1$ kHz.

The saturation voltages $V_{BE,\text{sat}}$ and $V_{CE,\text{sat}}$ are also of great interest in switching applications. $V_{BE,\text{sat}}$ is given as 0.15 volt minimum and 0.35 volt maximum, where $I_{c,\text{sat}}/I_B = 20$. $V_{CE,\text{sat}}$ at $I_{C,\text{sat}}/I_B = 40$ is given as 0.07 volt typical and 0.2 maximum.

The common-base alpha-cutoff frequency f_α, given at $V_{CB} = 5$ volts and $I_E = 1$ ma, is 5 MHz minimum, with a typical value of 14 MHz. Related to this parameter are C_{ob} and C_{ib}. C_{ob}, the common-base open-circuit output capacitance, is listed as 14 pf typical, 20 pf maximum. C_{ib}, the common-base open-circuit input capacitance, is listed as 13 pf typical. These, of course, are important when the transistor is operated at high frequencies or with pulses having fast rise times. Finally, the maximum collector-power dissipation is given as 150 mw.

Other parameters, which will be dealt with subsequently, are delay time t_d (0.07 μsec), rise time t_r (0.2 μsec), storage time t_s (0.7 μsec), and fall time t_f (0.4 μsec). In Chap. 8 many of these parameters will be derived from the characteristic curves, and their meaning will be more clearly defined.

QUESTIONS AND PROBLEMS

7-1   A transistor has an alpha of 0.95.   Determine its beta.

7-2   A transistor has an alpha of 0.87.   Determine its beta.

7-3   A transistor has an alpha of 0.97.   Determine its beta.

7-4 A transistor has a beta of 100. Determine its alpha.

7-5 A transistor has a beta of 200. Determine its alpha.

7-6 A transistor has a beta of 300. Determine its alpha.

7-7 A switching transistor is operating in saturation; its V_{CE} can be expected to be (low, moderate, high) in value.

7-8 A switching transistor is operating in cutoff; its V_{CE} can be expected to be (low, moderate, high) in value.

7-9 A switching transistor has an h_{FE} of 100. Its collector current is 1 ma, and its base current is 100 μa. Determine the transistor's operating mode.

7-10 A switching transistor has an h_{FE} of 50. Its collector current is 1 ma, and its base current is 100 μa. Determine the transistor's operating mode.

8
TRANSISTOR CHARACTERISTICS

This chapter is concerned with extending and deriving many of the parameters given in Chap. 7. The various characteristic curves will be used, where possible, to derive these parameters in an effort to provide a means to *visualize* their origin. Many of these can be extracted directly from the curves provided by the manufacturers and hence are readily obtainable.

In Chap. 7, an equivalent circuit of the transistor was used to explain many characteristics of a switching transistor. In this chapter, the manufacturer's curves will be used to extract the actual numerical values, giving the reader an appreciation for typical values.

Finally, the concept of *circuit limits* is given since many later examples will require these techniques to properly evaluate their circuit action. Our first concern, then, will be to examine briefly the collector family of curves.

8-1 THE TRANSISTOR AS A SWITCH

The collector family of curves is shown in Fig. 8-1. One ordinate of the graph represents collector current I_C, while the other is collector-to-emitter voltage V_{CE}. The running parameter (slanted lines) represents base

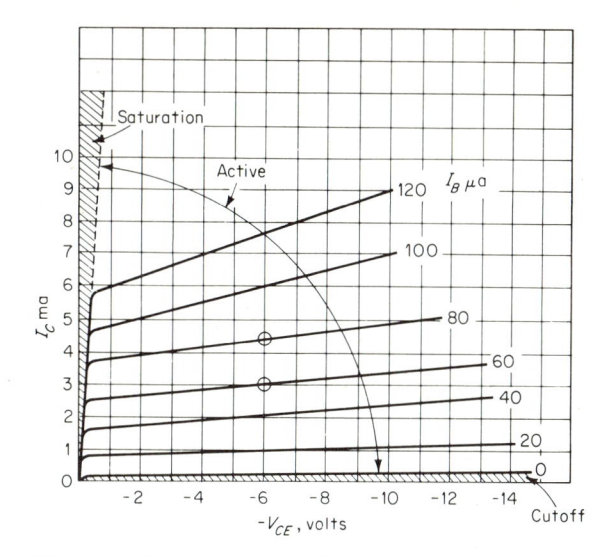

Fig. 8-1. Output-characteristic curves—regions of operation.

current I_B. From these curves the various currents and voltages that exist for different operating conditions can be read.

As an example, there is a point on the curves that is encircled on the 60-µa base-current line. If the transistor that these curves describe is operated with 60 µa of base current and if V_{CE} is 6 volts, 3 ma of collector current is flowing. This is read by projecting a line straight left from the intersection of $I_B = 60$ µa and $V_{CE} = 6$ volts. Any combination of collector current, base current, and collector-to-emitter voltage that the transistor can attain is to be found on the curves.

Knowing that $I_E = I_C + I_B$, the emitter current for any point on the curves can be determined. In the preceding example, emitter current is 3 ma + 60 µa, or 3.06 ma.

The curves can be used to derive characteristics of the transistor other than the currents and voltages. Since beta is I_C/I_B, it is possible to derive this important parameter by using the curves. Taking the same example as before,

$$\beta = \frac{3 \text{ ma}}{60 \text{ µa}} = \frac{0.003}{0.00006} = 50$$

By the same token, alpha is I_C/I_E. Therefore

$$\alpha = \frac{0.003}{0.00306} = 0.9804 \cong 0.98$$

Because these were derived from dc values of current, they are dc values of alpha and beta. Direct-current beta is often written h_{FE}.

Other characteristics, mentioned in Chap. 7, can be derived from these curves. For example, $I_{C,\text{sat}}$, $V_{CE,\text{sat}}$, $R_{CE,\text{sat}}$, and $I_{B,\text{sat}}$ can easily be extracted from the graph. The values lying along the slightly sloping line rising from the point of origin from which the base-current curves emerge are values of $V_{CE,\text{sat}}$. This line is called the *saturation line* representing values of I_B, I_C, and I_E attained by the transistor when in saturation. $V_{CE,\text{sat}}$ can be seen to vary between zero and about 0.4 volt for the values shown.

For any given value of collector saturation current, the required value of base current is also obtainable. If $I_{C,\text{sat}}$ is to be 5 ma, base current must be equal to, or greater than, 110 μa. This is derived from the point on the saturation line delineated by $I_C = 5$ ma, $V_{CE} = 0.4$ volt, and $I_B = 110$ μa (which must be estimated between 100 and 120 μa). From these, the corresponding values of certain other constants can be derived. α_{sat} and β_{sat} are occasionally required. β_{sat} is simply

$$\frac{I_{C,\text{sat}}}{I_{B,\text{sat}}} = \frac{0.005}{0.00011} = 45.5$$

and from this can be determined $\alpha_{\text{sat}} = 0.978$. Also,

$$R_{CE,\text{sat}} = \frac{V_{CE,\text{sat}}}{I_{C,\text{sat}}} = \frac{0.4}{0.005} = 80 \text{ ohms}$$

Normally, most cutoff values cannot be read from the curves since the value of leakage current cannot be shown to scale. To determine these, I_{cbo} must be determined from the tabulated information provided by the manufacturer and the following relationships must be evaluated:

$V_{CE,co} = V_{CC}$ (usually, but depends somewhat upon external circuit configuration)

$I_{C,co} = I_{cbo}, CB$ or I_{ceo}, CE

$R_{CE,co} = \dfrac{V_{CE,co}}{I_{C,co}}$

Other parameters relating to the emitter-base junction ($R_{BE,\text{sat}}$, $V_{BE,\text{sat}}$, etc.) must be derived from the input curves V_{BE} versus I_E.

In Chap. 7, the three regions of operation were illustrated, using a simple transistor circuit. The collector family of curves can also be used to describe these regions. Referring again to Fig. 8-1, a transistor operating in the center portion of the curves is said to be in the *active* region, as shown. If a transistor is to be used as an amplifier, it must be operated so as to never go outside this region.

The second possible region of operation is called *cutoff*. On the curves, this is the area near the abscissa where $I_C = I_{cbo} \cong 0$. I_{cbo} is the transistor leakage current, which is very similar to I_{co} in a diode. When a transistor is operated in the cutoff region, no collector, emitter, or base current flows and the transistor acts very much like an open switch. A great deal of the material in this book regarding transistors will obviously be concerned with this region of operation (as well as the saturation region) because of the switchlike action.

The third region in which operation is possible is the *saturation* region. This is the area of operation where maximum collector current flows and the transistor acts much like a closed switch. In this instance, it is said to be fully on since the collector current cannot be increased beyond this value. When a transistor is operated in the saturation region or cutoff, it not only cannot amplify, but the normal relationships between the various leg currents (α, β) do not hold true.

To correlate the curves to actual circuit conditions, consider Fig. 8-2, which is similar to Fig. 7-5. Figure 8-2a depicts operation in the active region, where collector voltage is something less than V_{CC} but not quite ground. In the active region, collector current is greater than base current by the factor beta, while emitter current is greater by $\beta + 1$. Since

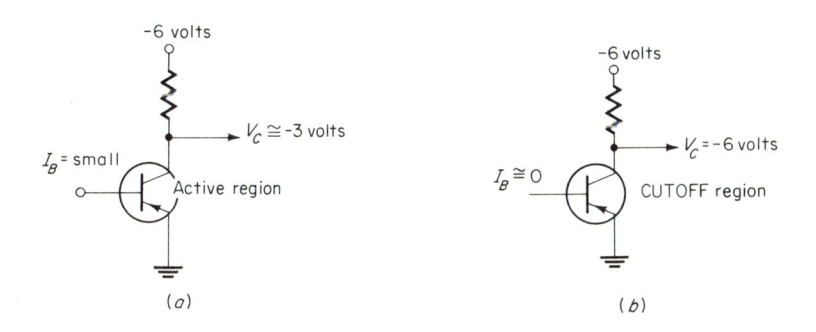

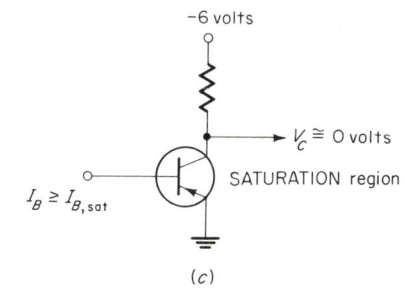

Fig. 8-2. Three regions of operation.

V_{CE} is approximately halfway between V_{CC} and ground, the transistor is capable of amplifying.

Figure 8-2b depicts the cutoff region, where, except for leakage current, I_C, I_B, and $I_E = 0$. Since there is no collector current, there is no voltage drop across the load resistor and therefore the collector voltage equals the supply voltage. It is apparent that the normal relationships between the leg currents cannot exist since there is essentially no current.

Finally, Fig. 8-2c shows the saturation region, where maximum current flows and the collector voltage is very nearly at ground. Again, because collector current cannot be increased, the transistor cannot amplify and acts much like a closed switch.

When a transistor is used in a switching circuit, the cutoff and saturation regions are, by far, the most important regions. In this application, a transistor is operated either fully on or fully off, and the only time it is in the active region is during the transition from off to on or on to off. The circuit action during these transitions will be discussed in detail in a later section.

Input and Temperature Curves

The collector output curves just described are extremely useful in determining how a transistor works in terms of its voltages and currents. There are other curves that reveal still more about how a transistor functions under a variety of conditions and environments. For example, the input curves indicate how the base-to-emitter voltage varies with changes in emitter current. Generally, as base current is caused to increase and therefore collector current increases, the base-emitter voltage will increase. However, this increase in voltage is not necessarily linear and so is best shown as a set of curves, an example of which is shown in Fig. 8-3.

This figure represents the input characteristics for a typical germanium transistor, with V_{BE} plotted against collector current. It is clearly revealed that as transistor current is caused to increase, base-to-emitter voltage also increases. For higher values of current the increase is almost linear, but at lower values it is not. Figure 8-4 illustrates the input characteristics of a typical silicon transistor. Note the inherently larger base-to-emitter voltage drop, which is typical of silicon transistors. Also given are other values for differing temperatures, which indicate that the base-emitter voltage is higher at lower temperatures and lower at higher temperatures.

Other sets of curves describe how transistor parameters vary with temperature. One such curve is shown in Fig. 8-5, which illustrates the variation of beta with corresponding changes in temperature. As the

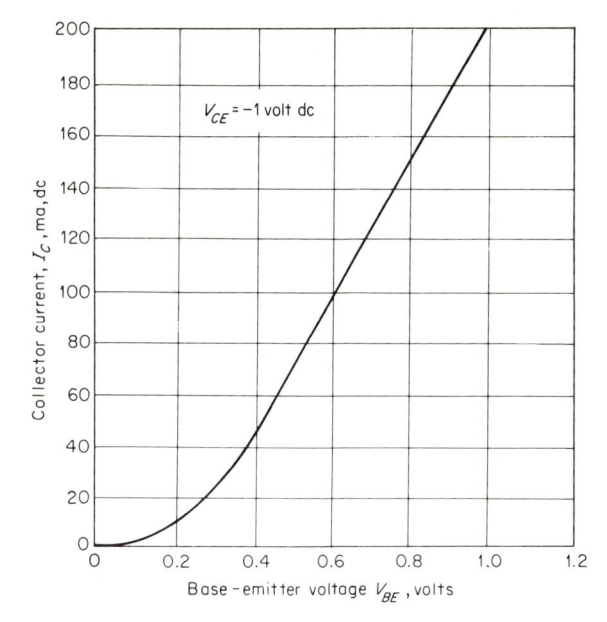

Fig. 8-3. Transistor input curves (Ge).

temperature increases, beta (h_{FE}) also increases. At about 100°C its value has about doubled. This, of course, must be considered in the design of transistor circuits.

Figure 8-6 illustrates how leakage current varies with temperature. This will be dealt with later in some detail.

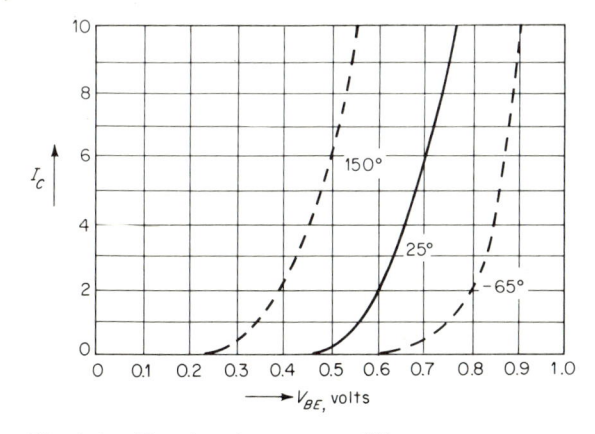

Fig. 8-4. Transistor input curves (Si).

116   PULSE AND SWITCHING CIRCUIT ACTION

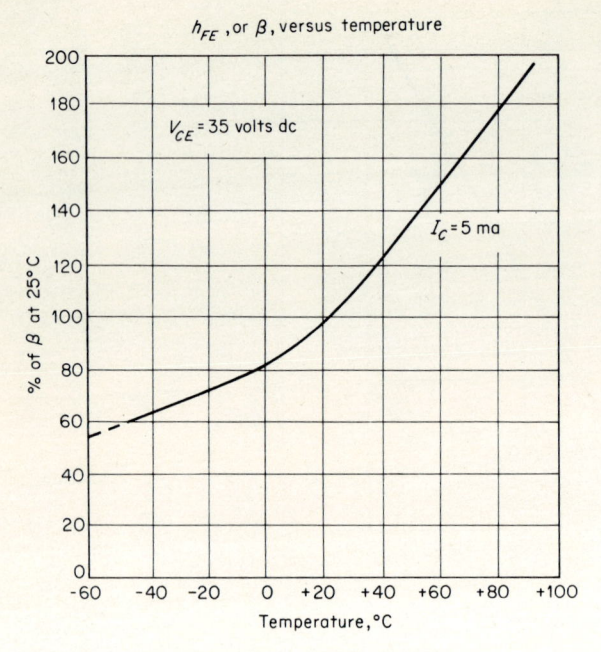

Fig. 8-5. Variation of beta with temperature.

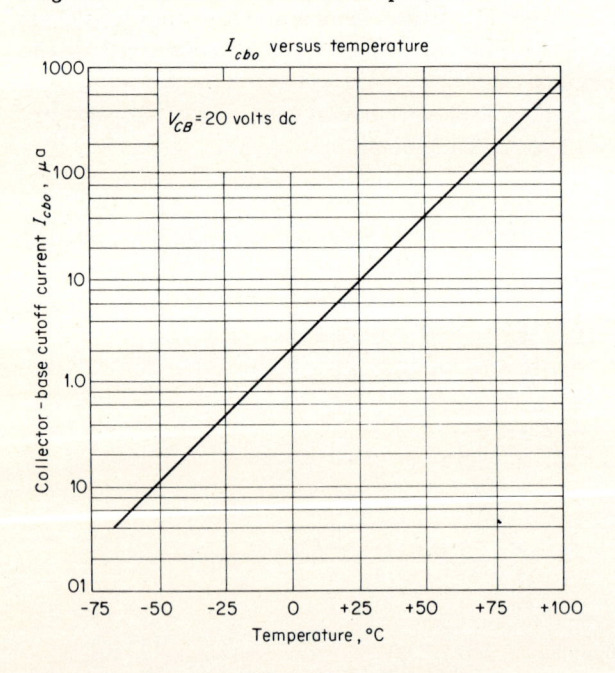

Fig. 8-6. Variation of I_{cbo} with temperature.

TRANSISTOR CHARACTERISTICS 117

Power-temperature derating curve

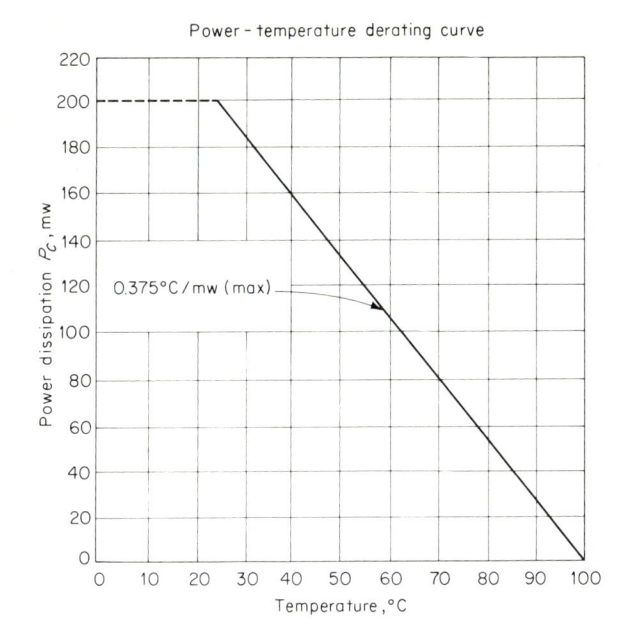

[Graph: Power dissipation P_C, mw vs Temperature, °C, showing a line from (25, 200) decreasing at 0.375°C/mw (max) to near (100, 0)]

Fig. 8-7. Power-derating curve.

The final curve to be shown here is given in Fig. 8-7. This is a typical derating curve, which indicates that at higher temperatures the collector dissipation P_C must be reduced from the normal value at 25°C to some lower value. Because $P_C = I_C \times V_{CE}$, this curve sets the limitation on the maximum collector current and voltage that can be allowed at some maximum temperature.

8-2 THE DC LOAD LINE

The dc load line is a device used to join both transistor and circuit data, which allows the circuit to be analyzed or evaluated in a very simple manner. The load line provides a convenient method of visualizing some of the more important circuit actions and is constructed on the collector family of curves, as illustrated in Fig. 8-8. Also shown is the circuit for which the load line was drawn.

The extremes of the load line represent certain basic limitations imposed on the transistor by the circuit itself. These are identified as limits 1 and 2 on the curves. Limit 1 is the maximum value which V_{CE} can attain *in this circuit* and is rather obvious since the supply voltage limits the maximum value in this circuit to 8 volts. That is, the transistor itself may

118 PULSE AND SWITCHING CIRCUIT ACTION

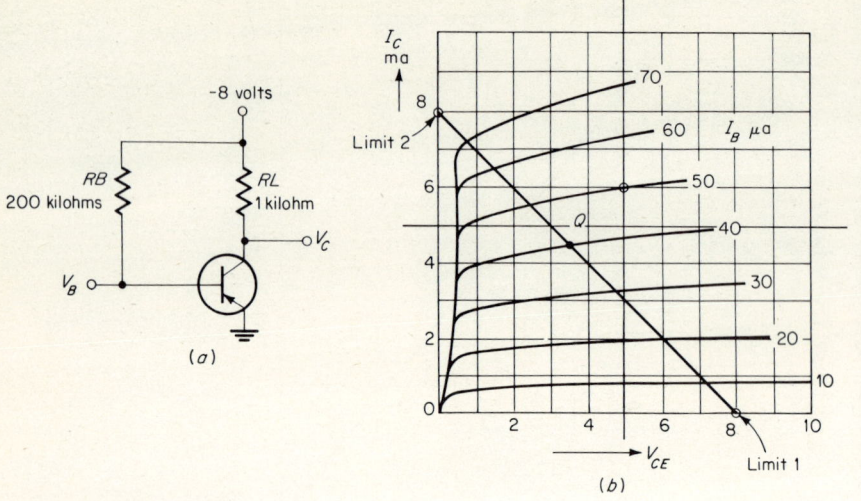

Fig. 8-8. A dc-load-line example.

be rated for, in this instance, 30 volts, but in this circuit $V_{CE,\max}$ can be no more than 8 volts.

Limit 2 specifies the maximum possible collector current, again *in this circuit*. The value of this limit is seen to be 8 ma and is determined as follows (use absolute value of $-V_{CC}$):

$$I_{C,\max} = \frac{V_{CC}}{RL} = \frac{8}{1{,}000} = 8 \text{ ma}$$

To construct the load line, then, the two limits are first found and a straight line is drawn between them. The transistor *must* operate at points that lie along this line.

The load line itself terminates on the ordinates of the graph. The transistor cannot attain operation directly on the ordinate but rather at points slightly away from it. Note Fig. 8-9, where the solid dots represent the calculated values while the circles represent the true values for the transistor.

In the case of limit 1, even though the transistor is in the off condition, a small current I_{cbo} still flows, which is shown as a point slightly above the absolute cutoff point. At limit 2, the true operation point is somewhat below the absolute maximum owing to the finite resistance of the transistor when in its saturated state. $V_{CE,\text{sat}}$ is often on the order of 0.1 to 0.3 volt, and thus collector-to-emitter voltage can never fall below this voltage. The slanting, vertical line from which the base-current values emerge to go

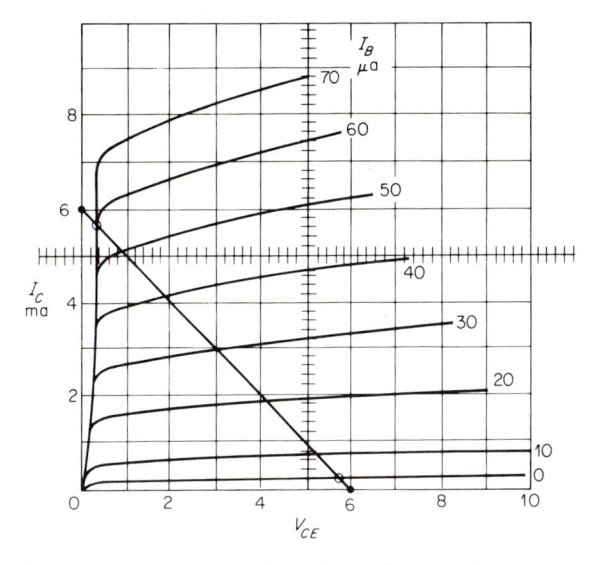

Fig. 8-9. Actual saturation and cutoff points of operation.

to the right has previously been termed the *saturation line* since the transistor must operate on this line when it is saturated.

It is interesting and informative to calculate the actual transistor resistance at these two extremes of operation. The saturation resistance $R_{CE,\text{sat}}$ for the unit represented by Fig. 8-9 is easily found:

$$R_{CE,\text{sat}} = \frac{V_{CE,\text{sat}}}{I_{C,\text{sat}}} = \frac{0.4}{5.6 \times 10^{-3}} = 71.4 \text{ ohms}$$

At the other extreme, the value of current cannot be accurately read, but from a data sheet one might find that I_{cbo} is 1 µa for this transistor:

$$R_{CE,co} = \frac{V_{CE,\text{max}}}{I_{cbo}} = \frac{6}{1 \times 10^{-6}} = 6.0 \text{ megohms}$$

These figures suggest the excellent qualities of the transistor for switching applications. The low on-resistance and the high off-resistance are the mark of a good switch. By special manufacturing processes, even lower values of $R_{CE,\text{sat}}$ and higher values of $R_{CE,co}$ can be attained.

8-3 THE BASIC SWITCHING CIRCUIT

A comparison of a linear amplifier and a typical switching circuit is illustrated in Fig. 8-10. The major characteristics of the linear circuit are

120 PULSE AND SWITCHING CIRCUIT ACTION

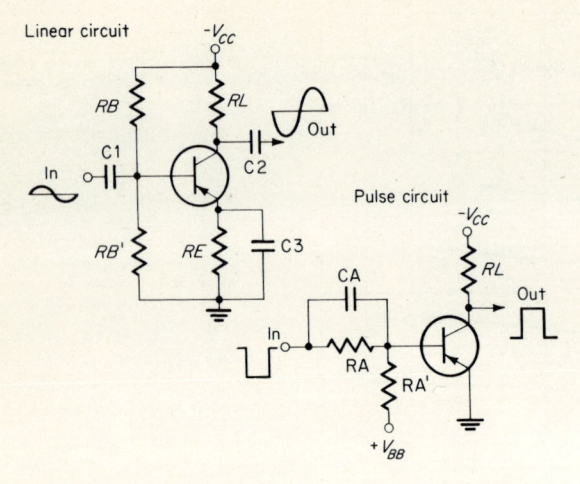

Fig. 8-10. Differences between the linear and pulse circuits.

small output signals, dc bias in the active region, and extreme linearity. The switching circuit, on the other hand, possesses characteristics that are just the reverse: very large input and output signals, no dc bias, as such, and extreme nonlinearity.

To begin to appreciate the overall circuit action of the switching circuit, it must first be understood that as ordinarily used the circuit does not amplify the input signal. This is not to say that it is not capable of amplification but only that in the ideal situation it does not amplify. This is easily seen when it is realized that the magnitude of the input signal often equals that of the value of V_{CC}. Hence, since the output of the circuit cannot exceed V_{CC}, the input and output are usually of the same magnitude.

One fundamental application of such a circuit, as normally used, is that of a shunt switch. If the transistor is off, there is no current through the load resistor RL and therefore no voltage drop across it. The output is therefore equal to V_{CC}. If the transistor is fully on (in saturation), the full V_{CC} drops across RL and the output is essentially at ground potential (0 volts). Thus, it is seen that these two conditions, saturation and cutoff, specify the only two allowable conditions of transistor operation. That is, in typical switching circuits the transistors are either in saturation or in cutoff. The only other allowable condition is the transition from one state to the other.

Obviously, during the transition from off to on (or on to off) the transistor must pass through the active region, and during this time it is capable of amplification. Hence, as will be seen, if the circuit is called upon to amplify, it will be found capable of this circuit action. The detailed explanation of such circuits will be given in Chaps. 9 and 10. At this point the basic description of circuit action will be confined to certain fundamental circuit responses.

Note the example in Fig. 8-11. This configuration is often encountered

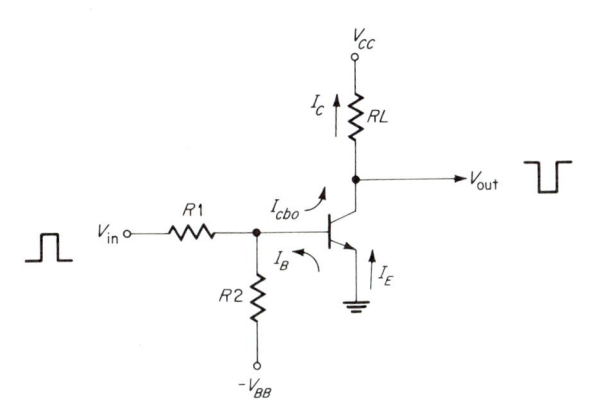

Fig. 8-11. An example of a switching circuit; leg currents and directions.

and will be used for the present example. The currents that flow in the transistor when it is conducting are shown, along with the input and output waveforms. First, assume that the input is at its most negative value, which in such a circuit is usually ground. The voltage divider, consisting of $R1$ and $R2$, is therefore connected between ground and $-V_{BB}$. Because of this, the base of the transistor must be some value more negative than ground. Since this is an *npn* transistor, a voltage such as this causes the base-emitter junction to be reverse-biased and the transistor is off. Collector current, then, is essentially zero, and therefore $V_C = V_{CC}$.

When the input goes more positive than ground by some amount, the base of the transistor also becomes more positive and the transistor is turned on. Hence, maximum collector current flows, and the voltage at the output line goes essentially to ground, typically 0.1 or 0.2 volt. Observing the input and output waveforms, note that they are 180° out of phase with one another. This is to be expected in view of the common-emitter configuration.

One of the currents shown is I_{cbo}, the collector-base leakage current. This current is temperature-sensitive and will give rise to improper circuit operation under adverse conditions. When the transistor is in the off state, the leakage current, if the temperature is high enough, can cause

122 PULSE AND SWITCHING CIRCUIT ACTION

the circuit to malfunction. Leakage current can increase to the point where an appreciable voltage drop appears across the load resistor RL. Thus, the transistor is no longer in the cutoff region but rather in the active region. For a switching circuit of this kind, this is the worst possible condition. In the design of switching circuits, this effect must be considered so as to make the circuit as independent of temperature as possible.

8-4 CIRCUIT LIMITS

The circuit action of a typical switching circuit (inverter) can be described in general terms quite simply. If the transistor is on, $V_C \cong 0$ volts, and if it is off, $V_C \cong V_{CC}$. The circuit shown in Fig. 8-12a, however, is not

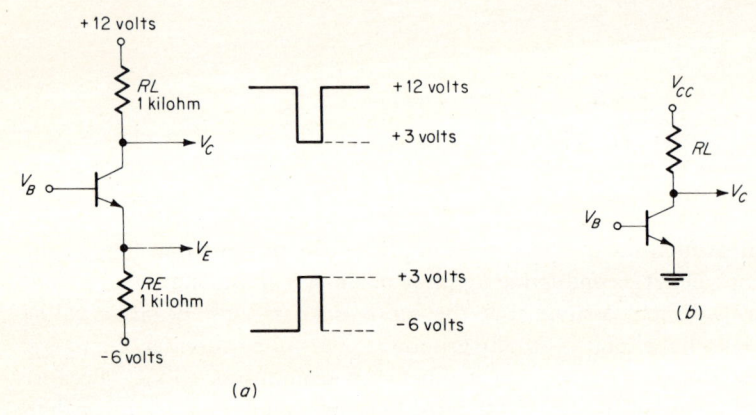

Fig. 8-12. Circuits used to develop the *circuit limits* concept.

nearly so simple to evaluate. Note that neither the emitter, the base, nor the collector leads are returned to ground. This removes much of the simplicity from the circuit and makes a complete circuit description more complex. Only by understanding the concept of *circuit limits* can this type of circuit be adequately analyzed.

The circuit of Fig. 8-12b will be used to introduce these basic ideas in terms that are already known. This circuit can be described in terms of its limits, but this is not usually done because of its simplicity. The usual description of circuit operation is implemented by realizing that the output lead is limited in its maximum excursion by V_{CC} and ground. These are, of course, limiting values beyond which the transistor cannot go.

In order to more accurately describe the circuit action, some new symbols are used that give new meaning to old ideas. For example, the maximum voltage to which V_C can rise is V_{CC}, but in terms of the circuit-limit concept this is called $V_{C,\max}$. This occurs, of course, when the transistor is cut off. The other limit is attained when the transistor is in saturation, in which case $V_{C,\min} = V_{CE,\text{sat}} \cong 0$ volts. Thus, the circuit limits of operation for the collector lead are $V_{C,\max}$ and $V_{C,\min}$. If the base and emitter are not connected to ground, they, too, will be found to have their maximum and minimum limits beyond which they cannot go. Normal operation, then, is either at or between these limits.

Carrying this idea one step further, consider Fig. 8-12a again. The voltage limits of this circuit, as well as the current limits, will now be found. This is accomplished in much the same manner as in the simpler circuit by alternately assuming the transistor to be in its two extreme states: cutoff and saturation.

Taking the cutoff condition as a first example, the circuit limits at each transistor lead will be determined. If the transistor is cut off, there will be no collector current, emitter current, or base current, excluding the leakage current I_{cbo}. Thus, $V_{C,\max} = 12$ volts, $V_{E,\max} = -6$ volts, and V_B must be more negative than -6 volts. Note that in the negative direction, V_B is not limited by any circuit action since the base-emitter junction is reverse-biased. The base can therefore go to any reasonable value less than the junction breakdown voltage. These values, then, specify the cutoff limits of the circuit.

The saturation limits of this circuit are not quite so easily determined. First, on the assumption that the transistor is in saturation (and therefore its internal resistance is nearly zero), the maximum collector current is found to be determined by the total applied voltage and the total series resistance between V_{CC} and $-V_{EE}$.

$$I_{C,\max} = \frac{V_{CC} + (-V_{EE})}{RL + RE} = \frac{18}{2 \text{ kilohms}} = 9 \text{ ma}$$

If it is assumed that α is very nearly 1, the emitter current is also 9 ma. That is, neither collector current nor emitter current can exceed 9 ma under any circumstances, and this therefore constitutes one of the circuit limits.

The maximum current limit will determine the voltage limits at the three transistor leads. The drop across RL will determine the collector saturation limit:

$$E_{RL} = I_{C,\text{sat}} \times RL = 0.009 \times 1000 = 9 \text{ volts}$$
$$V_{C,\min} = V_{CC} - E_{RL} = 12 - 9 = 3 \text{ volts}$$

This calculation suggests that when the transistor is in saturation, the collector voltage can be no more negative than +3 volts and *cannot* go completely to ground as might be expected. Knowing both $V_{C,\text{max}}$ and $V_{C,\text{min}}$ allows the *range* of collector voltage to be stated. In this circuit, the collector lead can range between V_{CC} and +3 volts as the two extreme limits. These are limits which cannot be exceeded!

If this circuit is to be used as a linear amplifier, the collector must never be allowed to approach either of these limits. As a pulse amplifier, these limits set the maximum possible excursion of the collector voltage. If the input is so large that the output tries to exceed the limits, the output waveform is said to be *clipped* and appears as a flat-topped waveform that is clipped at +12 and +3 volts.

Now, the emitter is also held to certain voltage limits beyond which it cannot go. One of these is −6 volts, as determined previously. The other limit is $V_{E,\text{min}}$ and is determined by the drop across the emitter resistor and the current through it, which is $I_{E,\text{max}}$. On the assumption that $I_{E,\text{max}} \cong I_{C,\text{max}}$, this drop is easily found:

$$E_{RE} = I_{E,\text{max}} \times RE = 0.009 \times 1000 = 9 \text{ volts}$$
$$V_{E,\text{min}} = -V_{EE} + E_{RL} = -6 + 9 = +3 \text{ volts}$$

Hence, if $V_{CE,\text{sat}}$ is disregarded, the emitter and collector are essentially at the same potential with respect to ground. (Actually, since the collector saturation voltage is often on the order of 0.1 volt or so, the collector is perhaps 0.1 volt more positive than the emitter. In many circuits this small value can be ignored.)

From the above discussion, it can be appreciated that the collector and emitter limits are a function of the series resistances and the power supplies. The base voltage V_B also has certain limits that determine to a large extent its circuit action. In the negative direction, the circuit itself imposes no limitation, as mentioned. The transistor itself determines any limitation in this direction, this being the reverse-biased breakdown voltage of the emitter-base junction.

However, in the direction that is more positive than −6 volts, a very definite limitation exists. The ultimate value of the maximum-excursion limit of V_B in this direction is influenced by, in part, $V_{BE,\text{sat}}$. Assuming that the base-emitter voltage at saturation for this transistor is 0.75 volt (a typical value for a silicon *npn* unit), the limit in this direction can easily be determined:

$$V_{B,\text{min}} = V_{E,\text{min}} + V_{BE,\text{sat}} = 3 + 0.75 = 3.75 \text{ volts}$$

Thus, if the base is at a voltage of 3.75 volts more positive than ground, the transistor is in saturation and the base cannot be made to go significantly farther in the positive direction. In the usual case, with normal component values, the foregoing statement is true. But, if a source having a low internal impedance is connected to the base, it is possible for the voltage here to be more positive than 3.75 volts. This would resolve into a simple dc circuit problem since the transistor would be so far into saturation that all junctions would be treated as essentially a short circuit. By Thevenizing the source resistance (or using a Millman equivalent), the value of voltage under these conditions can easily be found.

The concept of circuit limits is usefully applied to many circuits. The astable multivibrator and the Schmitt-trigger (both to be dealt with later) are two excellent examples that often require these principles before a meaningful analysis can be made.

QUESTIONS AND PROBLEMS

8-1 Refer to Fig. 8-1. Given that $I_B = 100$ μa and $I_C = 6$ ma, determine V_{CE} for this transistor.

8-2 Refer to Fig. 8-1. Given that $I_B = 120$ μa and $V_{CB} = 7.2$ volts, determine I_C for this transistor.

8-3 Refer to Fig. 8-3. For the transistor that this curve describes, find V_{BE} if $I_C = 100$ ma.

8-4 Refer to Fig. 8-3. For the transistor that this curve describes, find V_{BE} if $I_C = 180$ ma.

8-5 Refer to Fig. 8-3. Given that the emitter resistance $R_{EE} = V_{BE}/I_E$, find the approximate value of emitter resistance if collector current equals 100 ma ($\alpha \cong 1$).

8-6 Refer to Fig. 8-3. Given that the dynamic emitter resistance $r_e = \Delta v_{be}/\Delta i_e$, find the value of r_e if $\Delta i_c = 180 - 100 = 80$ ma.

8-7 Briefly describe the general characteristics of each of the three configurations: *CB, CE, CC*.

8-8 Refer to Fig. 8-12a. The 1-kilohm resistor in the collector circuit is to be changed to a 2-kilohm value. Determine the new limits of this circuit ($V_{CE,\text{sat}} = 0$).
(a) $V_{CE,\text{max}} =$
(b) $V_{CE,\text{min}} =$
(c) $I_{C,\text{max}} =$
(d) $I_{C,\text{min}} =$

8-9 Refer to Fig. 8-12b. Determine the new limits for this circuit if V_{CC} is changed to 5 volts and all values remain the same as drawn ($V_{CE,\text{sat}} = 0$).
(a) $V_{CE,\text{max}} =$
(b) $V_{CE,\text{min}} =$
(c) $I_{C,\text{max}} =$
(d) $I_{C,\text{min}} =$

8-10 Refer to Fig. 8-12a. The circuit is to remain as drawn, except that $-V_{EE}$ is to be changed from -6 to -12 volts. Determine the new limits of V_C and V_E ($V_{CE,\text{sat}} = 0$).
(a) $V_{C,\text{max}} =$
(b) $V_{C,\text{min}} =$
(c) $V_{E,\text{max}} =$
(d) $V_{E,\text{min}} =$

8-11 Refer to Fig. 8-12a. The circuit is to remain as drawn, except that V_{CC} is to be changed from 12 to 6 volts. Determine the new limits of V_C and V_E ($V_{CE,\text{sat}} = 0$).
(a) $V_{C,\text{max}} =$
(b) $V_{C,\text{min}} =$
(c) $V_{E,\text{max}} =$
(d) $V_{E,\text{min}} =$

8-12 Refer to Fig. 8-12a. The circuit is to remain as drawn, except that R_L is to be changed to a 3-kilohm resistor. Determine the new limits ($V_{CE,\text{sat}} = 0$).
(a) $V_{C,\text{max}} =$
(b) $V_{C,\text{min}} =$
(c) $V_{E,\text{max}} =$
(d) $V_{E,\text{min}} =$

9
TRANSISTOR SWITCHING CHARACTERISTICS

In the preceding chapter many switching characteristics of transistors were discussed and derived from published curves. We shall now begin to apply these to certain switching circuits and to see how they affect overall circuit action. The characteristics covered to this point have been, for the most part, *static* characteristics. Now, these ideas will be extended to include dynamic characteristics as well as dynamic circuit action.

9-1 TRANSISTOR SWITCHES

Fundamentally a transistor switch is operated so that when it is in the conducting mode (on), it is fully on, or saturated. Hence, collector current is maximum, and its value is set by the supply voltage and all series resistance in the collector and emitter leads. Conversely, when the transistor is operated in the off mode (nonconducting), essentially no current flows in the collector lead.

These two separate operating conditions are best described in basic

128 PULSE AND SWITCHING CIRCUIT ACTION

(a)

$$R_{CE1} = \frac{V_{CE}}{I_C} = \frac{0.15}{0.001} = 150 \text{ ohms}$$

$$R_{CE2} = \frac{V_{CE}}{I_C} = \frac{0.2}{0.002} = 100 \text{ ohms}$$

$$R_{CE3} = \frac{V_{CE}}{I_C} = \frac{0.3}{0.003} = 83 \text{ ohms}$$

$$R_{CE4} = \frac{V_{CE}}{I_C} = \frac{0.3}{0.004} = 75 \text{ ohms}$$

$$R_{CE5} = \frac{V_{CE}}{I_C} = \frac{0.35}{0.005} = 70 \text{ ohms}$$

$$R_{CE6} = \frac{V_{CE}}{I_C} = \frac{0.36}{0.006} = 60 \text{ ohms}$$

$$R_{CE7} = \frac{V_{CE}}{I_C} = \frac{0.37}{0.007} = 53 \text{ ohms}$$

(b)

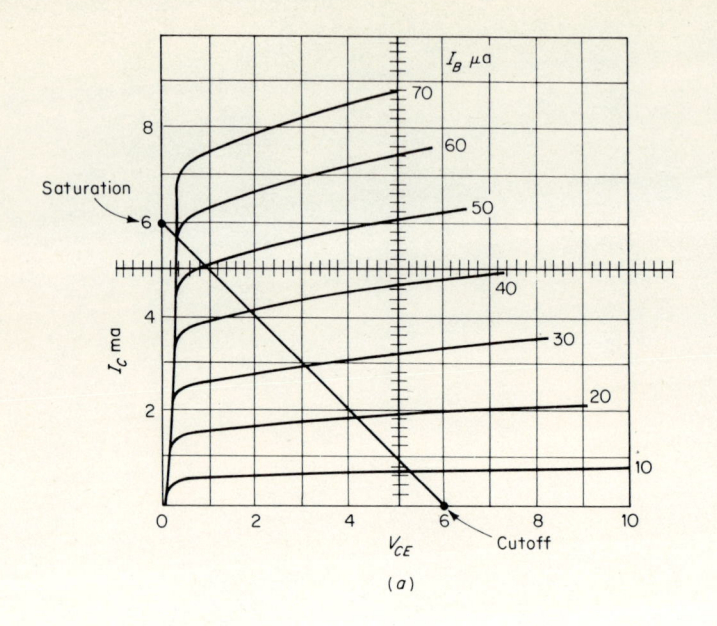

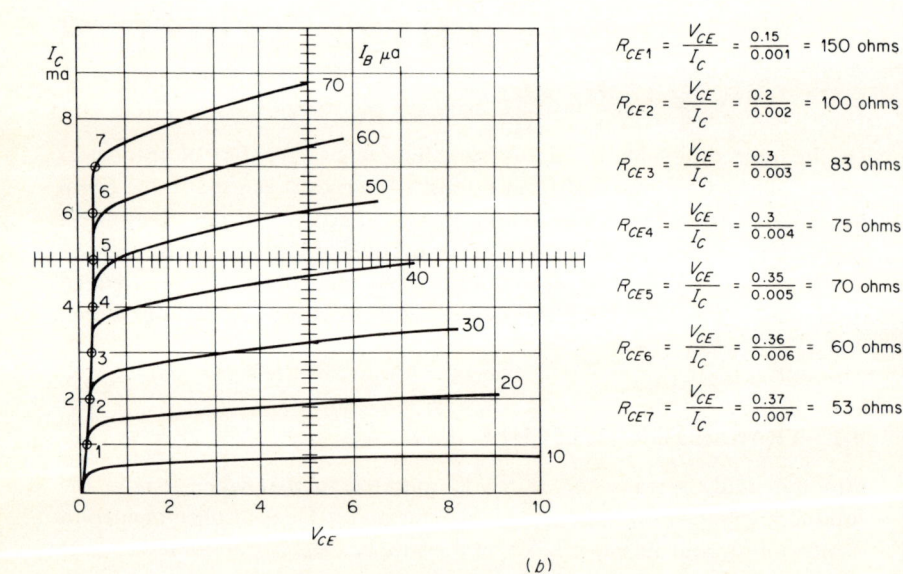

Fig. 9-1. (a) A load line for a typical switching circuit; (b) values of R_{CE} when transistor is in saturation.

terms by the dc load line constructed upon the collector curves. The upper limit, of course, represents the saturated condition, while the lower limit is the cutoff condition. This is illustrated in Fig. 9-1a. A circuit that functions as described by the load line is shown in Fig. 9-2. If

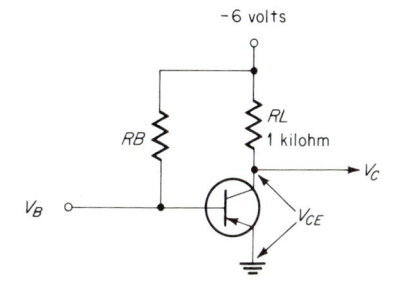

Fig. 9-2. A circuit for the load line of Fig. 9-1.

$-V_{CC}$ is -6 volts relative to ground and if $RL = 1$ kilohm, then the circuit response is described by the given load line.

Ignoring the circuitry in the base lead for a moment, if the base of the transistor is made positive with respect to ground, the transistor is off. Hence, with no current through the load resistor RL, there will be no voltage drop across it and V_{CE} is equal to -6 volts. This is indicated on the load line as the *cutoff point*.

Now, if the base is made negative with respect to ground, the transistor is turned on. As long as base current exceeds 60 μa, the transistor is in saturation and maximum collector current flows, limited in value only by $-V_{CC}$ and RL.

$$I_{C,\max} = \frac{V_{CC}}{RL} = \frac{6}{1 \times 10^3} = 0.006 \text{ amp} = 6 \text{ ma}$$

This, of course, assumes that the saturated transistor exhibits zero resistance, which is not quite true. Figure 9-1b illustrates the equivalent collector resistance R_{CE} during saturation. Seven points are shown, with the collector resistance of each point calculated. Because R_{CE} is small relative to the load resistor RL, the drop across it is also very small.

Thus, the transistor is operated between the limits of fully off and fully on. The only time it is in the active region is during a transition from off to on or vice versa. The degree of saturation can be explained in terms of *forced beta* β_F. The forced beta is the measured ratio of the true collector current to the true base current while the transistor is in saturation:

$$\beta_F = \frac{I_{C,\text{sat}}}{I_{B,\text{sat}}}$$

A transistor operated in the region that is just barely in saturation will have a forced beta of slightly less than normal. On the other hand, a transistor operated far into the saturation may have a forced beta of 10 or less.

The type of circuit shown in Fig. 9-2 is often called an *overdriven* amplifier, and this is very descriptive of circuit operation. The input of the transistor is driven so hard that it is either far into saturation or well beyond cutoff. The circuit is capable of amplifying, but in the usual case it does not do so. Since the signal source is in all probability operating from the same power supply, the signal voltage amplitude is as large as the collector supply voltage. Hence, e_o/e_i is usually 1. If, however, the input has been degraded for any reason and is less than V_{CC}, then the circuit *will* amplify. In this case the output will swing between V_{CC} and ground even if the input-voltage swing is considerably less.

To analyze the circuit operation some values must be assigned. Figure 9-3 shows a circuit with appropriate values. The operation of this circuit

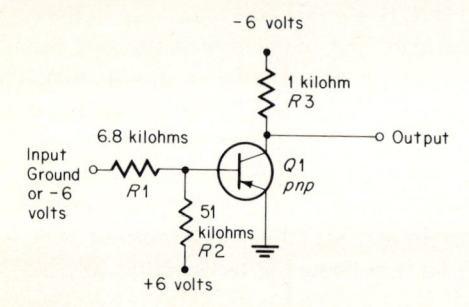

Fig. 9-3. A typical switching circuit example.

is relatively simple, as briefly explained in Chap. 8. The input can only be one of two conditions—either 0 (ground) or −6 volts. The circuit is arranged so that the transistor is turned either all the way on or all the way off. Since this is a *pnp* transistor, if the input is at ground, the transistor is off, while if the input is at −6 volts, the transistor is turned on. The voltage divider in the base circuit processes the signal so that it is suitable for the transistor. This is necessary because of the fact that if raw −6 volts were applied directly to the base of the transistor, it would be destroyed. Thus, the voltage divider serves to limit the current to a safe value and still provide positive turn-on and turn-off.

Another way of looking at the need for $R1$ and $R2$ is that the signal at the base should have some value of voltage less than the full supply voltage, and this reduced voltage should have an equivalent Thevenin's resistance much higher than that of the supply voltage. $R1$ and $R2$, then, synthesize a current-limited voltage for direct application to the base.

When the input goes to -6 volts, the base is driven very negative. How far negative it goes and how much base current is allowed to flow are functions of the values of the base resistors. Usually, the transistor is driven far into saturation when it is turned on. With the transistor very far into saturation, its internal resistance R_{CE} is very low—often between 10 and 20 ohms. Thus, the output voltage taken from the collector is nearly ground, typically -0.1 volt with a -6-volt input.

If the base-circuit input is at ground, or 0 volts, the base is driven somewhat positive, perhaps 1 volt. Since the transistor is *pnp* it is turned off. This means that there is no current through the load resistor, and therefore no voltage drop appears across it so that the output voltage is -6 volts. With an input level of 0 volts, the output is -6 volts, while with -6 volts in, the output is zero. The transistor is obviously inverting the signal. This is to be expected since it is a characteristic of the common-emitter configuration.

The general description above is useful, but it is desirable to be able to calculate some of the important values in the circuit. It is recommended that the reader study the following procedure carefully because many of the functions of the various parts become evident during the problem.

Choose one of the two possible input conditions, say, 0 volts. Mentally disconnect the transistor base from the junction of $R1$ and $R2$. (This prevents the transistor from possibly changing the values provided by the voltage divider. In this case, it will make little, if any, difference.) The voltage divider may be redrawn as shown in Fig. 9-4. Determine the

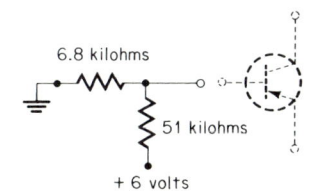

Fig. 9-4. The base voltage divider with a 0-volt input.

voltage at the junction of the two resistors in reference to ground. Since the emitter of the transistor is at ground, the voltage delivered to the base of the transistor, referenced to ground, will now be determined by the voltage divider. The voltage drop across $R1$ is

$$E_{R1} = \frac{R1}{R1 + R2} \times V_{BB} = \frac{6.8 \times 10^3}{(6.8 \times 10^3) + (51 \times 10^3)} 6$$

$$= \frac{6.8 \times 10^3}{57.8 \times 10^3} 6$$

$$= 0.118 \times 6 \cong 0.7 \text{ volt}$$

For the circuit condition shown, the junction of the two resistors can only be more positive than ground, and so it must be $+0.7$ volt with respect to ground. Reconnecting the base to the voltage divider will place $+0.7$ volt on the base. With the base more positive than the emitter, the *pnp* transistor is biased off and no collector current can flow.

Next, determine the output voltage. Since no collector current can flow, there is no current through the load resistor and therefore no voltage drop across it. The output must then be equal to -6 volts. Thus, a 0-volt input results in a -6-volt output.

With the input connected to -6 volts, Fig. 9-5, a new set of conditions will prevail. Again, mentally disconnecting the base lead to prevent upsetting the voltage divider, calculate the new voltage at the junction of $R1$ and $R2$:

$$E_{R1} = \frac{R1}{R1 + R2} \times V_T = \frac{6.8 \times 10^3}{57.8 \times 10^3} \times 12 = 0.118 \times 12 = 1.41 \text{ volts}$$

Since the drop across $R1$ is 1.41 volts, the junction of $R1$ and $R2$ must be more positive than -6 volts, or -6 volts $+1.41 = -4.6$ volts. With the junction of $R1$ and $R2$ at a potential of -4.6 volts referred to ground, it now becomes clear why the drop across $R1$ is not calculated with the transistor connected. With the base reconnected, the voltage at the junction will *not* be -4.6 volts, but will be about -0.2 volt instead. Keeping in mind the fact that when the emitter-base junction becomes forward-biased it conducts rather heavily (which indicates that the resistance of the junction is quite low), one can appreciate the fact that the base is effectively clamped to ground. Recall from Chap. 8, Fig. 8-3, that the base-emitter voltage must agree with the V_{BE}-I_C curve for the transistor; it is clear that the base-to-emitter voltage can never exceed a few tenths of a volt. This situation is illustrated in Fig. 9-5.

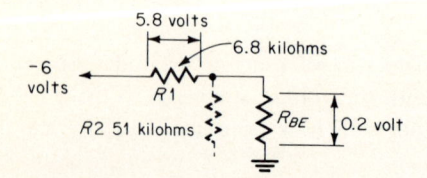

Fig. 9-5. The base voltage divider with -6-volt input.

Across the entire circuit, from ground to -6 volts, the total resistance consists of 6.8 kilohms and the resistance of the junction R_{BE}. If R_{BE} is low in value compared with the 6.8-kilohm resistor, we will make no large error in assuming that the current is limited by the 6.8-kilohm resistor only, in which case

$$I = \frac{E}{R} = \frac{6}{6.8 \times 10^3} = 0.88 \times 10^{-3} \text{ amp} = 0.88 \text{ ma}$$

Since 0.88 ma is flowing through R_{BE} and since 0.2 volt is dropped across it, R_{BE} must be equal to

$$\frac{E}{I} = \frac{0.2}{0.88 \times 10^{-3}} = 0.227 \times 10^3 = 227 \text{ ohms}$$

We can see that R_{BE} is quite small relative to the 6.8-kilohm resistor, and it is also clear why the base is at -0.2 volt.

Note particularly that when the base is connected to the junction of $R1$ and $R2$ (-6 volts at the input), the voltage divider now yields a much different value at the junction of $R1$ and $R2$. This difference can only be accounted for by the increased current through the voltage divider. This additional current is the base current itself. Since a larger current flows through $R1$, there must be a larger voltage drop across it and this is reflected as a more positive voltage at the junction of the two resistors.

The voltage drop across the emitter-base junction of a saturated germanium transistor is typically 0.2 volt (0.7 volt for silicon) and is often simply referred to as the *base-junction voltage*.

Now, determine the output voltage. Referring to Fig. 9-1a, if these curves are applicable to this transistor, it can be seen that it will require about 60 μa of base current to begin to saturate the transistor (that is, to operate near the upper extremity of the load line). We shall now approximate I_B. With the base disconnected, it was previously determined that the drop across $R1$ is 1.41 volts. The current through $R1$ is then

$$I_{R1} = \frac{E}{R1} = \frac{1.41}{6.8 \times 10^3} = 0.2076 \times 10^{-3} \text{ amp} \cong 208 \text{ μa}$$

With the base connected, however, the current through $R1$ is again E/R. But now $E = 5.8$ volts, and so

$$I_{R1} = \frac{5.8}{6.8 \times 10^3} = 0.853 \times 10^{-3} \text{ amp or } 853 \text{ μa}$$

The difference between the two values is very nearly the amount of base current that must be flowing:

$853 - 208 = 645$ μa

134 PULSE AND SWITCHING CIRCUIT ACTION

If the base current is larger than about 60 μa, the transistor is far into saturation and is said to be *full on*. The collector current must try to be βI_B, and this would cause a collector current of (if $\beta = 100$)

$$100 \times 643 \ \mu a \cong 64 \ ma$$

It is appreciated that this large amount of current simply cannot flow in the collector since Ohm's law indicates that the maximum current can only be

$$I_{max} = \frac{V_{cc}}{RL} = \frac{6}{1 \times 10^3} = 6 \times 10^{-3} = 6 \ ma$$

Since no more than 6 ma can flow but 64 ma would flow if it were allowed, there must be the full 6 ma in the collector circuit. The voltage drop across RL is then

$$E_{RL} = I_C \times RL = (6 \times 10^{-3})(1 \times 10^3) = 6 \ volts$$

The output must be 6 volts more positive than -6 volts and so must be 0 volts. The output lead is then firmly connected to ground through the heavily conducting transistor whose internal resistance is essentially zero! In this circuit, forced beta is

$$\beta_F = \frac{I_{C,sat}}{I_{B,sat}} = \frac{6 \ ma}{643 \ \mu a} = 9.33$$

Thus about ten times more base current flows than necessary.

As previously mentioned, it may not at first be apparent why the circuit is called an *amplifier*. Since the input is either 0 or -6 volts and the output is either -6 or 0 volts, no apparent amplification has been achieved. However, the input voltage is in reality a signal, and so it has been developed in perhaps a great deal of other circuitry. While passing through this other circuitry, it is quite possible that it will become badly degraded. In a case such as this, it might arrive at the input to this inverter amplifier as a -3-volt signal rather than the -6 volts that would be desirable. With -3 volts at the input, the voltage divider now sees the condition shown in Fig. 9-6. The voltage across the 6.8-kilohm resistor is

$$E_{R1} = \frac{6.8 \times 10^3}{57.8 \times 10^3} \times 9 = 1.06 \ volts$$

The junction of the resistors must be 1.06 volts more positive than -3 volts, or $-3 + 1.06 = -1.94$.

This is still sufficient negative voltage to drive the transistor into saturation, and the output will still be a firm ground. In this case, the transistor *is* amplifying since a change from 0 to −3 volts at the input will still yield a −6-volt to 0-volt swing at the output terminal.

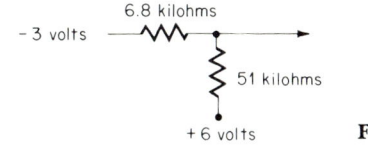

```
-3 volts ──6.8 kilohms──┬──────►
                        │
                        ≶ 51 kilohms
                        │
                     +6 volts
```

Fig. 9-6. Reduced-amplitude input.

To verify that with −3 volts at the input the transistor is truly in saturation, the base current will be determined for this circuit condition. This will be done using a different method than used heretofore to illustrate an alternate approach. Assume that R_{BE} is 250 ohms when the transistor is saturated. Millman's theorem provides a simple method of arriving at the correct figures. Referring to Fig. 9-5, the base voltage V_B is first determined ($e_{in} = -3$ volts):

$$V_B = \frac{e_{in}/R1 + V_{BB}/R2 + 0/R_{BE}}{1/R1 + 1/R2 + 1/R_{BE}} = \frac{-3/6800 + 6/51{,}000 + 0/250}{1/6800 + 1/51{,}000 + 1/250}$$

$$= \frac{-0.000441 + 0.000118}{0.000147 + 0.0000196 + 0.004} = \frac{-0.0003235}{0.0041667} \cong -0.078 \text{ volt}$$

$$I_B = \frac{V_B}{R_{BE}} = \frac{-0.078}{250} = -311 \text{ μa}$$

Because any value of base current in excess of 60 μa will cause saturation, the transistor is obviously still far into the saturation region, with only −3 volts at the input.

Cascaded Switching Amplifiers

When two or more pulse amplifiers are so connected that the output of the first is directed to the input of the second and that of the second is directed to the third, etc., the amplifiers are said to be *cascaded*. Such a circuit is shown in Fig. 9-7. This circuit is essentially no different than the single-stage amplifier. We can understand how the circuit functions if we work our way through it step by step in the same manner as was done for the single-stage amplifier. The calculations are indicated below in somewhat less detail than done previously to avoid unnecessary repetition.

Step 1 Assume a 0-volt input (ground).
    (a) V_B of $Q1 = (12 \times 10^3)/(45 \times 10^3) \times 6 = +1.60$ volts, and so $Q1$ is off and V_C of $Q1$ is at approximately −12 volts.

136 PULSE AND SWITCHING CIRCUIT ACTION

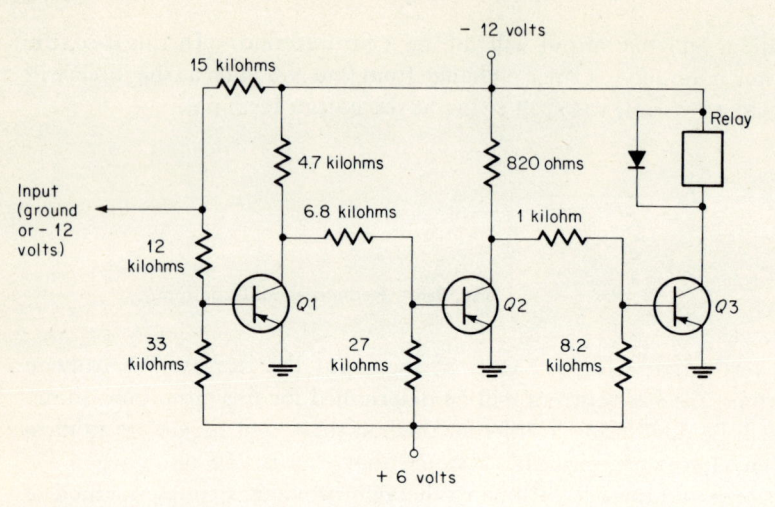

Fig. 9-7. Cascaded switching amplifier.

    (b) V_B of $Q2 = (4.7 \times 10^3 + 6.8 \times 10^3)/(38.5 \times 10^3) \times 18 = 5.38$ volts more positive than -12. $-12 + 5.38 = -6.63 \cong -6.6$ volts. $Q2$ is therefore in saturation. (Keep in mind the fact that when $Q2$ is in saturation, the base voltage is actually about -0.2 volt.)

    (c) V_B of $Q3 = (1 \times 10^3)/(9.2 \times 10^3) \times 6 = +0.652$ volt, and $Q3$ is off. Consequently, the relay is not energized.

Step 2 Assume a -12-volt input.

    (a) V_B of $Q1 = (12 \times 10^3)/(45 \times 10^3) \times 18 = 4.8$ volts more positive than -12. $-12 + 4.8 = -7.2$ volts; $Q1$ is in saturation, and V_{BE} is actually about -0.2 volt. The collector of $Q1 \cong 0$ volts.

    (b) V_B of $Q2 = (6.8 \times 10^3)/(33.8 \times 10^3) \times 6 = 1.2$ volts more positive than ground. Thus V_B is at $+1.2$ volts and $Q2$ is off. The collector of $Q2 = -12$ volts.

    (c) V_B of $Q3 = (1.82 \times 10^3)/(10.02 \times 10^3) \times 18 = 3.3$ volts more positive than -12 volts. $-12 + 3.3 = -8.7$ volts, and $Q3$ is in saturation. As a result, the relay is energized.

    In summary then, if the input is 0 volts, $Q1$ is off, $Q2$ is on, $Q3$ is off, and the relay is not energized. With a -12-volt input, $Q1$ is on, $Q2$ is off, and $Q3$ is on, providing energizing current for the relay.

Transistor Switching Characteristics

An ideal switch is one that has an infinite resistance when it is open and a zero resistance when it is closed. Also, it will have a way by which it can be opened or closed. As is known, the transistor can be used as a switch, and when so used the circuit is called a *switching circuit*. Transistors have many advantages over mechanical switches, such as the lack of contact bounce and arcing, very fast switching speed, and very high repetition rates.

There are no real disadvantages, but as stated earlier, the transistor is not a perfect switch, and so it will deviate from a perfect switch somewhat. For instance, a transistor operated in the off mode will always have some resistance, even though this may be several hundred thousand ohms or even megohms. The reason for this is that at normal temperatures, the reverse-saturation current, or leakage current, will flow across the reverse-biased collector junction. Any current flowing here is a deviation from a complete open circuit. This current I_{cbo} is partly temperature-dependent and at elevated temperatures can become quite large, causing a further reduction of the resistance at the collector junction.

This leakage current I_{cbo} actually has four separate components, each of which varies somewhat differently with temperature and applied voltage. Two of these components are just temperature-sensitive and so will increase only with an increase in the temperature. The other two are voltage-sensitive and will increase with increasing voltage at the collector junction. Thus, a transistor with a cutoff voltage applied to its base cannot turn off completely since a small current continues to flow. At normal temperatures and voltages, this leakage current is on the order of a few microamperes for germanium transistors and, with proper circuit design, is not troublesome. However, under the right conditions, a transistor that is supposed to be in cutoff might well be in the active region or even in saturation. A very high temperature or a load resistor of too great a value can cause this condition.

One other departure from a perfect switch is the case where a transistor is fully on. Ideally, this would exhibit a zero resistance, and even though the current is large there would be no voltage drop at all across the transistor. In practice, however, there is always a voltage drop across the transistor, which indicates that there is some resistance being encountered by the current. For good-quality switching transistors, this voltage drop is on the order of 0.1 volt and the internal resistance of the transistor is about 1 to 10 ohms. This must be considered when designing or analyzing transistor switching circuits. Since in a well-designed circuit these effects are minimized, we usually consider the transistor to be a perfect switch, keeping in mind that this is not necessarily true in all cases.

The preceding discussion considered the steady state or dc response of the transistor to switching conditions. Just as it is true that a transistor is not a perfect switch for dc conditions, we find that it is also not a perfect switch for transient conditions. Ideally, a high-speed switch should turn off and on in zero time. In other words, a perfect switch would activate instantly, with no time lag at all. This is literally impossible. It takes some time for a transistor to turn on and off, and the reasons for this must be considered.

One of these reasons is that the transistor must traverse the active region when making the transition from off to on, or on to off. This is illustrated in Fig. 9-8, where a load line is shown drawn upon the characteristic

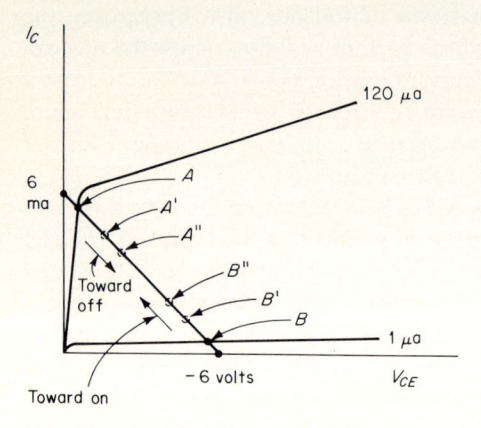

Fig. 9-8. On- and off-turning excursions along the load line.

curves of a transistor. The drawing shows two points, A and B, among others. Point A is the saturation point, while B is the cutoff point. If a transistor is operating at point B and is suddenly turned on, it cannot immediately jump to point A. Instead, it must move up the load line through B', B'', etc. While the operating point is moving up the load line, the transistor is in the active region and is therefore capable of amplifying. It takes some amount of time to get through the active region, and some reasons will be given for this later.

By the same token, if the transistor is turned off after having been turned on, it must again traverse the active region and again it takes some amount of time to do this. If the pulse-repetition rate is very slow, the turn-on and turn-off times can be neglected since these are on the order of microseconds. However, if the repetition rate is also on the order of microseconds, then the turn-on and turn-off times are of great importance.

Figure 9-9 shows a circuit with which the output pulse can be measured to compare it with the input applied to the base. This will show any time

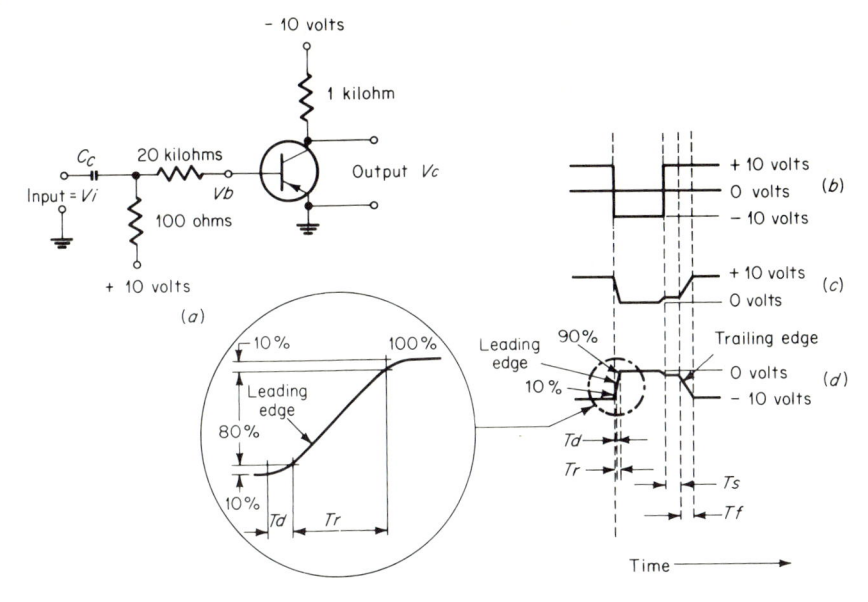

Fig. 9-9. Transistor departures from a perfect switch.

differences between the input and the output. Even though the input to the base is caused to change virtually instantaneously, the output lags by a fairly appreciable amount on both the leading and trailing edges.

The delay time Td is caused by several factors, the first of which is the fact that the collector-base junction and the emitter-base junction both exhibit appreciable capacitance, as does the circuit wiring. While the transistor is off, these capacitances are charged relative to ground, with $+10$ volts on the base and -10 volts on the collector. These capacitances must change their state of charge to about 0 volts at the base and 0 volts at the collector. They must change their state of charge through the resistances shown and therefore will take a certain period of time to reach the new state. Also, time must be allowed for the emitter current carriers to begin to diffuse across the base region. Adding further to the delay time are the facts that the emitter current is very small, the current gain is reduced, and a given change in base current will result in a smaller change in collector current than might be expected. This will delay the rise of collector current. The total effect of these factors is called the *delay time Td*.

The rise time Tr is a result of the base-current limitation; the larger the allowed base current, the shorter the rise time. In circuits where turn-on time is important, we would expect to find that the base-circuit resistance is low compared with another circuit. The rise time is the period of time the transistor is traversing the active region and is measured between the 10 and 90% points on the curve.

The storage time T_S is the delay the transistor exhibits before the collector current begins to drop in value. In other words, for some period of time after the base is presented a turn-off voltage, collector current continues to flow. The storage time can be quite appreciable under certain conditions. If the transistor has been on and if operation has been in the active region, the storage time is usually negligible. In many pulse applications, however, the transistors are operated in saturation, and in this case the storage time may approach several microseconds. The overall effect of T_S is to lengthen the output pulse, compared with the time duration of the input, for the circuit conditions illustrated.

A transistor is in saturation when the emitter junction and the collector junction are forward-biased and the base resistor and the load resistor are chosen so that the load resistor limits I_C rather than the transistor characteristics. A transistor is said to be lightly saturated if the base current is increased from the active region just to the point where V_{CE} is nearly zero. On the other hand, a transistor is heavily saturated when the base current is much larger than necessary to barely attain saturation. The further into saturation a transistor is driven, the longer the storage delay time.

The reason that T_S exists is that the saturated transistor behaves much differently in saturation than in the active region. With *both* junctions forward-biased, the *collector and the emitter* are emitting carriers into the base. The base region, which is very small relative to the other regions, becomes full of carriers. In a *pnp* unit, the carriers are holes; in an *npn* unit the carriers are electrons. Since the base region is lightly doped, most of the carriers can find nothing with which to combine and they swarm about within the confines of the base region as long as the transistor is in saturation. When the base is suddenly presented with a voltage such as to turn off the transistor, the carriers in the base must remain and will continue to combine. Since I_B continues to flow even after the base is turned off, I_C will continue to flow for the same length of time. The length of time these excess carriers are stored in the base is called the *storage time* (T_S). As soon as the stored carriers are swept away, the transistor starts to turn off. If the base resistor is made small in value in an attempt to decrease the storage time by providing a low-resistance path for sweeping away the stored charge, the beneficial effects are nullified owing to the fact that the same low-value resistor will allow the transistor to be driven further into saturation. The more highly saturated the transistor, the greater the stored charge, and the longer the delay time.

The initial current flowing in the base lead during the time the transistor is in saturation is called I_{B1}. The current that continues to flow after the turn-off time has occurred is called I_{B2} and is caused by the necessity of getting rid of the stored charge.

In Fig. 9-10a, a diagram of a transistor in saturation is shown, and in Fig. 9-10b the off-turning pulse is just applied to the base. The base region is seen to be filled with charge carriers (holes), and I_{B2} has started to flow. In Fig. 9-10c and d the stored charges are decreasing until finally they have disappeared and the transistor begins to turn off.

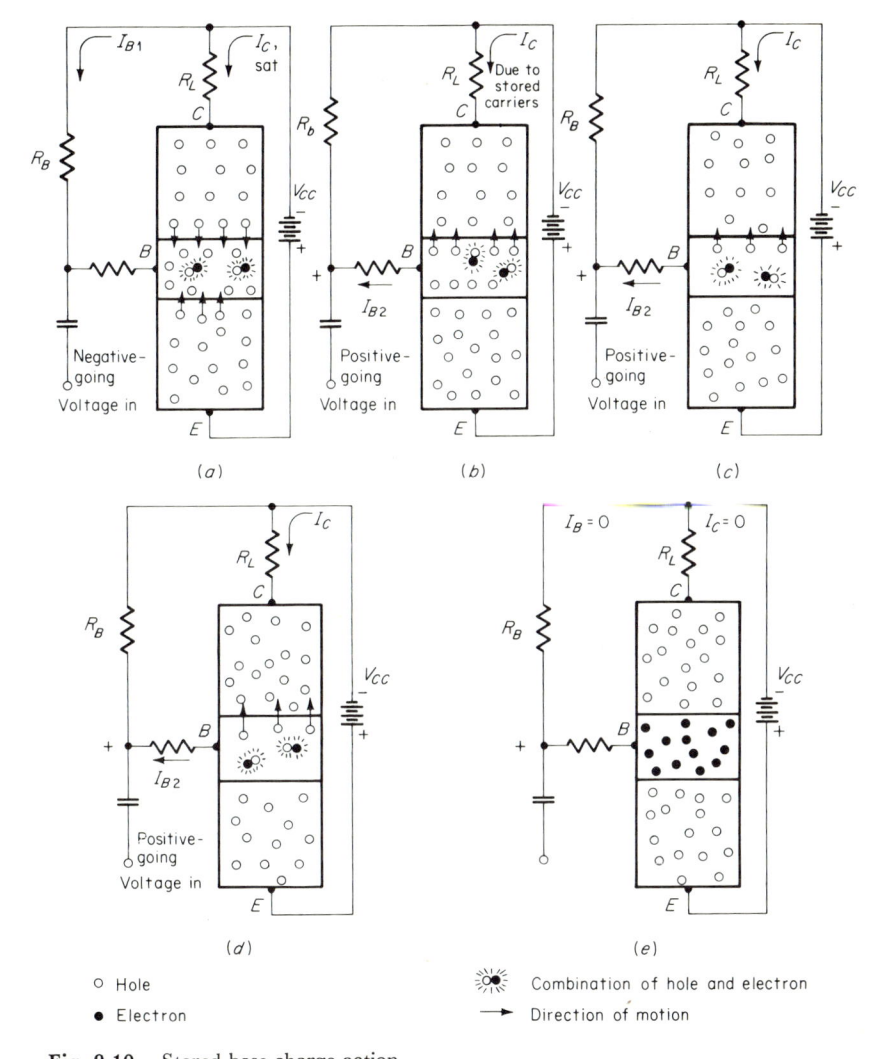

○ Hole
● Electron

Combination of hole and electron
→ Direction of motion

Fig. 9-10. Stored base-charge action.

The fall time T_f is analogous to the rise time in that the transistor goes through the active region during this period. T_f is measured from the 90% point on the curve to the 10% point.

Of all the preceding delay times, the storage time T_s is the worst offender. In extreme cases, T_s can very appreciably shorten or lengthen a pulse. If a transistor is normally off and then momentarily pulsed on and into saturation, the collector pulse will be longer in duration than the input pulse at the base. On the other hand, if a transistor that is normally saturated is briefly pulsed off, the storage-time delay will shorten the pulse and the output will be of shorter duration than the input. Thus, operating a transistor far into saturation, which accentuates T_s, can have seriously detrimental results.

Temperature Effects

Pulse circuits are usually less susceptible to temperature variations than linear amplifiers. This, however, does not mean that the question can be ignored. It simply means that there is less involved in the temperature stability of pulse circuits.

Switching circuits are usually operated in the on-off mode. There is, therefore, no such thing as a drift of the quiescent operating point. However, if an off-transistor experiences an increase in temperature, there will be a point where it comes on again even though the base is reverse-biased. If the transistor were already on, there would be no noticeable change as the temperature increased. But, when a signal voltage arrives at the base to turn the transistor off, it may remain on or partly on. Thus, the temperature characteristics of pulse circuits must be taken into consideration.

To see how I_{cbo} acts in a typical pulse circuit, it will be advantageous to investigate a circuit example. Suppose the circuit of Fig. 9-11 were to be used. In this circuit, the transistor is not biased firmly off, but instead the base is simply returned to ground. With no signal in (0 volts), the transistor is off but the base-emitter junction is not reverse-biased.

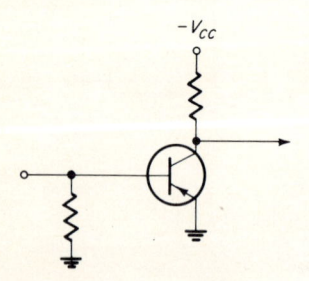

Fig. 9-11. Transistor with 0-volt bias.

Now I_{cbo}, the reverse saturation current, flows down the collector lead to the base of the transistor. If RB is large, the greatest part of I_{cbo} will flow on to the emitter and out to ground. This will result in amplification of this current, and beta times this value will also flow in the collector lead. If beta is 100, then 101 times I_{cbo} will flow in the collector and will cause a large drop across the load resistor. Thus, RL must be a very small value to minimize this effect.

One way of eliminating this amplification of current, at least to some degree, would be to insert a large resistor in the emitter of the transistor. This would tend to shunt more I_{cbo} out the base lead which would not be amplified. However, the emitter would be floating above ground, and in most circuits of this type this would be most undesirable.

Another solution to the problem (and this is the solution ordinarily used) is to firmly reverse-bias the emitter junction when the transistor is off. If V_B referred to ground is about 1 volt or more of reverse bias, essentially all of the saturation current will flow out the base lead. It will not be amplified by the transistor, and the amount of leakage current in the collector lead will be minimized. Such a circuit is shown in Fig. 9-12

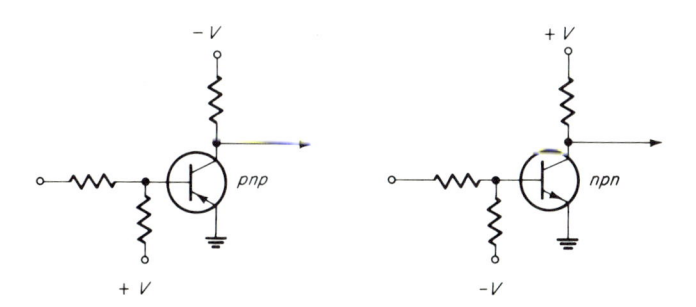

Fig. 9-12. Circuit arrangement to ensure positive forward and reverse bias.

for both *npn* and *pnp* transistors. This is the circuit configuration usually encountered. The voltage divider in the base circuit provides the necessary reverse bias that will allow I_{cbo} to flow out the base lead rather than out the emitter lead.

Power Dissipation

Normally, digital circuits such as pulse amplifiers do not dissipate a great amount of power. In the usual inverter, if the transistor is off, no collector current flows and the collector dissipation is zero, discounting I_{cbo}. If the transistor is saturated, V_{CE} is nearly zero, and so even though the collector current is large, the product of E and I is small. For instance, if a transistor is in saturation with as much as 200 ma flowing, the power dissipation might be, if V_{CE} is about 0.1 volt,

$$P_C = V_{CE} \times I_C = 0.1 \times 0.2 = 0.02 \text{ watt}$$

Since this transistor might well be rated for 200 mw, this is far below the rated maximum.

As long as the transistor remains cut off or saturated for relatively long periods of time between transitions from one extreme to the other, the power dissipation is usually far below rated maximum. However, if the transistor is switched from one extreme to the other very rapidly, then the power dissipation increases sharply. This is true because maximum power is generated by the transistor when it is operating in the center of its load line. This is illustrated in Fig. 9-13. When the transistor is switched very

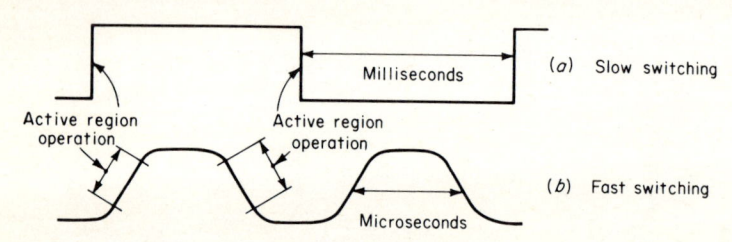

Fig. 9-13. The pulse repetition rate is a factor in transistor power dissipation.

rapidly from cutoff to saturation it remains in the active region for a much larger percentage of time and the average power increases. This is one limiting factor that determines the maximum repetition rate for a given circuit.

The above can be clearly shown by the use of a load line for a typical circuit. Figure 9-14 shows a load line that may be considered usual for a pulse inverter. Many switching circuits are designed so that the load line goes well above the maximum P_C curve since the length of time the transistor is in this region is normally small. But, if the transistor is caused to switch from one extreme to the other as fast as possible, it remains in the active region a large part of the time. Now, the collector-power dissipation becomes excessive and the transistor overheats.

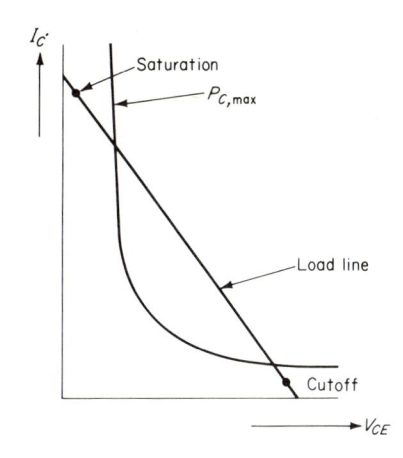

Fig. 9-14. Operation above $P_{C,\max}$ is permissible for some switching applications.

Switching Circuit Examples

One encounters few variations of the basic pulse-inverter circuit. Since in normal operation the transistor is either on or off, there are few varia-

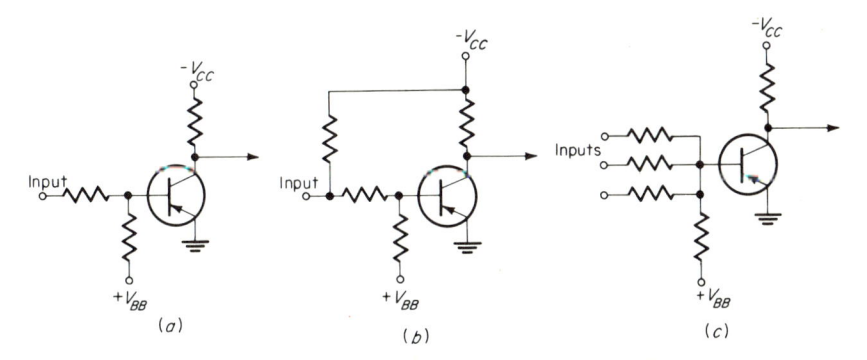

Fig. 9-15. Switching inverter circuit variations.

tions possible. In Fig. 9-15, we show several *pnp* switching circuits, each with slight differences. The major circuit variations come about because of differences in the resistor values.

Figure 9-16 shows an *npn* emitter follower, with the emitter returned through the load resistor to a negative supply and the collector grounded. The base will operate between $-V_{CC}$ and ground, with the emitter following the base voltage. If V_i is at $-V_{CC}$, the transistor is operating at cutoff. If V_i is at ground, the transistor is operating in saturation.

146　PULSE AND SWITCHING CIRCUIT ACTION

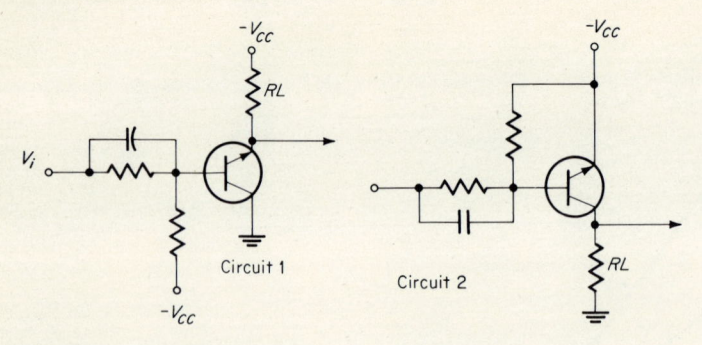

Fig. 9-16. *npn* emitter follower $-V_{CC}$.

Fig. 9-17. *npn* inverter $-V_{CC}$.

Figure 9-17 shows an *npn* inverter, again with the collector returned to ground. If V_B is at a voltage that is equal to or greater than $-V_{CC}$, the transistor is cut off, while if it is returned to ground through a suitable limiting resistance, the transistor is in saturation. The output is inverted with respect to the input.

Figure 9-18 shows an *npn* emitter follower using a positive supply voltage. The emitter will follow the base within the range of $+V_{CC}$ and ground. If V_B is at $+V_{CC}$, the transistor is in saturation, while if V_B is at ground, the transistor is cut off.

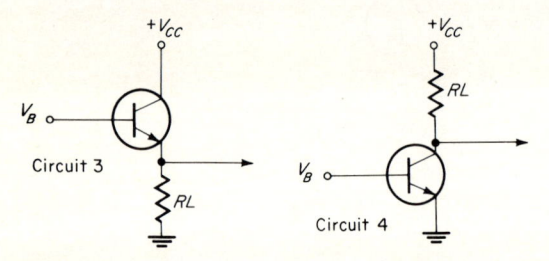

Fig. 9-18. *npn* emitter follower $+V_{CC}$.

Fig. 9-19. *npn* inverter $+V_{CC}$.

Figure 9-19 is an inverter using a positive supply. If V_B is more positive than ground, the transistor is on, while if V_B is more negative than ground, it is off. Again, the output is inverted with respect to the input.

A split-load amplifier is given in Fig. 9-20, using both positive and negative supplies. Whether the transistor is quiescently off or on in the active region depends upon the values of supply voltage, the two load

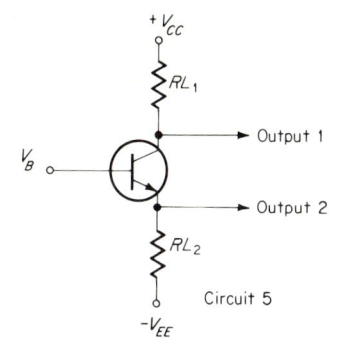

Fig. 9-20. *npn* split-load amplifier.

resistors, and the base voltage V_B. Output 1 is inverted, while output 2 is not.

Nonsaturating Techniques

In many circuit applications, operation in the saturated region is not only permitted but rather desirable. When a transistor is in saturation, the collector is returned to the emitter through a very low internal resistance, and if the emitter is grounded, the collector can be considered grounded also. With the transistor off, V_C will usually rise to the supply voltage. The collector will swing between V_{CC} and ground in this case, yielding very well defined voltage levels. The design of saturating circuits is relatively simple, and thus they are encountered quite often.

In high-speed circuitry, however, the storage time can be bothersome, and in this case operation is necessarily limited to the active region. A number of techniques are used to avoid operation in the saturation region. Probably the simplest way to accomplish this is with a diode clamp. In Fig. 9-21, the diode connects the collector to a voltage slightly more negative than ground and, if E_{clamp} is properly chosen, will prevent the transistor from entering the saturation region.

Suppose E_{clamp} is -1 volt and V_{CC} is -12 volts. If the transistor is off, V_C will rise to -12 volts and $D1$ is reverse-biased, and so it does not affect

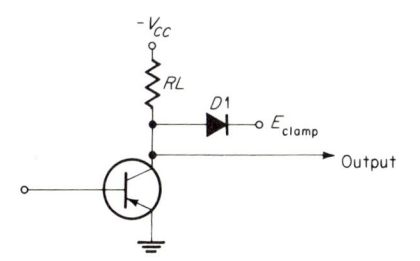

Fig. 9-21. *pnp* inverter with collector clamp.

circuit operation. If now the transistor is turned on, V_C goes toward ground, and if $D1$ were not in the circuit, the transistor would go into saturation and V_C would be about -0.1 volt. With the diode in the circuit and its cathode returned to -1 volt, as soon as the collector tries to go more positive than -1 volt, the diode becomes forward-biased and its internal resistance drops from very high to very low. Since the cathode of the diode is returned to -1 volt, the collector is prevented from going all the way to -0.1 volt. If the drop across the diode is 0.4 volt, the collector can only fall to $-1 + 0.4 = -0.6$ volt, and so cannot go into saturation. (V_B is about -0.2 volt and V_C is -0.6 volt, and so the collector junction is still reverse-biased. One condition necessary for saturation is that both junctions are forward-biased.)

When the base is again turned off, collector current will decrease in a time to Tf but there will be no storage time Ts. A disadvantage of this system is that when $D1$ conducts, the transistor collector current will rise to βIb, necessitating strict base-current limitation to avoid continuous operation in the region above maximum power dissipation.

A way of quickly getting rid of the stored-base charge, which will also reduce the storage time, is by overcompensating a direct-coupled voltage divider, as shown in Fig. 9-22. Here the transistor is allowed to saturate,

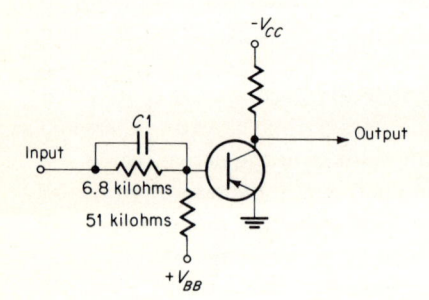

Fig. 9-22. A compensated voltage divider used in the base circuit.

and the capacitor ($C1$) provides the means to quickly dissipate the stored charge. If the input-source resistance is very low, the capacitor will allow the stored charge to dissipate itself through $C1$ rather than the 6.8-kilohm resistor. Since $C1$ is chosen to be relatively large, it exhibits practically zero opposition to fast changes and bypasses the current around the 6.8-kilohm resistor, thus speeding up the recovery from saturation.

Still another way to avoid the undesirable effects of Ts is to choose the values of the load and base resistors so that when the transistors are turned on, saturation is not quite reached. Since transistor characteristics are found to vary quite widely, this technique is seldom used unless it is feasible to hand-select the transistors for a beta that falls within a narrow range.

Also, inverse feedback can be used to stabilize the circuit and remove complete dependence of the circuit upon transistor characteristics. This however, makes for an unnecessarily complex circuit.

A final example of a way to reduce the storage time is shown in Fig. 9-23.

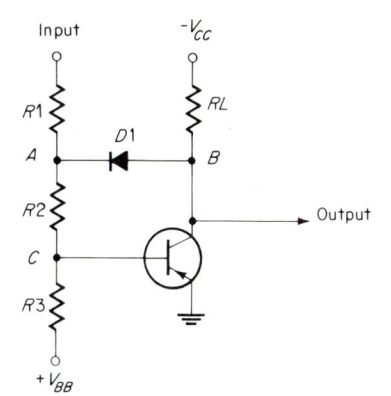

Fig. 9-23. A collector clamp used to prevent saturation.

In this instance, the base circuit is clamped to the collector by diode $D1$ under certain conditions. With the input at ground, the transistor is cut off and the diode is back-biased. When the input is driven negative, the transistor turns on, and as the collector falls toward ground (more positive) at some point $D1$ becomes forward-biased. If the input now swings still more negative, no additional base current can flow because it is being diverted into the diode and then into the collector circuits. Thus, by proper circuit design, the voltage at B can be made to be always more negative than that at C. Unless the collector can go more positive than the base, the transistor will not go into saturation. With the transistor kept out of saturation, any delay due to storage time is minimized.

By investigating typical voltages as they might exist in a circuit, we can better understand how the transistor is kept out of saturation. If point B is to be kept to -0.5 volt and if the diode drops 0.5 volt when conducting, point A must be -1 volt at this point. $R2$ is then chosen to put the base at about -0.3 volt. Thus, with these voltages existing, the transistor is not quite in saturation, and its turn-off time is quite short.

9-2 SWITCHING MODES

Transistor switching circuits are usually classed as one of three possible types, depending upon the operating *mode*. These three operating modes

150 PULSE AND SWITCHING CIRCUIT ACTION

are known as the *saturated mode*, the *current mode*, and the *avalanche mode*.

The switching amplifier shown in Fig. 9-3 is an example of saturated-mode operation. This type of circuit is by far the most widely used, primarily because of circuit simplicity and low-power consumption. However, when switching speed becomes increasingly important (as in high-speed equipment), the saturated-mode circuit suffers certain disadvantages as discussed earlier. To overcome these, the current-mode circuit is often used.

A basic current-mode circuit is shown in Fig. 9-24. This circuit, with

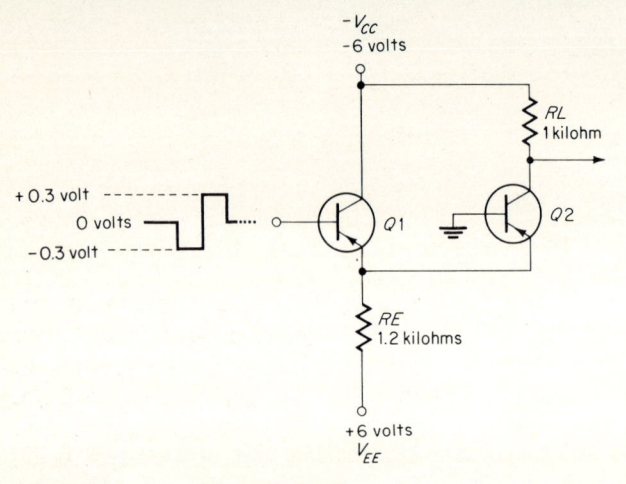

Fig. 9-24. A current-mode circuit.

proper values, eliminates one serious problem of the saturated-mode circuit. When a transistor is far into saturation and the base is suddenly presented a turn-off voltage, the transistor continues to conduct for some time, perhaps several microseconds. In high-speed circuitry this is highly undesirable. The current-mode circuit circumvents the problem by preventing saturation.

The basic circuit description, which will be discussed subsequently in more detail, is as follows. $Q1$ is an emitter follower and so is always in the active region if the signal does not exceed the power-supply limits. As shown, the signal is a very small voltage, and hence this requirement is easily met. (As will be seen, $Q1$ does turn off, but it never goes near saturation.)

With properly chosen components, if $Q1$ conducts, its emitter is, perhaps, a tenth of a volt more positive than the signal, or -0.2 volt. The emitter of Q2 is therefore at the same potential, and so Q2 is off (base slightly more negative than the emitter). The output is at $-V_{CC}$, owing to the lack of collector current through RL. Now, if the input is at the more positive level, the emitter of Q1 tries to follow. As soon as it becomes more positive than ground, by perhaps one-tenth of a volt, Q2 comes on (positive emitter is the same as a negative base). When Q2 is on, the emitter is clamped to two-tenths of a volt or so, and this allows Q1 to be turned off.

By properly choosing the values of RL and RE and their relationships to $-V_{CC}$ and V_{EE}, Q2 can be kept in its active region. With the values shown and with Q2 on, the output does not fall all the way to ground but to -1 volt. Thus, Q2 is still in the active region, and when the input returns to -0.3 volt, Q2 will be turned off very fast. Also, because Q2 is actually being operated in the common-base circuit, the high-frequency response is excellent and it will pass through the active region (on, toward off) very fast if the input signal is also fast. Thus, rise times in the nanosecond region can be faithfully reproduced.

This kind of circuit is said to operate in the current mode because, in effect, the current through RE is switched into either the emitter of Q2 or the emitter of Q1. The voltages produced are of less significance than the current, hence the name.

The third mode of operation is the so-called avalanche mode, which is seldom used. These circuits, while theoretically workable, do not find widespread usage in logic circuits because of instability problems asso-

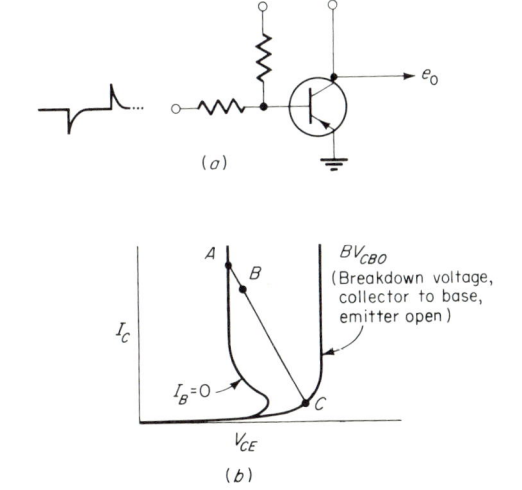

Fig. 9-25. The circuit and curves of an avalanche-mode circuit, along with the load line.

ciated with the negative-resistance avalanche region. The following brief discussion outlines the more significant operating characteristics.

When the maximum-rated collector-emitter voltage is exceeded, the transistor goes into avalanche conduction even though biased in the off condition. If V_{CC} is properly chosen, a small reverse bias on the base can keep the transistor conduction low, with V_{CE} just at the knee of the BV_{CBO} curve. A typical circuit and curves are given in Fig. 9-25. The off condition is shown as point C. If then a small negative trigger is applied to the input, the transistor "avalanches" and the operating point goes to A. When the input pulse disappears, the transistor then operates at point B.

The main attraction in this circuit is the very high speed with which the operating point switches from point A to point B. To return the circuit to point C again, a small positive trigger is applied as shown.

9-3 NONRESISTIVE LOADS

When transistor loading due to external circuitry is something other than resistive, the operation of the circuit is considerably altered. If the load should be highly capacitive, the transistor-output waveform is altered to the extent that voltage changes occur more slowly. Hence, in this instance the significant element that is altered is time, or rather the time response of the circuit. Examples of capacitive loading are evident throughout this text, and so no further mention should be necessary at this point.

However, if the loading is inductive, the transistor operation itself is altered because of the induced voltages caused by the inductor. Here, the actual points of operation that are possible for the transistor to attain are severely changed. Figure 9-26a, shows the static and dynamic load lines for the circuit of Fig. 9-26b. The maximum current for dc conditions $I_{C,\text{sat}}$ is indicated at the upper end of the dc load line. Note that for fast turn-on changes, the voltage induced by the inductor opposes the applied voltage (the source), and so little current flows at first, with an increase after the induced voltage begins to decay. Once current reaches its maximum value, it remains constant (V_{CC}/R_{total}), hence the induced voltage no longer exists.

As the transistor turns off, the collapsing lines of force cause an induced voltage that, in effect, increases V_{CC} perhaps several times over. Thus, the collector voltage is much larger than the value of the supply voltage ($e = L\,di/dt$). It is not uncommon to measure several hundred or even, in extreme cases, several thousand volts at this point.

In practical circuitry this effect *must* be prevented since the average

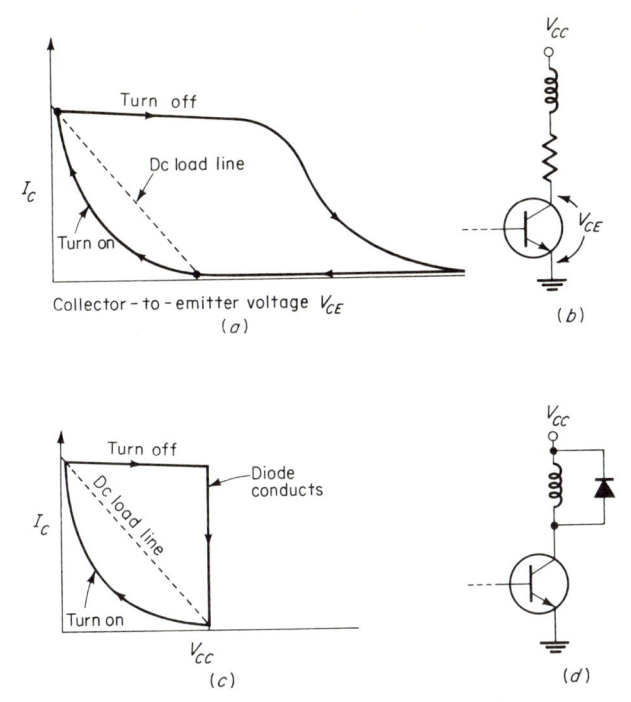

Fig. 9-26. The effects of inductance in the transistor circuit can destroy the transistor. Load lines (*a*) and (*b*) for the unprotected circuit and (*c*) and (*d*) for the protected circuit.

switching transistor has a maximum voltage rating of perhaps 30 volts. One way of doing this is illustrated in Fig. 9-26*c* and *d*. The diode is normally back-biased during *turn-on* and *full-on* operation. But when the transistor is turned off, the induced voltage is in a direction such as to forward-bias the diode and, in effect, the inductor is shorted and thus dissipates its energy in the diode. This prevents the collector from rising to more than V_{CC}, as shown by the dynamic load line. Other protection devices include various combinations of resistors, capacitors, back-to-back diodes, etc., all of which tend to dissipate the stored energy in themselves.

QUESTIONS AND PROBLEMS

9-1 A switching transistor is operated so that when it is in saturation, $V_{CE} = 0.2$ volt and $I_C = 4$ ma. Determine the value of collector-to-emitter saturation resistance.

9-2 A switching transistor is operated so that when it is in saturation, $V_{CE} = 0.3$ volt and $I_C = 1.5$ ma. Determine the value of collector-to-emitter saturation resistance.

9-3 A transistor is operated in such a manner that when it is cut off, its collector-to-emitter voltage is 8 volts while its maximum current when in saturation is 2 ma. What is the value of the load resistor RL?

9-4 A transistor is operated in such a manner that when it is cut off, its collector-to-emitter voltage is 8 volts while its maximum current when in saturation is 6 ma. What is the value of the load resistor RL?

9-5 A transistor is not the perfect switch that might be desired. Name the function that is the worst offender, in high-speed equipment, in this departure from perfection.

9-6 A certain transistor is known to suffer from minority-carrier storage. List several methods that might be used to alleviate this situation.

9-7 Refer to Fig. 9-3. The circuit is to be used as drawn, except for V_{CC}. Change this from -6 to -12 volts. Determine the value of $I_{C,\max}$.

9-8 Refer to Fig. 9-3. The circuit is to be used as drawn, except for V_{CC}. Change this from -6 to -8 volts. Determine $I_{C,\max}$.

9-9 Refer to Fig. 9-3. Using the circuit as drawn, find the value of input voltage to just barely turn the transistor on. Assume $V_{BE} = 0.2$ volt.

9-10 Refer to Fig. 9-3. Change the 51,000-ohm resistor to a 6800-ohm resistor, with all else remaining the same. Determine the value of input voltage to just barely turn the transistor on (assume $V_{BE} = 0.2$ volt).

9-11 Choose the correct answer. When a transistor is operating with $I_C = V_{CC}/RL$, it is said to be:
(a) In the active region
(b) In cutoff
(c) In saturation
(d) Amplifying linearly
(e) A class A amplifier

9-12 Choose the correct answer. When a transistor is operating with $V_{CE} = V_{CC}$, it is said to be:
(a) In the active region
(b) In saturation
(c) Amplifying linearly

(d) In cutoff
(e) A class A amplifier

9-13 The storage time of a transistor is made more severe by:
(a) Decreasing the base current
(b) Increasing the time constant in the emitter circuit
(c) Making the resistance in the base lead lower in value
(d) Increasing the base current

9-14 When operating a transistor alternately in cutoff and saturation, the storage time would be most likely to affect which of the following pulses?
(a) 100-msec pulse
(b) 1000-msec pulse
(c) 0.001-second pulse
(d) 1000-nsec pulse

9-15 Refer to Fig. 9-7. Considering the base-circuit voltage divider of Q1, determine the value of voltage at the input terminal to cause 0 volts at the base of Q1.

9-16 Refer to Fig. 9-7. Considering the base-circuit voltage divider of Q1, determine the condition (on or off) of Q1 if the voltage at the input is -3.33 volts.

9-17 Refer to Fig. 9-7. Considering the base-circuit voltage divider of Q1, determine whether Q1 is off or on if the voltage at the input is -1.5 volts.

9-18 Considering Questions 9-15, 9-16, and 9-17, determine the threshold at the input terminal to cause Q1 to just go from off to on or on to off. Assume $V_{BE} = 0.2$ volt.

9-19 Refer to Fig. 9-20. Briefly explain why V_C can never go to $-V_{EE}$ and V_E can never go to V_{CC}.

9-20 Refer to Fig. 9-20. Given that $V_{CC} = 10$ volts, $V_{EE} = -10$ volts, $RL_1 = 2000$ ohms, and $(RL_2) = 1000$ ohms, find the most negative value for V_C ($V_{C,\min}$).

10
AMPLIFIER SWITCHING CIRCUITS

In this chapter the principles previously presented will be applied to some practical circuits, with examples of each of the three basic configurations. Previously, the circuit examples were presented in their basic forms. Now they will be discussed in somewhat greater detail to allow a better understanding of actual circuit action.

10-1 THE CE (COMMON - EMITTER) CIRCUIT

The first circuit to be discussed is given in Fig. 10-1 and is seen to be a common-emitter inverter amplifier (switching amplifier) operating in the saturated mode. Before describing the circuit in detail, a brief discussion is desirable.

The input to the circuit is seen to be a pulse in the negative-going direction, making an excursion from perhaps -0.1 volt to something less negative than -6 volts and back to -0.1 volt. The signal has obviously been produced by another circuit very similar to the one shown, which has been on (saturated), then turned off to produce the pulse for some period of time, and then allowed to turn back on. The input terminal

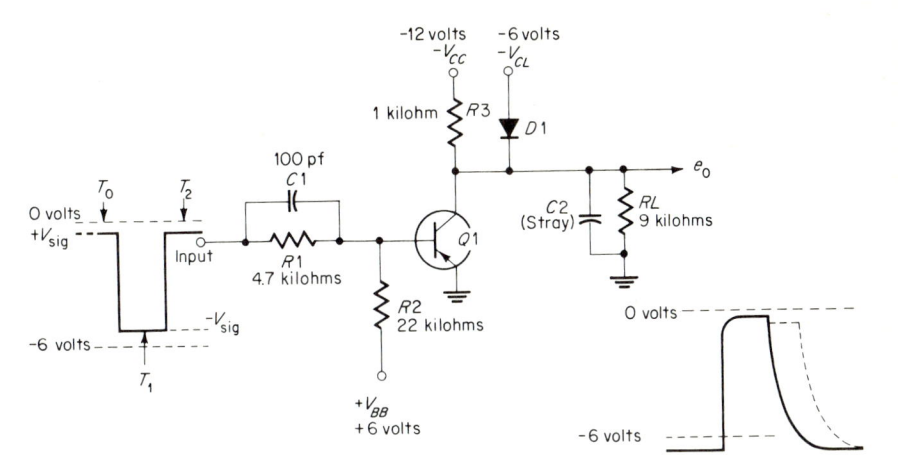

Fig. 10-1. Typical CE inverter-amplifier circuit.

shown may be assumed to be connected to the collector of this other inverter amplifier, called the *driver*.

During T_0, the interval prior to the start of the pulse, the driver is on and its collector is essentially at ground, about -0.1 volt. With a nearly 0-volt level at the input to $Q1$, the transistor is off since the drop across $R1$ causes the base to be somewhat more positive than ground. With $Q1$ off, its collector tries to rise to the value of $-V_{CC}$ but is held to the clamp voltage $-V_{CL}$ by the clamp diode $D1$. The output voltage is therefore at, perhaps, -6.5 volts, allowing 0.5-volt drop across $D1$.

When the input goes to its most negative level, during the interval labeled $T1$, the base of $Q1$ is driven to something more negative than ground, and $Q1$ is turned on. Its collector falls to perhaps -0.1 from -6.5 volts. Depending on the amount of shunt capacitance $C2$, the leading edge may be more or less rounded. Because the circuit is dc-coupled the transistor stays on for the duration of the input pulse.

At the instant the input waveform makes its positive-going excursion, $Q1$ begins to turn off. However, because it has been far into saturation, the stored charge in the base must be given time to dissipate and the transistor tends to stay on even after the input has returned to zero. This is indicated by the dotted line on the output waveform. However, because of the capacitor $C1$, called a *speedup capacitor*, the stored charge is removed much faster and the pulse is restored to its normal width if $C1$ is of the proper size. The collector of $Q1$ rises toward -12 volts but is clamped to the -6-volt supply by $D1$.

Circuit Analysis

With a general description of circuit action to act as a guide, the circuit will now be analyzed for more specific information. First, the static conditions for $Q1$, both off and on, will be verified. Then, a simplified transient analysis will be performed to determine certain dynamic aspects of the circuit.

With the input at -0.1 volt, the drop across $R1$ will determine if $Q1$ is actually off.

$$E_{R1} = \frac{R1}{R1 + R2} \times E_{total} = \frac{4.7 \text{ kilohms}}{4.7 \text{ kilohms} + 22 \text{ kilohms}} \times 6.1 = 1.074 \cong 1.1 \text{ volts}$$

Noting the direction of electron flow, the base must be more positive than -0.1 by 1.1 volt.

$$V_{B(Q1)} = -0.1 + 1.1 = 1.0 \text{ volt}$$

With the base at 1 volt more positive than the emitter, the *pnp* transistor is indeed in cutoff.

With $Q1$ in cutoff, there is no collector current through $R3$, the load resistor. However, the voltage divider consisting of $R3$ and RL will, of course, draw current. The resistor RL represents the ultimate load, which may consist of several additional inverters. Also, the diode $D1$ may or may not affect the operation at this point.

If $D1$ were disconnected, the voltage at the collector could never go more negative than

$$V_{C,\max} = \frac{RL}{R3 + RL} \times -12 = -10.8 \text{ volts}$$

With the diode connected, however, the cathode of the diode is returned to a source of -10.8 volts and hence must conduct. If a 0.5-volt drop is assumed, the collector must be resting at a potential of

$$-6 + (-0.5) = -6.5 \text{ volts}$$

If a curve of the diode characteristics is at hand, then the actual drop can be determined by first finding the diode current. The drop V_{D1} across $D1$, could then be read from the curve.

To determine the diode current, the Thevenin-equivalent circuit is used, as shown in Fig. 10-2. The equivalent source voltage has been determined previously as -10.8 volts. The source impedance is the parallel resistance of $R3$ and RL.

$$R_{th} = \frac{R3 \times RL}{R3 + RL} = \frac{1 \text{ kilohm} \times 9 \text{ kilohms}}{1 \text{ kilohm} + 9 \text{ kilohms}} = 900 \text{ ohms}$$

AMPLIFIER SWITCHING CIRCUITS   159

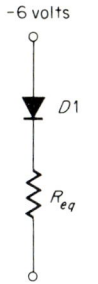

Fig. 10-2. Equivalent circuit for computing diode current.

The diode current I_D, then, is easily found:

$$I_D = \frac{E_{eq}}{R_{th}} = \frac{E_{source} - V_{CL}}{R_{th}} = \frac{(-10.8) - (-6)}{900} = \frac{4.8}{900} = 5.3 \text{ ma}$$

The diode drop can now be read from Fig. 10-3, which represents the forward characteristics of $D1$. At 5.3 ma, the diode drop agrees with the estimated value, as shown.

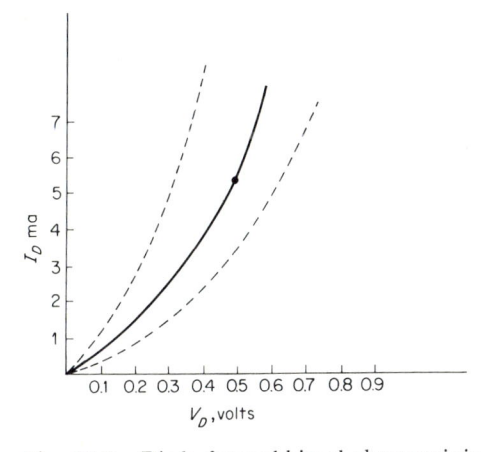

Fig. 10-3. Diode forward-biased characteristic curves.

Now, when the input goes to the other extreme, shown as about -5.5 volts, the net effect is to allow $Q1$ to go from cutoff to saturation. With -5.5 volts at the input, the voltage divider determines the base drive for $Q1$. Assume for a moment that the base is disconnected from the divider. The junction voltage E_J that would then appear between $R1$ and $R2$, referred to ground, is

$$E_{R1} = \frac{R1}{R1 + R2} \times E_{total} = \frac{4.7 \text{ kilohms}}{4.7 \text{ kilohms} + 22 \text{ kilohms}} \times 11.5 \cong 2 \text{ volts}$$

and $E_J = -5.5 + 2 = -3.5$ volts.

Hence, when the base is connected to the circuit, the current into the base will be large and the transistor is indeed fully into saturation. If the voltage divider tries to cause the base to be anything more negative than perhaps -0.3 volt with respect to ground, the transistor is on. The base, then, is clamped to -0.2 or -0.3 volt (germanium case) relative to ground.

To determine the base current at this time is only necessary to Thevenize the voltage divider as a source. The junction voltage (between $R1$ and $R2$) has been determined as -3.5 volts to ground. The source resistance looking back from the base of $Q1$ is

$$R_{th} = \frac{R1 R2}{R1 + R2} = \frac{4.7 \text{ kilohms} \times 22 \text{ kilohms}}{4.7 \text{ kilohms} + 22 \text{ kilohms}} = 3.87 \text{ kilohms}$$

$$I_B = \frac{E_{source}}{R_{th}} = \frac{3.5}{3.87 \text{ kilohms}} \cong 904 \text{ }\mu a$$

If it is desired to verify this value, this might be done in an alternate manner, as follows. The voltage across $R1$ is known, thus so is the current through it:

$$I_{R1} = \frac{E_{R1}}{R1} = \frac{-5.5 + 0.2}{4700} = \frac{5.3}{4700} = 1.13 \text{ ma}$$

The current through $R2$ is

$$I_{R2} = \frac{E_{R2}}{R2} = \frac{6 - (-0.2)}{22 \text{ kilohms}} = \frac{6.2}{22 \text{ kilohms}} = 0.282 \text{ ma}$$

The difference of these two currents is the base current:

$$I_B = I_{R1} - I_{R2} = 1.13 \text{ ma} - 282 \text{ }\mu a = 838 \text{ }\mu a$$

Because the two methods use slightly different approaches, they do not agree precisely. They are, however, quite close, the difference representing only a few tenths of one percent. Neither is precisely accurate since V_{BE} is only an assumed value and the actual resistance of the base-emitter junction has been ignored. Either, however, can be considered a very practical result.

There are several items of interest regarding the transient, or signal, response of this kind of circuit. One of these is the *trip-point* of $Q1$ with respect to the input signal when no speedup capacitor is used. That is, the input signal must be at some specified level to cause Q1 to start going off if it has been on, or to start going on if it has been off. Because of the voltage divider, this value of input voltage to just cause the transistor to trip may be either slightly below ground or it may be very much below ground for the circuit in Fig. 10-4a, which is similar to the preceding circuit except for the speedup capacitor.

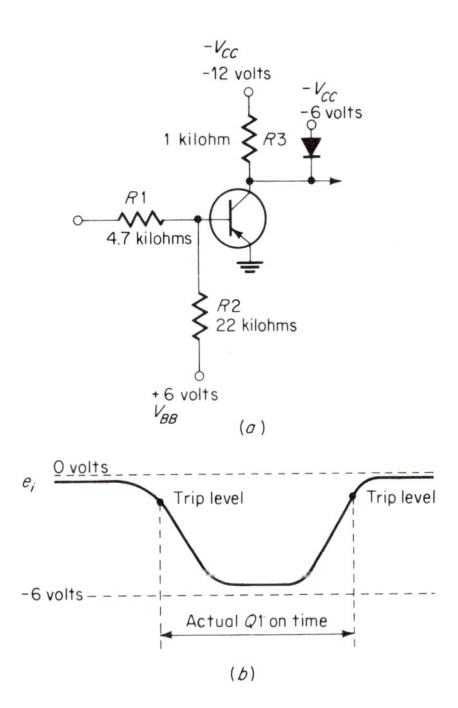

Fig. 10-4. A circuit and curves that illustrate the trip level.

By assuming a value of voltage at the base to just cause Q1 to trip, the input level necessary to produce this is easily found. If the base-emitter voltage is -0.2 volt during saturation, then we may assume that when V_B is -0.1 volt to ground, the transistor is beginning to trip. That is, if it has been on, it now begins to go off and vice versa.

To determine the trip level at the input terminal, the necessary drops across the resistors R1 and R2 are noted, along with the necessary voltage V_B at the junction of the two resistors. We know that the voltage at V_B must be -0.1 volt to ground. Also, the drop across R1 will stand in the same ratio to R1 that the drop across R2 stands to R2. This, of course, is a proportion:

$E_{R1}:R1::E_{R2}:R2$

Solving for E_{R1},

$$E_{R1}R2 = R1E_{R2} \quad \text{and} \quad E_{R1} = \frac{R1E_{R2}}{R2} = \frac{4.7 \text{ kilohms} \times 6.1}{22 \text{ kilohms}} = 1.3 \text{ volts}$$

The drop across $R1$, then, must be 1.3 volts to place V_B at the required -0.1-volt potential. The input voltage must then be

$$e_i = -E_{R1} + V_B = -1.3 + 0.1 = -1.2 \text{ volts}$$

Hence, as the input voltage changes from about ground to -1.2 volts, the transistor is just going from off to on. As the input goes from -5.5 to -1.2 volts, the transistor is just going from on to off. Figure 10-4b illustrates this case, and the actual "on" time of the transistor base is clearly shown.

The importance of this determination is fairly evident. If the value of $R1$ is too large or if $R2$ is too small (perhaps due to aging, heating, or a misread color code), the excursion of input voltage could conceivably be insufficient to turn the transistor on. This would be especially true if the signal happened to be degraded, as is normally the case when several diode gates are used.

A vital function is performed in a circuit such as shown in Fig. 10-1 by the speedup capacitor $C1$. As discussed earlier, the transistor suffers from minority-carrier storage when being turned off, and this causes the collector current to continue flowing even after the base is presented a turn-off signal. The measure of the length of time that collector current continues to flow is the storage time t_s.

The size of the capacitor is often determined empirically, using a temporary trimmer capacitor. Its value is usually chosen to provide overcompensation (see Chap. 3, Compensated Voltage Dividers). That is, when an exact value is determined that just allows t_s to be minimized, the permanent capacitor is chosen to be two to three times the critical value. This allows for capacitor, resistor, and transistor tolerances and insures both fast turn-on and turn-off times, which gives rise to a waveform at the base similar to that shown in Fig. 10-5.

The preceding discussion should not be construed to mean that the value of $C1$ cannot be calculated. However, the concept of charge-control theory is beyond the scope of this book. Also, because the significant parameters are quite different for each transistor, any determination is necessarily approximate at best. Additionally, because it is common practice to overcompensate, an exact determination is not necessary.

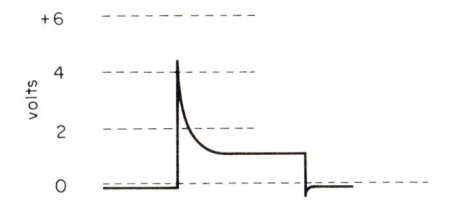

Fig. 10-5. Waveform showing the effects of overcompensation.

Usual values fall between about twenty-two and a few hundred picofarads, and, except for very high-speed equipment, nearly any given value will work for a single circuit. Note that when the speedup capacitor is used, the previously discussed trip level is altered since the capacitor acts as a coupler for the leading and trailing edges of the pulse and the input must go only a few tenths of a volt to affect the transistor. (For a simplified method of calculating for $C1$, see the design example at the end of this section.)

When several such circuits are cascaded, however, there is an additional consideration. Each speedup capacitor becomes a capacitive load upon the preceding inverter. This is shown in Fig. 10-6, where, with $Q2$ on and

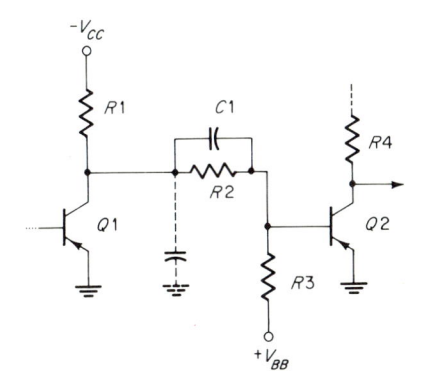

Fig. 10-6. The effects of $C1$ in a cascaded amplifier.

$Q1$ off, the right-hand plate of $C1$ is, in effect, connected to ground as indicated. Before the collector of $Q1$ can fall to ground, $C1$ must change its state of charge, which can increase the transition time appreciably. Hence, from this standpoint, $C1$ must be kept as small as possible while still helping to overcome the storage time of $Q2$. The choice of a value for $C1$, then, is complicated by conflicting requirements, and as always a good compromise is the best choice.

In connection with this particular circuit configuration, it is of some significance to realize that as the transistor is going from off toward on it has little trouble in driving the capacitive load. When, however, the transistor is being turned off, it cannot drive the capacitive load and the capacitor must recharge to the new condition through any resistive circuit available.

Referring again to Fig. 10-6, as $Q1$ is being turned on, $Q2$ is being turned off. $Q1$ can drive its collector to ground quite easily, with virtually no delay caused by $C1$. From the base of $Q1$, the speedup capacitor appears to be smaller by the current gain of the transistor. If $C1$ is a 100-pf capacitor and if $Q1$ has a beta of 100, $C1$ appears to be a 1-pf capacitance to the base of $Q1$. Hence, the collector of $Q1$ makes its excursion nearly as rapidly as it would if $C1$ did not exist.

However, when $Q1$ is being turned off, as V_{BE} approaches 0 volts, the transistor is off regardless of the collector voltage. Thus, $Q1$ can in no way assist in the recharge of $C1$ during this time. $C1$ must now charge to a value set by the voltage divider ($R1$, $R2$, $R3$, and the base-emitter junction of $Q2$). When $Q2$ comes on, of course, its base is clamped to ground. $C1$ then charges to the value set by $R1$ and $R2$, with a time constant determined by the Thevenin equivalent of $R1$ and $R2$ (and, of course, the value of $C1$). The collector of $Q1$, then, cannot recover its normal off level until $C1$ completes its recharge. This can definitely decrease the resolution of the circuit, i.e., the maximum repetition rate. The larger $C1$, the poorer the resolution, and the fewer the pulses per second it can handle.

Base-Emitter Protection

A slight variation of the basic circuit is shown in Fig. 10-7. In this instance, a diode is used to protect the transistor from excessive base-to-emitter voltage. Many transistors have a maximum V_{BE} of, perhaps, 2.5 volts when reverse-biased. If the peak value of the waveform at the

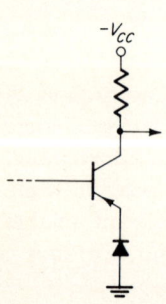

Fig. 10-7. Base-emitter protection.

AMPLIFIER SWITCHING CIRCUITS 165

base exceeds this, the transistor will be damaged. The diode shown will prevent this by also becoming reverse-biased. This allows the base and emitter to float together with the input signal so that the difference of potential between them never exceeds the rated value. Hence, large signals can be used with transistors having restricted V_{BE}.

Threshold Voltages

Under certain conditions, the circuit requirements specify a certain *threshold* voltage. That is, the transistor in question must be presented a certain input that exceeds the threshold (or trip level) before any action results. One method of accomplishing this is shown in Fig. 10-8, along with the appropriate waveforms.

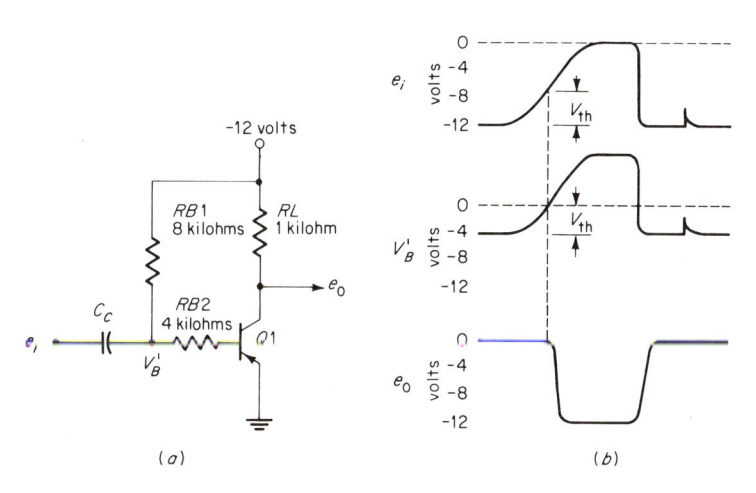

Fig. 10-8. A circuit and waveforms to illustrate the threshold voltage set by $RB1$, $RB2$, and C_C.

The input signal e_i makes a normal excursion from -12 volts to ground and back to -12 volts. Note that the input must rise to at least -8 volts before the transistor comes on. As long as the voltage at V'_B remains more negative than ground, $Q1$ stays on. The circuit therefore is said to have a 4-volt threshold, which gives it excellent noise immunity. Note the small noise spike which, even though it is 3 volts in amplitude, does *not* give a false output.

Very briefly, the circuit is analyzed as follows. Normally, $Q1$ is on and so the base current flowing through $RB1$ and $RB2$ causes some voltage to appear at V'_B. Ignoring the base-to-emitter drop, this is easily determined:

$$V'_B = \frac{RB2}{RB1 + RB2} \times -12 = \frac{4 \text{ kilohms}}{12 \text{ kilohms}} \times -12 = -4 \text{ volts}$$

This is indicated on the waveforms as the static level at V'_B prior to the input pulse. The input capacitor C_C plays an important role in overall circuit operation. Its right-hand plate sees a -4-volt level with respect to ground. The left-hand plate, however, is at -12 volts, being held there by the input signal.

As the input begins to fall toward ground at a rate much faster than the capacitor's time constant, the output plate of $C1$ follows the input plate. This must be true since the capacitor cannot change its state of charge instantly. As long as the time constant of C_C and the associated resistance is 100 times or more greater than the rise time of the pulse, the output plate of C_C follows exactly the input plate. This, of course, must occur if the net voltage from plate to plate is to remain the same. Just prior to the input pulse, the left plate is at -12 volts and the right plate at -4 volts for a total plate-to-plate voltage of $+8$ volts. Just as the input gets to 0 volts, the right plate has risen to $+8$ volts. The capacitor still has 8 volts from plate to plate.

If the input were to remain at ground for a long time, point V'_B would eventually go back to -4 volts, of course, since there is nothing but the capacitor charge to keep it at $+8$ volts. However, since the input falls back to -12 volts quite soon, the capacitor holds V'_B at $+8$ volts during the required interval.

Figure 10-9 illustrates the capacitor voltages as the input goes from

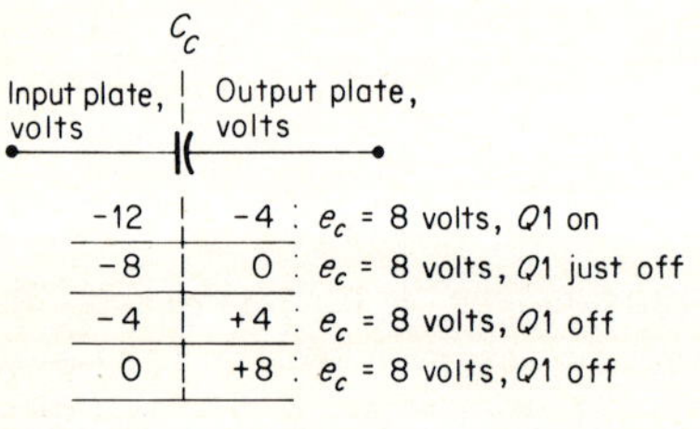

Input plate, volts	Output plate, volts	
−12	−4	e_c = 8 volts, $Q1$ on
−8	0	e_c = 8 volts, $Q1$ just off
−4	+4	e_c = 8 volts, $Q1$ off
0	+8	e_c = 8 volts, $Q1$ off

Fig. 10-9. Capacitor action during a positive excursion.

-12 to 0 volts referred to ground. The fact is emphasized that if the rise time of the input pulse is small enough, the capacitor cannot change its state of charge by any significant amount. Note that the transistor begins to go off just as the output plate arrives at 0 volts. This, of course, is the threshold voltage at the base and at V'_B. The threshold voltage at the input is -8 volts, as indicated.

Overvoltage Protection

Another method of overvoltage protection is shown in Fig. 10-10. In this instance, an additional resistor is used to prevent excessive collector-to-emitter voltage when $Q1$ is turned off. If $-V_{CC}$ is -100 volts and if $R1$ and $R2$ are equal in value, then $V_{CE,\max}$ can never rise to more than

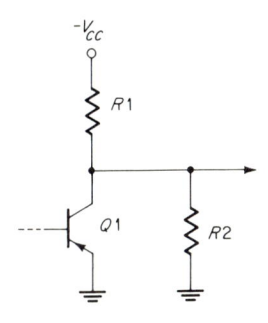

Fig. 10-10. One method of protecting the transistor from excessive V_{CE}.

-50 volts. Hence a transistor rated at 60 volts maximum will function quite well, with a 10-volt safety factor.

Fan-out Considerations

Fan-out is defined as the number of multiple loads on a given inverter and is illustrated in Fig. 10-11. In this example $Q1$ has a fan-out of four

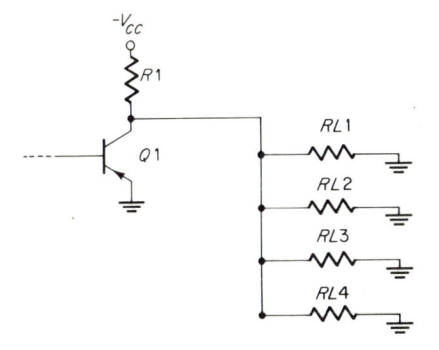

Fig. 10-11. Fan-out, showing four separate loads.

separate loads, each of which represents, perhaps, the total input resistance of another inverter. Although determining the maximum number of loads allowed is properly a design problem, certain considerations are of importance to us.

168   PULSE AND SWITCHING CIRCUIT ACTION

Of prime importance is the maximum value of collector voltage. Because of the voltage-divider action when $Q1$ cuts off, its collector cannot rise completely to $-V_{CC}$. This, of course, is one major consideration in the maximum number of external loads allowed. To allow minimum loss, the resistor $R1$ should be made as small as possible while the external loads should be made as large as possible, consistent with other requirements.

A simple example will clarify the loading on $Q1$ imposed by the external loads. If $-V_{CC}$ is -10 volts, $R1$ is 1 kilohm, and each external load is 8 kilohms, the collector voltage at $Q1$ is determined as follows:

$$R_{eq} = \frac{1}{1/RL1 + 1/RL2 + 1/RL3 + 1/RL4} = 2 \text{ kilohms}$$

$$V_{C,\max} = \frac{R_{eq}}{RL + R_{eq}} \times -V_{CC} = \frac{2 \text{ kilohms}}{3 \text{ kilohms}} \times (-10) = -6.67 \text{ volts}$$

With this circuit, then, the collector can never rise to more than -6.67 volts, even though the supply voltage is -10 volts.

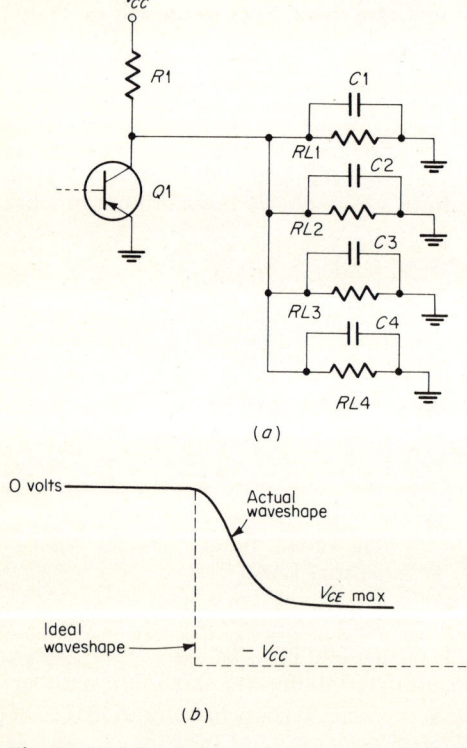

Fig. 10-12. Fan-out, showing highly capacitive loads.

An additional facet is often encountered. If speedup capacitors are used, as illustrated in Fig. 10-12a, the load is no longer completely resistive and can become highly capacitive. In this instance, not only does $V_{CE,\max}$ become less than $-V_{CC}$ to a considerable extent, but because of the capacitive load it may require some time to attain its maximum value. Hence, the waveform at the collector will be exponential in form, as suggested by the waveform in Fig. 10-12b. If the fan-out is large, the delay caused by the need to charge the capacitors can be very appreciable.

The time constant of this circuit can be approximated by adding together the values of $C1$ through $C4$ since they will charge in parallel. If the four resistors are equal, R_{eq} is any one value divided by 4. The time constant can then be determined by the Thevenin equivalent shown in Fig. 10-13.

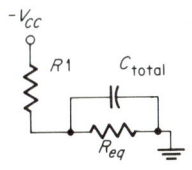

Fig. 10-13. An equivalent circuit for capacitive loads.

The ultimate voltage to which the capacitors will charge is the ultimate dc drop across R_{eq}. The Thevenin-equivalent source resistance is the parallel resistance of $R1$ and R_{eq}. Hence, the following relationship is reasonably valid for this circuit:

$$t_c = R_{th} \times C_{Total}$$

This, of course, does not consider the junction activity of the other transistors. That is, the RC combinations are not directly connected to ground as shown, but instead they are connected to the base of a transistor. Only when that particular transistor is on, is this wire actually at ground, and then only to the extent that the base itself is clamped to ground potential.

Silicon npn Inverter

Any transistor having a base-emitter voltage much greater than 0.1 or 0.2 volt will differ somewhat from the previously given examples. Nearly any silicon npn will have a V_{BE} of roughly 0.7 volt with the current values found in the usual circuits. If a data sheet is at hand, the V_{BE} can be accurately read from the sheet. If not, and if the unit is known to be a silicon npn, the typical value of 0.7 volt will be found to be a reasonable assumption.

170 PULSE AND SWITCHING CIRCUIT ACTION

The fact that V_{BE} is significantly greater for some transistors implies that in some instances the circuit requirements are somewhat different. The circuit in Fig. 10-14 will illustrate this point. Inverter Q1 is conventional in every respect, except perhaps for RL, the load resistor.

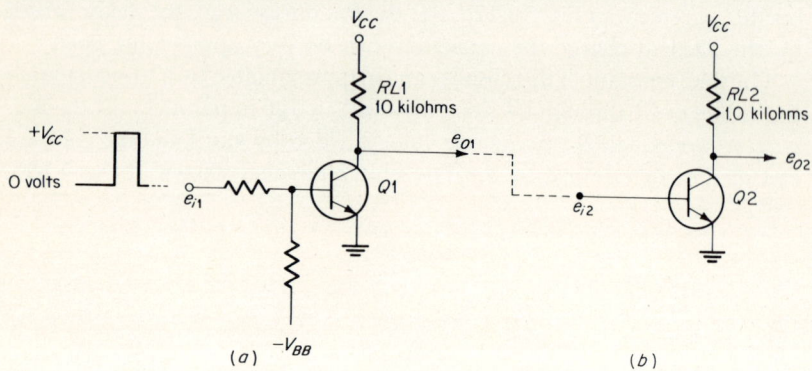

Fig. 10-14. A circuit sometimes used with silicon *npn*'s.

Because the base of Q2 is directly connected to the collector of Q1 without benefit of any series-limiting resistance, the 10-kilohm resistor serves not only as the dc load for Q1 but also as the base-current-limiting resistance for Q2.

Briefly, the circuit functions as follows. With the input to Q1 at ground, the base of Q1 is negative to some extent and Q1 is off. Its collector tries to rise to $+V_{CC}$ but is clamped by the base of Q2 to about $+0.7$ volt above ground. Q2 is heavily on, and its collector is essentially at ground, perhaps $+0.1$ to $+0.2$ volt.

When the input rises to $+V_{CC}$ as shown, Q1 is turned on and its collector falls to $+0.1$ or $+0.2$ volt above ground. Because the base-emitter junction of Q2 requires at least $+0.5$ volt to turn Q2 on, it must therefore be off. Note that the base-emitter junction of Q2 is not reverse-biased in the usual sense. However, since it is far from being forward-biased, Q2 is off and its collector is at $+V_{CC}$. When the input returns to ground, Q1 is turned off, thus turning Q2 back on.

The base current I_{B2} for Q2 is simply the current through $RL1$ when Q1 is off. Allowing 0.7 volt dropped across the base-emitter junction if V_{CC} is 10 volts, there must be 9.3 volts across $RL1$ during this time:

$$I_{B2} = \frac{E_{RL1}}{RL1} = \frac{9.3}{10 \text{ kilohms}} = 0.93 \text{ ma} = 930 \text{ }\mu\text{a}$$

The collector current for Q2, assuming it is in saturation, is also easily found (allow 0.2 volt across Q2):

$$I_C = \frac{E_{RL2}}{RL2} = \frac{9.8}{10 \text{ kilohms}} = 0.98 \text{ ma} = 980 \text{ μa}$$

It can be appreciated that by allowing for the proper V_{BE} of the transistor in question, the circuitry is not drastically different in its basic operation. The main point of difference is that the greater V_{BE} will sometimes allow simplification of the circuit since the need for actually reverse-biasing the junction is relaxed.

However, in connection with this, it should be noted that in some instances such a circuit still requires reverse bias but for a different reason. Besides simply turning a transistor off, reverse-biasing the emitter-base junction allows operation at much higher temperatures. The reverse voltage at the base forces all the leakage current out of the base and hence none will cross the junction to become amplified. By not reverse-biasing this junction, leakage current can become a problem. This, of course, is a function of the transistor itself. Some silicon units have leakage currents about the same magnitude as germanium units. In these cases, the need to reverse-bias the junction during off times is apparent.

There are, however, many silicon transistors with *very* low leakage currents (a few nanoamperes), and these will not necessarily require such treatment. Again, the manufacturers' data sheets are the final authority in determining a particular transistor's suitability for a certain circuit configuration.

Mixed Transistor Types

There are certain circuit conditions that require a firm connection to some point other than ground. If, for example, *pnp* transistors in the inverter circuit are used, the connection of the output wire to $-V_{CC}$ is through the load resistor. Only when the transistor is in saturation is there a *firm* connection, and since the emitter is usually at ground, this firm connection will be to ground.

When it is necessary to provide a firm connection to the negative supply, then the circuit shown in Fig. 10-15 can be used. Note that Q1 is *pnp* while Q2 is *npn*. Using this circuit, then, will allow direct connection to the −6-volt supply when Q2 is driven to saturation and a connection to ground through the 1-kilohm resistor when Q2 is off.

Briefly, the circuit action is as follows. Assume the base of Q1 is driven more positive than ground. Q1 is off, and its collector tries to rise toward −12 volts. However, as it rises to more than −6 volts in the negative

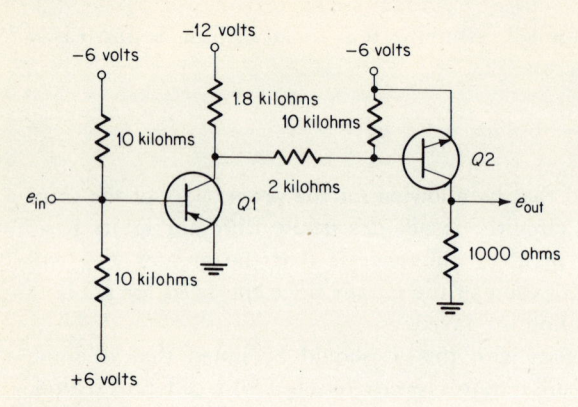

Fig. 10-15. A circuit showing one method of using mixed types.

direction, the base of $Q2$ sees a negative voltage with respect to its emitter and $Q1$ is off. The actual base voltage is determined by the three resistors between the -6-volt and -12-volt supplies.

When the base of $Q1$ is presented a negative signal, $Q1$ turns on and its collector falls to essentially ground. The base of $Q2$ now sees a voltage considerably more positive than -6 volts, as determined by the 10- and 2-kilohm resistors, and so $Q2$ turns on. With $Q2$ in saturation, its collector is clamped to its emitter, which is returned to -6 volts. Hence, the output is firmly connected to -6 volts. Several examples are given in Fig. 10-16. Each illustrates a firm connection to the various power-supply possibilities. Each is self-explanatory and will not be discussed further.

Electric Inverter

In logic circuits (see Chap. 11), it is often necessary to transfer a negative voltage to a positive voltage or vice versa. That is, if a signal of -6 volts exists, it may be necessary to invert this to $+6$ volts, for example. A circuit to accomplish this is given in Fig. 10-17 and is often called an *electric inverter*. Instead of operating between a power-supply value and ground, it operates between two power supplies, $+6$ and -6 volts.

The basic operation of the circuit follows. With the input at 0 volts, the base sees a voltage slightly positive with respect to ground by the drop across the 2.2-kilohm resistor. This is very negative with respect to $+6$ volts, to which the emitter is tied, and so $Q1$ is in saturation. The base is clamped to, perhaps, $+5.8$ volts, allowing 0.2 volt across the base-emitter

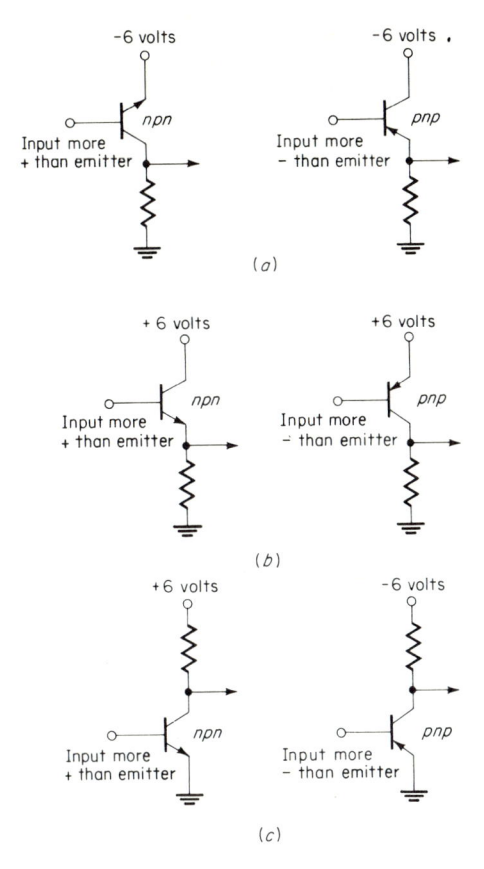

Fig. 10-16. Several circuits showing methods of connecting the output line to various parts of the circuit. (a) Firm connection to -6 volts; (b) firm connection to $+6$ volts; (c) firm connection to ground.

junction. With $Q1$ saturated, the output is firmly held to $+6$ volts. If the input now goes to $+6$ volts, the base-emitter junction is no longer forward-biased and $Q1$ turns off, which allows the collector to rise to -6 volts, as shown. When the input falls back to ground, the collector returns to $+6$ volts.

Operation of the circuit can be made more reliable by inserting the diode in the emitter, as suggested in Fig. 10-17b. Without the diode, it is not possible to firmly reverse-bias the emitter-base junction when the input goes to $+6$ volts. With the diode, however, the emitter is returned to perhaps $+5$ volts if the diode drop is on the order of 1 volt. Now the

174 PULSE AND SWITCHING CIRCUIT ACTION

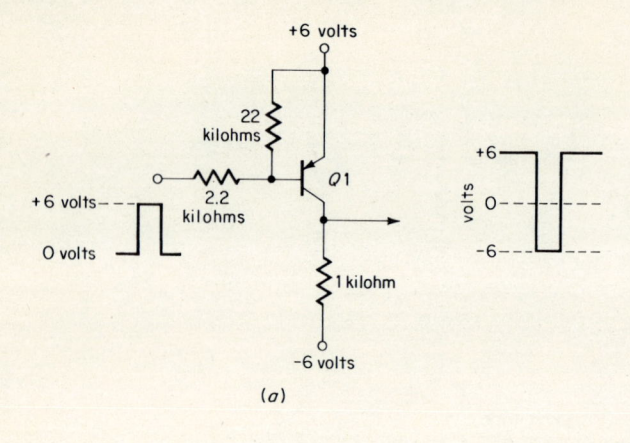

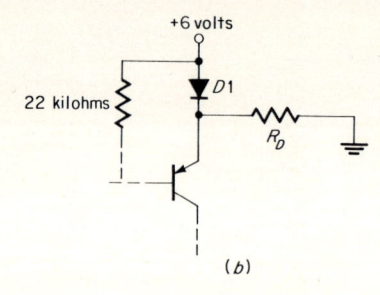

Fig. 10-17. The circuit of the electric inverter.

transistor base can be reverse-biased, and the transistor is definitely in cutoff. Of course, now the output can rise only to +5 volts, but this would seldom be any problem.

Using a diode in this manner is often called a *local-diode power supply*. Of course, the diode is not really supplying anything but a constant voltage drop, but the name is descriptive of the circuit function. The purpose of R_D is to keep the diode conducting at all times.

Current-mode Circuit

Because operation in the current mode is quite different than saturated-mode operation, it requires more explanation than given previously. A typical current-mode circuit is shown in Fig. 10-18.

Basically, the circuit is useful in high-speed switching circuits, and its advantages lie in eliminating the storage-delay time. The circuit values are chosen to permit operation in the active region rather than in the saturation region when the transistor is on. With the circuit shown, this is quite easy to do. As the input is caused to be slightly more negative than ground, $Q1$ is on and a current I_{E1} is flowing in the emitter. The collector current flowing in $RL1$ is sufficient to cause the collector voltage to fall near, but not to, ground. Hence, $Q1$ is in the active region, and $Q2$ is off.

Now, when the input goes slightly more positive than ground, $Q1$ is turned off and its emitter current is switched to $Q2$. The collector of $Q2$ falls from V_{CC} toward, but not completely to, ground. Therefore, $Q2$ is still in its active region. When the input returns to some negative value, $Q1$ comes back on while $Q2$ is turned off again.

By carefully analyzing the circuit for the various conditions, the foregoing general description can be verified. First, consider the condition where the input is slightly negative with respect to ground, perhaps

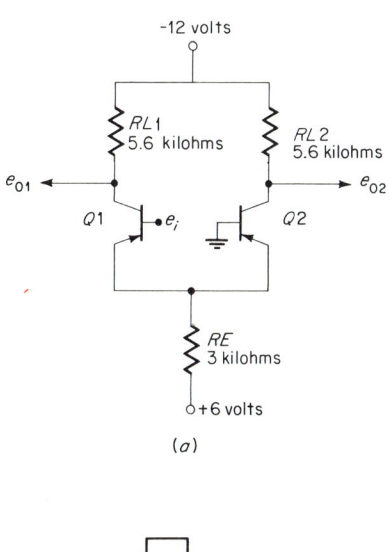

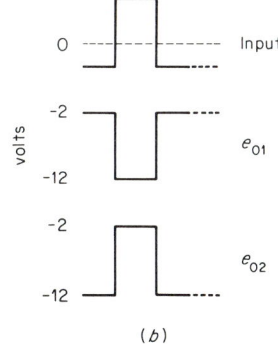

Fig. 10-18. A current-mode circuit and waveforms.

−0.4 volt. $Q1$ is on, and some amount of emitter current is flowing. Allowing 0.2 volt across the base-emitter junction, I_{E1} is easily found:

$$I_{E1} = \frac{V_{EE} - V_E}{RE} = \frac{6 - 0.2}{3 \text{ kilohms}} = \frac{5.8}{3 \text{ kilohms}} = 1.93 \text{ ma}$$

If the transistor's beta is 100, its alpha is 0.99 and thus its collector current is αI_E:

$$I_{C1} = \alpha I_{E1} = 0.99 \times 1.93 \text{ ma} = 1.91 \text{ ma}$$

Knowing the current through $RL1$ will allow the drop across it to be determined, as well as V_{C1}.

$$E_{RL1} = I_{C1} \times RL1 = 1.91 \text{ ma} \times 5.6 \text{ kilohms} \cong 10.7 \text{ volts}$$
$$V_{C1} = -V_{CC} + E_{RL1} = -12 + 10.7 = -1.3 \text{ volts}$$

Thus, with $Q1$ on and the collector voltage at −1.3 volts while V_B is −0.4 volt, the transistor is still in the active region; i.e., the collector junction is reverse-biased. At the same time $Q1$ is on, $Q2$ should be off. If V_{EQ1} is at about −0.2 volt, then V_{EQ2} must be −0.2 volt also. Since the base of $Q2$ is directly grounded, it is 0.2 volt more positive than the emitter and $Q2$ is indeed off.

When the input goes to, perhaps, +0.5 volt, the emitter of $Q1$ tries to follow along. When the emitters are about 0.1 volt more positive than ground, $Q2$ begins to conduct, and with the emitters at +0.2 volt, $Q2$ is full on (for this circuit full on is something less than saturation). As the base of $Q1$ continues on to +0.5 volt, the emitters are *clamped* to about +0.2 volt. Hence, $Q1$ turns off as $Q2$ turns on.

Now, the current in R_E is flowing in $Q2$. With the emitter of $Q2$ at about +0.2 volt, the emitter current is

$$I_{E2} = \frac{V_{EE} + V_E}{RE} = \frac{6 + 0.2}{3 \text{ kilohms}} \cong 2.07 \text{ ma}$$

This is not significantly different than I_{E1}, with the actual difference being only about 100 μa. Completing the analysis for I_C, E_{RL}, and V_C will allow a comparison of the two states:

$$I_{C2} = \alpha I_E = 0.99 \times 2.07 \cong 2.05 \text{ ma}$$
$$E_{RL2} = I_{C2} \times RL2 = 2.0 \text{ ma} \times 5.6 \text{ kilohms} = 11.2 \text{ volts}$$
$$V_{C2} = -V_{CC} + E_{RL} = -0.8 \text{ volt}$$

Again, the collector junction is reverse-biased, and so $Q2$ is still in the active region. Note that, in effect, the current in the $Q1$ emitter circuit has been switched to the $Q2$ emitter circuit when the input goes positive.

AMPLIFIER SWITCHING CIRCUITS   177

As the input returns to its negative value, the current is switched back to Q1 again.

This particular circuit has certain advantages, one of which is the complementary outputs. That is, a single input can, if necessary, yield two opposing outputs. The output e_{o1} is out of phase with the input while e_{o2} is in phase. Hence, either or both outputs may be used, depending upon the circuit requirements.

Design Example

As a final example of the common-emitter switching circuit, a simplified design procedure will be shown so as to emphasize many of the circuit actions and limits. The circuit to be designed is given in Fig. 10-19a.

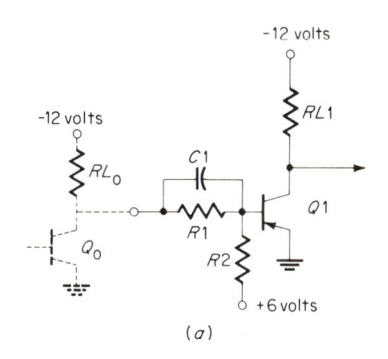

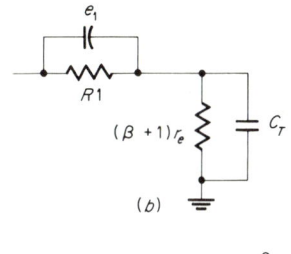

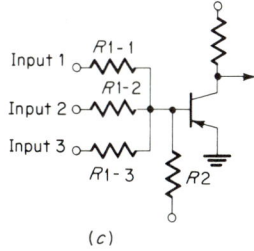

Fig. 10-19. Circuit configurations for the design example.

178 PULSE AND SWITCHING CIRCUIT ACTION

The proper value of each component will be determined, consistent with the design limitations dictated by the overall circuit requirements.

First, however, a few stipulations must be set forth, such as the power-supply values and the choice of transistor. The power supplies provide −12 and +6 volts, while the transistor is to be a 2N711 medium-speed germanium switching transistor. From the manufacturer's specification sheet, the transistor parameters needed are extracted. These are listed in Table 10-1. Also, the maximum operating temperature is to be 65°C.

Table 10-1
Transistor Characteristics for the Design Example

2N711 *Characteristics at* 25°C

$h_{FE,\max} = 150$ $t_{s,\max} = 200$ nsec
$h_{FE,\min} = 20$ $t_{s,\text{typ}} = 90$ nsec
$V_{CE,\max} = 12$ volts $I_{cbo,\max} = 3$ μa
$V_{BE,\max} = 1$ volt $I_{ebo,\text{typ}} = 100$ μa
$I_{C,\max} = 50$ ma $V_{CE,\text{sat}} = 0.2$ volt
$P_{C,\max} = 300$ mw $V_{BE,\text{sat}} = 0.3$ to 0.5 volt
$t_{\text{off}} = 80$ nsec for $\beta_F = 10$ at $I_C \cong 10$ ma

Step 1   Determine maximum leakage current $I_{cbo'}$ at 65°C. (I_{cbo} doubles every 10°C.)

Leakage current at 25°C = 3 μa
35°C = 6 μa
45°C = 12 μa
55°C = 24 μa
65°C = 48 μa
$I_{cbo'}$ = 48 μa

Step 2   Choose $V_{BE,\text{off}}$.

$V_{BE,\max} = +1$ volt

Set $V_{BE,\text{off}}$ to +0.75 volt.

Step 3   Determine $RL_{\max}$ and $RL_{\min}$. The leakage current at 65°C will determine the maximum value of RL, while the maximum values of allowable power and current will determine the minimum value of RL.
    (a) Choose the maximum allowable drop across RL (ΔE_{RL}) due to the increase in leakage current. ΔE_{RL} is to be no more than 0.5 volt.
    (b) Calculate $RL_{\max}$.

$$RL_{\max} = \frac{\Delta E_{RL}}{I_{cbo'}} = \frac{0.5}{48 \ \mu a} = 10{,}417 \text{ ohms}$$

Nearest standard value: ±5% = 10 kilohms; $RL_{\max}$ = 10 kilohms.

(c) Calculate RL_{min}.

$$RL_{min} = \frac{V_{CC}}{I_{C,max}} = \frac{12}{50 \text{ ma}} = 240 \text{ ohms}$$

(d) Choose RL to be less than 10 kilohms and more than 240 ohms. RL is to be 1.0 kilohm.

Step 4 Calculate the value of I_B to firmly saturate $Q1$, which determines the value of $R1$.

(a) $I_{B,sat,min} = \dfrac{I_{C,sat}}{h_{FE,max}} = \dfrac{12 \text{ ma}}{150} = 80 \text{ }\mu\text{a}$

(b) $I_{B,sat,max} = \dfrac{I_{C,sat}}{h_{FE,min}} = \dfrac{12 \text{ ma}}{20} = 600 \text{ }\mu\text{a}$

(c) Saturated base current $I_{B,sat}$ should be arbitrarily set at two to four times the maximum value to insure saturation under worst-case conditions.

$$I_{R1} = I_{B,sat,max} = 2 \times 600 \text{ }\mu\text{a} = 1.2 \text{ ma}$$

(d) Calculate forced beta B_f.

$$B_f = \frac{I_{C,sat}}{I_{B,sat}} = \frac{12 \text{ ma}}{1.2 \text{ ma}} = 10$$

(e) $R1$ plus the load resistor of the previous stage ($RL_0 = 1$ kilohm) is R_T.

$$R_T = R1 + RL_0 \qquad R1 = R_t - RL_0$$

$$R_T \cong \frac{V_{CC}}{I_{R1}} = \frac{12}{1.2 \text{ ma}} = 10 \text{ kilohms}$$

$$R1 = 10 \text{ kilohms} - 1 \text{ kilohms} = 9 \text{ kilohms}$$

Nearest standard value: $R1 = 9.1$ kilohms.

Step 5 Calculate the value of $R2$. The entire voltage divider, RL_0, $R1$, and $R2$, is across a total of 18 volts. However, when $Q0$ is on, its load resistor is, in effect, not in the applicable circuit. Considering that the collector of $Q0$ is at ground when it is on, the voltage divider $R1$ and $R2$ should place $+0.75$ volt on the base of $Q1$.

$E_{R1}:R1::E_{R2}:R2$

$E_{R1} \times R2 = R1 \times E_{R2}$

$$R2 = \frac{R1 \times E_{R2}}{E_{R1}} = \frac{9.1 \text{ kilohms} \times 5.25}{0.75} = 63.7 \text{ kilohms}$$

Nearest standard value; $R2 = 62$ kilohms.

180 PULSE AND SWITCHING CIRCUIT ACTION

Step 6 Calculate the value of the speedup capacitor
(a) Calculate the value of the saturation base-to-emitter resistance.

$$R_{BE,\text{sat}} = \frac{V_{BE,\text{sat}}}{I_{B,\text{sat}}} = \frac{0.3}{1.2 \text{ ma}} = 250 \text{ ohms}$$

(b) The turn-off time is given as 80 nsec. This represents an equivalent capacitance, where T = total charge time, C_t = transistor junction capacitance, R = junction resistance.

$$T = 5(RC_t) \qquad \therefore C_t = \frac{T}{5R}$$

$$C_{t,\text{max}} = \frac{T}{5R_{BE,\text{sat}}} = \frac{80 \text{ nsec}}{5 \times 250} = \frac{80 \times 10^{-9}}{1.25 \times 10^3} = 64 \text{ pf}$$

This value is actually somewhat higher than the minimum required since R_{BE} does not, of course, remain constant. As the transistor begins to turn off, R_{BE} increases in value, and while $Q1$ is in the active region, approaches the value of $(\beta + 1)r_e$. For typical transistors, this is on the order of 3000 ohms. Using this value of resistance in the above formula gives a different effective capacitance:

$$C_t = \frac{T}{5R} = \frac{80 \text{ nsec}}{5(3 \text{ kilohms})} = 5.3 \text{ pf}$$

This represents a minimum value of the effective transistor capacitance. A value halfway between the two values (29 pf) is a good estimate of needed capacitance. This, however, must now be related to the size of $C1$. The applicable network is shown in Fig. 10-19*b*, and the following relationship will determine the value of $C1$ to just compensate the voltage divider:

$$R1C1 = (\beta + 1)r_e \times C_T$$

$$C1 = \frac{(\beta + 1)r_e \times C_T}{R1} = \frac{3 \text{ kilohms} \times 29 \text{ pf}}{9.1 \text{ kilohms}} = 9.56 \text{ pf} \cong 10 \text{ pf}$$

Hence, within the accuracy of the foregoing assumptions, the minimum value of $C1$ necessary to assure fast turn-off is 10 pf. In practice, two to four times this value would be used to make certain of overcompensation. Since 33 pf is a standard size, this would make an excellent choice; therefore $C1 = 33$ pf.

Step 7 Determine the minimum loading on $Q1$. The number of external loads allowed on $Q1$ is a function of the maximum allowed current through RL ($Q1$) when $Q1$ is off. A current that is too large will cause an appreciable drop across RL and the net effect is as though $Q1$ were on and in its active region. This, of course, must be avoided in so far as possible.

(a) Choose maximum drop across RL (ΔE_{RL}) to be allowed due to loading.

$\Delta E_{RL} = 0.5$ volt

(b) RL and the total effective dc load form a voltage divider, and thus the effective load R_{load} is easily determined:

$RL:0.5$ volts::$R_{\text{load}}:11.5$ volts

$$R_{\text{load}} = \frac{RL \times 11.5}{0.5} = 23 \text{ kilohms}$$

Thus, a single load can be no less than 23 kilohms; two loads (in parallel) must be 46 kilohms each; four loads must be 4×23 kilohms, or 92 kilohms each.

Step 8 Calculate for the worst-case input condition, on the assumption that $Q1$ is being fed from several identical inverters. The maximum number of inputs, as indicated in Fig. 10-19c, is determined by the minimum value of V_{BE} allowed to insure good turn-off.

(a) Assume $+0.2$ volt is to be the minimum V_{BE} for turn-off conditions. Then, R_t is the total parallel resistance of $R_{1-1}, R_{1-2},$ and R_{1-3}. Finally the collector sides of all $R1$'s are -0.2 volt, which is the $V_{CE,\text{sat}}$ of the driving transistors.

$E_{Rt}:R_T::E_{R2}:R2$

$$R_T = \frac{E_{Rt} \times R2}{E_{R2}} = \frac{0.4 \times 62 \text{ kilohms}}{5.8} \cong 4.28 \text{ kilohms} \cong 4.3 \text{ kilohms}$$

(b) Since each R_1 is equal to 9.1 kilohms, only two inputs are allowed:

$$N_{\text{inputs}} = \frac{R_X}{4.3 \text{ kilohms}} = \frac{9.1 \text{ kilohms}}{4.3 \text{ kilohms}} \cong 2.117$$

where R_X is the value of any one R_1. The next-smaller integer (2.0) is the maximum number of allowed inputs. Thus, one of the three inputs *must be removed*, or other means used to satisfy the design requirements, possibly a complete redesign.

Step 9 Determine the power rating of each resistor.

(a) $RL:(I_{C,\text{sat}})^2 \times RL = P_{RL}$

$P_{RL} = I^2 R = (12 \text{ ma})^2 \times 2 \times 10^3 = 0.288$ watt

Therefore, RL should be no less than a ½-watt resistor to assure cool operation. (P_{rated} should be at least $2 \times P_{RL}$.)

(b) $R1:(I_{R1,\text{max}})^2 \times R1 = P_{R2}$

$P_{R1} = (I_{B,\text{sat}})^2 \times R1 = (1.2 \text{ ma})^2 \times 9.1 \text{ kilohms} = 0.0131$ watt

A ½-watt resistor is more than enough in this case.

(c) $R2: (I_{R2,\max})^2 \times R2 = P_{R2}$

$$P_{R2} = (I_{\max})^2 \times R2 = \left(\frac{E_{R2\max}}{R2}\right)^2 \times R2 = \left(\frac{6.2}{62 \text{ kilohms}}\right)^2 \times 62 \text{ kilohms}$$

$$= 0.00062 \text{ watt}$$

Again, this resistor dissipates practically no power, and hence the ½-watt size is quite adequate.

10-2 THE CB (COMMON - BASE) SWITCHING CIRCUIT

Although the common-base circuit is used less frequently than the CE circuit, it nevertheless occupies an important place in the scheme of things. Basically, a common-base circuit is used where either there is to be no signal inversion or the high-frequency response of a CE circuit is not adequate. The circuit of Fig. 10-20 will serve to illustrate the basic properties of this circuit.

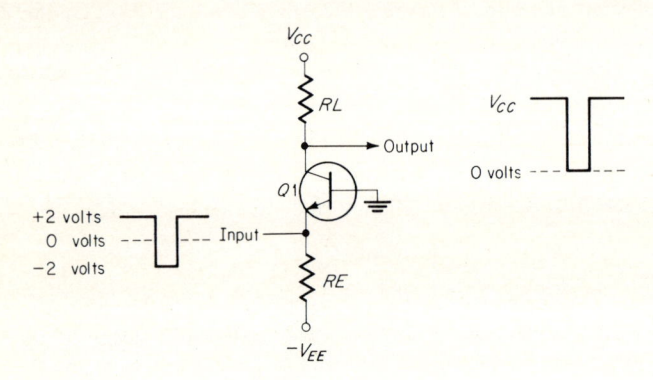

Fig. 10-20. The CB switching circuit.

Consider the case where an input operates between the levels of +2 and −2 volts and assume that Q1 is a silicon *npn* transistor. During the time the input is at a level of +2 volts, the emitter is 2 volts more positive than the base. In other words, the base is 2 volts more negative than the emitter and Q1 is off. The collector, then, is at V_{CC} since there is no current through RL. Thus, a positive input voltage results in a positive output voltage.

Now, when the input falls to −2 volts, the base is more positive than the emitter and Q1 turns on. With properly chosen component values, Q1 can be assumed to be in saturation. Hence, the collector falls to

AMPLIFIER SWITCHING CIRCUITS   183

ground, essentially. Actually, $V_{CE,\text{sat}}$ is on the order of 0.2 volt, and the collector cannot be exactly ground but is nevertheless very close. Obviously, a negative-going pulse at the input results in a similar output, and the lack of inversion is clearly evident. As the input returns to +2 volts, the output returns to V_{CC}.

To describe the circuit action more concisely, Fig. 10-21 shows basically the same circuit with component values given. First, consider the input

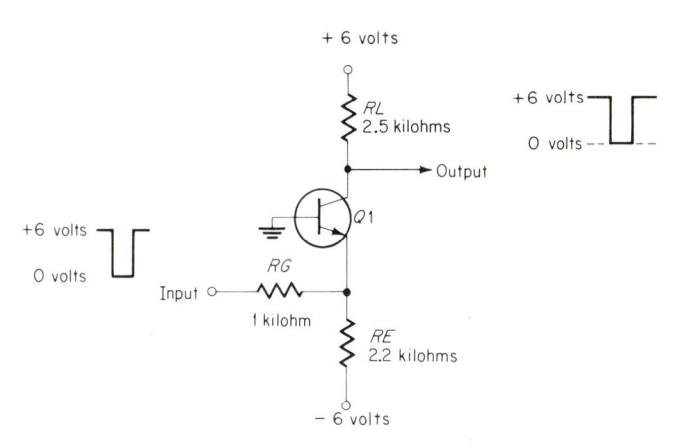

Fig. 10-21. A CB circuit used for an analysis example.

voltage to be at the +6-volt level. Because the emitter is at something more positive than the base, $Q1$ is off. The actual emitter voltage referred to ground is set by the voltage divider.

$$E_{RG} = \frac{RG}{RG + RE} \times E_{\text{total}} = \frac{1 \text{ kilohm}}{3.2 \text{ kilohms}} \times 12 = 3.75 \text{ volts}$$

$$V_E = +V_{CC} - E_{RG} = +6 - 3.75 = +2.25 \text{ volts}$$

Thus, with the emitter at +2.25 volts with respect to the base, the transistor is indeed off. The collector, then, is at a level of +6 volts.

Now, with the input at 0 volts (ground), the transistor sees a different set of conditions. The voltage at the emitter will *try* to become some value more negative than ground, as set again by RG and RE. However, if the emitter tries to go more negative than ground by, perhaps, 0.75 volt, $Q1$ is turned on and the emitter is clamped to about −0.75 volt. The voltage as set by RG and RE tries to be

$$E = \frac{RG}{RG + RE} \times -V_{EE} - \frac{1 \text{ kilohm}}{3.2 \text{ kilohms}} \times -6 \cong -1.88 \text{ volts}$$

The more negative this point tries to go, the harder the transistor is turned on, and with V_E actually clamped to -0.75 volt approximately, the emitter current is essentially determined by the current through RE:

$$I_{RE} = I_E = \frac{E_{RE}}{RE} = \frac{6 - 0.75}{2.2 \text{ kilohms}} \cong 0.0024 \text{ amp} = 2.4 \text{ ma}$$

Because I_C and I_E are very nearly equal, the drop across RL may now be found:

$$E_{RL} = I_C \times RL = 2.4 \text{ ma} \times 2.5 \text{ kilohms} \cong 6 \text{ volts}$$
$$V_C = V_{CC} - E_{RL} = 6 - 6 = 0 \text{ volts}$$

Obviously, $Q1$ is in saturation, and the output will be firmly held to ground. As in the case of common-emitter circuit, the input need not be a full 6-volt excursion. The circuit is capable of voltage amplification if necessary, and so if the input is degraded somewhat, the output will still be a full 6-volt pulse. Because of the inherently good frequency response, the output will exhibit essentially the same rise times as does the input within the limitation set by the transistor itself.

10-3 THE CC (COMMON - COLLECTOR) SWITCH

The common-collector circuit, or emitter follower, is often encountered in digital circuitry, although strangely enough it is seldom used as a true switch. Most often, it is used in a way that keeps it in the active region all the time. When used in this manner, it cannot be properly referred to as a *switch*. However, the surrounding circuitry is normally used in the true switching mode, and so it is convenient to classify the emitter follower as a *switching circuit* when used in conjunction with true switches.

On the other hand, a transistor is sometimes used in the common-collector configuration as a true switch, being at various times in the active region, the cutoff region, or the saturation region. In this instance it cannot be truly called an *emitter follower* since in saturation or cutoff the emitter is incapable of following the base. Again, because it appears to be an emitter follower when viewed schematically, it will here be classed in general terms as an emitter follower regardless of its true mode of operation.

Consider first, the circuit in Fig. 10-22, which is classed as a true emitter follower. The input signal will be assumed to operate between the levels of ground, and -6 volts. As shown, the output is a nearly exact replica of the input, except for the slight offset caused by the voltage drop across the base-emitter junction. For a typical germanium transistor this is about 0.1 volt, and for certain silicon *npn* units this is on the order of 0.7 volt.

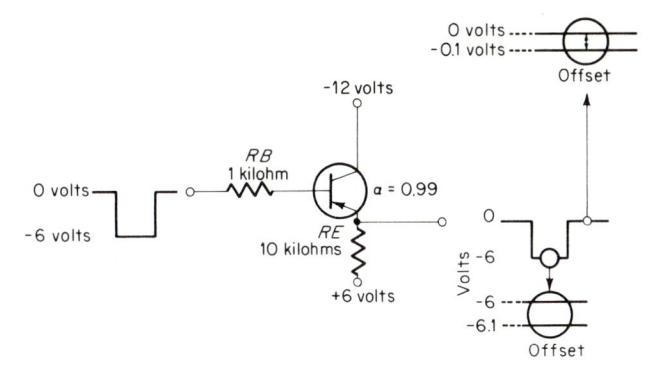

Fig. 10-22. A CC circuit (emitter follower) used in a switching application.

Except for this offset, the output is usually considered to be an exact replica of the input. Hence, it is easy to understand why the circuit is often called the *emitter follower*.

Common-collector Circuit Analysis

To understand the basic operation of the emitter follower and why the emitter follows the base, the following discussion is offered. The first consideration to be made when describing the circuit operation (and this is true of any circuit analysis procedure) is to determine the maximum circuit limits. Once these are determined, then all operations must lie between these limits. The limits of interest in this circuit are the most positive value of V_E, called $V_{E,\max+}$, the most negative value of $V_E(V_{E,\max-})$, as well as the emitter currents at these two extremes.

$V_{E,\max+}$ can only be $+6$ volts since the emitter cannot go more positive than the supply voltage to which it is returned. In order to attain this limit the base must be more positive than $+6$ volts or at least no more negative than $+6$ volts. Hence, if V_B is at or above $+6$ volts, the transistor is in cutoff and no current is flowing (exclusive of leakage current).

The same reasoning allows the value of $V_{E,\max-}$ to be found. The collector is returned to -12 volts, and if the transistor is to become saturated, the base must be made to be more negative than the collector (or at least *as* negative as the collector) to insure that the collector-base junction is forward-biased. If the base is at -12 volts, the emitter will also be nearly at -12 volts. Hence $V_{E,\max-}$ is very nearly -12 volts, the difference being attributed to the collector-to-emitter resistance when saturated, $V_{CE,\text{sat}}$, typically 0.1 to 0.2 volt.

The following two limits have been defined. If the transistor is cut off, V_B and therefore V_E are at $+6$ volts, while if the transistor is at the edge of saturation, V_B and V_E are approximately at -12 volts. *The area of operation between cutoff and saturation is the active region.* Because the emitter voltage is free to change (because of RE), it will follow any voltage change on the base within these limits.

That this is true is evident from a consideration of the requirements of the transistor itself. Assume that both base and emitter of Fig. 10-22 are at precisely $+6$ volts. The emitter-base junction is not forward-biased, and so no base or emitter current flows. Now, if the base is made, perhaps, 0.2 volt more negative than $+6$ volts ($+5.8$ volts with respect to ground), the emitter-base junction becomes forward-biased. Now, base current flows, and hence emitter current flows.

Because emitter current flows, there is a voltage drop across the emitter resistor RE. This causes the emitter voltage to become more negative (less positive). But, this is the same direction that the base is made to go, thus the emitter appears to follow the base. It does this because as the base is caused to go more negative, the transistor is caused to turn on harder and the emitter is caused to go more negative by the increased emitter current. The actual difference between the two voltages is simply the drop across the base-emitter junction V_{BE}. Because this varies slightly between saturation and cutoff, the output-voltage swing is very slightly less than the input-voltage swing, by perhaps a few hundred millivolts. That is, if $V_B = 5.7$ volts, $V_E \cong 5.75$, while if $V_B = -11$ volts, $V_E = -10.5$ volts, a difference of about 500 mv. Thus, the output is just noticeably smaller if the input-voltage swing is nearly as large as the circuit limits. However, if the input is only a fraction of the limits (perhaps 2 or 3 volts), the output is smaller by so small an amount that it would be difficult to measure.

To analyze the circuit for its static conditions is simplicity itself. One must first define the voltage at the base since there is no biasing network as such, and if the base is left open-circuited as shown, the transistor, of course, will be off. Assume first that the base is returned to ground. Assuming no measurable drop across RB, the base and emitter are nearly at ground also. However, because of the junction drop, the emitter will actually be slightly more positive than the base. Allowing 0.1 volt for the junction drop will place the emitter at a potential of $+0.1$ volt. Therefore, if V_E is $+0.1$ volt, across RE a known potential exists:

$$E_{RE} = V_{EE} - V_E = +6 - 0.1 = 5.9 \text{ volts}$$

Now, the current through RE, which must be emitter current, is easily found.

$$I_{RE} = I_E = \frac{E_{RE}}{RE} = \frac{5.9}{10 \times 10^3} = 0.59 \text{ ma}$$

Only two other values might be of concern: I_C and I_B.

$I_C = \alpha I_E = 0.99 \times 0.59 \times 10^{-3} = 0.584$ ma

$I_B = \dfrac{I_E}{\beta + 1} = \dfrac{0.59 \times 10^{-3}}{101} = 5.8 \ \mu a$

The previous assumption that there was no noticeable drop across RB will now be confirmed:

$E_{RB} = I_B \times R_B = (5.8 \times 10^{-6})(1 \times 10^3) = 5.8$ mv

Hence, only 0.0058 volt is dropped across RB, and with typical measurement devices this would not be a significant amount. The only reason for the inclusion of RB is to limit base current should an accidental short circuit occur. Otherwise, it accomplishes practically nothing during normal circuit operation. Often, RB is actually the load resistor of a previous stage. To determine the transistor conditions if the input rises to -6 volts, simply perform the previous calculations substituting the new value. Again, allow 0.1 volt for junction drop.

$E_{RE} = V_{EE} + (-V_E) = +6 + (-5.9) = 11.9$ volts

$I_{RE} = I_E = \dfrac{E_{RE}}{RE} = \dfrac{11.9}{10 \times 10^3} = 1.19$ ma

$I_C = \alpha I_E = 0.99 \times 1.19 \times 10^{-3} \cong 1.18$ ma

$I_B = \dfrac{I_E}{\beta + 1} = \dfrac{1.19 \times 10^{-3}}{101} \cong 11.8 \ \mu a$

$E_{RB} = I_B \times R_B = (11.8 \times 10^{-6})(1 \times 10^3) = 11.8$ mv

As can be appreciated, determining the transistor conditions for these two values is quite straightforward. It should be noted that the foregoing conditions cannot be called *static*, or *quiescent*, conditions in the usual sense. (However, since the input can only be one of the two given values, these are the only two static conditions possible, keeping in mind the fact that the input terminal is permanently connected to some other device and thus cannot be other than 0 or -6 volts.)

The preceding discussion concerned a circuit that is operated continually in the active region. It is quite possible to use the common-collector configuration in the cutoff or saturation modes, as illustrated in Fig. 10-23. The advantage of this kind of operation is that when the output is at -6 volts, it is *firmly* connected to the -6-volt supply through the saturated transistor. If the external load is highly capacitive, this is a real advantage since the capacitance can charge to -6 volts *very* quickly through the virtually zero resistance of the saturated transistor.

The circuit operation can be explained rather easily. If the input is at a 3-volt level, the base is more positive than the emitter, which can go no

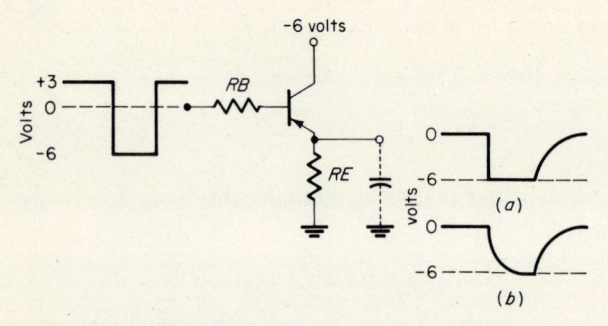

Fig. 10-23. An emitter follower operated so that it is either in cutoff or saturation. The waveforms show circuit action (a) without capacitive loading and (b) with capacitive loading.

more positive than ground ($V_{E,\max+}$), and the transistor is off. No emitter current results in no drop across RE, and hence the output is at ground.

Now, when the input goes to -6 volts, the collector-base junction becomes forward-biased (or nearly so) and the output wire is firmly clamped to the supply terminal. With a capacitive load, the leading edge will be similar to the waveform of Fig. 10-23a, being a very fast transition from 0 to -6 volts. If the transistor were not allowed to saturate, the leading edge would appear more rounded, as shown in the waveform in Fig. 10-23b. As long as the transistor is in the active region, its collector-base resistance R_{CB} is some appreciable amount and the capacitor-charge time is determined by the resistance value through which it must charge.

Note that both waveforms show a very rounded fall time, from -6 volts back to ground. Figure 10-24 shows one method of improving the fall time of such a circuit. The added diode will allow an alternate discharge path for capacitor current when the transistor is turned off. The only

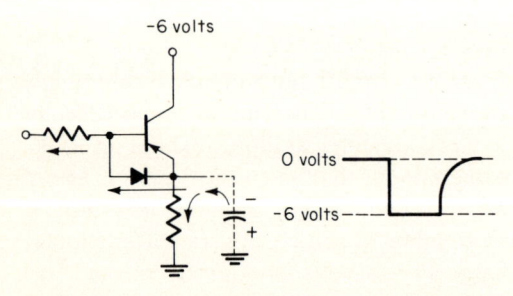

Fig. 10-24. One method of improving the pulse shape of the emitter follower.

limitation is that *RB* and any other series resistance back to ground must be relatively low in this path if appreciable improvement is to be gained. Note that usually a small rounding of the fall time is still apparent. However, much improvement is evident and is gained with a minimum of addition circuitry.

It is often necessary to know the output resistance of an emitter follower. Recalling from basic transistor theory that the output resistance of the emitter follower is low, it will now be found that it is probably much lower than one might anticipate. Consider the circuits shown in Fig. 10-25.

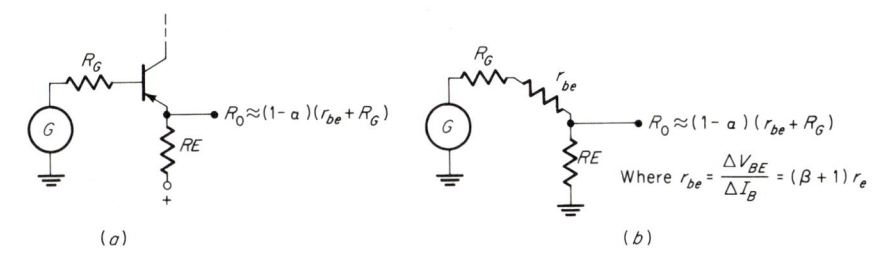

Fig. 10-25. The output resistance of the emitter follower.

Figure 10-25a indicates the items that must be known to determine this quantity. R_G is the total resistance to ground external to the base and is often primarily the generator (internal) resistance of the preceding stage. The dynamic base-emitter resistance r_{be} is determined by the slope of the V_{BE}-I_B curves for the transistor. If the curves are not available, the typical value of 500 to 1500 ohms may be used with small error.

To see what a typical value of R_O might be, assume that R_G is 1 kilohm. (Normally R_E is so much larger than R_O that is can be ignored, although it is possible that the two would be similar in value. In this case, R_O and R_E would be considered to be in parallel and the true output resistance would be smaller than either.) The value of R_O can now be determined. Assume that h_{fe} is 100 ($\alpha = 0.99$).

$$R_O = (1 - \alpha)(r_{be} + R_G) = (1 - 0.99)(500 + 1000) = 0.01 \times 1500 = 15 \text{ ohms}$$

Thus, the true output resistance of such a circuit is quite low, with typical values much lower than might be expected from a cursory inspection of the circuit components.

QUESTIONS AND PROBLEMS

10-1 Refer to Fig. 10-1. The input terminal is connected to a dc source of -3.6 volts. Determine the state of $Q1$ (off, on, active region).

10-2 Refer to Fig. 10-1. The input terminal is connected to a dc source of -2.5 volts. Determine the state of $Q1$ (off, on, active region).

10-3 Refer to Fig. 10-1. The input terminal is connected to a dc source of $+2.5$ volts. Determine the state of $Q1$ (off, on, active region). Assume $V_{D1} = 0.3$ volt.

10-4 Refer to Fig. 10-1. $-V_{CC}$ is to be changed to -10 volts, and RL and $C2$ are to be removed, but all else is to remain as shown. Determine e_o if $Q1$ is off and if $V_{D1} = 0.8$ volt.

10-5 Refer to Fig. 10-1. $-V_{CC}$ is to be changed to -15 volts, but all else is to remain the same. Assume V_{D1} is 1.0 volt when the diode conducts. Determine e_o if $Q1$ is cut off, disregarding RL and $C2$.

10-6 Refer to Fig. 10-1. Using the circuit as drawn, briefly explain the effect on the circuit action if $C1$ were open or removed.

10-7 Refer to Fig. 10-3. For the diode which these curves represent, determine the static diode resistance if $V_D = 0.3$ volt.

10-8 Refer to Fig. 10-3. For the diode which these curves represent, determine the static diode resistance if $V_D = 0.6$ volt.

10-9 Refer to Fig. 10-3. Given that $r_{df} = \dfrac{\Delta V_D}{\Delta I_D}$; for the diode which these curves represent, determine the dynamic resistance of the diode if diode current is varied from 2 to 3 ma.

10-10 Refer to Fig. 10-3. Given that $r_{df} = \dfrac{\Delta V_D}{\Delta I_D}$; for the diode which these curves represent, determine the dynamic resistance of the diode if diode current is varied from 4 to 6 ma.

10-11 Refer to Fig. 10-4. Change $R1$ from 4700 to 9100 ohms. Determine the trip level evident at the input. Assume $V_{BE} = 0.25$ volt.

10-12 Refer to Fig. 10-4. Change $R1$ from 4700 to 22,000 ohms. Determine the trip level evident at the input. Assume $V_{BE} = 0.25$ volt.

10-13 Refer to Fig. 10-8. Changing only $RB2$ from 4000 to 6000 ohms, determine the threshold voltage at e_i if $V_{BE} = 0.25$ volt and if e_i starts at 0 volts and progresses in the positive direction. Assume that $Q1$ does not turn off until $V_B = 0$ volts.

10-14 Refer to Fig. 10-8. Changing only $RB2$ from 4000 to 6000 ohms, determine the threshold voltage at e_i if $V_{BE} = 0.25$ volt and if e_i starts at -12 volts and progresses in the positive direction. Assume that $Q1$ does not turn off until $V_B = 0$ volts.

10-15 Refer to Fig. 10-12. $R1$ is to be a 2000-ohm resistor while $RL1$ to $RL4$ are to be 4000 ohms each. Disregard $C1$ to $C4$. Determine $V_{C,max}$ if $-V_{CC} = 10$ volts.

10-16 Refer to Fig. 10-12. $R1$ is to be a 2000-ohm resistor while $RL1$ to $RL4$ are to be 1000 ohms each. Disregard $C1$ to $C4$. Determine $V_{C,max}$ if $-V_{CC} = -10$ volts.

10-17 Refer to Fig. 10-21. Change only RL from 2500 to 1000 ohms. Determine the output V_C when the input is at a 0-volt level. Assume $V_{BE} = 0.75$ volt.

10-18 Refer to Question 10-17. For the same circuit and conditions, find a value of RL (other than 1000 ohms) to just put $Q1$ in saturation. Assume $V_{BE} = 0$ volts when $Q1$ is in saturation.

11
LOGIC AND LOGIC CIRCUITS

11-1 NUMBER SYSTEMS

The decimal number system is so widely known and used that the fact that other methods are possible is not often considered. But an electronic digital system very frequently uses a method of handling numbers that is not decimal. Any number system is possible, but in electronic equipment, probably the most widely used is the binary system.

In order to understand the binary number system, it is necessary to first investigate thoroughly the more familiar decimal system so that a close comparison between the two is possible. Once this is accomplished, it will be a simple matter to consider binary numbers by themselves and to manipulate them according to the rules of arithmetic. When binary numbers can be easily added and subtracted, it is quite certain that one understands the subject sufficiently well.

The first step, then, is to investigate the organization of the decimal system. When these ideas are firmly fixed in mind, binary numbers will be compared with them and the similarity between systems noted. Then, the arithmetic manipulation of binary numbers will complete the discussion.

The Decimal Number System

A basic property of any number system is the *base* or *radix* of the system. In the case of the decimal system, the radix is 10; that is, there are 10 separate symbols or digits used to represent quantity, and these are 0, 1, 2, 3, 4, 5, 6, 7, 8, 9. Because it is often necessary to indicate quantities greater than 9, the digits are arranged by columns, each column having a different *weight* or multiplying factor. The columnar structure for the decimal system is shown in Fig. 11-1 for several values both greater and smaller than 1.

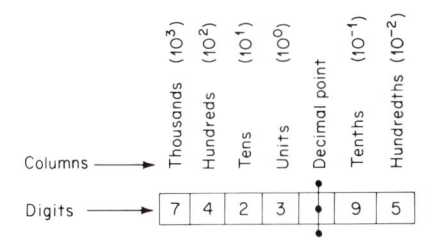

Fig. 11-1. Decimal-column position.

The digits inserted in each column carry the weight of that column multiplied by the digit itself. For instance, the digit 3 in the units column means 3×1. The digit 4 in the hundreds column means 4×100, or 400. Hence, the total number as shown can be represented as

$$(7 \times 1000) + (4 \times 100) + (2 \times 10) + (3 \times 1) + (9 \times 0.1) + (5 \times 0.01)$$
$$= 7000 + 400 + 20 + 3 + 0.9 + 0.05 = 7423.95$$

The column-positional weight, then, is necessary to allow numbers larger than 9 or smaller than 1 to be accounted for without the necessity for inventing new symbols. For example, 10 could be written as X, while 11 might be *, and so on. However, each new quantity would require a new symbol. Such a system would rapidly become unwieldy.

In a decimal system, however, the changeover from one column to the next-higher-order column is accomplished by the *carry*. That is, by increasing a 9 in the units column by 1, the units column is returned to zero while the carry is placed in the next-higher-order column. Hence, the result of adding 1 to 9 is 10, which, of course, is a two-digit number. Note that the 1 in the tens column has a total weight of 10 rather than simply 1. This operation is indicated in the following expression, where cr = carry:

$$\begin{array}{r} \text{cr} \\ 09. \\ (+)01. \\ \hline 10. \end{array}$$

This process may be reversed where necessary. In subtraction it is often necessary to *borrow, which is exactly opposed to a carry*. A carry is the addition of a unit from a certain column to the next-higher-order column. A borrow is the addition of a unit from a certain column to the next-lower-order column. Because a 10 in one column becomes a 1 in the next-higher-order column on a carry, the same 1 in the next-higher-order column becomes a 10 in the next-lower-order column on a borrow. Hence, the borrow and carry are mutually equivalent, but opposite, functions.

To illustrate the carry and borrow operations the following examples are worked out and are self-evident. (Note that the carry is treated as a 1, while the borrow is treated as a 10.)

$$\begin{array}{r} \text{bbb} \\ 099.87 \\ +080.42 \\ \hline 180.29 \end{array} \qquad \begin{array}{r} 3 \\ +10 \\ 180.\cancel{4}2 \\ -60.33 \\ \hline 120.09 \end{array}$$

The Binary Number System

With the decimal number system thoroughly understood, it will be a simple matter to correlate it with the binary number system. The binary number system uses but two digits 0 and 1, as contrasted to the ten digits of the decimal system. Because of this, the weight of each column is different from that of the decimal system. The radix for this system, therefore, is 2.

An example of a binary number and the columnar weighting is shown in Fig. 11-2. The units column has the same meaning as in the decimal system. However, because there are just two digits used, only a 0 or 1 may

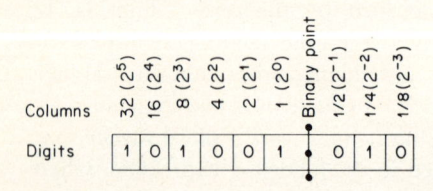

Columns	32 (2^5)	16 (2^4)	8 (2^3)	4 (2^2)	2 (2^1)	1 (2^0)	Binary point	1/2 (2^{-1})	1/4 (2^{-2})	1/8 (2^{-3})
Digits	1	0	1	0	0	1	•	0	1	0

Fig. 11-2. Binary-column position.

be used. If it is necessary to add a 1 to a 1 already in a column, the carry is used as before to implement this function.

The number shown in Fig. 11-2 is 101001.010. This, of course, will have a decimal equivalent. The meaning of this binary number can be specified by adding together the column weights for every column where a 1 appears, as was done earlier for the decimal system:

$$(1 \times 32) + (0 \times 16) + (1 \times 8) + (0 \times 4) + (0 \times 2) + (1 \times 1) + (0 \times \tfrac{1}{2})$$
$$+ (1 \times \tfrac{1}{4}) + (0 \times \tfrac{1}{8}) = 32 + 8 + 1 + \tfrac{1}{4} = 41\tfrac{1}{4}$$

Because the digits to the right of the binary point are seldom used in practice, they will not be discussed further.

When a given binary column is filled (contains a 1) and when the digit is to be incremented by 1 (added to by 1), the methods used are the same as in the case of the decimal system. That is, the column in question is returned to zero and a carry is placed in the next-higher-order column. As an example, the following binary number is to be incremented by 1. As before, the carry (cr) becomes a 1 in the next-higher-order column:

```
cc
0011   (number)
   1   (add 1)
 100
```

A borrow may be performed just as in the decimal system. A binary digit moved from a more significant column becomes a borrow, and in the new column it carries a weight greater than before by the radix of the system. This, of course, is the same as in the case of the decimal system. As an example, a 1 in the third column is to be moved to the second column.

```
        1
0100   00 + 0
        1
```

This peculiar notation is necessary because a 1 moved one column to the right is now weighted twice its previous value. Twice 1 is 2, but there is no digit 2 in the binary system. Hence, we write two 1s, which, of course, means the same thing. These, then, will be treated as individual digits, as will be seen in the section covering binary addition.

Binary-to-Decimal Conversion

The process of converting a binary number to its decimal equivalent is rather simple and has been illustrated previously. The column weights for each 1 appearing in the number are noted and are added to all other column weights containing a 1.

Decimal-to-Binary Conversion

Converting a decimal number to its binary equivalent is illustrated in Table 11-1 and is easily accomplished by the following method. The

Table 11-1
Decimal-to-Binary Conversion

```
         MSD
 (1) 342.     Decimal Number.
 (2) 256.     Next lower binary number in the series
              1, 2, 4, 8, etc.
 (3) ┌ 86.  1 Difference = successful subtract.
 (4) │ 128.   Next lower binary number below (2)
              above.
 (5) └ 86.  0 Difference - NON-successful subtract.
 (6)   64.    Next lower binary number below (4)
              above.
 (7) ┌ 22   1 Difference = successful subtract.
 (8) │ 32     Next lower binary number below (6)
              above.
 (9) └ 22   0 Difference = NON-successful subtract.
(10)   16     Next lower binary number below (8)
              above.
(11) ┌  6   1 Difference = successful subtract.
(12) │  8     Next lower binary number below (10)
              above.
(13) └  6   0 Difference = NON-successful subtract.
(14)    4     Next lower binary number below (12)
              above.
(15)    2   1 Difference = successful subtract.
(16)    2     Next lower binary number below (14)
              above.
(17)    0   1 Difference = successful subract.
(18)    1     Next lower binary number below (16)
              above.
        0   0 Difference - NON-successful subtract.
        LSD
```

number to be used will be 342. First the number itself 342 is written. Then, the number that is just smaller than 342 in the binary series of numbers is written below, as shown in Table 11-1. To determine the step 2 number, simply inspect the numbers 1-2-4-8-16-32-64-128-256-512, etc. Then choose the one that is nearest to 342 but smaller, and this is, of course, 256.

Then, in step 3, subtract 256 from 342 and write the difference. This is a successful subtract, and so a 1 is written in the box opposite the remainder. In step 4, the binary number just smaller than 256, which is 128, is written below the remainder, and again a subtraction is attempted. In this instance, however, 128 will *not* successfully subtract from 86. Therefore, this is not a successful subtraction, and a 0 is written in the box.

Next, the remainder (86) is brought down and the next-lower binary number is subtracted from it; in this case 64 is subtracted from 86. This is a successful operation, and so a 1 is written in the box. This operation is carried out to its conclusion as shown, writing a 1 wherever a successful subtraction is carried out, and a 0 where the operation is unsuccessful. Then, the binary number is written from the MSD toward the LSD, and the binary equivalent of 342 is 101010110. To prove that this is the proper number:

256 128 64 32 16 8 4 2 1
 1   0   1   0   1 0 1 1 0
and
256 + 64 + 16 + 4 + 2 = 342

Thus, the fact that the two numbers are mutually equivalent is proved.

Binary Arithmetic

It is just as feasible to operate arithmetically upon binary numbers as upon decimal numbers. Basically, any arithmetic function that can be performed upon a decimal number can also be performed upon a binary number although perhaps not conveniently. The process of addition and subtraction can be easily performed using binary numbers, and examples of each will subsequently be shown. Fortunately, arithmetic operations beyond addition and subtraction using binary notation, while possible, are seldom, if ever, encountered.

Binary Addition

The process of addition using binary numbers is basically identical to that using decimal notation if allowances are made for the number of digits used in each system. The rules for addition may be summarized as shown in Table 11-2a.

The first example states that 0 plus 0 is 0. Then, 1 plus 0 is 1, and, of course, 0 plus 1 is also 1. The fourth example, however, states that 1 plus 1 results in 0 plus a carry to the next-higher-order column. The fifth example states that a carry (from a lower-order column) plus 0 plus 0 yields 1. Next, a carry plus 1 plus 0 yields 0 plus a carry, with the next example producing the same results. Finally, a carry plus two 1s is 1 plus a carry. Hence, all possible combinations are shown for a two-digit addition.

Table 11-2
Examples of the Binary Addition and Subtraction Rules

a

	(1)	(2)	(3)	(4)
	0	1	0	cr 01
	0	0	1	01
	─	─	─	─
	0	1	1	0

	(5)	(6)	(7)	(8)
	cr 00	cr cr 01	cr cr 00	cr cr 01
	00	00	01	01
	─	─	─	─
	1	0	0	1

b

(1)	(2)	(3)	(4)
			1
			br+
			1
00	01	01	10
−00	−00	−01	−01
───	───	───	───
00	01	00	br1

Several examples of binary addition follow. Each example is proved by converting to decimal as shown.

$$\begin{array}{rl} 0010 =& 2 \\ 0101 =& 5 \\ \hline 0111 =& 7 \end{array} \qquad \begin{array}{rl} 0111 =& 7 \\ 0011 =& 3 \\ \hline 1010 =& 10 \end{array}$$

$$\begin{array}{rl} 0011010 =& 26 \\ 1010111 =& 87 \\ \hline 1110001 =& 113 \end{array} \qquad \begin{array}{rl} 1011010 =& 90 \\ 1111111 =& 127 \\ \hline 11011001 =& 217 \end{array}$$

Binary Subtraction

Again, there is little difference in the process of subtracting either binary or decimal numbers if account is taken of the number of digits involved. The rules of binary subtraction are summarized, as shown in Table 11-2*b*. Each of these can be described in words to emphasize the important points:

Rule 1. This states that 0 minus 0 equals 0.

Rule 2. This states that 1 minus 0 equals 1.

Rule 3. This states that 1 minus 1 equals 0.

Rule 4. This states that 0 minus 1 equals 1 plus a borrow from the next-higher-order column.

Rules 1 to 3 are self-explanatory but rule 4 may require explanation. A digit borrowed from a next-higher-order column, when placed in the lower-order column, has a value greater than before by the radix of the system. Therefore, the borrow results in a 2 in the original column. But the only digits allowed are 0 and 1; there is no digit 2 in the binary system. Hence, the borrowed digit becomes 1 + 1, which of course is a legitimate way to express 2 in binary. Then, the 1 in the subtrahend is subtracted from one of the 1s in the minuend leaving a 1 to be brought down in the answer.

Several examples of binary subtraction follow, and it is recommended that they be studied carefully. Again, each is proved as before.

```
  1101 = 13          1001 = 9
 -1001 =  9         -1000 = 8
 ─────              ─────
  0100 =  4          0001 = 1

    1                   1 1
   +1                   + +
   1+                   0 1 1
   0 0 1
  1̸1̸011 = 27         1̸0010 = 18
 -01111 = 15         -00100 =  4
 ──────             ──────
  01100 = 12          01110 = 14
```

Note that to indicate a borrow a slanted line is drawn through the 1. A 1 through which a slanted line is drawn must be considered 0, of course. In the decimal system we reduce the value of the digit being borrowed from by 1 (borrow 1 from 6 leaves 5), but in the binary system the value of 1 less than the only digit allowed (1) is 0.

Commonly Used Codes

Generally speaking, a code is simply a system of symbols by means of which meaningful communication can be effected. At this time, our main concern is only with numeric codes used in the arithmetic sections of digital paraphernalia.

In the past many coding systems have been used for representing numbers. A code sometimes used is the decimal code, which has the advantage that everyone is familiar with it. While the decimal system of notation is not often thought of as a code, it most certainly is one. It is not used as often as others, but nevertheless it does an admirable job in some instances. Because everyone is already familiar with decimal notation, very little need be said about it. It is used ordinarily when the information to be processed arrives at the processing point in serial form. That is, the electrical form of the digital information is a pulse, and hence the pulses arrive in serial fashion, spread out in time, one after the other.

An example of the use of the decimal system for arithmetic operations is in a typical electronic calculator. This equipment, in one instance, uses a *dynamic memory* system wherein the digital information is kept circling around the memory loop. This loop consists of an acoustic delay line and a set of decimal counters that accept pulses from the line, process them (add, subtract, shift, etc.), and then return them to the memory loop at the proper instant. In this system a decimal 7 is represented by seven pulses; a 3 is represented by three pulses, and so on. The actual coding, then, is very simple.

Another coding method much used is the straight binary code, where all numbers are coded in binary fashion. Hence: 1 is 0001; 7 is 0111; 15 is coded as 1111; and 105 is represented as 1101001. This system has the disadvantage that the large numbers require a great deal of hardware to hold and manipulate these numbers.

To simplify the hardware considerably, a modified combination of both decimal and binary notations is often used. This method is called *binary coded decimal*, or simply BCD. In this system the individual decimal digits are coded in binary notation and are operated upon singly. Thus, the only binary codes allowed are those that represent the decimal digits 0 to 9. A multiple-digit number is written just as the decimal counterpart but with each digit coded in binary. For example, 742 written in BCD is

0111   0100   0010
(7)   (4)   (2)

This system has the advantage that only four elements of hardware are

Table 11-3
The Creeping Code

Decimal	Creeping Code
0	00000
1	00001
2	00011
3	00111
4	01111
5	11111
6	11110
7	11100
8	11000
9	10000
10	(1)00000

Table 11-4
Other Coding Systems

a

Decimal	Binary	Gray Code
0	0000	0000
1	0001	0001
2	0010	0011
3	0011	0010
4	0100	0110
5	0101	0111
6	0110	0101
7	0111	0100
8	1000	1100
9	1001	1101
10	1010	1111
11	1011	1110
12	1100	1010
13	1101	1011
14	1110	1001
15	1111	1000
16	10000	11000

b

Decimal	Octal	Excess-3	Biquinary
0	0000	0011	01–00001
1	0001	0100	01–00010
2	0010	0101	01–00100
3	0011	0110	01–01000
4	0100	0111	01–10000
5	0101	1000	10–00001
6	0110	1001	10–00010
7	0111	1010	10–00100
8	–	1011	10–01000
9	–	1100	10–10000

needed to operate upon the number because, as normally used, each digit is operated upon at different times. A single four-stage register, consisting of four flip-flops, can therefore handle any size number, no matter how many digits, since it need contain only one digit at a time.

Another code occasionally encountered is the *Gray code*. This code is similiar to the binary code in some respects, but the counting progression is quite different. Table 11-4*a* shows this code; note that only *one bit at a time* is changed.

Still another code quite often used is the *Johnson code*, often known as the *creeping code*. It is similiar to the Gray code in that only one bit changes at a time, and this is useful in high-speed applications. Table 11-3 shows this code for several decimal values. Note that this code is particularly well adapted for a scale-of-10 counting arrangement and is often used in a *switch-tail ring counter*. (This counter is described in Chap. 13.)

Many other coding systems are now used or have been used in the past. Among these, are the *octal code*, the *excess*-3 *code*, and the *biquinary code* in addition to many other special-purpose codes used in specific instances. These codes are illustrated in Table 11-4*b*.

11-2 LOGIC SYMBOLOGY

Typically, logic circuitry often seems complex simply because there are so many individual components. In the early days of electronics, a full and complete schematic diagram of, for example, a six- or seven-tube radio could easily be put on paper. Even the schematic of a relatively complex TV set posed no problems. Nearly all electronic technicians were trained in reading and analyzing schematic drawings, and many became quite proficient at this.

Occasionally, one encountered a block diagram of the equipment, but this was usually a simplification used only for introductory purposes to allow a better view of the overall function. In typical nondigital representation, the schematic diagram is essential because each component is different from all others and serves a different purpose. Because there is little, if any, redundancy, the block-diagram approach has little use.

In digital circuitry, however, the opposite conditions prevail. A particular circuit (an inverter, for example) may be used in dozens and dozens of places. Each inverter is identical to all others and serves the same general purpose. Hence, a block-diagram approach is quite sensible and in the case of larger equipment is absolutely necessary. A detailed schematic drawing of a large computer would be so large, or would occupy so many pieces of paper, that its use by a troubleshooter would be virtually impossible. Hence, the need for a simplified representation exists, but it must be sufficiently detailed to be as useful as the schematic.

A drawing of digital equipment, in block form, is called a *logic diagram*. The symbols used to represent the various logic functions are called *logic symbols*. Ideally, the use of a standard set of logic symbols should result in complete simplification. Unfortunately, this ideal has yet to be even approached. No one set of symbols has been universally approved, and reviewing this problem industry-wide, there are nearly as many sets of symbols as there are companies using them. Even within one company, several sets of symbols may be used, leading to great confusion when cross-training of personnel becomes necessary. To illustrate this problem, Fig. 11-3 shows some of the symbols used to implement one function, that

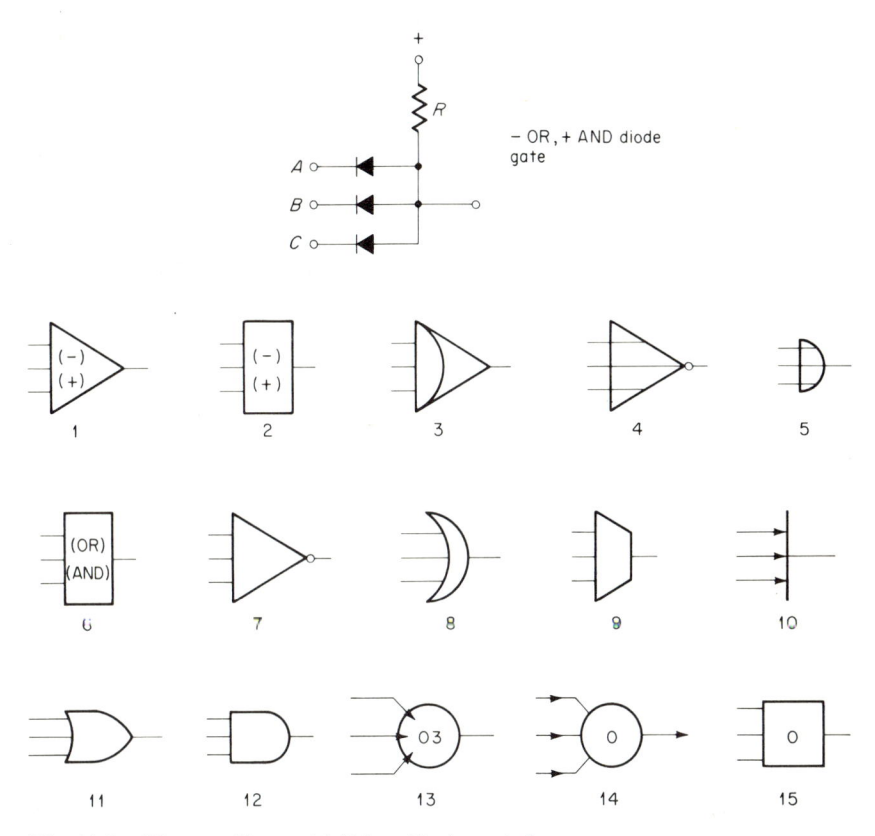

Fig. 11-3. The appalling multiplicity of logic symbols.

of the −OR, +AND function. In each instance, the circuit is identical, but the multiplicity of logic symbols is appallingly evident.

To be sure, a set of standards does exist: Mil-Std 806B. However, of equal importance, the American Standards Association has a different standard (ASA-Y32.14) which is quite different in many respects. Because of the many conflicting ideas, we will use here the Mil-Std 806B symbols since, as of this writing, it seems to be slightly preferred to any others. And even this standard will require slight modification to encompass circuits, or functions, not included in it.

To begin the study of logic symbols, one must first be aware of the function of each circuit to be represented. It will be considered that the reader has no knowledge of this subject. Thus, a detailed description of each of the major functions follows. The main functions fall into four major classes: logic gates, multivibrators, amplifiers, and miscellaneous circuit functions. In this section only the logic functions will be discussed, with the actual circuit implementation discussed elsewhere.

Logical Gates

The AND function is often called the *decision-making* circuit. It provides a means of activating a circuit only with a certain condition existing. Any condition other than the correct one will not allow the circuit to become active. A very elementary AND function is exemplified in Fig. 11-4a1 by two switches (A and B) in series with a source and a lamp. If both switches are open, the lamp is out. If one or the other switch is closed but not both, the lamp will not light. Only when switches A and B are closed will the lamp be energized.

Thus, a switch in the open condition is the inactive condition, while with the two switches closed, the active condition is existent. The switches are two-state devices, or binary devices. Often, the states of binary devices are represented as follows:

0 = inactive state
1 = active state

This allows the simplification of a statement regarding circuit conditions. Rather than writing, "The switch A is open and switch B is closed," it is necessary to write only

$A = 0$ $B = 1$

A truth table can be constructed for the AND function as shown in Table 11-5. The table is interpreted as follows: "The lamp L will light

Table 11-5
AND Truth Table

B	A	L
0	0	0
0	1	0
1	0	0
1	1	1

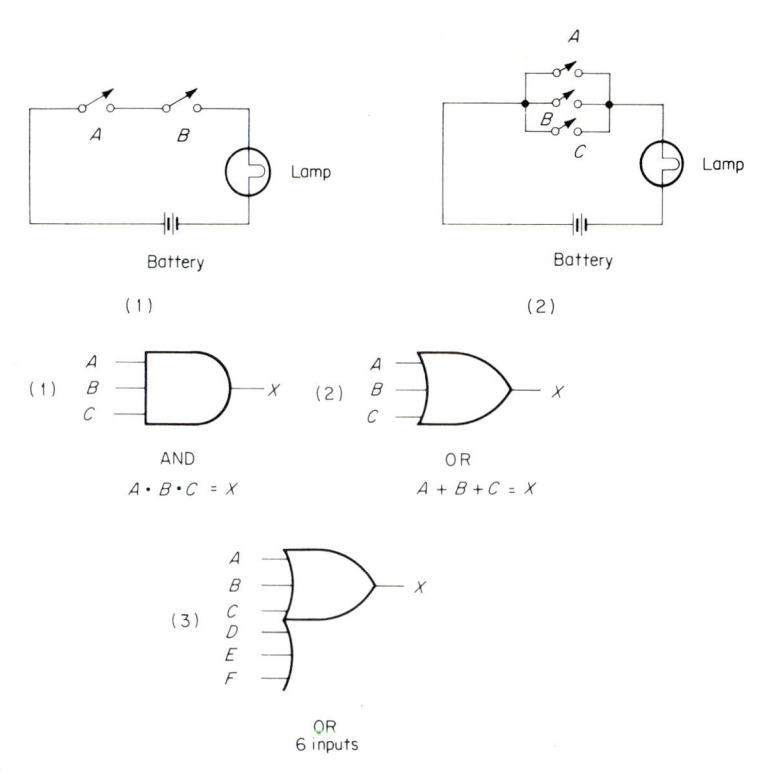

Fig. 11-4. (a) Basic AND/OR functions using switches; (b) the AND/OR logic symbols.

only when switches A AND B are in the active (closed) state." Similar truth tables are used throughout industry to easily describe logic functions of otherwise complex, or involved, circuits.

Figure 11-4b1 illustrates the logic symbol for an AND function. The three wires A, B, and C are inputs, while the single wire X is the output. Without regard to the inner workings at this time, the function of the circuit is explained as follows: "If the inputs, A AND B AND C are active, the output is active. If *any* input is inactive, the output is inactive." This entire idea can be simplified by using the expression $A \cdot B \cdot C = X$, where the dot ($\cdot$) represents the word "AND." The expression implies that a logic 1 appears at A AND B AND C, and therefore a logic 1 also appears at X, the output. That is, the logic expressions *are always written for the logic* 1 *condition at the output.* The truth table for this gating circuit is very similar to the previous one. Note that all possible combinations are shown in Table 11-6.

Table 11-6
Input Combinations for a Three-variable AND Gate

C	B	A	X
0	0	0	0
0	0	1	0
0	1	0	0
0	1	1	0
1	0	0	0
1	0	1	0
1	1	0	0
1	1	1	1

To implement this AND function, the actual circuit is designed to operate on two different voltage, or current, levels. As an example, logic 1 might be +6 volts while ground (0 volts) is logic 0. In another piece of equipment, logic 1 could be −12 volts, while ground (0 volts) is logic 0. The only basic requirement for choosing the two levels is that they are sufficiently different to be easily distinguished by the electronic circuits. That is, using +1 and +2 volts for the two levels might be a poor idea, for the circuitry necessary to distinguish them would be more or less complex and there is always a chance that one might be mistaken for the other. Once the two levels are chosen and assigned their logical meaning, then for this particular equipment they will never change.

As an example, using the AND gate in Fig. 11-4b1, if logic 1 = −12 volts and logic 0 = 0 volts, the output will be −12 volts only if A AND B AND C are at −12 volts. If *any* input falls to ground, the output will also fall to ground.

The OR gate shown in Fig. 11-4b2 is somewhat similar to the AND gate, but the logic function is opposite. With this circuit, a logic 1 at A OR B OR C will yield a logic 1 at the output. Again, a truth table will specify the complete action for any combination of inputs.

Again using the switch analogy, Fig. 11-4a2, if three switches were to be placed in parallel to each other but in series with a source and a lamp, it is easily seen that closing any switch will activate the lamp. If the switches are labeled A, B, C, the entire OR function can be simplified:

$$A + B + C = X$$

where + = OR.

Using the standard abbreviation for OR (+), this reads: "A logic 1 at A OR B OR C yields a logic 1 at X."

Both the AND- and OR-gate logic symbols shown have three inputs. If more than three inputs exist, they are drawn as shown in Fig. 11-4b3, thus avoiding a crowding of the lines.

The symbols in Fig. 11-5 introduce a new idea to the basic AND and OR functions. The NOT function is sometimes difficult to comprehend at first. As an example, consider the gate in Fig. 11-5a. This is called a

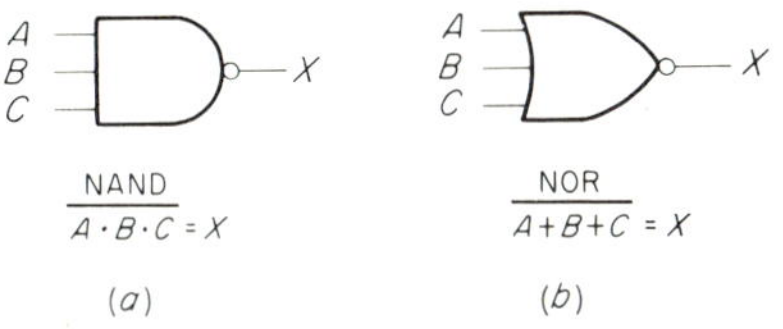

NAND
$\overline{A \cdot B \cdot C} = X$
(a)

NOR
$\overline{A + B + C} = X$
(b)

Fig. 11-5. The NAND/NOR symbols.

NAND gate, being a contraction of NOT-AND. Note the small bubble o at the output wire. As used here, this represents a logic inversion. (This usage is at variance with Mil-Std 806B.)

Logic inversion, as such, is a simple matter to consider. Such a circuit changes a logic 0 at the input to a logic 1 at the output. By the same token, a logic 1 at the input is inverted to become a logic 0 at the output. This is usually implemented by a simple common-emitter-amplifier circuit, used in conjunction with the gate itself.

The NAND circuit, then, is simply the AND gate followed by an inverter. To show that the signals have been inverted at the output, the bar symbol (⁻) is used. Thus, to write the expression for this gate, the entire AND condition for the input is shown with the bar overall. Thus, $\overline{A \cdot B \cdot C} = X$ for the NAND function. This is read: "Not $A \cdot B \cdot C$," or "$(A \cdot B \cdot C)$ barred." This is interpreted as: "If any input is at logic 0, the output will be logic 1. If all inputs are at logic 1, the output will be at logic 0."

To explain this idea further, some voltage levels will be used. Assume that logic 1 is −6 volts and logic 0 = ground (0 volts). The AND gate itself will yield a 1 out if A AND B AND C are at logic 1. However, between the gate output and X there is an inverter. Thus, a 1 at the input to the inverter becomes a 0 at the output.

Now, to obtain a logic 1 at X, certain conditions must prevail at A, B, and C. A logic 1 at X will only occur when the inverter input is at logic 0, or ground, and so the inputs to the gate must be such as to yield logic 0 at the gate output. Because of the fact that any input to an AND gate at logic 0 yields an output of logic 0, this is the input requirement at A or B or C. In terms of voltage: If A is at ground, $X = -6$ volts; if B is at ground, $X = -6$ volts; and if C is at ground, $X = -6$ volts. If all are at -6 volts, the output at X is ground.

Note that the NAND gate has similar characteristics to the OR function. If A is not 1, OR if B is not 1, OR if C is not 1, X will be at logic 1. It will be determined later that any logical gate has a dual function. The AND gate often performs as an OR: the OR gate often performs as an AND, etc. To avoid ghastly confusion, the writing of expressions of logic ($A \cdot B \cdot C = X$, for example) are always written for a *logic* 1 *at X;* then, the proper input conditions to yield $X = 1$ are written.

The NOR gate, shown in Fig. 11-5b, is similar to the NAND gate in certain respects. It consists of a simple OR gate followed by an inverter. As before, the bar symbol is shown in the logic expression for the input condition: $\overline{A + B + C} = X$. This is read: "Not $A + B + C$," or "($A + B + C$) barred." This expression is interpreted as: "If all inputs are at logic 0, the output X will be logic 1. If any input is at logic 1, the output X will be logic 0."

If a logic 1 appears at A OR B OR C, the OR gate output itself will be at logic 1, which will be inverted to logic 0 at X. Hence, to have a logic 1 at X, there must be logic 0 at the inverter input, which requires that A AND B AND C be at logic 0.

Again, note that the NOR gate has similar characteristics to that of the AND function. As mentioned, this duality of gates in general will be discussed in detail later. Our present purpose is simply to become acquainted with the various logic symbols and the logic statements they represent.

A special case of the AND function is the resistor-capacitor gate, sometimes called the *ac gate.* Here, ac simply refers to a voltage change, usually a pulse, at one input. The symbol sometimes used is shown in Fig. 11-6, along with the basic schematic. Because this circuit performs

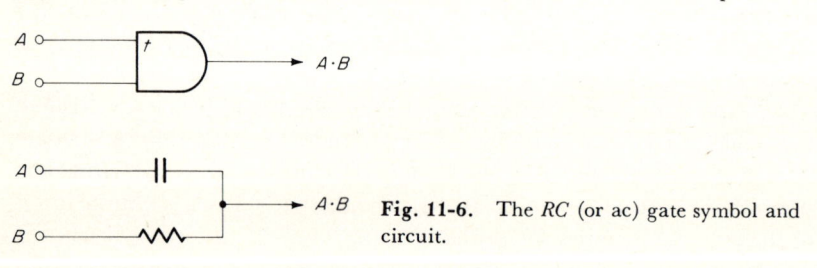

Fig. 11-6. The *RC* (or ac) gate symbol and circuit.

the basic AND function, the AND symbol is used and t is used to denote the *transition* or capacitor input. (This circuit has been briefly discussed earlier.)

Multivibrators

There is probably a greater number of logic symbols representing *bistable multivibrators*, or *flip-flops*, than any other circuit. Although it is true that flip-flops can take several forms, basically they perform identical logical functions. Other multivibrators are the one-shot, the astable, and the Schmitt-trigger.

A bistable multivibrator is a switching circuit, usually composed of two transistors in its basic form, that has two distinct states of being, either of which is stable. That is, when caused to be in one of its two states, it will remain in that state until instructed to change, at which time it reverts to the opposite state. Such a device is obviously useful in a binary-oriented system.

Without regard to the inner workings at this time, the flip-flop states are given names to allow one to be distinguished from the other. Each such circuit has two possible outputs that are evident on two wires. The nature of the flip-flop is such that if a logic 1 is evident on either output wire, logic 0 *must* appear on the other. When the flip-flop is transferred to the opposite state, the wire that was at logic 1 reverts to logic 0 and the wire that was at logic 0 changes to logic 1. A second instruction (usually a pulse) to transfer will result in the flip-flop again reverting to its original condition.

One of these conditions is called the *set*, or *logic 1*, condition, while the opposite condition is called the *clear*, *logic 0*, or *reset*, condition. In the usual case, which condition is set and which is reset is arbitrary on the first choice. But, once this is decided, it is maintained thereafter in the interests of clarity.

Figure 11-7 illustrates the general symbol for a *JK* flip-flop. The two

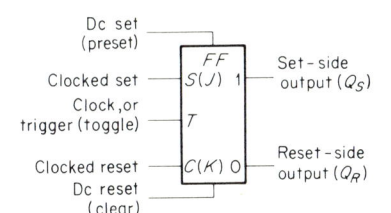

Fig. 11-7. The symbol for the standard *JK* flip-flop with edge-triggered inputs.

output wires are shown on the right and are labeled the "set-side output" and the "reset-side output." They are also known as the 1 and 0 outputs, respectively. If the set-side output Q_S is at logic 1 (and therefore the reset-side is at logic 0), the flip-flop is said to be in the *set condition*, or, simply, *set*. On the other hand, if the reset-side wire Q_R is at logic 1 (and the set-side output at logic 0), the flip-flop is said to be in the *reset condition*. There are no other possibilities for the basic circuit with no input applied; the flip-flop is either set or reset.

The other wires shown (five in number) are the various inputs. The function of each is clearly labeled. A logic 1 at the preset input (dc set) will set the flip-flop. A logic 1 at the clear input (dc reset) will reset the flip-flop. These inputs are simply dc levels and usually act on the flip-flop to the exclusion of the other inputs; that is, they often take precedence over the clocked (or pulsed) inputs, which is a function of the electronic design of the circuit.

The clocked-set input will set the flip-flop with a clock pulse evident at this wire. To reset the flip-flop via a pulse, the clocked reset is used. It should be evident that if a flip-flop is in the set condition, an input (logic 1) at any set input can accomplish nothing. Also, if it is already reset, an input (logic 1) at any reset input does nothing. The set and reset inputs, then, are perfectly straightforward.

The T input, however, operates differently. Regardless of the state of the flip-flop, it is caused to make a transition to the opposite state. If originally set, a pulse at the T input will reset the flip-flop, while if originally reset, a pulse at T will set it. This input is variously called the *clock input*, the *trigger input*, the *toggle input*, the *transfer input*, or the *complement input*.

The basic diagram of Fig. 11-7 is often modified by the omission of any unused (or nonexisting) inputs. Several examples are given in Fig. 11-8. Figure 11-8*a* illustrates a flip-flop with T and C inputs. A pulse at T will complement the flip-flop, while a dc level at C will clear, or reset, it.

Figure 11-8*b* shows a flip-flop that is known as a *latch*. Only dc levels are used as inputs, with each performing its indicated function. While the dc inputs to most flip-flops can perform the latch function, they seldom, if ever, are called upon to do so. The latch function occurs when both the S and C inputs are at the proper dc level. When this occurs, Q_S and Q_R are at the *same* logic level (logic 0 or 1, depending on circuit design). This is the only instance where the outputs of a flip-flop are not complementary.

Figure 11-8*c* shows a complementary flip-flop that requires a pulse, not only at the T input (to complement the flip-flop) but also at the J input (set) and the C input (clear). Still other variations of the basic circuit are possible, with multiple S, C, or T inputs used to actuate the flip-flop for various logic conditions or at different times.

The standard flip-flop of Fig. 11-7 is usually constructed from discrete components. Hence, such items as coupling or differentiating capacitors

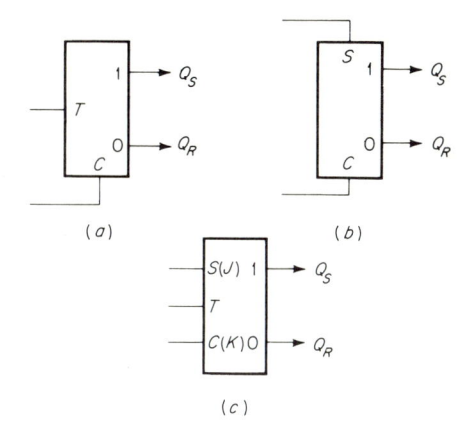

Fig. 11-8. Various combinations of flip-flop inputs.

can easily be designed into the circuit. When it becomes necessary to construct a flip-flop using integrated-circuit techniques, then direct coupling is used because adequate-size capacitors cannot easily be incorporated. To compensate for the inability to provide for capacitively coupled inputs, the circuit functions are implemented differently. Hence, because the flip-flop itself is different, the symbolization is somewhat different. This is not to be construed to mean that capacitors are not formed by integrated techniques. Values on the order of 100 pf are commonly used where necessary, but direct coupling is often more advantageous. (See Chap. 17.)

Figure 11-9a shows one symbol for an IC flip-flop. The set and clear inputs operate on dc levels, but the J input (set) and the K input (clear) operate on dc levels that do not of themselves affect the flip-flop. The truth table also shown indicates the operation of the logic inputs. If both J and K are at logic 0, a pulse at the clock input CK will not alter the setting of the flip-flop. However, if $J = 0$ and $K = 1$, the flip-flop will reset on the clock. If $K = 0$ and $J = 1$, it will set on the clock. (In both cases, if the flip-flop were already in the state toward which the inputs indicate a change, no change occurs, of course. That is, if the flip-flop is already set and a set signal occurs, no change is evident at the flip-flop.)

Finally, if both J and K are at logic 1, the clock pulse will complement the flip-flop and it makes a transition to the opposite state. Using this kind of circuit, it is important to realize that, in effect, CK AND J, or CK AND K, are ANDed. Either or both J or K must be present along with the clock to produce any effect upon the flip-flop. A very simple

212 PULSE AND SWITCHING CIRCUIT ACTION

(a)

t_n		t_n+1
J	K	Q
0	0	Q_n
0	1	Q_R
1	0	Q_S
1	1	$\overline{Q}_n$

Where: t_n = bit time before clock
t_n +1 = bit time after clock
Q_n = FF state before clock
$\overline{Q_n}$ = Q_n complement
Q_R = FF reset
Q_S = FF set

(b)

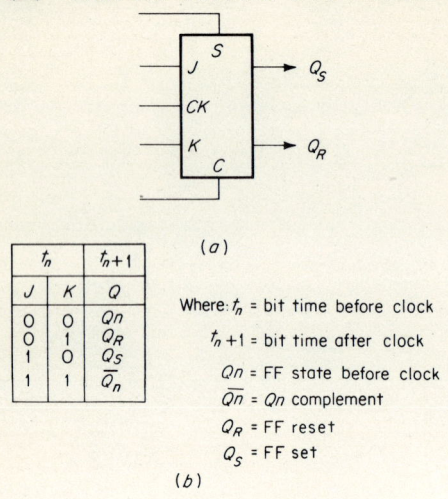

Fig. 11-9. The clocked *JK* flip-flop.

addition to the basic flip-flop allows multiple *J* or *K* inputs. This is shown in Fig. 11-10. It is important to realize that the inputs are ANDed; that is, *J*1 AND *J*2 AND *J*3 must be at logic 1 before a clock allows the flip-flop to set. Also, to reset, *K*1 AND *K*2 AND *K*3 must be at logic 1. Then, the flip-flop resets on the clock. A circuit of this type is very convenient, for the input gates perform much of the logic, leaving fewer external gates necessary. Except for the input-gating, this circuit performs exactly like the previous circuit.

A slightly different function is performed by the *D*-type flip-flop, shown in Fig. 11-11. Here, a logic 1 (dc level) at *D* will, on the clock, set the flip-flop. A logic 0 at *D* will, again on the clock, clear, or reset, the flip-flop. Additional inputs may also be used.

A special form of multivibrator is shown in Fig. 11-12. This is variously known as the *one-shot multivibrator*, or the *single-shot*. Very simply described,

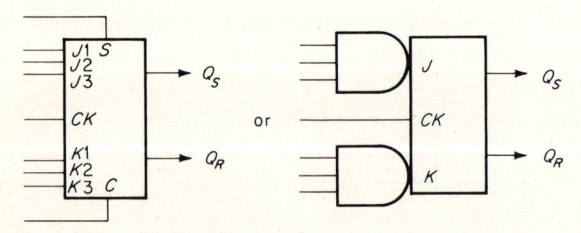

Fig. 11-10. A *JK* flip-flop using input gates operating on the clock pulse.

LOGIC AND LOGIC CIRCUITS   213

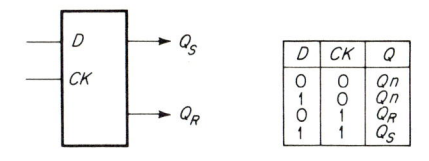

D	CK	Q
0	0	Q_n
1	0	Q_n
0	1	Q_R
1	1	Q_S

Fig. 11-11. The D-type flip-flop.

the one-shot, when triggered with a pulse, produces a pulse at the output, the width of which is determined by the internal constants of the multivibrator. Because the circuit returns to its quiescent state of its own accord, there is no need for a reset input.

The final multivibrator symbol is shown in Fig. 11-13, and this is the Schmitt-trigger circuit. This circuit is responsive to changes in the dc

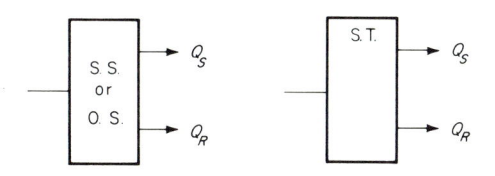

Fig. 11-12. The one-shot multivibrator.

Fig. 11-13. The Schmitt-trigger symbol.

level at the input, the *trip* point being set by circuit design. If the input voltage is more negative than some critical value (the trip point), the flip-flop is in one of its two states, perhaps reset. If then the input goes more positive than the critical value, the circuit makes a transition to the opposite state where it stays as long as the input remains above the trip point. Then, as the input voltage is allowed to go more negative, the circuit returns to its original condition. It is often used as a *squaring* circuit, with a slowly changing input converted to a square wave with very fast rise and fall times.

The final few symbols used for various purposes are shown collectively in Fig. 11-14. Figure 11-14a illustrates the symbols for current or voltage amplifiers, both inverting and noninverting, and the symbol for a time-delay device, such as an LC delay line, an acoustic (or sonic) delay line, or a mercury column. The amount of delay is usually a part of the symbol. The symbol for an oscillator is shown next, and this represents any kind of oscillator, linear or otherwise. An approximate replica of the output waveform is usually shown, along with the frequency. Finally, the electric inverter is a circuit that, for example, converts a negative logic 1 to a positive logic 1. That is, if -12 volts is logic 1 in a given system, occa-

Noninverting   Inverting

Current or voltage amplifier symbol

(a)

Time delay (b)

Oscillator (c)

Electric inverter (d)

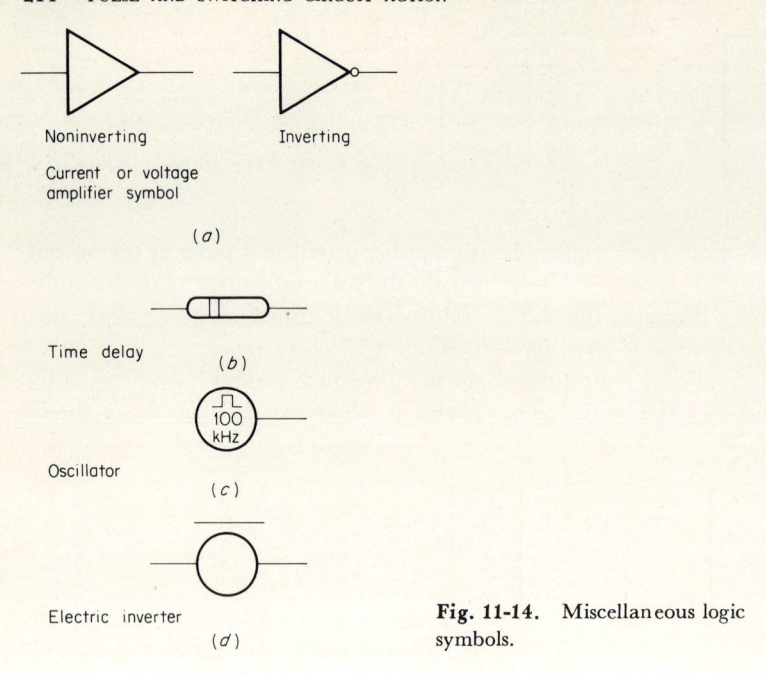

Fig. 11-14. Miscellaneous logic symbols.

sionally it is necessary to change this to, perhaps, +12 volts, which is still to be considered logic 1; the electric inverter accomplishes this kind of function.

11-3 BOOLEAN ALGEBRA

A digital computer performs its separate functions by being able to execute individual steps in sequential order. Each of these separate steps is accomplished by a signal called an *instruction*, which may be composed of many different conditions. Each of the conditions can be described in true-false language.

It will be recalled that in the binary number system there are only two digits, 0 and 1. Boolean algebra is a method of simplifying the notation of circuits and conditions that use the binary number system. This notation is particularly well adapted to circuits having but two discrete states. Any device having only two possible conditions can be represented by Boolean algebra.

Boolean logic is based upon the idea that a statement is either true or false. A true statement has a value of 1, while a false statement has a value of 0. Since digital circuitry uses signals that are of two levels, the symbols 1 and 0 are ideal for use in reference to these circuits. If a signal is false, or 0, it is considered absent, and if a signal is true, or 1, it is considered present. An alternate way of viewing this is to consider the logic level 0 to be a signal to prevent an action while the logic level 1 is the signal that allows the action. Hence, logic 0 is a form of *inhibit* signal while logic 1 is an *enable* signal.

Boolean algebra uses the set of symbols from the preceding section, some of which are already familiar. These symbols are the + sign, the × sign or ·, the parentheses (), and the bar, or not, symbol, as in $\bar{A}$. The + sign is the symbol for OR, the × sign, or the centerline dot · are the symbols for AND, the parentheses are used as signs of grouping, and the bar symbol is used to denote an inverse relationship. Some of these symbols have been used before but have not been used to manipulate logic expressions.

An example of the use of these symbols will clarify their meaning:

$A \cdot B \cdot C = A$ AND B AND C $A + B + C = A$ OR B OR C
$A \times B \times C = A$ AND B AND C $ABC = A$ AND B AND C
$\bar{A}$ = not-A = inverse of A = A-bar

An expression such as $A \cdot B \cdot C = X$ is called a Boolean equation. It has essentially the same meaning as an algebraic equation but relates to a logical circuit rather than just a numerical relation. Such an equation is read: "If a logic 1 is at inputs A AND B AND C, there will be a logic 1 at the output X." As in algebra, the equals sign = denotes that both sides of the expression are mutually equivalent. In order to clearly understand how logic circuits function especially in relation to several other similar circuits, certain rules exist that must always be followed.

Boolean Algebra Laws

There are several laws that pertain to the manipulation of the expressions of Boolean algebra. These must be understood in order to perform the proper operations upon the given expressions or to resolve a given circuit configuration to its Boolean equivalent.

The first of these laws is the *commutative law*, which states that the results of the AND or OR connectives are not affected by the sequence of the logical elements that they connect.

Commutative law:

$A \cdot B = B \cdot A$
$A + B = B + A$

In other words, the logical elements may appear in any order, and the result is the same. As long as the same connective is used, the position, or order, of the logical elements does not affect the result. This can be proved by the diagram for the statement, as shown in Fig. 11-15a. Because of the fact that both signals must exist simultaneously, the order in the written expression is unimportant.

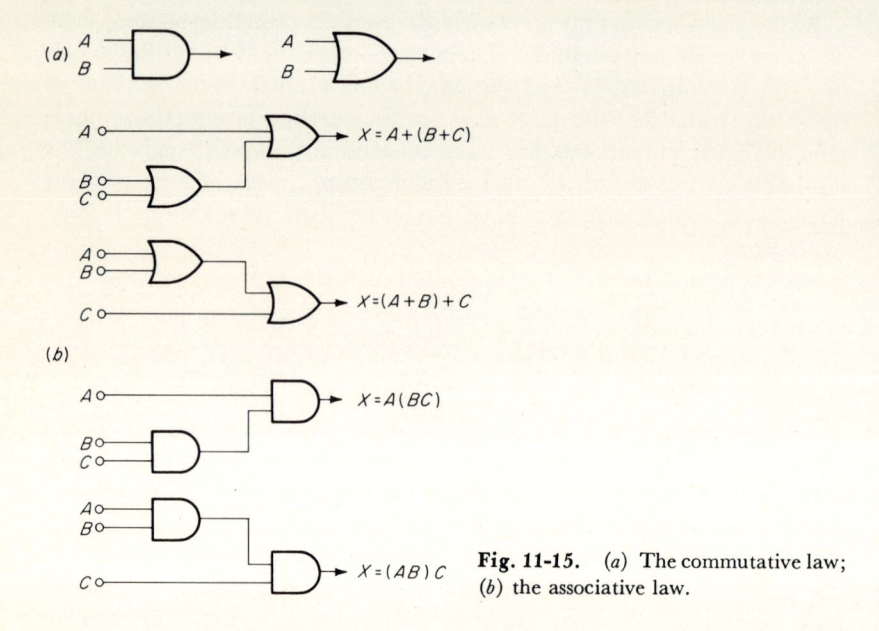

Fig. 11-15. (a) The commutative law; (b) the associative law.

Another law is the *associative law*, which states that the logical elements may be grouped in any combination provided they are connected by the same sign.

Associative law:

$$A + (B + C) = (A + B) + C$$
$$A \cdot (B \cdot C) = (A \cdot B) \cdot C$$

Using AND or OR gates, the implementation of the associative law is shown in Fig. 11-15b.

The *distributive law* is based upon the characteristics of the symbols, or connectives, and can be best explained by example.

Distributive law:

$$AB + AC + AD = A(B + C + D) \quad \text{and} \quad (A + B)(A + C) = A + BC$$

The distributive law is often used to simplify Boolean equations. The functional diagram of the two foregoing equations, Fig. 11-16, will serve to show how the redundant element can be eliminated.

Note particularly that Fig. 11-16a is needlessly complex, while Fig. 11-16b is simpler and still produces exactly the same result. If these particular gates were diode gates, Fig. 11-16a requires thirteen components while Fig. 11-16b needs only nine. Likewise, Fig. 11-16c requires

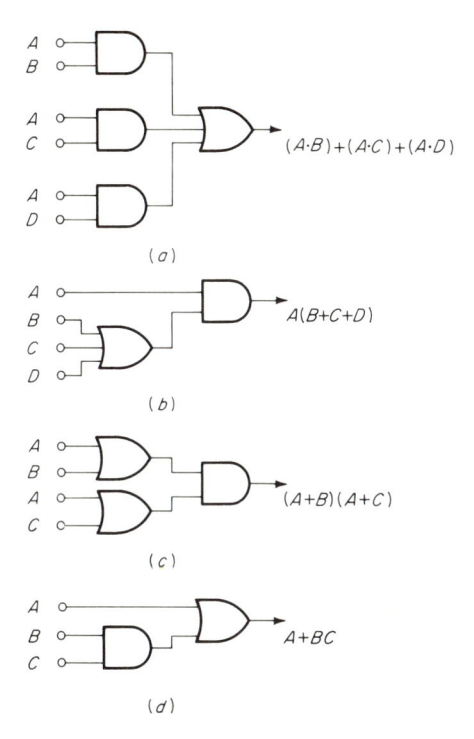

Fig. 11-16. Circuits used to describe the distributive law.

seven components while Fig. 11-16d needs six. By simplifying the expressions, we can often save components and thus eliminate unnecessary cost. Later, simple ways of simplifying logical expressions will be presented.

The *law of NOT* is a singular feature of Boolean algebra that is unique. The NOT function is indicated by a bar over the symbol and it means *the inverse of* the symbol itself. For example, we can say that if a signal appears at a wire called A, there must be no signal on a wire called $\bar{A}$ (A-bar or A-NOT). The symbol itself represents one particular condition while the bar-symbol represents the opposite condition.

The final law to be discussed is often called *DeMorgan's theorem* and describes the method of handling the NOT function when two or more elements appear in the equation.

DeMorgan's theorem:

$$\overline{(A \cdot B)} = \bar{A} + \bar{B}$$
$$\overline{(A + B)} = \bar{A} \cdot \bar{B}$$

Note that a bar over a compound expression *changes or inverts the logic function.* $\overline{A \text{ AND } B}$ becomes $\bar{A}$ OR $\bar{B}$. This must be true since the bar over the entire expression is also over the connective.

$$\overline{A \cdot B} = \bar{A} + \bar{B}$$

The inverse of A is $\bar{A}$, the inverse of AND is OR, and the inverse of B is $\bar{B}$. Thus, an AND function in the original equation, when transposed according to DeMorgan's rules, becomes an OR function. Also, an OR expression barred becomes an AND expression by the same rule.

A few examples will serve to illustrate how the above manipulative rules are related to the actual circuit. Consider Fig. 11-17. When an

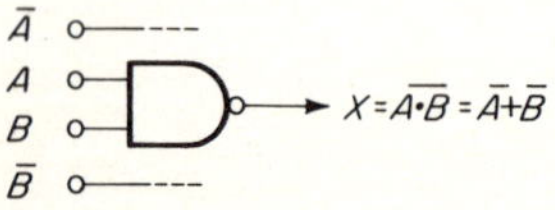

$X = \overline{A \cdot B} = \bar{A} + \bar{B}$

Fig. 11-17. A circuit (NAND gate) to illustrate DeMorgan's law.

inverter is used in connection with a gate, DeMorgan's rules are particularly applicable. In the first example, let us set logic 1 equal to -6 volts and logic 0 to ground (0 volts). Thus, to yield a valid output of the AND gate (input to the inverter), A AND B must be at -6 volts. The circuit output X will then be at ground because of the inverter. To yield a logic 1 at X requires that EITHER A or B be at ground. Thus, the wire labeled $\bar{B}$ has -6 volts while B is at ground. Or, if wire $\bar{A}$ is at -6 volts, A is at ground. It may often occur that there is no real wire to be labeled $\bar{A}$ or $\bar{B}$, but we write the expressions as if there were. In other words, *the expression is always written for the wire where logic 1 is or would be.*

The gate in the first example can be a negative OR or a positive AND gate. In this case, it is acting as a positive OR since a ground (0 volts) at either input results in a logic 1(-6 volts) at the output. Note that the output equation is always written for a logic 1 at this point in the circuit.

A second example is shown in Fig. 11-18, where again logic $1 = -6$ volts and logic $0 =$ ground. The gate itself, shown as an OR gate, must in this case be actually performing an AND function, as indicated by the equations. To yield a logic 1 at X (-6 volts), the inputs A and B must *both* be at ground. Thus, the logic 1 on the input pairs of wires is at $\bar{A}$ AND $\bar{B}$. Six volts at either gate input will result in a logic 1 into the input of the inverter, giving logic 0 at X.

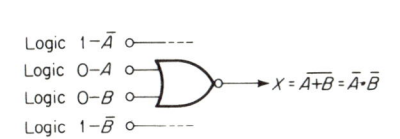

Logic $1-\bar{A}$
Logic $0-A$
Logic $0-B$
Logic $1-\bar{B}$

$X = \overline{A+B} = \bar{A} \cdot \bar{B}$

Fig. 11-18. DeMorgan's law using NOR logic.

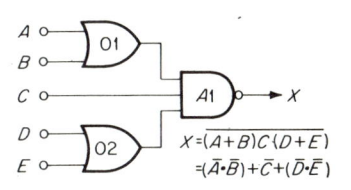

$X = \overline{(A+B)C(D+E)}$
$= (\bar{A} \cdot \bar{B}) + \bar{C} + (\bar{D} \cdot \bar{E})$

Fig. 11-19. Complex logic circuit.

A third example, given in Fig. 11-19, shows a slightly more complex circuit, with the appropriate equations. To write the equation for this circuit, write up to, but not including, the inverter: $(A + B)C(D + E)$. This signal, passing through the inverter, becomes inverted in its entirety, and so a bar is placed over the entire expression:

$$\overline{(A + B)C(D + E)}$$

To expand this equation by DeMorgan's rule, separate each item, properly barred, and invert the connective signs:

$$\overline{(A + B)} \cdot \bar{C} \cdot \overline{(D + E)} = (\bar{A} \cdot \bar{B}) + \bar{C} + (\bar{D} \cdot \bar{E}) = X$$

This is the true equation for the circuit and is really saying this: To have a valid output:

1. There must not be signals A and B.
2. OR, there must not be the signal C.
3. OR, there must not be the signal D and E.

Again, because of the inverter, the gates labeled $O1$ and $O2$ are actually acting as AND gates, as indicated by the final equation. Also the $A1$ gate is really an OR gate. But keep in mind the fact that the $O1$ and $O2$ gates are acting as AND gates for the *absence* of signals, and gate $A1$ is an OR gate for the *absence* of signals.

That some of the foregoing ideas are confusing is easily understood. It may, at first, be of help to realize that it really is of little importance to put a label on a gate. Knowing the voltage levels of the signals and the required level at the output will allow one to see what the input condition

220   PULSE AND SWITCHING CIRCUIT ACTION

must be to effect a valid output. In the foregoing instance, if logic 1 is negative, a negative output will occur when A and B are 0 (ground), C is 0, or D and E are 0, and whether we call the gates AND or OR is of little practical importance.

A final example is shown in Fig. 11-20, which incorporates an inverter

$$X = \overline{(A+B)(C+D)} + E = \overline{A} \cdot \overline{B} + \overline{C} \cdot \overline{D} + E$$

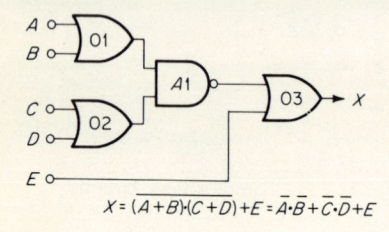

Fig. 11-20. A circuit incorporating an inverter within the system.

within the system of gates rather than at the output. The only significant difference in this set of gates is that the gate $O3$ is not influenced by the inverter. The bar does not extend over the E symbol or its connective ($+$), and hence the signal E is not affected by the inverter.

In the Appendix there appears a summation of 22 Boolean algebra theorems, along with a brief description of each and an illustration showing the actual hardware implementation. The reader should study each of these carefully, for they allow a comprehensive understanding of logic circuits.

Karnaugh Maps

A variation of a truth table, known as a *Karnaugh map*, can be employed as a useful tool in the simplification of logic networks. That this is often necessary will become evident in the following examples. As will be seen, a very real saving can be effected by minimizing the number of components necessary to perform a given logic operation.

First, consider the logic diagram in Fig. 11-21, where a relatively simple logic network is shown. The first step in attempting to simplify this

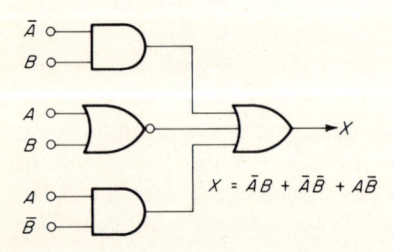

$$X = \overline{A}B + \overline{A}\overline{B} + A\overline{B}$$

Fig. 11-21. A logic circuit for use with the Karnaugh map.

expression is to write the logic expression for logic 1 at the output. Then a Karnaugh map, or truth table, is constructed for a two-variable circuit, as shown in Fig. 11-22. Each block is then filled in only if this is a condition for an output. That is, each block represents one condition for an output; Fig. 11-22b shows a block filled in for the input condition

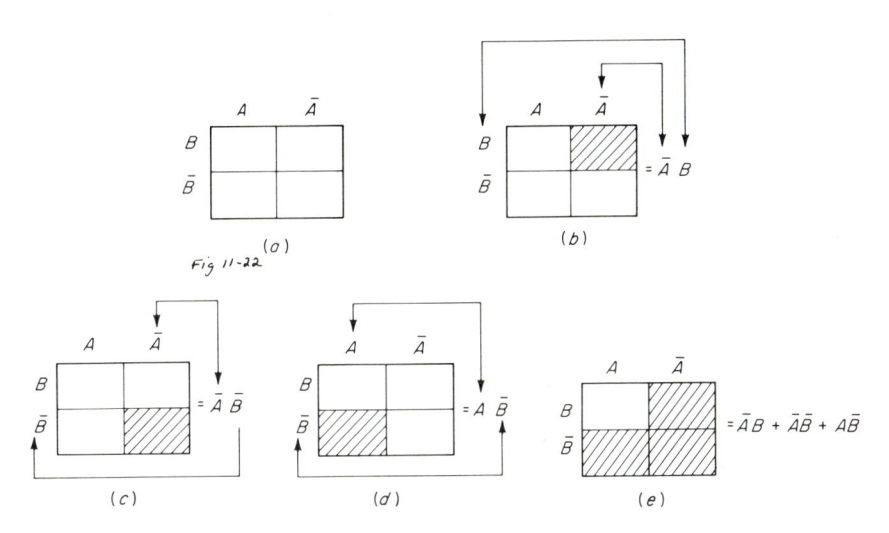

Fig. 11-22. The two-variable Karnaugh map. The circuit example of Fig. 11-21 is entered on the map in (e).

$\bar{A}B$. Likewise, the other two input conditions are shown in Fig. 11-22c and d. All three results are combined in the final form of the Karnaugh map as shown in Fig. 11-22c. In this form, the table allows the original network to be simplified (if, of course, simplification is possible) by removing redundant or needlessly repetitive signals of the form $A + \bar{A} = 1$ and $A\bar{A} = 0$.

To use the truth table, inspect it carefully for *adjacent* blocks in either a vertical or horizontal direction. Write the expression for every two adjacent blocks, first in one direction and then in the other.

In the vertical direction the two adjacent filled-in blocks result in the following equation and its alternate form, derived by use of the distributive law:

$$\bar{A}\bar{B} + \bar{A}B = \bar{A}(B + \bar{B})$$

But, since either B or $\bar{B}$ will always be present (see Appendix, Theorem 8), *this is a redundant term* and can be eliminated, leaving only $\bar{A}$. By inspecting the original expression, note that both B and $\bar{B}$ appear and hence are redundant. In such simple equations the alternate form is often not necessary since the redundant item can be found by inspection if the other element is identical in each expression.

Now, in the horizontal direction, the two adjacent blocks yield the following expression: $\bar{A}\bar{B} + A\bar{B} = \bar{B}(A + \bar{A})$. Since either A or $\bar{A}$ will always occur, $A + \bar{A}$ is eliminated, leaving only $\bar{B}$. Again, this can be determined by inspection in such simple equations (see Appendix, Theorem 7).

Combining the two conditions that were left above yields $\bar{A} + \bar{B}$. The original equation and the simplified form are, then, mutually equivalent:

$$\bar{A}B + \bar{A}\bar{B} + A\bar{B} = \bar{A} + \bar{B}$$

Not only is the logic expression greatly simplified, but the hardware necessary to provide implementation is equally simplified. Figure 11-23

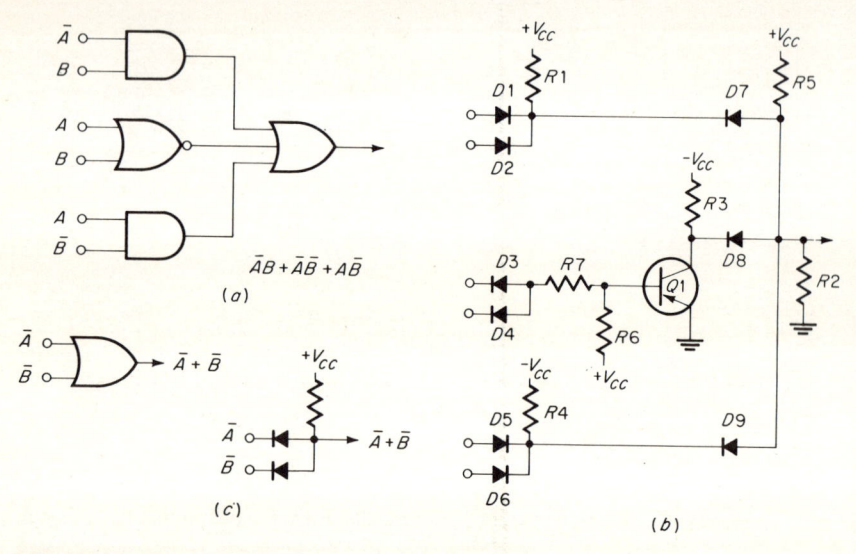

Fig. 11-23. (a) An original logic circuit before simplification; (b) electronic equivalent of the original circuit; (c) the simplified results.

illustrates this point quite clearly, both logically and electronically. Not always is such a drastic reduction in the number of parts possible, but even a small saving can be quite worthwhile.

Using a Karnaugh map for three variables is possible by arranging the table so that any one variable can be combined with all others. Such a table is shown in Fig. 11-24, along with a circuit to be simplified. In this instance, rather than fill in the square, a number is inserted relating to the number adjacent to a part of the output expression. Hence, square 1 represents the first part of the equation $AB\bar{C}$, and so on, for all four parts of the equation.

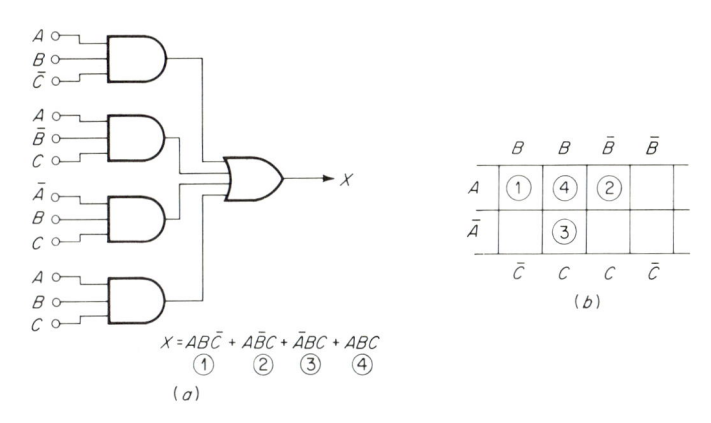

Fig. 11-24. (a) Three-variable circuit before simplification; (b) the equation entered on the map.

To remove the redundancies, look for adjacent blocks; write the parts of the equation represented by the two blocks and simplify by removing the redundant element. In this instance, blocks 1 and 4 yield the following simplification:

$AB\bar{C} + ABC = AB(C + \bar{C}) = AB$

Blocks 4 and 2 yield the following simplification:

$ABC + A\bar{B}C = AC(B + \bar{B}) = AC$

Finally, in the vertical direction, blocks 4 and 3 can be simplified:

$ABC + \bar{A}BC = BC(A + \bar{A}) = BC$

Hence, the simplified form of the original expression is as follows:

$X = AB + AC + BC$

Applying the distributive law further simplifies the result:

$X = AB + AC + BC = A(B + C) + BC$

224   PULSE AND SWITCHING CIRCUIT ACTION

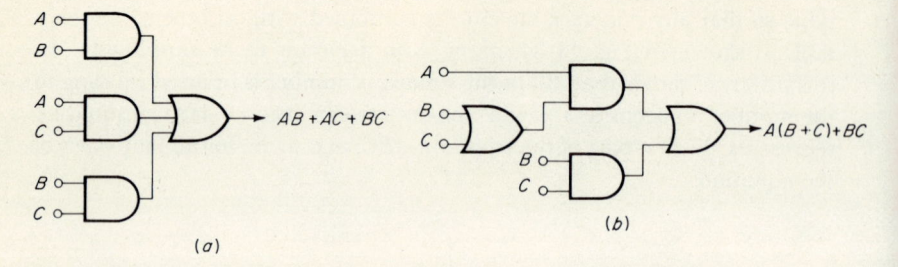

(a) [circuit producing $AB + AC + BC$]

(b) [circuit producing $A(B+C)+BC$]

Fig. 11-25. The circuit in Fig. 11-24 after simplification.

The simplified logic diagram derived from the map appears in Fig. 11-25a and can be seen to be much less complex than the original. Figure 11-25b is only slightly less complex than Fig. 11-25a. Finally, a Karnaugh map for four variables is shown in Fig. 11-26, along with the circuit it represents. Because there are four variables, sixteen squares are required and each must overlap with the other three variables.

The map, as shown, is arranged to simplify the equation $\bar{A}BCD + A\bar{B}CD + ABCD + \bar{A}\bar{B}CD$. Each block that describes a part of the equation is shaded. Block 1 is $ABCD$; block 2 is $A\bar{B}CD$; block 3 is $\bar{A}BCD$; and block 4 is $\bar{A}\bar{B}CD$. Now, blocks 1 and 2 combine to form the following

$X = ABCD + A\bar{B}CD + \bar{A}BCD + \bar{A}\bar{B}CD$
     ①   ②   ③   ④

(a)

(b)

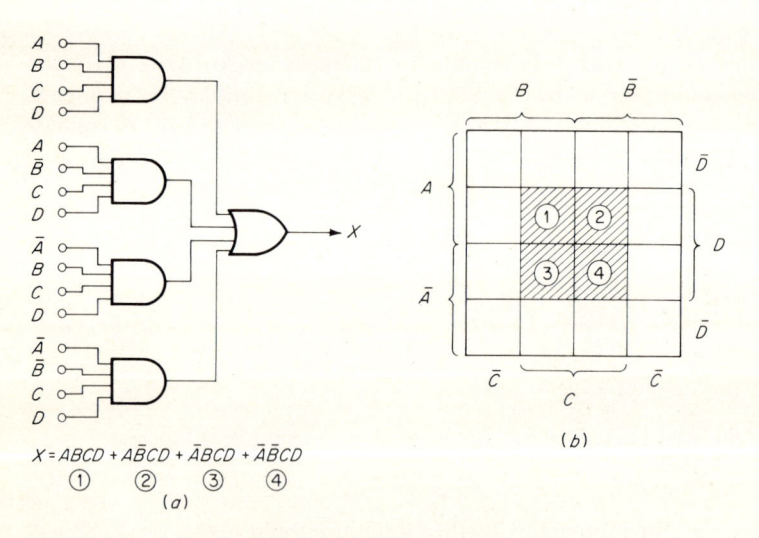

Fig. 11-26. A four-variable circuit with corresponding map.

expression:

$ABCD + A\bar{B}CD = ACD(B + \bar{B}) = ACD$

Obviously either B or $\bar{B}$ must exist at all times and so may be eliminated, leaving ACD. Blocks 3 and 4 form

$\bar{A}BCD + \bar{A}\bar{B}CD = \bar{A}CD(B + \bar{B}) = \bar{A}CD$

Again B or $\bar{B}$ is dropped, leaving $\bar{A}CD$. Blocks 1 and 3 can be expressed as

$ABCD + \bar{A}BCD = BCD(A + \bar{A}) = BCD$

Blocks 2 and 4 can be expressed as

$A\bar{B}CD + \bar{A}\bar{B}CD = \bar{B}CD(A + \bar{A}) = \bar{B}CD$

Combining the first two results,

$ACD + \bar{A}CD = CD(A + \bar{A}) = CD$

and the last two,

$BCD + \bar{B}CD = CD(B + \bar{B}) = CD$ and $CD + CD = CD$

Hence, CD is the simplified form of $\bar{A}BCD + A\bar{B}CD + \bar{A}\bar{B}CD + ABCD$.

Finally, it should be noted that in using the Karnaugh map the ends are to be considered as adjacent also. That is, the blocks to be considered are:

1. Blocks horizontally adjacent
2. Blocks vertically adjacent
3. Blocks on the extreme ends of the same horizontal row or vertical column

Logic Simplification: Alternate Method

When the number of variables to be simplified is greater than four, it becomes quite complicated to use the Karnaugh map. While formats for five- and six-variable maps are available, they are unwieldy and cumbersome owing to the large number of squares needed. Instead, an alternate method exists which will accomplish the same thing and do it much more easily.

The same principle upon which the Karnaugh map operates can be readily applied to logic simplification without need for the map itself. To explain this, it is first necessary to understand what is accomplished by locating horizontally or vertically adjacent blocks on the map. The map itself, because of its organization, allows the comparison of *each* term of a logic equation to each *other* term, looking for combinations that contain *identical parts of a term*, except one.

As an example, consider the two terms $A\bar{B}C\bar{D} + ABC\bar{D}$. Note that parts A, C and $\bar{D}$ are identical in both terms, with only the B part appearing in both the barred (inverted) and unbarred (normal) forms. Because all parts *but one* are identical, with the fourth part (B and $\bar{B}$) appearing in both forms, these terms of the equation *would appear in adjacent blocks on the map*.

As verification, Fig. 11-27 shows this clearly. Also, note that $\bar{A}BCD + \bar{A}\bar{B}\bar{C}D$ are *not* adjacent because *more than one* variable appears

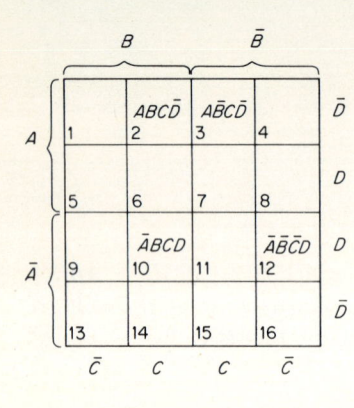

Fig. 11-27. A Karnaugh map verifying that $A\bar{B}C\bar{D} + ABC\bar{D}$ are adjacent and that $\bar{A}BCD + \bar{A}\bar{B}\bar{C}D$ are *not* adjacent.

both barred and unbarred. Once this principle is understood, it is a simple matter to simplify a logic equation by comparing each term with *all* others. The following examples will serve to illustrate the method.

Example 1

Simplify the equation $\bar{A}B\bar{C}D + \bar{A}BCD + \bar{A}BC\bar{D} + \bar{A}\bar{B}\bar{C}\bar{D}$.

Step 1 Assign a number to each term.

$$\bar{A}B\bar{C}D + \bar{A}BCD + \bar{A}BC\bar{D} + \bar{A}\bar{B}\bar{C}\bar{D}$$
$$\ \ (1) \qquad\ \ \ (2) \qquad\ \ (3) \qquad\ \ \ (4)$$

Step 2 Using the following sequence, compare each term with all others: 1 and 2, 1 and 3, 1 and 4, 2 and 3, 2 and 4, 3 and 4.

1 and 2: $\bar{A}B\bar{C}D + \bar{A}BCD = \bar{A}BD(C + \bar{C})$
1 and 3: $\bar{A}B\bar{C}D + \bar{A}BC\bar{D} = \bar{A}BC(D + \bar{D})$
1 and 4: $\bar{A}B\bar{C}D + \bar{A}\bar{B}\bar{C}\bar{D}$ (no change)
2 and 3: $\bar{A}BCD + \bar{A}BC\bar{D}$ (no change)
2 and 4: $\bar{A}BCD + \bar{A}\bar{B}\bar{C}\bar{D}$ (no change)
· 3 and 4: $\bar{A}BC\bar{D} + \bar{A}\bar{B}\bar{C}\bar{D} = \bar{A}C\bar{D}(B + \bar{B})$

LOGIC AND LOGIC CIRCUITS   227

Step 3   Note which of the terms of the original equation took part in a simplification in step 2.   Comparing terms 1 and 2 resulted in a simplification, as did 1 and 3, and 3 and 4.   Hence, all four terms participated in a simplification.   Any term that does *not* take part in a simplification must remain unchanged and will appear unchanged in the final result.

Step 4   Combine all simplifications, including any unchanged terms (none in the example).

$$\bar{A}BD + \bar{A}B\bar{C} + \bar{A}\bar{C}\bar{D}$$

Step 5   Note whether further simplification is possible.   Repeat steps 1, 2, 3, and 4, if possible, for the equation in step 4.   (In the example, none is possible.)

Step 6   If necessary, prove the result by the Karnaugh map.   Referring to Fig. 11-27, the squares numbered 9, 10, 13, and 16 represent the equation in step 1.   Using the map to simplify the equation will result in the same reduced equation as in step 4 above.

Step 7   Applying the distributive law gives the following result.

$$\bar{A}BD + \bar{A}B\bar{C} + \bar{A}\bar{C}\bar{D} = \bar{A}(BD + B\bar{C} + \bar{C}\bar{D})$$

Example 2
Another example using a five-variable equation will now be given.   Simplify the equation $\bar{A}B\bar{C}D\bar{E} + \bar{A}BCD\bar{E} + \bar{A}\bar{B}\bar{C}D\bar{E} + \bar{A}\bar{B}\bar{C}\bar{D}\bar{E}$.

Step 1   Assign a number to each term.

$$\bar{A}B\bar{C}D\bar{E} + \bar{A}BCD\bar{E} + \bar{A}\bar{B}\bar{C}D\bar{E} + \bar{A}\bar{B}\bar{C}\bar{D}\bar{E}$$
$$\ \ (1)\ \ \ \ \ \ \ \ \ \ (2)\ \ \ \ \ \ \ \ \ \ (3)\ \ \ \ \ \ \ \ \ \ (4)$$

Step 2   Compare each term with all others.

1 and 2: $\bar{A}B\bar{C}D\bar{E} + \bar{A}BCD\bar{E} = \bar{A}BD\bar{E}$
1 and 3: $\bar{A}B\bar{C}D\bar{E} + \bar{A}\bar{B}\bar{C}D\bar{E} = \bar{A}\bar{C}D\bar{E}$
1 and 4: $\bar{A}B\bar{C}D\bar{E} + \bar{A}\bar{B}\bar{C}\bar{D}\bar{E} =$ (no change)
2 and 3: $\bar{A}BCD\bar{E} + \bar{A}\bar{B}\bar{C}D\bar{E} =$ (no change)
2 and 4: $\bar{A}BCD\bar{E} + \bar{A}\bar{B}\bar{C}\bar{D}\bar{E} =$ (no change)
3 and 4: $\bar{A}\bar{B}\bar{C}D\bar{E} + \bar{A}\bar{B}\bar{C}\bar{D}\bar{E} = \bar{A}\bar{B}\bar{C}\bar{E}$

Step 3   Terms 1, 2, 3, and 4 all took part in a simplification.

Step 4   Combine all simplified terms.

$$\bar{A}BD\bar{E} + \bar{A}\bar{B}D\bar{E} + \bar{A}\bar{B}\bar{C}\bar{E}$$

Step 5   No further simplification is possible (except to apply the distributive law).   The original equation, using a circuit having a total of 20 parts has been reduced to one having only 12 parts, thus effecting a considerable saving of hardware.

228 PULSE AND SWITCHING CIRCUIT ACTION

To graphically illustrate the advantages of simplification, Fig. 11-28 shows both circuits of Example 2 before and after simplification. The saving of hardware is quite apparent. If these are diode gates, nine diodes and one resistor have been removed and yet the logic function is identical. This method can be used for any number of variables, with the only difference being in the sequence of comparison (step 2).

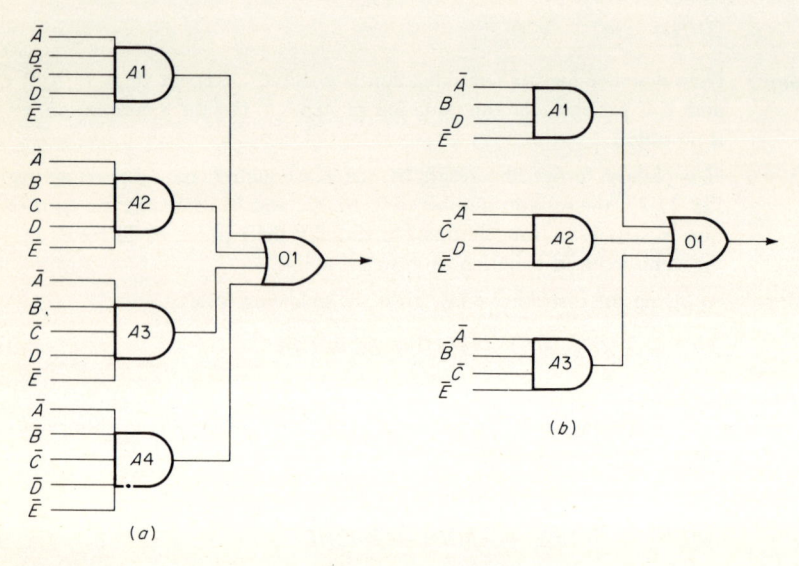

Fig. 11-28. The results of simplification: (a) before; (b) after.

For seven variables, the sequence is (read downward)

1-2	2-3	3-4	4-5	5-6	6-7
1-3	2-4	3-5	4-6	5-7	
1-4	2-5	3-6	4-7		
1-5	2-6	3-7			
1-6	2-7				
1-7					

From this can be derived the sequence for any number of variables.

Example 3

A further example of a more complex equation will now be worked out to show the total method to be used for complete simplification. Simplify the equation

$$ABC DE + \bar{A}BCDE + A\bar{B}CDE + \bar{A}\bar{B}CDE + AB\bar{C}DE + \bar{A}B\bar{C}DE$$
$$\quad (1) \qquad\quad (2) \qquad\quad (3) \qquad\quad (4) \qquad\quad (5) \qquad\quad (6)$$

1 and 2: $BCDE$
1 and 3: $ACDE$
1 and 4: (no change)
1 and 5: $ABDE$
1 and 6: (no change)
2 and 3: (no change)
2 and 4: $\bar{A}CDE$
2 and 5: (no change)
2 and 6: $\bar{A}BDE$
3 and 4: $\bar{B}CDE$
3 and 5: (no change)
3 and 6: (no change)
4 and 5: (no change)
4 and 6: (no change)
5 and 6: $B\bar{C}DE$

Note that all terms participated in at least one simplification, thus none of the original terms need be used. Now, combine all reduced terms again.

$BCDE + ACDE + ABDE + \bar{A}CDE + \bar{A}BDE + \bar{B}CDE + B\bar{C}DE$
 (1)   (2)   (3)   (4)   (5)   (6)   (7)

1 and 2: (no change)
1 and 3: (no change)
1 and 4: (no change)
1 and 5: (no change)
1 and 6: CDE
1 and 7: BDE
2 and 3: (no change)
2 and 4: CDE
2 and 5: (no change)
2 and 6: (no change)
2 and 7: (no change)
3 and 4: (no change)
3 and 5: BDE
3 and 6: (no change)
3 and 7: (no change)
4 and 5: (no change)
4 and 6: (no change)
4 and 7: (no change)
5 and 6: (no change)
5 and 7: (no change)
6 and 7: (no change)

Again, all terms participated, so simply rewrite the reduced expressions.

$CDE + BDE + CDE + BDE$

Hence, since two of the four parts are identical to the other two, the final expression is

$$CDE + BDE = DE(C + B)$$

The great reduction in input requirements is caused by the fact that the original equation had very nearly the maximum number of redundancies possible. Such a contrived situation probably would not frequently exist in practice, although the next problem is a prime example of such an instance.

Example 4

As a final example of logic simplification using this method, we shall consider a practical application that could very well be actually used. The circuit shown in Fig. 11-29 is a *binary counter*, which has not been discussed up to this point. Chapter 13 deals with counters in detail, and at this time only a brief introduction is necessary.

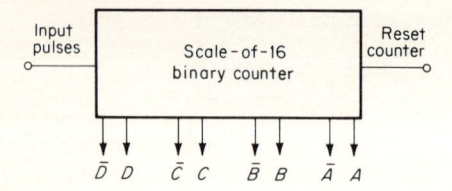

Fig. 11-29. A scale-of-16 counter showing the output wires.

A binary counter is a circuit consisting of several bistable multivibrators that, when suitable pulses are present at the input, advances the state of the output wires in binary fashion. By connecting AND gates to the output wires, a given condition can be sensed so as to know when the given condition has been reached by the counter.

The reset condition of the counter is with logic 1 at $\bar{A}\bar{B}\bar{C}\bar{D}$; logic 0 therefore is at $ABCD$. The state of the counter can be specified by stating the output wires that have a logic 1 on them. The following tabulation shows all possible conditions of the counter:

Counter State	Output Wires
Reset, or 0 =	$\bar{A}\bar{B}\bar{C}\bar{D}$
1 =	$A\bar{B}\bar{C}\bar{D}$
2 =	$\bar{A}B\bar{C}\bar{D}$
3 =	$AB\bar{C}\bar{D}$
4 =	$\bar{A}\bar{B}C\bar{D}$
5 =	$A\bar{B}C\bar{D}$

$$6 = \bar{A}BC\bar{D}$$
$$7 = ABC\bar{D}$$
$$8 = \bar{A}\bar{B}\bar{C}D$$
$$9 = A\bar{B}\bar{C}D$$
$$10 = \bar{A}B\bar{C}D$$
$$11 = AB\bar{C}D$$
$$12 = \bar{A}\bar{B}CD$$
$$13 = A\bar{B}CD$$
$$14 = \bar{A}BCD$$
$$15 = ABCD$$
$$16, \text{reset, or } 0 = \bar{A}\bar{B}\bar{C}\bar{D}$$

If, for example, the reset state of the counter must be known to develop a signal for use elsewhere that is at logic 1 only when the counter is in the reset state, the AND gate of Fig. 11-30 would be used. The input wires would simply

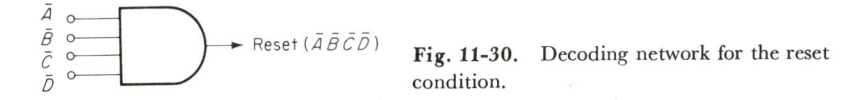

Fig. 11-30. Decoding network for the reset condition.

be connected to $\bar{A}$ AND $\bar{B}$ AND $\bar{C}$ AND $\bar{D}$, and only when the counter is in the reset condition will the AND gate have a logic 1 output. Any, or all, of the other possible conditions can be similarly sensed by using a separate gate for each condition. The gate inputs, then, will connect to the appropriate counter output wires for the specified condition.

With this brief description of circuit operation as a basis for further discussion, let us proceed to the logic simplification problem concerning this circuit arrangement. Assume that when the counter has advanced to the 11th, 12th, 13th, 14th, or 15th states, a signal must be developed to so indicate. That is, as long as the counter is at a condition representing 0, 1, 2, 3, 4, 5, 6, 7, 8, 9, or 10, the added gating circuit must yield a logic 0 out. But, when the 11th to 15th states are attained, the gating network must yield a logic 1.

The most straightforward way of accomplishing this is indicated in Fig. 11-31. The output, labeled X, is a signal that, in effect, indicates that the counter's condition is greater than 10 but less than 16 (or reset). This gating system would function as desired, but if these are to be diode gates they would require a total of 31 components: 24 diodes and 6 resistors.

To verify that there may be certain redundancies in the original expression it will now be subjected to the simplification process.

Step 1   The original logic equation is

$$\underset{(1)}{AB\bar{C}D} + \underset{(2)}{\bar{A}\bar{B}CD} + \underset{(3)}{A\bar{B}CD} + \underset{(4)}{\bar{A}BCD} + \underset{(5)}{ABCD}$$

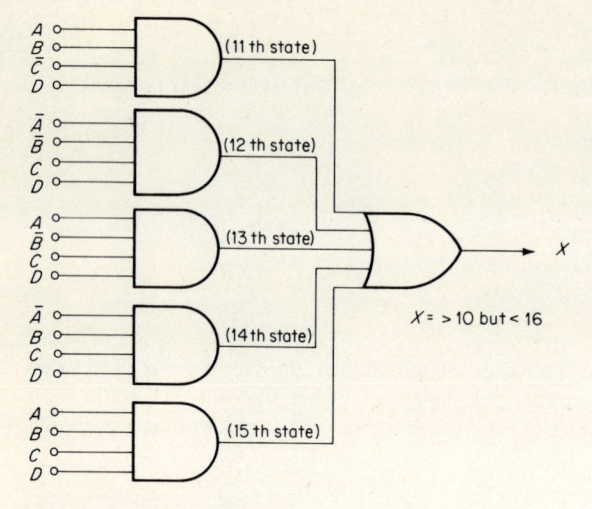

Fig. 11-31. Decoding network for the 11th to 15th states.

Step 2 Compare each term with all others.

  1 and 2: no change
  1 and 3: no change
  1 and 4: no change
  1 and 5: *ABD*
  2 and 3: $\bar{B}CD$
  2 and 4: $\bar{A}CD$
  2 and 5: no change
  3 and 4: no change
  3 and 5: *ACD*
  4 and 5: *BCD*

Step 3 Rewrite (all terms participated).

  $ABD + \bar{B}CD + \bar{A}CD + ACD + BCD$
  (1)   (2)   (4)   (4)   (5)

Step 4 Further reduce if possible.

  1 and 2: no change
  1 and 3: no change
  1 and 4: no change
  1 and 5: no change
  2 and 3: no change
  2 and 4: no change
  2 and 5: *CD*

3 and 4: CD
3 and 5: no change
4 and 5: no change

Step 5 Again rewrite. [Note that term 1 above (ABD) did not participate.]

$ABD + CD + CD$

Obviously, one CD term is unnecessary. The simplified result is therefore

$ABD + CD$

The necessary hardware to implement this is illustrated in Fig. 11-32, where the number of components that have been eliminated is quite clear.

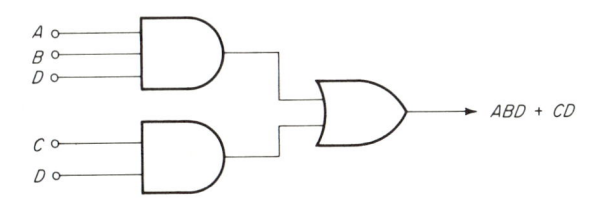

Fig. 11-32. The decoding network in Fig. 11-31 after simplification.

ORed AND Gates

The Karnaugh map is usually applied to AND gates that are ORed together. The reverse situation is just as feasible, of course. A circuit of OR gates that are ANDed is given in Fig. 11-33. This circuit contains certain redundancies that will be removed.

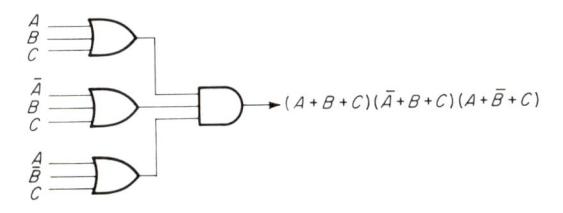

Fig. 11-33. An example of ANDed OR gates.

The Karnaugh map operates on the principle that a certain combination of signals, such as $A + \bar{A}$, always equals 1. The simplified system previously shown, which does not use the map, also is based upon this premise. In the case of a series of OR gates that are ANDed, the principle is somewhat different. An unnecessary signal appearing at the input represents a condition that can *never* appear, such as $A \cdot \bar{A}$. Obviously, these signals can never appear simultaneously and so can be eliminated.

In the circuit in Fig. 11-33, the process of simplification is identical to that given previously. The only difference is the *reason* for eliminating unnecessary signals. Note that such a signal cannot be termed *redundant* in this instance since it is not unnecessarily repetitious.

There are several ways to simplify this circuit, two of which will be given. The first method makes use of the fact that a Boolean equation can be inverted without losing its identity. Hence, the first step is to invert the entire expression using DeMorgan's theorem. This results in a configuration that can easily be operated upon by previous methods.

Step 1   Invert the entire expression.

$$\overline{(A + B + C)(\bar{A} + B + C)(A + \bar{B} + C)} = \bar{A}\bar{B}\bar{C} + A\bar{B}\bar{C} + \bar{A}B\bar{C}$$

Step 2   Simplify (simplified method or Karnaugh map).

$$\bar{B}\bar{C} + \bar{A}\bar{C}$$

Step 3   Reinvert to restore original circuit.

$$\overline{\bar{B}\bar{C} + \bar{A}\bar{C}} = (B + C)(A + C)$$

Hence, the circuit is reduced to a much simpler configuration. However, this can be further reduced, which will be done following the next discussion.

The second method consists in the comparison of each set of input conditions, looking for an impossible set of signals.

Step 1
$$(A + B + C)(\bar{A} + B + C) = B + C(A \cdot \bar{A}) = B + C$$
Step 2
$$(A + B + C)(A + \bar{B} + C) = A + C(B \cdot \bar{B}) = A + C$$
Step 3
$$(\bar{A} + B + C)(A + \bar{B} + C) = (\bar{A} + B + C)(A + \bar{B} + C)$$

Because all three groups, or terms, took part in the simplification, step 3 is ignored and the two results are combined.

Step 4

$$(A + B + C)(\bar{A} + B + C)(A + \bar{B} + C) = (A + C)(B + C)$$
$$= AB + BC + AC + CC$$

This can be further reduced by the application of the basic theorems. Note that C is common to three of the terms. Hence, this can be written as follows, recognizing that $CC = C$:

$AB + C(A + B)C$

Because C must be present in any case, except for AB, the $(A + B)$ term accomplishes nothing. The final expression, then, is

$AB + C$

Each of these methods is closely related and will produce a valid simplification.

Simplification of Complex Logic

In practical circuitry, it is seldom the case that all OR gates or all AND gates comprise the input section. Usually, AND and OR functions are intermixed, and this slightly complicates the simplification of the network. It becomes necessary, therefore, to formulate a simplification method for any combination of AND or OR functions.

A single group of OR gates is reduced (or simplified) by the use of several basic theorems. Consider the circuit given in Fig. 11-34. In

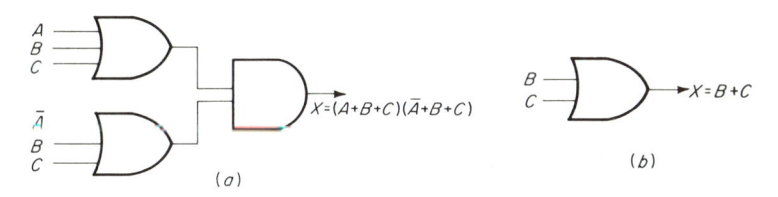

Fig. 11-34. (a) ANDed ORs to be simplified; (b) result of simplification.

this instance, the unnecessary element $A \cdot \bar{A}$ is always equal to 0 and therefore can be eliminated. A method of proving this relationship is to logically sum the two expressions:

$$\frac{\begin{array}{c} A + B + C \\ \bar{A} + B + C \end{array}}{0 + B + C} \text{ logical sum (ANDed)}$$

Any signal that, in such combination, results in a 0 can be eliminated, leaving only $B + C$. The simplified circuit is shown in Fig. 11-34b.

In more complex networks, the redundant, or unnecessary, element (or elements) is not as clearly evident. The circuit given in Fig. 11-35 is

236 PULSE AND SWITCHING CIRCUIT ACTION

[Circuit diagram with inputs $\bar{A}, B, C, D$; $\bar{A}, B, C, \bar{D}$; $\bar{A}, \bar{B}, C, D$; $\bar{A}, \bar{B}, \bar{C}, \bar{D}$; $\bar{A}, B, C, \bar{D}$ feeding four AND gates into an OR gate (31 components)]

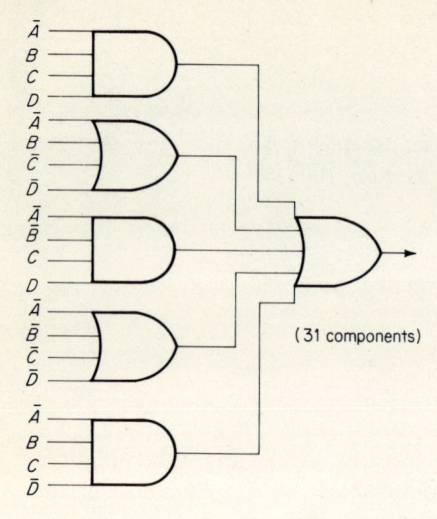

Fig. 11-35. A circuit to implement a complex AND/OR equation.

such a network. To determine if any redundancies exist and therefore remove them, the following procedure can be applied.

Step 1   Convert the OR gates to equivalent AND gates by inverting all ORed signals by DeMorgan's theorem.

$$(\bar{A}BCD) + (\overline{\bar{A} + B + \bar{C} + \bar{D}}) + (\bar{A}\bar{B}CD) + (\overline{\bar{A} + \bar{B} + \bar{C} + \bar{D}}) + (\bar{A}BC\bar{D})$$

$$= (\bar{A}BCD) + (A\bar{B}CD) + (\bar{A}\bar{B}CD) + (ABCD) + (\bar{A}BC\bar{D})$$

Step 2   Simplify by the usual means.

$$\bar{A}CD + \bar{A}BC + \bar{B}CD$$

Step 3   Check all input signals for availability. (Because of the inverting process above, there is a possibility that a signal will appear that is not available at the input to the original circuit.)

     Available   Not Available
     $\bar{A}$ A
     $\bar{B}B$
     $\bar{C}C$
     $\bar{D}D$

The signal A does not appear in step 2. If it did, that term would have to be reinverted or an inverter placed at the appropriate input, whichever is most economical in terms of component cost.

LOGIC AND LOGIC CIRCUITS   237

Step 4   Perform a logical summation of the two original OR-gate inputs.

$$\bar{A} + B + \bar{C} + \bar{D}$$
$$\bar{A} + \bar{B} + \bar{C} + \bar{D}$$
$$\overline{\bar{A} + 0 + \bar{C} + \bar{D}}$$

The *B* part can be removed.

Step 5   Add $\bar{A}$ and $\bar{D}$ to the result given in step 2. This results in the simplified circuit and is shown in Fig. 11-36.

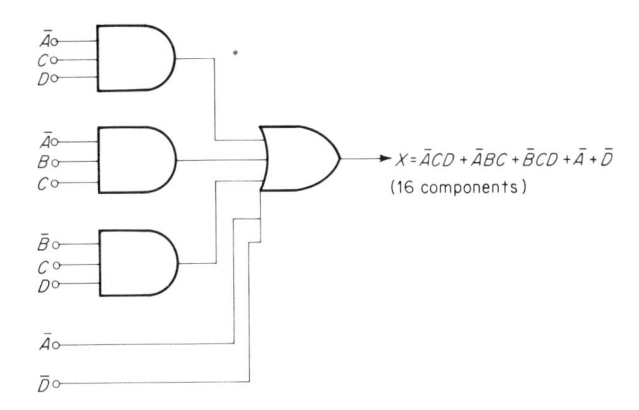

$$X = \bar{A}CD + \bar{A}BC + \bar{B}CD + \bar{A} + \bar{D}$$
(16 components)

Fig. 11-36. Final result of simplifying the complex AND/OR equation.

A final example using NAND/NOR gates is shown in Fig. 11-37. This circuit is simplified much as before by applying the rules in a logical sequence.

Step 1   Write the output equation.

$$\overline{(\overline{ABC})(\overline{\bar{A} + \bar{B} + \bar{C}})(\overline{A\bar{B}C})} = (ABC) + (\bar{A} + \bar{B} + \bar{C}) + (A\bar{B}C)$$

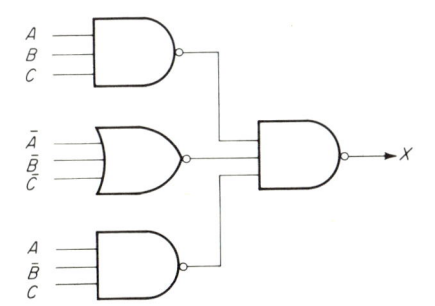

Fig. 11-37. NAND/NOR network simplification problem.

Step 2   Invert the OR function.

$$ABC + ABC + A\bar{B}C$$

Step 3   Remove one *ABC* term.

$$ABC + A\bar{B}C$$

Step 4   Simplify

$$AC(B + \bar{B}) = AC$$

The simplest form, then, is *AC*. This is implemented as shown in Fig. 11-38, using the appropriate NAND/NOR components.

$\bar{A}$ ○———⟫○——→ $X = \overline{(\bar{A} + \bar{C})} = AC$ **Fig. 11-38.** NAND/NOR simplification results.
$\bar{C}$ ○———

11-4   LOGIC CIRCUIT DESIGN

As a final example of the principles set forth in this chapter, two circuits will be designed, which not only will allow a reiteration of foregoing principles and methods, but also will yield a good understanding of this particular class of circuits. These will be the half-adder and full-adder circuits.

An adder is a circuit that, in conjunction with other circuitry, allows a digital computer to perform arithmetic. Such a circuit allows one binary bit to be summed with another to produce a result, or summation. If the principles of binary addition are not remembered fully, it is recommended that the reader review Sec. 11-1.

To illustrate the action of a simple half-adder circuit, consider Fig. 11-39. Figure 11-39*a* illustrates binary addition as would be done by hand. There are two methods of accomplishing this, called *parallel-bit summation* and *serial-bit summation*. Using parallel-bit summation, all four bits are acted upon at the same time, with the sum immediately evident at the four output wires, as suggested by Fig. 11-39*b*. On the other hand, with serial-bit addition (Fig. 11-39*c*), the bits are summed in sequence, with the least significant digits summed first, etc. This requires, of course, that the bits that have already been summed must be stored in some device until all additions are completed. Then, the complete sum, with any appropriate carries, is available at the storage device (usually a shift register, see Chap. 13).

To keep the discussion as simple as possible, only serial-bit addition

LOGIC AND LOGIC CIRCUITS   239

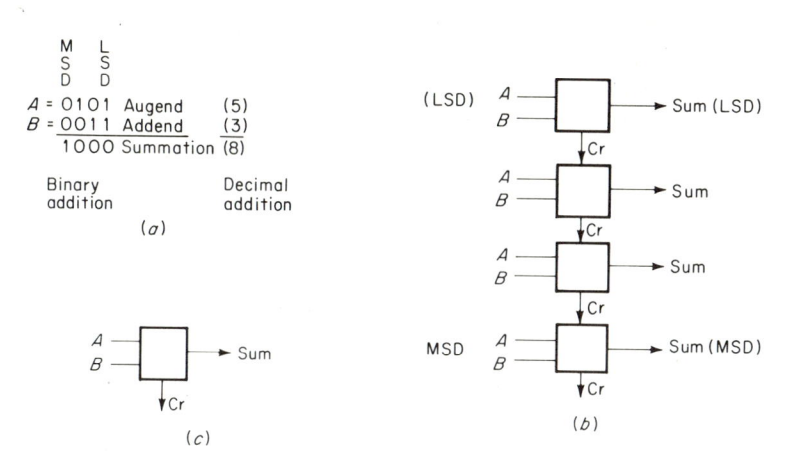

```
      M    L
      S    S
      D    D
A  = 0101  Augend      (5)
B  = 0011  Addend      (3)
     1000  Summation   (8)

     Binary            Decimal
     addition          addition
              (a)
```

(b) Parallel diagram:
- (LSD) A, B → box → Sum (LSD), Cr down
- A, B, Cr → box → Sum, Cr down
- A, B, Cr → box → Sum, Cr down
- MSD A, B, Cr → box → Sum (MSD)

(c) A, B → box → Sum, with Cr down

Fig. 11-39. The half-adder function. (*a*) Binary addition; (*b*) parallel-bit summation; (*c*) serial-bit summation.

will be considered. Thus, the bits shown in Fig. 11-39*a* will be used to verify that the circuit we shall design will indeed produce a correct sum. Each group will be summed individually, starting with the least significant bits. First, however, the half-adder will be designed, followed by the full-adder, and it is the latter that is used to construct a complete adder. The reason for this is simply that the half-adder has no provision for accepting a binary carry from the preceding bits, whereas the full-adder does have this provision. As evident in the example, several binary carries are produced, hence the final circuit must have this provision.

To begin the design of the half-adder, some convenient method must be found to specify the requirements of such a circuit. Recalling the rules of addition from Sec. 11-1, a truth table can be constructed such as shown in Fig. 11-40. The two inputs A and B represent the binary bits to be

Input		Output		Input
A	B	S	C_o	Equation
0	0	0	0	$\bar{A}\bar{B}$
1	0	1	0	$A\bar{B}$
0	1	1	0	$\bar{A}B$
1	1	0	1	AB

Fig. 11-40. Truth table for the half-adder.

summed. If both bits are at logic 0, no sum is to be produced, of course, as well as no carry-out C_o. When either bit is at logic 1, a sum must be produced but no carry-out. However, if *both* bits are at logic 1, then the sum output must be zero but a carry must be produced. (This carry must be stored for use when the next set of bits arrives at the input, which must be able to accommodate it.)

The input equation specifies the condition of the inputs that is required for the specified output. Simply drawing these as AND gates, properly ORed together, will result in a usable circuit, shown in Fig. 11-41a. A

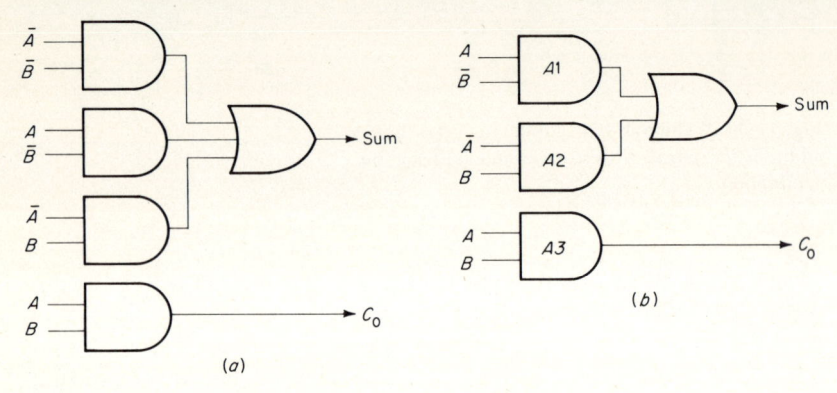

Fig. 11-41. (a) Half-adder designed from the truth table; (b) simplified half-adder circuit.

little thought will reveal that the gate with the input $\bar{A}\bar{B}$ is unnecessary since with both A and B at zero the other two gates will yield the required zero out. Hence, the simplification shown in Fig. 11-41b is the final circuit.

To verify the action of such a circuit, assume all possible input conditions and note the outputs. If the input condition is $\bar{A}\bar{B}$, no sum results and hence no sum or carry-out is produced. If the signal $A\bar{B}$ exists, gate $A1$ results in a 1 at the sum output, with no C_o. Also, the same output condition occurs for $\bar{A}B$. Finally, if the input bits are both at logic 1 (AB), the sum output is zero but the gate $A3$ yields a 1 at C_o. (Note that the output C_o must be separate from the sum output.) This circuit, then, performs the function of addition if there is no carry at the input from a preceding summation.

To allow for the possibility of a previously generated binary carry at the input C_i, the full-adder circuit is used. The principles involved in building up such a circuit are similar to those just used for the half-adder. The first step consists in constructing the truth table, including the input carry C_i and the output carry C_o. As before, the rules for producing the sum are given in Sec. 11-1 and provide the information necessary to construct the truth table. The truth table for the full-adder is given in Fig. 11-42, where it is evident that there are now three input conditions

Input			Output		Output Equation
A	B	C_i	S	C_o	
0	0	0	0	0	$\bar{A}\bar{B}\bar{C}_i$
1	0	0	1	0	$A\bar{B}\bar{C}_i$
0	1	0	1	0	$\bar{A}B\bar{C}_i$
1	1	0	0	1	$AB\bar{C}_i$
0	0	1	1	0	$\bar{A}\bar{B}C_i$
1	0	1	0	1	$A\bar{B}C_i$
0	1	1	0	1	$\bar{A}BC_i$
1	1	1	1	1	ABC_i

Fig. 11-42. Truth table for the full-adder.

A, B, and C_i. The equations are also given, and simply selecting the equations that result in the sum output equal to 1 will yield a part of the circuit. The equation for the sum output S is, then,

$$S = A\bar{B}\bar{C}_i + \bar{A}B\bar{C}_i + \bar{A}\bar{B}C_i + ABC_i$$

A moment's reflection will verify the validity of this equation. A logic 1 at A OR B OR C_i should result in a 1 at the sum output, as should a 1 at $A \cdot B \cdot C_i$. These are the conditions according to the rules for binary addition for a logic 1 at S.

Now, to yield a signal at the carry-out output, the table is again inspected and the equation built up of the necessary conditions:

$$C_o = AB\bar{C}_i + A\bar{B}C_i + \bar{A}BC_i + ABC_i$$

As before, these can be verified as to their validity.

Now, to implement these equations, simply draw out each one in terms of the proper AND or OR functions specified. Figure 11-43 shows the completed circuit. To use this circuit for the original problem illustrated in Fig. 11-39a, apply the inputs in sequential fashion, starting at the least significant digit. A 1 at A AND B results in the condition $AB\bar{C}_i$. Thus gates $A1$ to $A4$ yield no output, and $S = 0$. However, gate $A7$ gives an output, and so $C_o = 1$. This, of course, is a correct summation. Now,

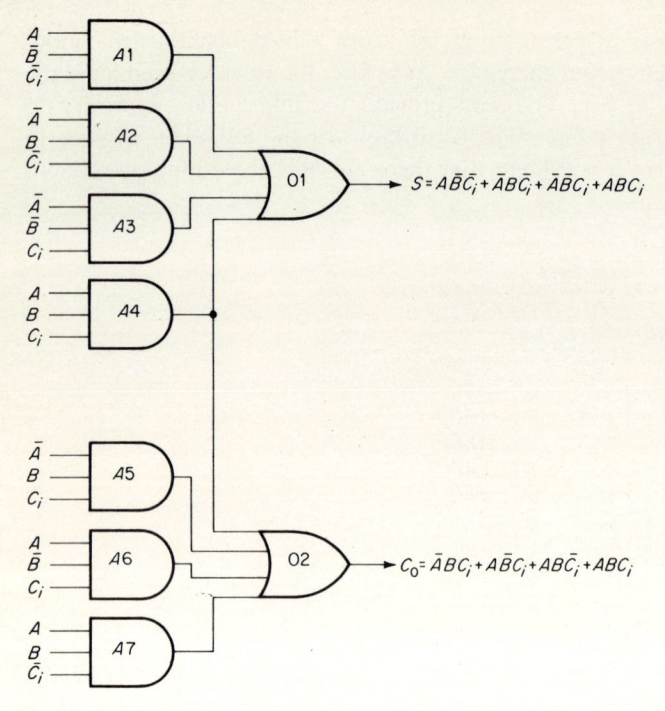

Fig. 11-43. Circuit for the full-adder.

the C_o signal is stored temporarily, to be used when the next set of binary digits arrives at the inputs. It then becomes the C_i signal for these new digits.

Continuing, the next most significant digits are $A = 0$, $B = 1$, and $C_i = 1$ (from the last C_o). The input condition is, then, $\bar{A}BC_i$. Hence, gate $A5$ yields an output at C_o, but $S = 0$, as it should. Again, C_o at logic 1 is stored temporarily for use with the next binary digits. When the third set of digits arrives, the input condition is $A\bar{B}C_i$, which allows $A6$ to give an output at C_o. Again, $S = 0$. Finally, the last set of digits are zeros, but C_i is a logic 1. Thus, the input condition is $\bar{A}\bar{B}C_i$, and hence $A3$ yields a 1 at S, while a 0 appears at C_o. The binary sum then, is 1000. This is, of course, an 8 in decimal notation, which is the correct sum of 0101 (5) and 0011 (3).

It should be noted that if the sum is greater than 9, additional circuitry is necessary to change the simple binary sum to a binary-coded-decimal representation *plus a decimal carry* to the next *set* of BCD digits to be summed. This, however, is beyond the intent of this book. Also, inspecting the full equation for the carry-out C_o, it appears that some redundancy may exist.

By applying the simplifying procedures explained earlier, this expression can be reduced to

$$AB + AC_i + BC_i = A(B + C_i) + BC_i$$

which represents a considerable saving in components.

11-5 LOGIC CIRCUITS

In this section the implementation of the previous logic networks will be discussed. That is, the actual electronic hardware used to perform the various logic functions will be presented, with emphasis placed upon the *electrical* activity of each component and its influence upon the others, while at the same time integrating the circuit discussion with the logic functions.

Diode Gates

There are two possible diode-gate-circuit configurations used. These are arbitrarily designated as X and Y gates. The reason for this is that each circuit possesses a dual logical meaning. That is, the X circuit may provide either an AND function or an OR function; also, the Y circuit may yield either an AND function or an OR function. We cannot, therefore, classify them exclusively as either AND or OR gates.

The X gate is shown in Fig. 11-44, and the number of diodes shown is typical, although up to eight or ten inputs may be used if necessary. The

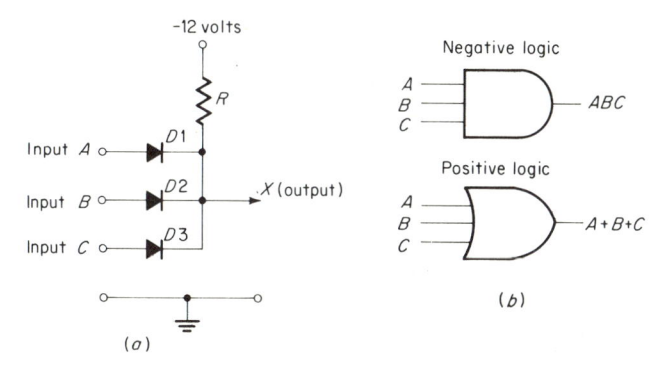

Fig. 11-44. (a) X-gate circuitry provides negative AND or positive OR functions; (b) logic diagrams for the X gate.

limiting factors are the leakage current of the diodes at the highest expected temperature and the diode junction capacitance. If this is to be called an *AND gate*, an output will be produced only when *all* inputs have a valid signal. On the other hand, if it is to be called an *OR gate*, a valid signal appearing at *any* input will result in a valid output signal.

In order to define the X gate as an AND gate, we must define the voltages of the two logical signals. Logic 1 is some *negative* value (say, -6 volts), and logic 0 is a more positive value (say, 0 volts, or ground). In terms of voltage, -6 volts at A AND -6 volts at B AND -6 volts at C will give -6 volts at the output. The Boolean equation for this circuit, which is another way of saying the same thing, is $A \cdot B \cdot C = X$.

If it is desired to define the X gate as an OR gate, logic 1 must be the more *positive* value and logic 0 the more negative value. Using the voltage values from the previous example (logic $1 = 0$ volts and logic $0 = -6$ volts), $A + B + C = X$.

The electrical operation of the circuit is simplicity itself. If a negative voltage appears at A AND B AND C, the output is negative. The same idea can be stated another way: If either A OR B OR C is at a more positive voltage, the output is more positive. By assigning values, the electrical operation can be determined. Assume logic 1 is -6 volts and logic 0 is ground.

If $A \cdot B \cdot C$ are at ground, all diodes are conducting and the output is essentially at ground. If the input to A goes to -6 volts, leaving B and C at ground, diode $D1$ becomes reverse-biased and the -6 volts at A can in no way influence the output. If B goes to -6 volts, $D2$ also becomes reverse-biased and the output is still zero, clamped to ground by the forward-biased $D3$. Now, if C also goes to a -6-volt value, all three diodes are again in conduction and the output is -6 volts. With all three inputs at -6 volts, the output is -6 volts. If, then, any input goes to ground, that diode becomes the controlling factor and the output goes to ground, back-biasing the remaining diodes that are still at -6 volts.

In analyzing a diode gate, it is important to realize that a certain diode, or group of diodes, will control the output by back-biasing all other diodes. To see how this controlling action occurs, consider Fig. 11-44a, which is defined as an X gate. From an electrical viewpoint, each individual diode will conduct if the anode side is more positive than V_{CC} (in this instance, -12 volts). This is only true, however, if the other two diodes are figuratively removed by open-circuiting their input side.

Keeping in mind the fact that in the actual circuit all three inputs are connected to a signal voltage (in other words, *not* open-circuited), it will be found that the input-signal voltages determine the output-voltage level. To ascertain the electrical action of the circuit, assume that the two signal levels are ground and −6 volts. (The logical meaning of the signals will not be considered now; only the electrical action is of present importance.)

First, assume that all three inputs are at a potential of 0 volts, or ground. All diodes have their anodes more positive than the potential to which the cathodes are returned. Hence, all diodes are conducting about equally, and, neglecting the voltage drop across the diodes, the output voltage is also at ground. [Actually, if the inputs are at precise ground (0 volts), the output will be more positive than ground by the drop across the diodes, perhaps 0.5 volt.]

Now, when the input A goes to −6 volts, with the others still at ground, a very significant factor is to be considered. With 0 volts at inputs B and C, diodes 2 and 3 are connected in a circuit with a total potential of 12 volts (−12 volts to ground). However, diode 1, with −6 volts at A, is connected in a circuit that has a 6-volt total potential difference (−6 to −12 volts). Hence, diode 1 loses its influence on the circuit and becomes back-biased since $D2$ and $D3$ retain control and hold the output at, or nearly at, ground. $D2$ and $D3$ are conducting, and $D1$ is back-biased, with −6 volts at its anode and essentially ground at its cathode.

Now, if one of the other inputs goes to −6 volts, say, $D2$, it simply becomes back-biased, with $D3$ continuing to conduct, thus still clamping the output to ground. When all three inputs become −6 volts, again all three diodes conduct and the output is now also at −6 volts (neglecting the diode drop, as before). Only with −6 volts at A AND B AND C will the output also be at a potential of −6 volts. With the circuit in this condition, each diode is connected in a circuit that has a total potential difference of 6 volts.

If, now, any one of the diode inputs is allowed to go to ground potential, it will then have a total difference across its circuit of 12 volts. Hence, it immediately assumes control, and the output goes to ground, which, of course, will cause the other diodes to become back-biased. In other words, if A OR B OR C is at ground, the output is at ground. Note that the output assumes the same potential as the *most positive* input voltage (neglecting diode drop).

The X gate is recognized by the fact that it is a diode gate that has inputs that look into the diode anodes. The polarity of the signals does not affect the description of the circuit, except to the extent that if logic 1 is negative, the circuit is properly called a *negative AND gate* and if logic 1 is positive, it is called a *positive OR gate*.

Although ground is usually one of the two logic signals, it does not necessarily have to be. Suppose the X gate is used in a system where +6 volts = logic 1 and −6 volts = logic 0. The circuit will function as a positive OR gate regardless of the actual levels chosen. Or if logic 0 is

+10 volts and logic 1 is −3 volts, it will nevertheless function as a negative AND. Several possibilities are shown in Fig. 11-45 for the X gate, where 0 volts is ground.

− AND		+ OR	
Logic 1, Volts	Logic 0, Volts	Logic 0, Volts	Logic 1, Volts
−6	0	−6	0
0	+6	0	+6
+3	+9	+3	+9
−6	−3	−6	−3
−12	−6	−12	−6

Fig. 11-45. Possible voltage levels for the X gate.

The other kind of diode gate, called a Y gate is similar to the X gate, except for an inverse logic function. The Y gate is shown in Fig. 11-46 in a typical configuration. Depending upon how the logic signals are defined, the Y gate can be called a *positive AND* or a *negative OR* gate. To define the Y gate as a negative OR gate, the logic signals must be negative for logic 1 and positive for logic 0; a logic 1 at A OR B OR C yields a logic 1 at the output X. If logic 1 is −6 volts and logic 0 is ground, −6 volts at either A OR B OR C gives −6 volts at the output.

On the other hand, if the Y gate is to be used as a positive AND gate,

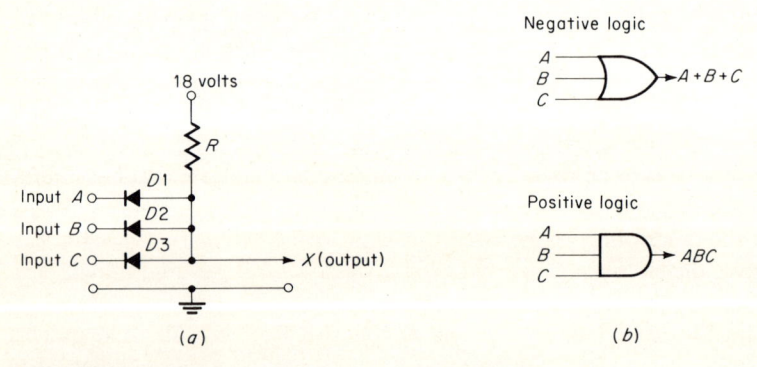

Fig. 11-46. (a) Y-gate circuitry provides negative OR or positive AND functions; (b) logic diagrams for the Y gate.

logic 1 must be the more positive of the two signals, while logic 0 is negative. Thus, a positive level is needed at A AND B AND C to give a logic 1 at the output. If logic 1 is ground and logic 0 is -6 volts, a ground at A AND B AND C is necessary to yield a ground, or logic 1, at the output.

The electrical operation of the Y gate is as straightforward as that of the X gate. If logic 1 is -6 volts and logic 0 is ground, the gate is operating as a negative OR gate. That is, a -6-volt signal at A OR B OR C will give a -6-volt signal out. If all inputs are at ground, the output is at ground since all diodes are conducting equally. If only input A goes to -6 volts, $D1$ becomes more heavily conducting (12-volt circuit) and controls the output, which must become -6 volts. Diodes $D2$ and $D3$ are reverse-biased (6-volt circuit), and their input can in no way influence the output. When input B goes to -6 volts, too, nothing changes except that $D2$ conducts rather than being reverse-biased. The output, of course, remains at -6 volts. In other words, it requires only one input to produce an output, and more than one input accomplishes nothing significant.

To operate the circuit as a positive AND gate simply requires redefining the logic levels. If logic 1 is, say, ground and logic 0 is -6 volts, an output is produced by a logic 1 at A AND B AND C. The output is negative for any other combination of inputs. Again, ground is usually one of the two logic levels, although any other set of levels, one more positive than the other by a significant amount, can be used. Figure 11-47 shows a few possibilities.

$+ AND$		$- OR$	
Logic 1, Volts	Logic 0, Volts	Logic 1, Volts	Logic 0, Volts
0	-6	-6	0
6	0	0	6
-6	-12	-12	-6
12	6	-3	3
8	-8	-8	8

Fig. 11-47. Possible voltage levels for the Y gate.

Diode-Transistor Gates

The diode-transistor gate combines the advantages of the isolation afforded by the diode gates with the amplification provided by an inverter. The logic is performed by the diodes, while the transistor simply inverts the signal, yielding the NAND/NOR function.

The circuit is shown in Fig. 11-48. With all three inputs at ground, the voltage divider cuts off the transistor, and the output is $-V_{CC}$. To cause the output to change to ground (to turn the transistor on) all three inputs must go negative. The diode gate, then, is recognized as an X gate—a negative AND or a positive OR gate. ($R1$ and the diodes are considered the gate proper.)

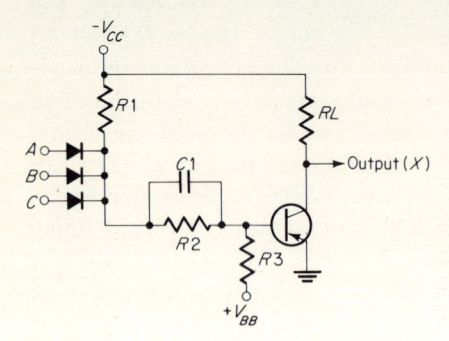

Fig. 11-48. Diode-transistor logic (DTL).

If logic 1 is -6 volts and logic 0 is ground, the output equation is $\overline{A \cdot B \cdot C} = X$. If logic 1 is ground and logic 0 is -6 volts, the output equation is $\overline{A + B + C} = X$. The capacitor around $R2$ is a speedup capacitor used to enhance switching speed where necessary. The voltage-divider values are chosen to provide the proper drive to the base and assure firm cutoff. Typical values might be $R1 = 12$ kilohms, $R2 = 6.8$ kilohms, $R3 = 22$ kilohms, $C1 = 330$ pf, and $RL = 12$ kilohms. $-V_{CC}$ might be -12 volts and V_{BB} perhaps $+6$ volts.

Resistor-Transistor Gates

A typical NOR/NAND circuit is shown in Fig. 11-49. It appears similar to the diode-transistor gate discussed previously but, in fact, is somewhat different in operation. The gating is performed by resistors, and, in this case, the gate itself ($R1$, $R2$, $R3$, and $R4$) is a negative OR or a positive AND gate, depending upon the logic definition. If all inputs are at zero (ground), the divider action of the resistors places a positive voltage on the base of the *pnp* transistor and it is off. Thus the output is at $-V_{CC}$. Now, if any input goes negative, the transistor turns on and the output

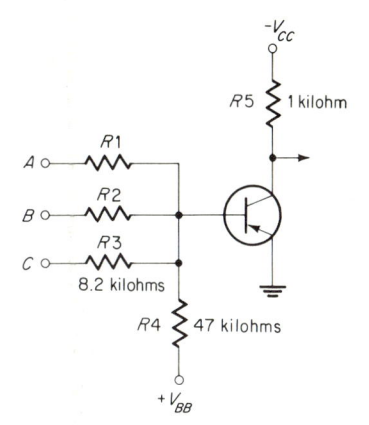

Fig. 11-49. Resistor-transistor logic (RTL).

falls to ground. If more inputs go negative, the output remains at ground. If A OR B OR C goes negative, the output is at ground (NOR); conversely, if A AND B AND C are at ground, the output is at $-V_{cc}$ (NAND).

The resistors at the input must be chosen carefully to ensure the described operation without fail. If $R1$, $R2$, and $R3$ are 8.2-kilohm resistors and $R4$ is 47 kilohms, a maximum of six inputs can be used; if seven inputs or more are applied to the transistor, it will not turn off reliably under all operating conditions. However, there is seldom a need for more than six inputs, and so this fact usually presents no problem.

A disadvantage of the circuit is the possibility of cross-talk between inputs. That is, input A is fed a voltage change; this can be evident at input B since there is nothing but the resistors to isolate one signal from another. If the input signal is sufficiently negative to turn on the transistor, the base is clamped to ground and no cross-talk can occur in this condition.

Resistor-Capacitor (*RC*) Gates

The fourth kind of gate often seen in use is the so-called *RC gate*. The operation of this gate was described briefly in Chap. 3. It is somewhat different from the previously discussed gates. As used, the *RC* gate always has but two inputs, one to the resistor and one to the capacitor. This deceptively simple circuit is shown in Fig. 11-50 and consists of only one resistor and one capacitor. It will be necessary to recall the information from Chap. 3 in order to fully appreciate the ac gate. (Some of the following material represents a repetition of the discussions in Chap. 3.)

There are two kinds of *RC* gates, the $+RC$ gate and the $-RC$ gate, and

250   PULSE AND SWITCHING CIRCUIT ACTION

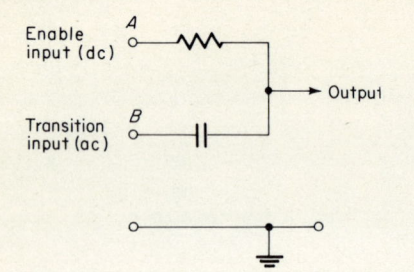

Fig. 11-50. *RC*, or ac, gate.

we shall discuss them in this order. The $+RC$ gate is schematically no different from the $-RC$ gate, as we shall subsequently see. The difference lies in the way the two of them are used in the circuit. Consider first a system where logic 1 is -6 volts and logic 0 is ground. At either input, then, there might be one or the other of these. At the enable input, a level, or dc voltage, is applied. At the transition, or ac input, a pulse is applied. A valid output will be defined as a voltage *different* from either of the two logic levels. This must not confuse the reader! A logic 1 at the input is -6 volts, as previously stipulated. A logic 1 at the output will be, for the $+RC$ gate, a *positive* voltage referred to ground. As we know from Chap. 3, it is possible to obtain a positive pulse from a capacitor even though no positive source appears in the circuit. Under certain conditions, then, a positive pulse may appear at the output, and this is the valid output signal.

By assigning the mentioned voltages to the circuit, its basic operation can be described in suitable terms. If the enable input A is set at -6 volts, the state of charge of the capacitor is determined partly by this voltage. If sufficient time is allowed, the right plate of the capacitor must be charged to -6 volts. At the same time, the ac input, say, has been at ground for some time when along comes a relatively short negative pulse, as shown in Fig. 11-51*a*. With the values of R and C properly chosen, the output must be as shown in Fig. 11-51*b*. At no time does the output go more positive than ground, and the gate is said to have been *disqualified* or *inhibited*. That is, it did not produce a valid output.

If the enable input is at ground, a different condition exists. Now, the right plate of the capacitor is at ground, as shown in Fig. 11-51*d*. When the signal goes negative by 6 volts, the right plate is also driven negative by 6 volts from ground. The right plate, however, must return to ground since the resistor (enable) input is at ground. Now, when the input goes from -6 volts back to ground, the right plate of the capacitor must also go 6 volts more positive than it was. Since it was at ground just before the transition, it is driven to $+6$ volts momentarily, returning exponentially

LOGIC AND LOGIC CIRCUITS **251**

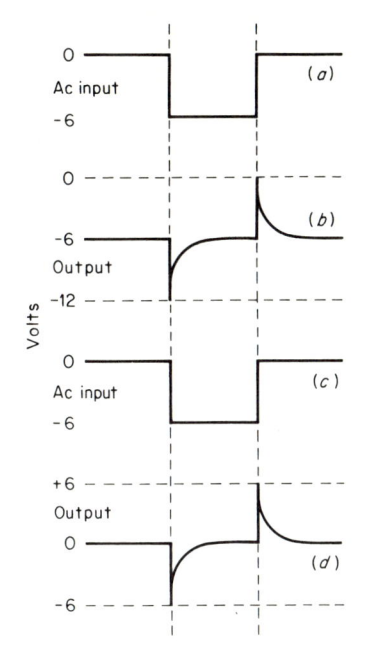

Fig. 11-51. *RC*-gate waveforms.

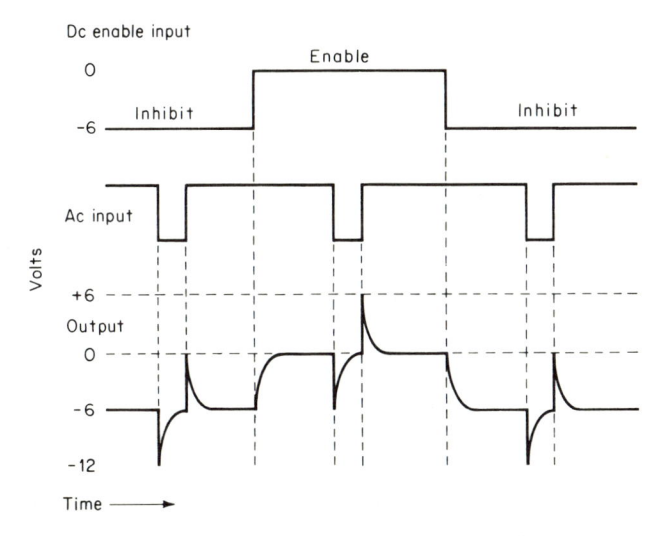

Fig. 11-52. *RC*-gate waveforms for the enable and inhibit conditions.

252 PULSE AND SWITCHING CIRCUIT ACTION

to ground. The gate is said to be *enabled*, and the +6-volt spike is the valid output produced by the positive-going transition of the ac input. The condition necessary to produce a valid output could be stated as follows: If the enable input has been at ground for some time *and* a positive transition occurs at the ac input, a valid output will occur.

Figure 11-52 shows the two conditions where a $+RC$ gate is first inhibited, then enabled, and then inhibited again. Note that it takes some time for the enabled condition to take place. Also, when the dc input goes negative, some time is necessary to reestablish the negative level at the output. This, of course, is owing to the fact that as the dc input changes, the capacitor must charge or discharge through R to the new voltage. Thus the time constant of C and R is most important in this circuit.

Figure 11-53 shows what happens if the input pulses occur at the same

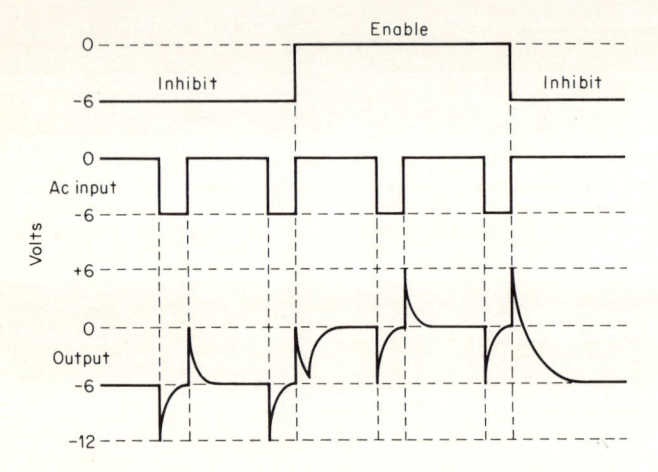

Fig. 11-53. RC gate with coincident transitions.

time as the enabling input. If the enable input has been negative for some time, it is inhibiting the gate even though a transition occurs in step with the enabling level. Likewise, if the pulse to the dc input is just going negative after having been enabled (0 volts), a transition at this time in the positive direction is enabled and an output results even though the inhibit input (dc) is negative at this instant.

The $-RC$ gate is schematically identical to the $+RC$ gate. The only difference between the two gates lies in which of the spikes is to be used. The $-RC$ gate uses the most negative spike as the valid output. The appropriate dc voltage at the enable input must be used, of course. The negative spike can be used to turn on a *pnp* transistor or to turn off an *npn* transistor.

Common-load Gates: Direct-coupled Transistor Logic (DCTL)

There are four variations of common-load gates. The justification for their use lies in the economy of parts employed. That is, where they are applicable, several diodes and resistors may be eliminated. The first of these circuits appears in Fig. 11-54. If Q1 and Q2 are off, the output is at

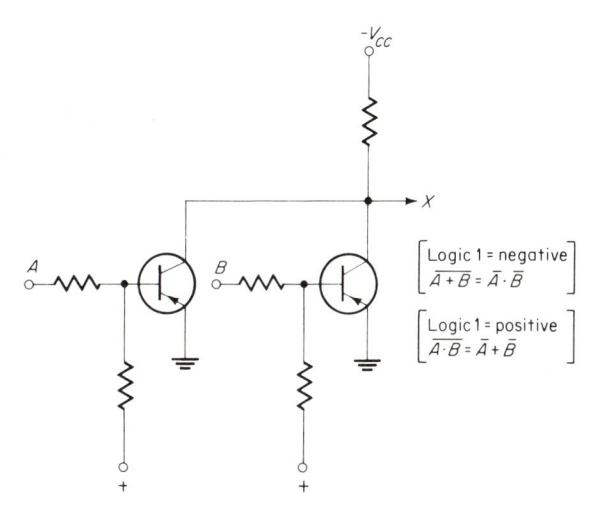

Fig. 11-54. Common-collector-load gate, parallel connection.

$-V_{CC}$. If either transistor comes on, the output falls to ground. Thus the transistors invert the logic function at the bases, and, depending upon the logic definition, the circuit could be described as an *inverted negative OR* (*NOR*) or an *inverted positive AND* (*NAND*) gate. If logic 1 is the more negative of the two logic levels, the output equation is $\overline{A + B} = X$, or $\overline{A} \cdot \overline{B}$. If logic 1 is the more positive of the two levels, the output equation is $\overline{A \cdot B} = X$, or $\overline{A} + \overline{B}$, which is an identical statement. Note that if *npn* transistors are used, the logic definitions and the actual relative voltage levels would be reversed.

Another possible connection is shown in Fig. 11-55. Here, the transistors are in series, and both must be on to cause the output to be 0 volts. If either is off, the output is $-V_{CC}$. Again, this circuit would be called an

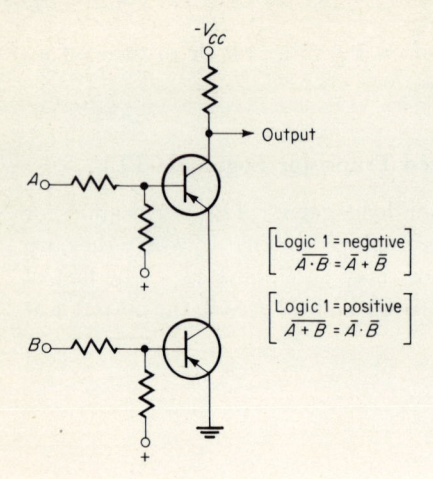

Fig. 11-55. Common-collector-load gate, series connection.

$$\left[\begin{array}{l} \text{Logic 1 = negative} \\ \overline{A \cdot B} = \bar{A} + \bar{B} \end{array} \right]$$

$$\left[\begin{array}{l} \text{Logic 1 = positive} \\ \overline{A + B} = \bar{A} \cdot \bar{B} \end{array} \right]$$

inverted negative AND or an *inverted positive OR*. If logic 1 is equal to a voltage more negative than logic 0, the output equation is

$$\overline{A \cdot B} = \bar{A} + \bar{B} = X$$

If logic 1 is equal to a voltage more positive than logic 0, the output equation is

$$\overline{A + B} = \bar{A} \cdot \bar{B} = X$$

If, in the case of the last two circuits (common-collector gates), the load resistor is taken from the collector and put in the emitter, two other possible combinations will result. (These are still known as *common-load gates*.) One of these circuits is shown in Fig. 11-56. This circuit is not as straightforward as some of those previously discussed. In this case, a truth table will help to understand the circuit, and it is seen that it is a

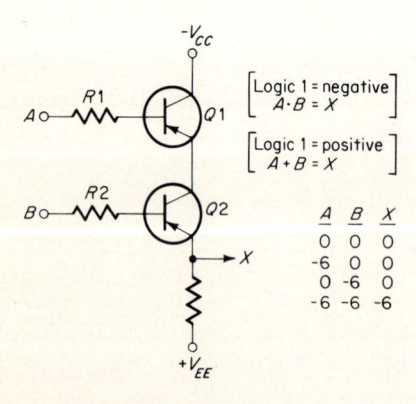

$$\left[\begin{array}{l} \text{Logic 1 = negative} \\ A \cdot B = X \end{array} \right]$$

$$\left[\begin{array}{l} \text{Logic 1 = positive} \\ A + B = X \end{array} \right]$$

A	B	X
0	0	0
-6	0	0
0	-6	0
-6	-6	-6

Fig. 11-56. Common-emitter-load gate, series connection.

negative AND or a positive OR gate. Assume V_{CC} is -6 volts and V_{EE} is $+6$ volts. If A and B are at ground, the output is at ground. Q1 acts as an emitter follower, while Q2 is in saturation since its collector is 0 volts. Now, if A is -6 volts and B is still zero, the emitter of Q1 is at -6 volts and Q1 is in saturation, providing a -6-volt collector supply for Q2. The base of Q2 is still at 0 volts, as is its emitter. Q2 is not now in saturation but is simply acting as an emitter follower operating in its active region.

If A is 0 volts and B is -6 volts, the emitter of Q1 is at ground and this is the collector supply for Q2. With B at -6 volts, the collector of Q2 is far into saturation, and the collector, base, and emitter are at ground. The full 6-volt input at B is dropped across the series base resistor, and the base current is large. Thus the emitter of Q2 is at ground. Only when both A and B are at -6 volts, will the output be -6 volts. In this case, the emitter of Q1 will be at -6 volts and provides the supply voltage for the collector of Q2. With B at -6 volts, the emitter of Q2 can, and will, go to -6 volts. Note that the circuit does not invert the logical sense of the signal as is the case with common-collector-load gates.

The last of the common-load gates is shown in Fig. 11-57. In this

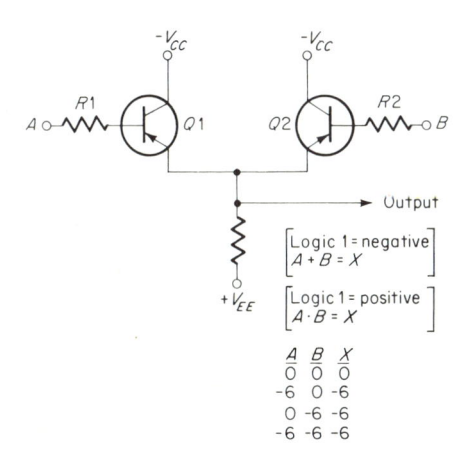

Fig. 11-57. Common-emitter-load gate, parallel connection.

circuit the transistors are in parallel with a common-emitter-load resistor. Again, the truth table will be helpful in describing all possible combinations of inputs. If $V_{CC} = -6$ volts, $V_{EE} = +6$ volts, and if logic 1 is the more negative of the two logic signals, the circuit is a negative OR gate. On the other hand, if logic 1 is the more positive of the logic signals, the circuit is a positive AND gate.

In this circuit both transistors act as emitter followers. If A is at 0 volts and B is at 0 volts, the output is 0 volts. If A goes to -6 volts and B stays at 0 volts, the output goes to -6 volts. With the emitter of $Q2$ at -6 volts and its base at ground, its base is 6 volts more positive than its emitter and it is cut off. (This circuit will afford excellent isolation between signal sources because of this.) Likewise, if B is at -6 volts and A is at ground, $Q1$ is cut off.

It should be pointed out that any common-load gate may have more than two inputs, and, considering this possibility, they will still function as previously described. Keep in mind the fact that any of the preceding circuits using transistors may be constructed using *npn* transistors. It will be instructive for the student to analyze each in terms of *npn* transistors to make certain that both kinds of circuits are understood. To do this, simply redraw each circuit using *npn* transistors. Then, use $+V_{CC}$ as the collector supply, $-V_{EE}$ as the emitter supply, and $-V_{BB}$ as the base supply. The signal levels will be ground (0 volts) and $+V_{CC}$. Then analyze each circuit both electrically and logically.

QUESTIONS AND PROBLEMS

11-1 Express the decimal number 27 in binary.

11-2 Express the binary number 101101 in decimal.

11-3 Express the decimal number 537 as a binary number.

11-4 Add the following binary numbers. Prove your answers.

(a) 11010011
    10010111

(b) 001101
    101111

11-5 Subtract the following binary numbers. Prove your answers.

(a) 1000010
    0100011

(b) 1110001
    0011111

11-6 Express the following number in BCD: 287.

11-7 Express the following numbers in the creeping code using 5 bits.
(a) 3
(b) 7
(c) 9

11-8 Using the Gray code, express 1101 in decimal form.

11-9 Draw the symbol for a three-variable AND gate.

LOGIC AND LOGIC CIRCUITS   257

11-10   Draw the symbol for a three-variable OR gate.

11-11   Draw the symbol for a three-variable NAND gate.

11-12   Draw the symbol for a three-variable NOR gate.

11-13   Draw the symbol for a two-variable RC (ac) gate.

11-14   Draw the symbol for a flip-flop with clocked set-reset inputs, T input, and dc set-reset inputs.

11-15   Draw the symbol for a JK flip-flop with clock input and S and C inputs.

11-16   Draw the symbol for a JK flip-flop with three-variable J and K inputs integral with the flip-flop.

11-17   Draw the symbol for (a) a noninverting amplifier and (b) an inverting amplifier.

11-18   Using DeMorgan's theorem, invert the following expressions:

(a) $A \cdot B$ (b) $A + B$
(c) $\overline{A \cdot B}$ (d) $\overline{A + B}$

11-19   Refer to Fig. 11-19. All inputs are to be changed to bar, or NOT, signals. Write the simplest output equation.

11-20   Refer to Fig. 11-19. The input signals D and E are to be changed to bar, or NOT, signals. Write the simplest output equation.

11-21   Refer to Fig. 11-20. Invert (change to bar signals) the C and D inputs. Write the output equation in simplest form.

11-22   Refer to Fig. 11-20. Invert all input signals except E. Write the output equation.

11-23   Refer to Fig. 11-21. Change the output OR gate to a NOR gate. Write the output equation. (Do not simplify.)

11-24   Refer to Fig. 11-21. Change the upper AND gate to a NAND gate. Write the output equation. (Do not simplify.)

11-25   Given the following Boolean expression, which of the networks shown corresponds to the equation $X = A + \bar{B} + \bar{C}$?

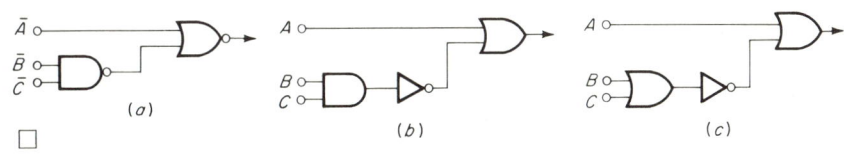

11-26 Remove any redundancies from the following expression:
$\bar{A}B\bar{C}D + \bar{A}BCD + \bar{A}B\bar{C}\bar{D} + \bar{A}\bar{B}\bar{C}\bar{D}$

11-27 Remove all redundancies from the following equation:
$\bar{A}BC\bar{D} + ABC\bar{D} + \bar{A}\bar{B}\bar{C}D + A\bar{B}\bar{C}D$

11-28 Remove all redundancies from the following equation:
$(\bar{A}BCD) + (\bar{A} + B + \bar{C} + \bar{D}) + (\bar{A}\bar{B}CD) + (\bar{A} + \bar{B} + C + \bar{D}) + (\bar{A}BC\bar{D})$

11-29 Remove all redundancies from the following equation:
$ABCDEF + \bar{A}BCDEF + A\bar{B}CDEF + \bar{A}\bar{B}CDEF + AB\bar{C}DEF + \bar{A}B\bar{C}DEF$

11-30 Remove all redundancies from the following equation:
$ABC\bar{D} + \bar{A}BC\bar{D} + ABCD + \bar{A}BCD$

11-31 Remove all redundancies from the following equation:
$\bar{A}B\bar{C}D + \bar{A}BCD + \bar{A}B\bar{C}\bar{D} + \bar{A}\bar{B}\bar{C}\bar{D}$

11-32 Remove all redundancies from the following equation:
$AB\bar{C}\bar{D} + A\bar{B}\bar{C}\bar{D} + \bar{A}\bar{B}\bar{C}\bar{D} + \bar{A}B\bar{C}\bar{D}$

11-33 Write the correct Boolean expression for the following gating network:

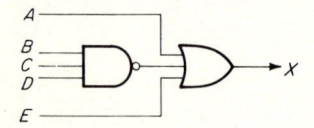

11-34 Write the correct Boolean expression for the following gating network:

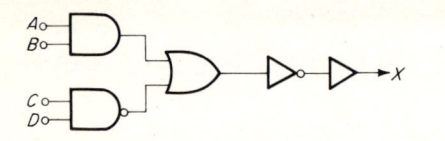

12
MULTIVIBRATORS

This chapter is concerned with a class of circuits called *multivibrators*, which is a subdivision of the larger class of oscillators. Oscillators possess the property of *regeneration*, which allows a circuit to act as a generator of ac waveforms. Regeneration, otherwise known as *positive feedback*, is a property of a circuit that allows it to produce a *self-sustaining output*. That is, the oscillator will produce an ac output of its own accord, with nothing more than the application of the normal dc power.

Such oscillator circuits as Armstrong, Hartley, Colpitts, etc., produce a sinusoidal-output waveform.[1] Multivibrators, on the other hand, either produce a square-wave output or are used to provide special effects necessary to the operation of pulse and digital circuits. Both of these functions, however, have the property of regeneration, which gives them their special characteristics. For example, the astable multivibrator produces a continuous output that consists of a train of square waves. The monostable multivibrator yields a single pulse at its output upon the receipt of an input-trigger pulse. Other similar circuits, all of which will be dealt with in detail, produce still different results.

[1] For a circuit description of these oscillators, see Henry C. Veatch, "Transistor Circuit Action," McGraw-Hill Book Company, New York, 1968.

As previously stated, all oscillators, whether linear or pulse, possess the common property of regeneration, and this attribute causes them to function as they do. The following section discusses the property known as regeneration, and this is followed by a detailed description and analysis of the individual circuits.

12-1 INTRODUCTION TO REGENERATION

An oscillator is a circuit that produces an ac output by virtue of providing its own input. As suggested above, there are many different kinds of oscillator circuits, and each is dependent upon the requirement that there is amplification somewhere in the circuit. Hence, most oscillators are simply amplifiers, with a portion of the output fed back to the input. A block diagram is shown in Fig. 12-1 that is helpful in understanding these interesting and useful circuits.

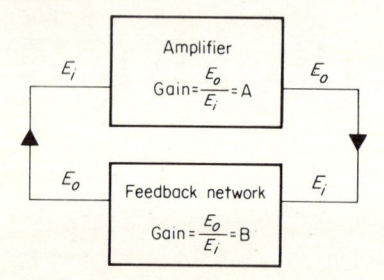

Fig. 12-1. Block diagram of an oscillator.

The output of the amplifier is seen to be presented back to its input through the feedback network. The main requirement here is to ensure that the feedback signal is applied *in phase* with any signal already at the input. Thus, the feedback circuit must often provide some phase shift if the output and input are not exactly in phase. If the amplifier circuit is a common-emitter amplifier, which provides 180° of phase shift, the feedback circuit must provide an additional 180° shift. On the other hand, if the amplifier is a common-base circuit, no additional phase shift is required since the output and input are already in phase with each other.

Another important requirement is that the loop gain $A \times B$ must be greater than unity (1). The gain of the feedback circuit is usually less than 1, and so the gain of the amplifier must be appreciable to overcome this loss. This is usually written as: $\text{gain}_{osc} = A \times B > 1$. If power gain is being considered, this relation is called the *Barkhausen criterion*.

Usually, however, it is enough to simply say that the requirement for oscillation is met if the voltage gain is greater than 1.

In a practical circuit, an active device such as a transistor provides the amplification, while passive components such as resistors, capacitors, and inductors provide the feedback network. As mentioned, the feedback network may or may not provide some degree of phase shift but almost always provides some loss.

The foregoing description of the general oscillator pertains, for the most part, to oscillators such as the Hartley, Colpitts, etc. While much of this discussion also relates to any regenerative circuit, it will be found that there are some differences between sinusoidal oscillators and the multivibrators. The major area of difference is in the feedback network, which in the case of the multivibrators usually consists of an active device, as will be seen, rather than passive components.

12-2 RELAXATION OSCILLATORS

A relaxation oscillator uses one or more capacitors to cause oscillation and usually produces a square-wave output. This interesting circuit is one of a class of *RC* oscillators called *astable multivibrators*. The name is derived from the fact that there is no stable state, and the circuit oscillates because it is continually hunting for a stable state. As in any conventional oscillator, there is regeneration, and in this case the phase shift necessary to produce regeneration is provided by an additional transistor. Hence the circuit generally encountered uses two transistors, one to give 180° of phase shift and the other to yield the remaining 180°. As usually designed, both transistors provide gain, and thus the loop gain is very high.

Simple Astable Multivibrators

A relaxation oscillator may be defined as one in which the waveform changes abruptly and periodically to produce a nonsinusoidal waveform, usually a square wave. The circuit operates because of the time delay of capacitive charge or discharge through a resistor.

A typical astable multivibrator circuit is shown in Fig. 12-2. Note the closed loop; it is this that provides the regeneration and gain necessary to cause oscillation. This feedback loop is emphasized in Fig. 12-3, where the same circuit is redrawn somewhat. The output of $Q1$ is fed to the input of $Q2$, while the output of $Q2$ is returned to the input of $Q1$. This forms the closed loop, and the 360° phase shift occurs because each transistor provides 180°.

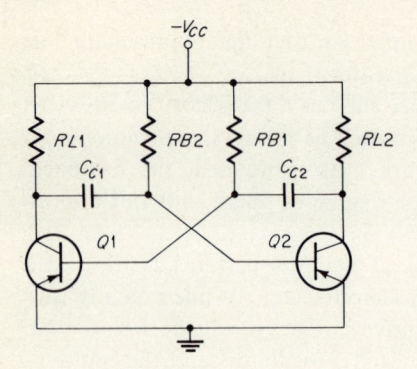

Fig. 12-2. An astable (free-running) multivibrator.

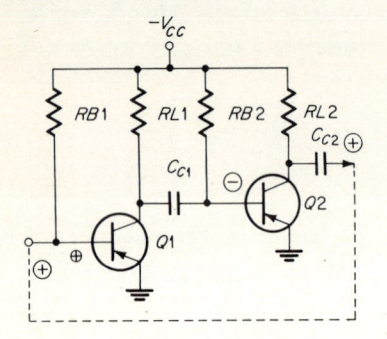

Fig. 12-3. The regenerative closed loop.

Note, also, that both transistors are operated in the common-emitter configuration; hence the loop voltage gain can be made very high. The smallest change at any point in the circuit is amplified, inverted, reamplified, and reinverted. It thus appears at the starting point greatly amplified, as suggested by the polarity signs in the figure.

To appreciate, first, how the circuit works, a simplified drawing is given in Fig. 12-4, where the transistors are replaced by simple switches. One stipulation must be made to cause the simplified circuit to perform like the actual circuit. If switch 1 is open, switch 2 *must* be closed. If switch 1 is closed, switch 2 *must* be open.

As drawn, switch 1 is closed and switch 2 is open. The waveforms shown indicate the conditions at A and B as the switches are alternately transferred. Initially, with switch 1 closed, a large current flows from $-V_{CC}$ through the load resistor and out to ground. The full $-V_{CC}$ is dropped across the resistor, and point B is therefore at ground, or 0 volts. At the same time, point A is at $-V_{CC}$ since, with switch 2 open, there is no current flow through the resistor and therefore no voltage drop across it. When the switches transfer, point B will rise to $-V_{CC}$ while point A will fall to ground, or 0 volts. As the switches are continually transferred, the

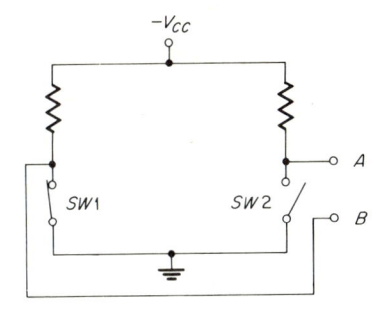

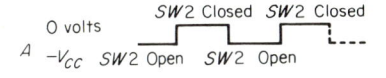

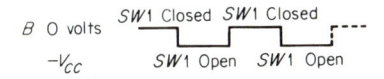

Fig. 12-4. Simplified equivalent circuit of a multivibrator with waveforms.

output from either wire, A or B, is a square wave, each being 180° out of phase with the other.

The multivibrator circuit shown in Fig. 12-2 operates in the same manner. One transistor conducts, while the other is cut off. Then the situation reverses, automatically, at a frequency determined by the time constants in the circuit. When power is applied to this circuit, the transistors alternately conduct and cut off and will produce a train of square waves for as long as power is applied.

To more fully appreciate the details of operation in this circuit, we shall investigate the circuit shown in Fig. 12-5. This is the same circuit shown

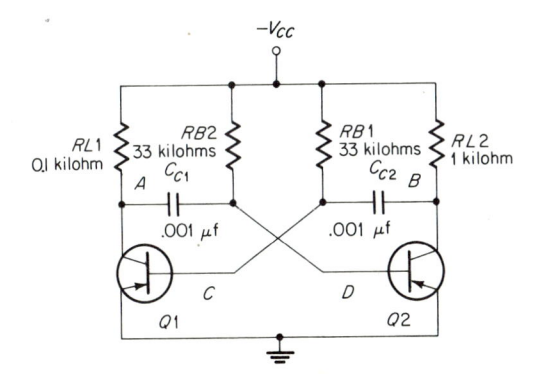

Fig. 12-5. An astable multivibrator circuit.

264 PULSE AND SWITCHING CIRCUIT ACTION

in Fig. 12-2, but with component values added. To start the discussion, it will be assumed that $Q1$ is on while $Q2$ is off. At this point it is not known just what is holding $Q2$ off, but as the following description unfolds, this will become clear.

If $Q1$ is on, its collector is essentially at ground. (A typical switching transistor might have a saturation voltage, $V_{CE,\text{sat}}$, equal to about 0.1 volt. Hence V_C is about -0.1 volt, actually.) With $Q2$ off, its collector is essentially at $-V_{CC}$, which might, for example, be -12 volts. The right-hand plate of capacitor C_{C2} is at a potential of -12 volts, and since $Q1$ is saturated, its base is clamped nearly to ground, with a typical value of -0.3 volt. The left plate of C_{C2} is therefore clamped essentially to ground also. C_{C2} is charged nearly to 12 volts, plate to plate, and as long as $Q2$ is off, it will remain so charged.

Now something (we shall later see just what) will allow $Q2$ to begin to conduct. A very rapid regenerative switch occurs, and $Q2$ goes far into saturation very quickly. But C_{C2} has a 12-volt charge that it cannot get rid of quickly, and with $Q2$ all the way on, the collector side of C_{C2} is at ground. Since the 12-volt charge still remains, with the left plate more positive than the right plate, the left plate must now be at $+12$ volts. To see how this can be, consider Fig. 12-6.

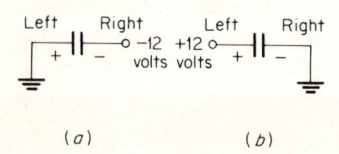

Fig. 12-6. Capacitor action at the instant of transition.

In Fig. 12-6a the capacitor is charged as shown, which represents the condition just before the transition. The left plate of C_{C2} is at ground, while the right plate is at -12 volts. Figure 12-6b indicates the condition of the capacitor just after the transition. With the right plate of the capacitor suddenly grounded by $Q2$ going into conduction, the left plate must rise to $+12$ volts since the capacitor must retain the 12-volt charge for a brief instant. With $+12$ volts at the base of $Q1$ (*pnp*), the transistor turns off. It is clear that the reason why one transistor turns off is that the other one comes on and the capacitor develops the voltage that accomplishes the turn-off.

Now, with $Q1$ off, the length of time it will remain off is very important. Since there is no $+12$-volt supply in the circuit, it can be appreciated that eventually the base of $Q1$ must become negative. However, the coupling capacitor C_{C2} will keep the base of $Q1$ positive while it recharges to the new condition, and during this time $Q1$ is held off. The base of $Q1$ will gradually go from +12 volts toward −12 volts as the capacitor recharges through $RB1$ to the power-supply voltage.

Figure 12-7 shows this occurrence graphically. The waveform represents the base voltage $Q1$ at this instant of time with reference to ground.

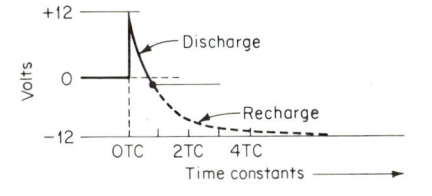

Fig. 12-7. Base-voltage waveform due to capacitor action.

The sharp transition, at $0TC$, is the time $Q1$ goes from on to off. The positive voltage is clearly seen. Now, as time progresses, C_{C2} begins to recharge toward −12 volts. The discharge curve is exponential, and the base of $Q1$ is held positive with respect to ground for some period of time. At one point on the curve, indicated by a dot, the base of $Q1$ is allowed to become slightly negative, and at this point in time, $Q1$ comes back on. The capacitor, then, is never allowed to finish recharging to the new condition. As soon as the base junction becomes forward-biased, the left plate of the capacitor is clamped to ground. The solid portion of the curve, labeled "discharge," represents the time that $Q1$ is held off.

At the instant that $Q1$ comes back on, $Q2$ is turned off in exactly the same manner. C_{C1} performs the same function for $Q2$ as does C_{C2} for $Q1$. The continual switching back and forth continues as long as power is applied. The waveforms at both collectors and bases are shown in Fig. 12-8 and they verify this fact. Note that the waveforms for $Q1$ and $Q2$

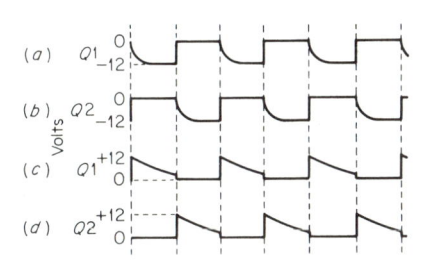

Fig. 12-8. Collector and base waveforms for the astable multivibrator.

are 180° out of phase. This clearly shows that if one transistor is on, the other must be off. Also, the exponential fall of base voltage during the time each transistor is held off is evident, along with the small step caused as the transistor comes on. (This step is shown slightly exaggerated.)

To determine the time each transistor is held off, the time constant t_c of the base resistor and the coupling capacitor is found.

$t_c = RC = 33$ kilohms $\times 0.001$ µf $= 33$ µsec

Figure 12-7 indicates that each transistor is held off for a period of time T somewhat less than one complete time constant. The following relationship will yield the exact length of time that each transistor is held off:

$T = RC \ln \dfrac{E_{max}}{E_{max} - e_c} = 33$ µsec $\ln \dfrac{24}{24 - 12.1}$

$\cong 33$ µsec $\ln 2 = 33$ µsec $\times 0.694 = 23$ µsec

where E_{max} = difference of potential from most positive to most negative, $+12$ to -12 volts

e_c = change in potential across capacitor during time of interest

To account for one complete cycle, two of the above periods must be considered, or 46 µsec. Finally, the frequency of oscillation may be determined as follows:

Frequency (Hz) $= \dfrac{1}{\text{period (sec)}} = \dfrac{1}{46 \times 10^{-6}} = 0.022 \times 10^6 = 22$ kHz

This multivibrator, then, will oscillate at a frequency of 22 kHz, and will generate 22,000 square waves per second.

The above calculation is valid only for a symmetrical circuit, that is, if the two coupling capacitors and the two base-biasing resistors are equal in value. If they are not equal in value, the period that each base is held off must be determined separately and added. This becomes the total period, which is then used to determine the frequency of oscillation.

Self-starting Multivibrators

One of the disadvantages of the circuit shown in Fig. 12-2 is that often, when power is first applied, both transistors tend to go toward saturation at the same time. If this occurs, the circuit cannot begin to oscillate. Power must be removed and then reapplied until the circuit begins to oscillate. In commercial equipment, this must be avoided at all costs. The circuit shown in Fig. 12-9 is frequently used to overcome this shortcoming.

If a circuit can be devised that will prevent both transistors being in saturation together while still allowing one or the other to saturate, the circuit will be self-starting and will then oscillate under any condition. The circuit shown has this attribute and will be found to be a most reliable circuit. It is impossible for both transistors to saturate simultaneously in

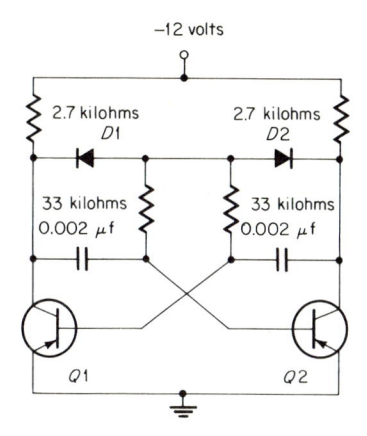

Fig. 12-9. Self-starting circuit using diodes.

this circuit, and the two diodes that have been added help to produce this circuit function.

Just how the diodes work to prevent saturation of both transistors at the same time is best understood by assuming that both transistors *are* saturated, even though this could not happen. If this were the case, the small voltage drops across the transistor junctions would have to be considered, for they are quite significant in the analysis of this circuit. Each collector would be at about -0.1 volt, referred to ground, and each base would be approximately -0.3 volt if both $Q1$ and $Q2$ were in saturation. The anodes of $D1$ and $D2$ are returned to the bases of the transistors, and so they would be at a potential of -0.3 volt. Their cathodes, however, are returned to the collectors, and so must be at -0.1 volt. Now, -0.1 volt is more positive than -0.3 volt, and both diodes are back-biased.

Close inspection of the base circuits reveals that all base current must come from the -12-volt supply *through the diodes*. If the diodes are back-biased, there can be no base current; if there is no base current, the transistors cannot be in saturation. Thus, with the diodes present, the situation where both transistors saturate at the same time is impossible.

With the circuit oscillating and if $Q1$ is off, $Q2$ is conducting and $D2$ is back-biased. However, $D1$ is forward-biased and provides base current for $Q2$. Conversely, with $Q2$ off, $D1$ is back-biased but $D2$ provides base current for $Q1$.

Except for the self-starting feature and one other minor point, this circuit functions exactly the same as the one shown in Fig. 12-5. The other minor difference is that the coupling capacitor must recharge, not only through RB, but also through RL. Thus the time constant in this circuit is determined by the capacitor and by $RB + RL$.

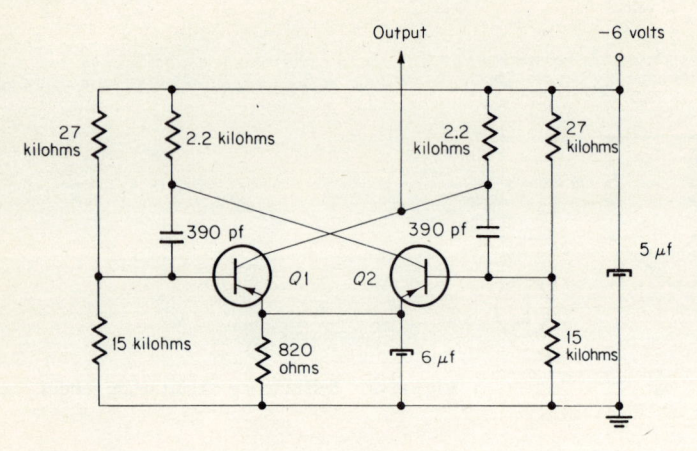

Fig. 12-10. Self-starting circuit using a common-emitter resistor and base-circuit voltage divider.

An alternative approach to the self-starting feature is shown in Fig. 12-10. The reason for the self-starting ability of this circuit is not quite as obvious as in the example of Fig. 12-9. To analyze this circuit properly, it will be necessary to recall some of the material presented earlier regarding the voltage limits at the base and emitter leads. This is an excellent example of the application of these principles to other than linear amplifier circuits.

In this circuit, the emitters are lifted up from ground by the common-emitter resistor. Also, the bases are biased by a voltage divider, in a manner not unlike that of the universal circuit. The circuit works much like any other astable multivibrator as far as its basic function is concerned. The coupling capacitors, 390 pf, serve two purposes. First, they provide the ability to turn off the proper transistor at the proper time, as before. Second, they help, along with the base resistors, to set the operating frequency. As will be seen, the frequency of this oscillator is nominally 50 kHz. This allows the pulse width of 10 μsec to be produced, with a repetition rate of 50 kHz. Thus the time for one complete cycle is 20 μsec.

To make clear how the self-starting feature works, the circuit will be analyzed in some detail. Then the frequency of oscillation will be determined for the circuit values shown in Fig. 12-10. When the circuit is operating normally, the emitters are lifted above ground by the voltage drop across the emitter resistor. Since only one transistor will conduct at a time, the drop across this resistor will be determined by the emitter current of either one of the saturated transistors.

The resistors in either base circuit are chosen to place into saturation the transistor that is on, but care is taken to prevent overdriving the transistor by too much. That is, the transistor that is on is in saturation, but is not driven into saturation any further than necessary. Again, with properly chosen values, the situation with both transistors in saturation is, actually, impossible, but the action of the circuit is easier to understand if it is assumed to be possible. If both transistors become saturated, the current through the emitter resistor will double, of course. Now twice the current through the emitter resistor will result in twice the voltage drop, and the emitters are driven much more negative. Driving the emitters more negative results in the same circuit action as if the bases were driven more positive. With the circuit values shown in the schematic diagram, however, if both transistors were saturated, this would cause them to be cut off, which is obviously impossible.

If the transistors begin to conduct more heavily together, their respective collector currents are reduced and the transistors will remain in the active region. Because the loop gain is very high, they cannot remain in the active region for long and oscillations will again resume. Hence the circuit is indeed self-starting and will not fail to function even if the transistors are artificially saturated. That is, if the collectors and emitters are shorted together with, say, clip leads, the instant the clip leads are removed the circuit will begin to oscillate. On the other hand, the circuit shown in Fig. 12-5 will often remain saturated and will usually refuse to oscillate if treated in this fashion.

Let us now verify that the preceding general discussion is accurate. The first step in analyzing a circuit such as this is to determine the level of voltage at the emitter for normal operating conditions. On the assumption that only one transistor is operating at a time, the emitter current is determined by the applied voltage in the collector-emitter circuit of either transistor.

$$V_E = \frac{RE}{RL + RE} \times (-V_{cc}) = \frac{0.82 \text{ kilohm}}{2.2 \text{ kilohms} + 0.82 \text{ kilohm}} \times (-6)$$
$$= -1.63 \text{ volts}$$

A dc voltmeter, or dc-coupled oscilloscope, would then read -1.63 volts on the emitter when the multivibrator is functioning properly. This voltage is maintained at a constant value by the emitter-bypass capacitor.

When either transistor is on, the base-biasing network determines the transistor condition. It is to be expected that the biasing resistors will drive the base voltage to something more negative than -1.63 volts, then, to assure that the transistor is truly in saturation. The voltage which the voltage divider tries to impress upon the base is simply a function of the ratio of the two resistors and the applied voltage.

$$V_B = \left(\frac{RB'}{RB + RB'}\right) \times (-V_{cc}) = \frac{15 \text{ kilohms}}{27 \text{ kilohms} + 15 \text{ kilohms}} \times (-6)$$
$$= -2.14 \text{ volts}$$

Hence the voltage divider tries to put -2.14 volts at the base, but the emitter cannot go more negative than -1.63 volts. Accounting for the drop across the junction, about 0.2 volt, the base cannot be more negative than -1.83 volts. The emitter clamps the base to this voltage, and the transistor is indeed in saturation. When a transition occurs and one of the transistors is driven off, its base is lifted to something more positive than -1.83 volts, and the transistor is turned off. It is held off by the charge of the appropriate coupling capacitor for some period of time, after which it is allowed to turn on again.

Referring again to Fig. 12-10, the self-starting feature can now be described in actual terms. Normally, the emitter voltage is -1.63 volts, referred to ground. The bases of the transistors, when in the on condition, are clamped to about -1.83 volts, considering the junction drop. If both transistors should come on together, the emitter would rise to 2 times -1.63 volts, or -3.26 volts. If this could happen (and it could not), the emitters would be at -3.62 volts, while the bases would be at -2.14 volts. The bases would be more positive than the emitters by 1.48 volts, and the transistors would be cut off. Obviously, the transistors cannot be saturated and cut off at the same time. Hence, as both transistors try to conduct more heavily, the bases can go as negative as -2.14 volts. If we assume 0.2 volt across the junction, the emitters will then be at -1.94 volts, and both units would be in the active region where oscillation must again begin.

An important feature that must be considered by the designer is to set the limits to exactly the right values to ensure self-starting. In the case of the circuit just described, the designer would first set the proper value of collector load resistor and emitter resistor to be consistent with other circuit requirements. Once a value for RL and RE is determined, he chooses values for RB and RB' to set the voltage at the bases to the proper value to ensure self-starting. He knows that, normally, V_E is -1.63 volts in the present circuit (Fig. 12-10) and that if both transistors saturate, it will rise to twice this, or -3.26 volts. He then chooses the ratio of RB to RB' to set V_B so as to fall as near as possible halfway between these two limits. Halfway between -1.63 and -3.26 volts is -2.45 volts. With standard 5% resistors being used, he can come very close to this value. In the circuit example, note that the divider is providing -2.14 volts to the transistor base and this is well within the requirements of this circuit. If the base resistors are not properly chosen and if they apply a voltage to the base that is more negative than, in this case, -3.26 volts, the circuit will not be self-starting and will be no more dependable than the simple circuit first shown in Fig. 12-2.

Finally, the operating frequency is of some importance. This, of course, is related to the time constant in each base circuit. This calculation is made in much the same way as for the circuit in Fig. 12-5 but is complicated by several factors. For one thing, the value of resistance to be used with each capacitor is not completely self-evident. In this circuit, the capacitor can charge or discharge by way of either base resistor RB or RB'. Hence these two resistors appear to be in parallel as far as the charge time of the capacitor is concerned. A further complicating factor is the voltages toward which each coupling capacitor charges and discharges. The voltage divider in the base circuit determines one of these limits, while the power supply $-V_{CC}$ determines the other.

Because this is a symmetrical circuit, only the conditions at one transistor need be determined. The circuit conditions for $Q1$ will be determined first, and this will allow the overall operating frequency to be found. The coupling capacitor connecting the collector of $Q2$ to the base of $Q1$ will be the one to which we shall direct our attention. The waveform at the base of $Q1$ is shown in Fig. 12-11, referred to ground. The voltage

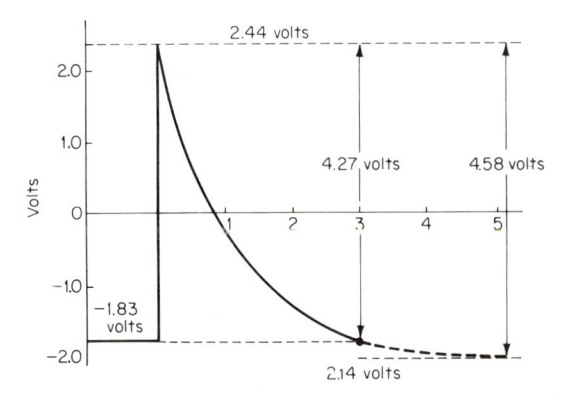

Fig. 12-11. Self-starting multivibrator waveform (base) showing maximum excursion limits.

levels given have already been determined in the previous discussion. The most negative excursion shown is -2.14 volts. This is the voltage set by the voltage divider at the base and appears to the capacitor as a source. Another level shown is -1.83 volts, and this is simply the voltage at either base when that transistor is on. The emitter voltage is normally -1.63 volts; allowing 0.2 volt across the base-to-emitter junction, the base must be $(-1.63) + (-0.2) = -1.83$ volts.

The most positive voltage on the drawing is 2.44 volts. This is the voltage to which the base will rise at the instant of transition and is arrived at as follows. Assume Q1 is on. The voltage *across* the coupling capacitor is determined by the voltage at the base of Q1 and the voltage at the collector of Q2. If Q1 is on, its emitter is at -1.63 volts, and so its base is approximately at -1.83 volts. The collector of Q2 is at a potential equal to the supply voltage, or -6 volts. The difference between these two values is the voltage across the capacitor when Q1 is on.

$$E_C = -V_{CC} + V_{BQ1} = -6 + 1.83 = -4.17 \text{ volts}$$

Thus the voltage across the capacitor is 4.17 volts, with the base side at -1.83 volts with respect to ground. Now, at the instant of the transition, the collector of Q1 falls to about -1.73 volts, with respect to ground, allowing 0.1 volt from collector to emitter. This is a 4.27-volt excursion in the positive direction $(-6.0 + 1.73 = -4.27)$. Hence, if one capacitor plate goes rapidly 4.27 volts in the positive direction, the other plate must also go 4.27 volts in the positive direction since the capacitor cannot change its state of charge instantly. The base of Q1 is thus driven to a value that is 4.27 volts more positive than it was.

$$V_B = -1.83 + 4.27 = 2.44 \text{ volts}$$

The base of Q1, then, is driven to 2.44 volts above ground. This is shown in Fig. 12-11 as the most positive excursion of the waveform. These voltages will allow the determination of the period that one transistor is held off.

$$T_{1/2 \text{ period}} = RC \ln \frac{E_{\max}}{E_{\max} - e_c}$$

where $RC = \dfrac{27 \text{ kilohms} \times 15 \text{ kilohms}}{27 \text{ kilohms} + 15 \text{ kilohms}} (390 \times 10^{-12}) = 3.76 \text{ }\mu\text{sec}$

$$T_{1/2 \text{ period}} = 3.76 \text{ }\mu\text{sec} \ln \frac{4.58}{4.58 - 4.27} = 3.76 \text{ }\mu\text{sec} \ln 14.77$$

$$= 3.76 \text{ }\mu\text{sec} \times 2.69 \cong 10 \text{ }\mu\text{sec}$$

One transistor is off for 10 μsec, and this represents one half-cycle. Since we are dealing with a symmetrical circuit, the time, or period, for one complete cycle is twice this value, or 20 μsec; from this we can find the frequency.

$$F = \frac{1}{T_{\text{period}}} = \frac{1}{20 \times 10^{-6}} = 50 \times 10^3 \text{ Hz} = 50 \text{ kHz}$$

The calculation of the frequency of oscillation, then, is not overly difficult, although somewhat detailed.

Nonsymmetrical Multivibrators

The circuit diagram of a nonsymmetrical multivibrator is shown in Fig. 12-12. Note that the value of one base resistor is 33 kilohms while the other

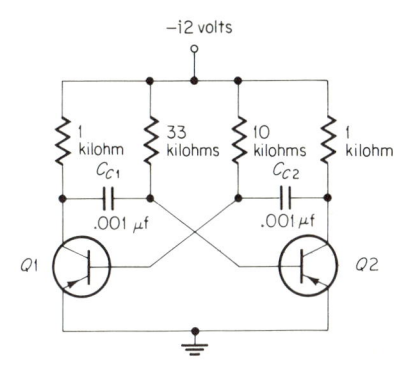

Fig. 12-12. A nonsymmetrical multivibrator.

is 10 kilohms. Since both coupling capacitors are equal in value, this will give rise to unequal time constants in the base circuits. The period of time that each transistor is held off, then, will be different.

To find the frequency of oscillation for this circuit, the time constant for each base circuit is calculated.

$t_{c1} = RB2 \times C_{C1} = 33$ kilohms $\times 0.001$ μf $= 33$ μsec
$t_{c2} = RB1 \times C_{C2} = 10$ kilohms $\times 0.001$ μf $= 10$ μsec

The sum of these two periods represents the time constant for both base circuits.

$T_{period} = t_{c1} + t_{c2} = 43$ μsec

The frequency of oscillation can now be determined. First, the actual off time must be determined, and this is relatively simple for this circuit. E_{max} for each capacitor is 24 volts since the emitters are at ground and the total voltage to which each capacitor will charge is approximately two times the supply voltage.

$$T_{total} = T_{period} \ln \frac{E_{max}}{E_{max} - e_c} = 43 \text{ } \mu\text{sec} \ln \frac{24}{24 - 12.1}$$
$$= 43 \times 0.694 = 29.84 \text{ } \mu\text{sec}$$
$$F_{osc} = \frac{1}{T_{total}} = \frac{1}{29.84 \text{ } \mu\text{sec}} = 33.5 \text{ kHz}$$

Typical waveforms for a circuit such as this are given in Fig. 12-13 where the lack of symmetry is clearly evident.

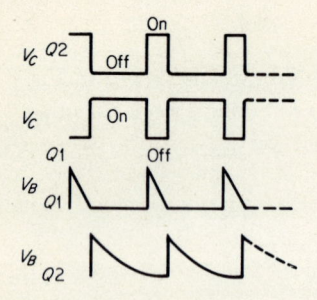

Fig. 12-13. Nonsymmetrical waveforms.

12-3 BISTABLE MULTIVIBRATORS

The bistable multivibrator is widely used in pulse and digital circuitry. Otherwise known as a *flip-flop*, this circuit forms the backbone of electronic calculators, computers, and many other forms of digital equipment.

Fundamental Circuit Action

Basically, the flip-flop is a two-transistor circuit that exhibits *bistable* characteristics; that is, it will exist in one of its two states, either of which is stable. For example, in the circuit shown in Fig. 12-14, if Q1 is on, Q2 is off and the circuit will remain in this state or condition until forced to

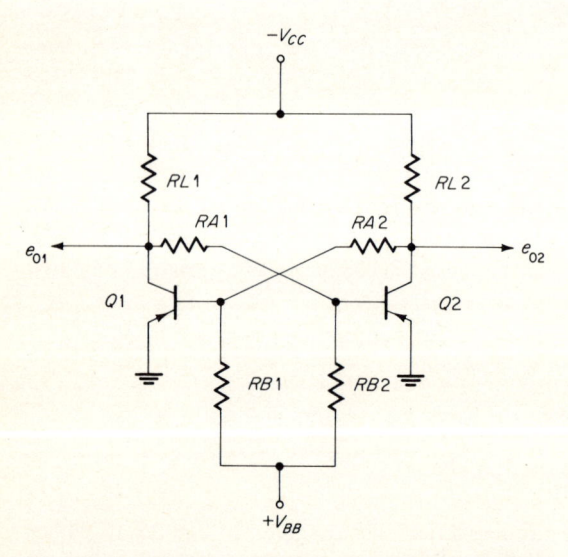

Fig. 12-14. The basic bistable multivibrator (flip-flop).

change. If, however, Q2 is caused to turn on (by means of a suitable input signal not indicated on the drawing), Q1 will turn off and again the circuit will remain in this new condition until instructed to revert to the original condition.

To verify the bistable nature of such a circuit, consider Fig. 12-15,

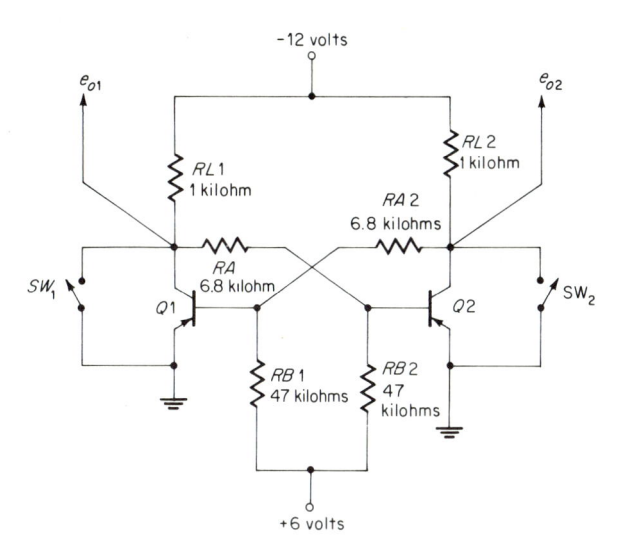

Fig. 12-15. A flip-flop with an elementary triggering arrangement.

which has circuit values added plus two switches $SW1$ and $SW2$. These switches, which need only be momentary contacting, will allow the circuit to be transferred to either of its two states: Q1 on, Q2 off; and Q1 off, Q2 on.

First, assume that Q1 is on. It will be verified later that when either transistor is on, it is far into saturation. With Q1 on and saturated, the circuit is so designed that Q2 is firmly off. This, too, will be verified. With Q1 in saturation, its collector-to-emitter voltage $V_{CE,\text{sat}}$ is, perhaps, 0.1 volt. Hence, the collector can be considered to be held firmly at ground potential since the emitter is at ground. The voltage at the base of Q2 is therefore determined by the voltage divider, consisting of the attenuation resistor $RA1$ and the base resistor $RB2$, and the voltage across this divider.

$$V_{BQ2} = \frac{RA1}{RA1 + RB2} V_{BB} = \frac{6800}{6800 + 47{,}000} 6 = 0.758 \cong 0.76 \text{ volt}$$

Hence, when Q1 is on, the base of Q2 is held to +0.76 volt. This, of course, keeps Q2 in the off condition. Because of the direct coupling from the collectors to the bases, these conditions can be expected to remain with no tendency to change. That is, as long as Q1 remains on, Q2 will remain off.

We shall now verify that Q1 is indeed in saturation. With Q2 in cutoff, its collector rises nearly to $-V_{CC}$, or -12 volt. This allows the base of Q1 to "see" the -12-volt supply through 7.8 kilohms. The approximate base current actually flowing is a function of the Thevenin-equivalent voltage and resistance as viewed from the base. The open-circuit voltage at the base (with no base current flowing) is determined by the voltage divider.

$$V_{th} = \left(\frac{RL2 + RA2}{RL2 + RA2 + RB1} \times V_{total}\right) - V_{CC}$$

$$= \left(\frac{1000 + 6800}{1000 + 6800 + 47{,}000} \times 18\right) - 12$$

$$= (0.1423 \times 18) - 12 = 2.56 - 12 \cong -9.44 \text{ volts}$$

$$R_{th} = \frac{(RA2 + RL2)RB1}{RA2 + RL2 + RB1} = \frac{7800 \times 47{,}000}{54{,}800} \cong 6700 \text{ ohms}$$

On the assumption that the emitter-base junction of the transistor exhibits negligible resistance, the base current can now be determined:

$$I_{B,sat} \cong \frac{|V_{th}|}{R_{th}} = \frac{9.44}{6{,}700} = 1.41 \text{ ma}$$

To determine if the transistor is truly in saturation it is only necessary to find the worst-case base current that is required to barely cause saturation. If the transistor manufacturer specifies the minimum dc beta as 20, then the circuit must provide at least the following value of base current:

$$I_{B,min,sat} = \frac{I_{C,sat}}{h_{FE,min}} = \frac{V_{CC}/RL}{20} = \frac{12 \text{ ma}}{20} = 600 \text{ }\mu a$$

Hence, the transistor only requires 600 µa (or less, depending on the actual beta) to just saturate Q1 *if* its beta is the smallest guaranteed value. However, the circuit provides about two and four-tenths times more base current than necessary to make certain that the on-transistor is *firmly* on. Another way of stating this is to use forced beta:

$$\beta_{forced} = \frac{I_{C,sat}}{I_{B,sat}} = \frac{12 \text{ ma}}{1.41 \text{ ma}} = 8.51$$

Thus, with forced beta significantly smaller than the minimum published value for this transistor, it must be firmly in saturation. As before, this is a stable condition and nothing tends to change. With Q1 on and in saturation, Q2 is off and this represents one of the two stable states. To cause the circuit to change to the other state (make a transition) either the off-transistor must be made to turn on or the on-transistor must be made to turn off. The switches shown in Fig. 12-15 are a rather crude (but effective) means of causing a transition. With the circuit in the preceding condition (Q1 on, Q2 off), switch 2 can be operated briefly, causing the circuit to make a transition.

With $SW2$ depressed (closed) for an instant, ground appears through the switch at the collector of Q2. The base of Q1 therefore suddenly goes to $+0.76$ volt and Q1 turns off. With Q1 off, its collector rises toward $-V_{CC}$ and Q2 turns on. This action can occur in a time that is on the order of a microsecond or less; hence when the switch is restored to its normally open position, the flip-flop has long since been transferred. Now, with Q2 on and Q1 off, the circuit is in the other of its two stable states, where it will remain until caused to transfer again.

The foregoing discussion confirms the static conditions occurring during either of the two stable states, but the mechanism of the transition itself has yet to be investigated. To best appreciate the circuit action during the time one transistor is going off and the other is coming on, one must recall that a single transistor is capable of amplification. In Fig. 12-16, this is clearly evident. The circuitry in which Q1 is placed is seen to be a

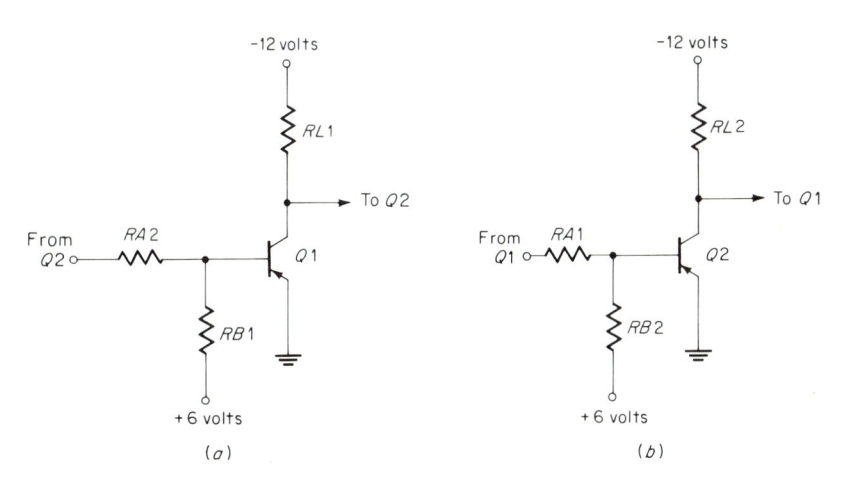

Fig. 12-16. The individual parts of the flip-flop. Each transistor (a) and (b) is in a simple common-emitter circuit.

simple common-emitter configuration as is that of $Q2$. Each, then, is able to provide both current and voltage amplification.

However, these transistors are not, of course, biased as linear amplifiers. Instead, they are switching amplifiers and, as such, are usually considered to operate either in the on or off modes. But, as mentioned in an earlier chapter, for a transistor to go from the off state to the on state or vice versa it *must traverse the active region*. Only during this time can the transistor provide any amplification. Because of the *closed-loop* connection, the overall voltage gain is very high, actually the product of the individual stage gains. If the voltage gain of each transistor is, for example, 200, the loop gain is $200 \times 200 = 40,000$. Therefore, during the time in which both transistors are in the active region, the total amplification is very large.

Because the transistor circuit is wired as it is, the circuit is capable of regeneration if the loop-voltage gain is greater than 1. As mentioned, a regenerative loop exists if a signal voltage at any point in the circuit can return to this same point amplified and *in phase* with the original signal. In the circuit of Fig. 12-14, a positive-going voltage at the base of $Q1$ is amplified and inverted and is presented to the base of $Q2$ as a larger negative-going voltage. This is again amplified and inverted and delivered to the base of $Q1$ as a much larger positive-going signal, reinforcing the original. In this way, a transistor just starting to go from saturation into the active region is given a tremendous push and is forced all the way to cutoff very rapidly. The other transistor, of course, is similarly forced from cutoff through the active region into full saturation. Because of the very large voltage gain, the voltages at the collectors change *very* rapidly. For example, a medium-speed alloy transistor in such a circuit can make a transition in less than 0.09 μsec, while a high-speed silicon-planar unit requires on the order of a few nanoseconds. The main limiting factors are the shunt capacitances, due to wiring as well as the transistor junctions, that must be recharged to the new condition, and the storage time of the transistor.

Before such a regenerative transition can occur, *both* transistors must be in the active region to insure that the *loop gain* is greater than 1. It would serve no purpose to simply cause the on-transistor to go slightly into the active region if the voltage at its collector does not rise from ground far enough to begin to turn the other transistor on. *Both* transistors must be far enough into the active region to insure that the overall voltage gain is greater than 1 to initiate the transition. Once started, the transition will continue of its own accord, regardless of further outside influences. Because of this action, it is only necessary to apply a short pulse to the flip-flop just sufficient to start the switching transition, and the circuit will complete the action even in the absence of further signal.

The basic circuit shown in Fig. 12-15 is useful for illustrative purposes only. The use of a mechanical toggle switch is seldom encountered. However, it should be noted at this point that if the switches are replaced with transistors used as pulse inverters, then the circuit is a very real one and is frequently used in commercial equipment. This method of causing a transition is probably the simplest and most straightforward. As will presently be seen, more complex ways of causing the flip-flop to make a transition are commonly used also.

To simplify the discussion and analysis of such a circuit, it is convenient to assign a logical sense to the outputs, as was done in Chap. 11. Figure 12-17 indicates the logic diagram for a flip-flop shown with a minimum of

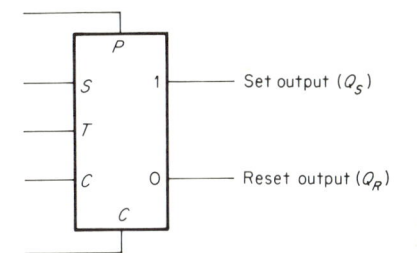

Fig. 12-17. A logic diagram of a flip-flop.

inputs. The flip-flop is said to be set if a logic 1 appears at the set output, while it is reset if a logic 1 appears at the reset output. The inputs to either set or reset the flip-flop are also shown and will be described subsequently.

Multivibrator Triggering

A flip-flop can be caused to make a transition by the application of a pulse of suitable amplitude and duration. The only requirements are that the pulse be applied to a part of the circuit that will produce the desired action and that the pulse has sufficient energy to lift the loop gain to something greater than 1. The pulse may be injected into the collector circuit, the emitter circuit, or the base circuit. The preference appears to be to use the base circuit for this, although one encounters the other methods occasionally. Because of this we shall stress base-circuit triggering, but examples of other methods will be shown.

One of the reasons for triggering by way of the base is that the transistor is most sensitive to an input at this point. Hence, the circuit can be made to be exceptionally reliable with a minimum amount of input energy. Also, because of the voltage requirements of the base, the input need only be a relatively small voltage, especially if applied directly to the base connection.

The circuit given in Fig. 12-18 is probably the most widely used triggering input. A suitable pulse applied to the terminal labeled "T input" will cause a transition regardless of the state of the flip-flop. Such an input is called a *toggle*, *trigger*, or *transition*, input, and hence is abbreviated T. The action of the T input has previously been described in terms of logic. If the flip-flop is reset, a pulse at the T input will set it; if initially set, the pulse will reset the flip-flop. As explained earlier, if a logic 1 exists at the set output Q_S, the flip-flop is said to be set. If a logic 1 exists at the reset output Q_R, it is said to be reset. A pulse at the T input, then, changes the logical state of the flip-flop to the opposite state.

To understand how the circuit works it is necessary to realize that the components making up the T input actually comprise a pair of ac gates, discussed in detail in Chap. 3. The three components R_{G1}, D_{G1}, and C_{G1} form the resistor-capacitor ac gate for the reset side of the flip-flop. By the same token, R_{G2}, D_{G2}, and C_{G2} are the gating components for the set side.

This input configuration is often called a *self-gated*, or *self-enabled*, T input. Note that the enabling resistors return to the collector of the transistor on the corresponding side of the flip-flop. Because of this connection, *the state of the flip-flop itself* determines which capacitor will be allowed to present a signal to its base. This greatly improves circuit reliability, and if properly designed the circuit cannot fail to make a transition with a proper input pulse.

A further consideration of this method of triggering is that it is the on-transistor that is turned off rather than the reverse, which allows for a certain degree of noise immunity. That is, owing to the fact that the base of a transistor that is in saturation is firmly clamped to ground, it requires an amount of energy in excess of some critical value to begin to turn the transistor off. Because the impedance level to ground is so low at the base, typically 200 to 400 ohms, any noise or other electrical disturbance cannot easily cause a transition inadvertently.

Still referring to Fig. 12-18, assume that $Q1$ is on and therefore $Q2$ is off. Also, let $-V_{CC} = -12$ volts and $+V_{BB} = +6$ volts. The input will be a negative-going 12-volt pulse operating between 0 (ground) and -12 volts. As stated, $Q1$ is on and $Q2$ is off; thus, resistor R_{G1} is returned to ground, and R_{G2} is returned to -12 volts through $RL2$. Recalling from Chaps. 3 and 11 the action of such a gate, it can be appreciated that C_{G2} is inhibited while C_{G1} is enabled, or permissed. Thus, as indicated on the waveforms, when a pulse arrives, the base of $Q1$ will receive a positive

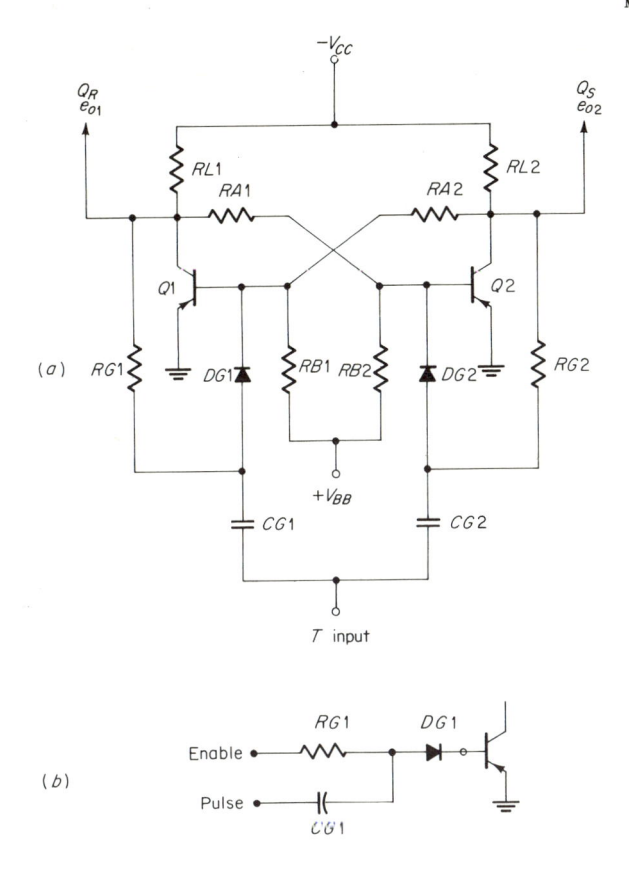

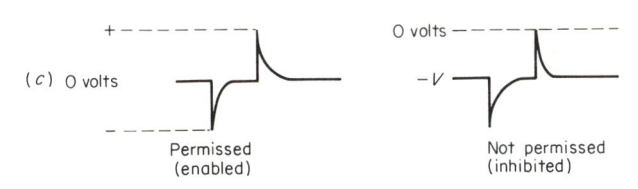

Fig. 12-18. (a) Flip-flop with a T input; (b) RC (ac) gate for $Q1$; (c) waveforms for permissed and inhibited conditions.

pulse on the trailing edge of the signal pulse through D_{G1}; this is more than sufficient to turn off $Q1$. As its collector rises toward $-V_{CC}$, $Q2$ is forced into saturation and the circuit has made a transition. If another pulse is then presented to the T input, C_{G2} is now enabled since $Q2$ is on. Also, C_{G1} is inhibited because $Q1$ is off. Hence, this pulse (the second one) will

282 PULSE AND SWITCHING CIRCUIT ACTION

be applied through D_{G2} to the base of $Q2$, turning it off. As before, when its collector rises toward $-V_{CC}$, $Q1$ comes on and the flip-flop has returned to its original condition.

An alternate method of base-circuit triggering is shown in Fig. 12-19. The input pulse is applied through C_T to the *steering* diodes, and on the

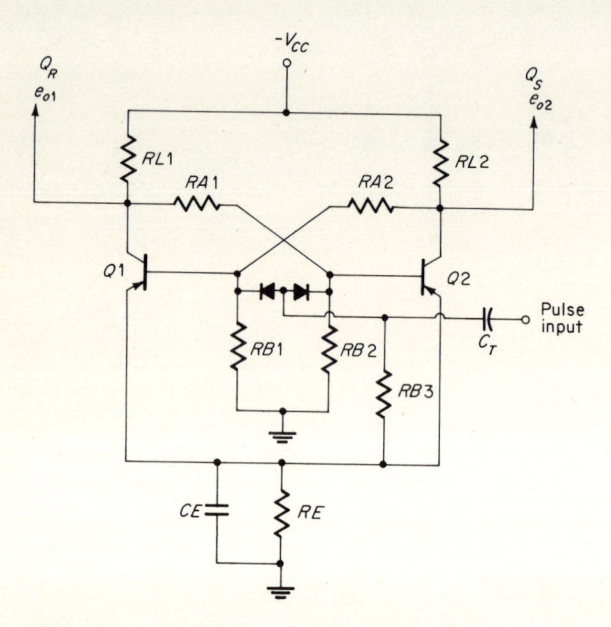

Fig. 12-19. An alternate method of base-circuit triggering.

positive-going transition the on-transistor is turned off. As soon as the loop gain exceeds 1 the flip-flop takes over and completes the transition.

Figure 12-20 shows a circuit that allows the circuit to be triggered at the emitter. In this arrangement, the input pulse must be of very short duration, sufficient only to start the regenerative action. Additionally, the input current requirement is quite large due to the low value of RE.

Collector triggering is illustrated in Fig. 12-21. As a positive pulse is applied, the diode connected to the off-transistor conducts, lifting the collector more positive and turning the other transistor on, effecting a transition.

As a final example of triggering methods, a composite circuit, using many different techniques, is shown in Fig. 12-22. First, note the transistors $Q3$ and $Q4$. These are collector clamps and operate in much the

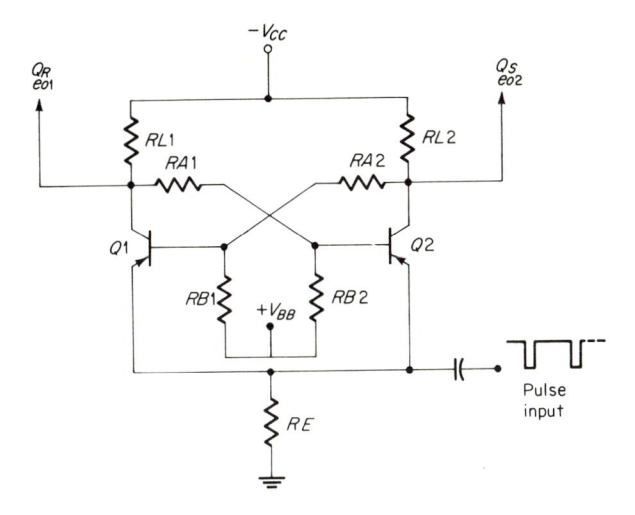

Fig. 12-20. A flip-flop with emitter triggering.

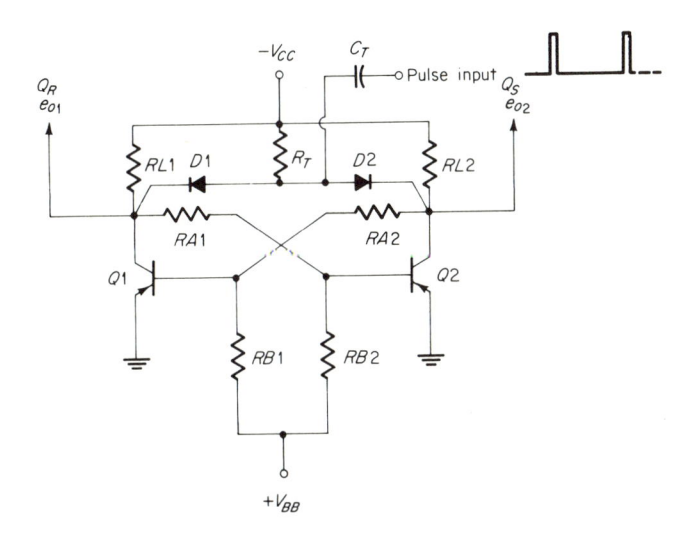

Fig. 12-21. A flip-flop with collector triggering.

same manner as the mechanical switches first shown in Fig. 12-15. The circuit is labeled in terms of negative logic. If $Q2$ is off, the flip-flop is set, while if $Q1$ is off, it is reset. Besides the T input, separate, externally enabled set-reset ac gates are provided, as well as set-reset dc inputs directly at the bases.

Briefly, the circuit functions as follows. Assume the flip-flop is reset

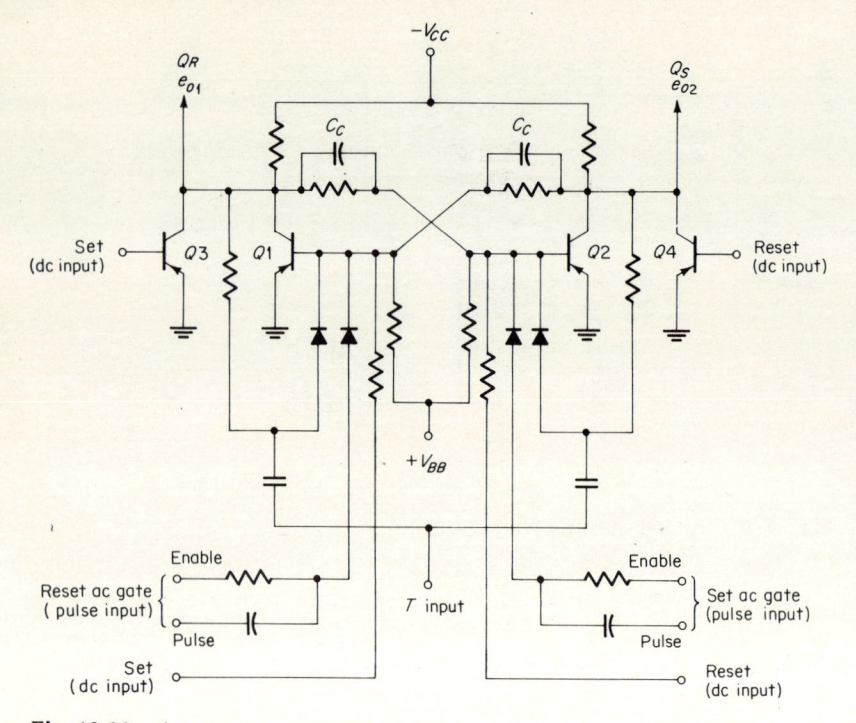

Fig. 12-22. A composite circuit showing several different methods of triggering.

(Q1 off, Q2 on). Logic 0 (ground) appears at the set output (e_{o2}). There are four ways in which the flip-flop can be transferred to the set condition. First, a pulse at the T input will toggle the circuit, and since the flip-flop is reset before the pulse, it becomes set. Second, a negative pulse at either set dc input (base input or collector input) will cause the circuit to become set. Finally, if the set ac gate is enabled (resistor returned to ground), a pulse at the capacitor will set the flip-flop.

It should be noted that with the flip-flop reset, a pulse or dc level at any reset input can in no way affect the circuit. (This, of course, is not true of the toggle input.) With the flip-flop set, an input signal at a reset input will cause the flip-flop to become reset.

One final point regarding this circuit should be mentioned. Note the capacitors connected across the attenuation, or cross-coupling, resistors. These are called the *speedup capacitors*. Without these capacitors, the turn-off speed of a saturated transistor is determined to a large extent by the storage time. The capacitors serve to minimize storage time and so speed up the transistor's turn-off time. Depending upon circuit requirements, their size might range from ten to perhaps several hundred picofarads. Without these capacitors, the flip-flop's ability to respond to a high-speed pulse is severely limited. On the other hand, if they are made too large, they themselves require an excessively long time to return to a static charge (long time constant) and they will again limit the fastest rate at which the circuit can respond.

A circuit variation used to improve the rise time, among other things, is shown in Fig. 12-23. Note the addition of two diodes at each collector. These provide a clamping action, and hence are known as *collector-clamping diodes*. In addition to improving the rise time of the collector waveform, they provide the additional benefit of a direct connection from a power

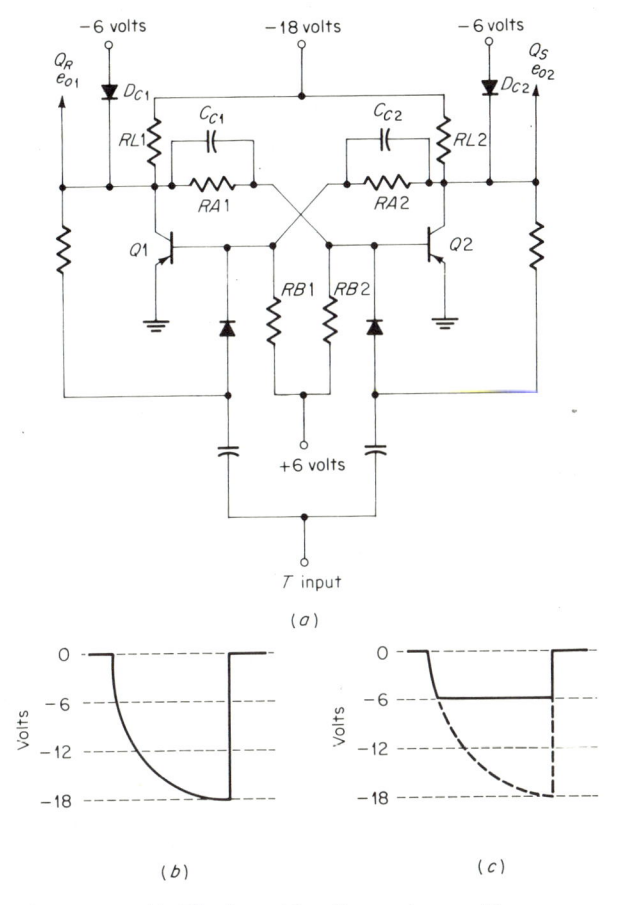

Fig. 12-23. (*a*) Flip-flop with collector clamps; (*b*) waveform without clamps; (*c*) waveform with clamps.

supply to the load through a heavily conducting diode. Hence, a predominately capacitive load need not be charged through a relatively large load resistor ($RL1$ or $RL2$) to a relatively large voltage.

The diodes function very simply. If $Q1$ is on, for example, its collector is essentially at ground. The cathode of D_{C1}, then, is 6 volts more positive than its anode and it is reverse-biased. When the flip-flop makes a transition, the collector of $Q1$ starts toward -18 volts and without the clamping diodes would yield the waveform shown in Fig. 12-23b. However, with the diode, as the collector voltage begins to exceed -6 volts by any significant amount, D_{C1} becomes forward-biased and clamps the collector to -6 volts. (In practice, of course, a small drop appears across the diode and the collector voltage differs from -6 volts by this small amount.) The waveform in Fig. 12-23c clearly indicates the rise-time improvement.

The waveforms for a flip-flop reveal much about its basic circuit action. A typical circuit is shown in Fig. 12-24, and the corresponding waveforms are shown in Fig. 12-25. The input pulse presented to the T input is seen to be a square wave, operating between ground and -6 volts. This is differentiated by the input capacitors and on the positive-going edge of the

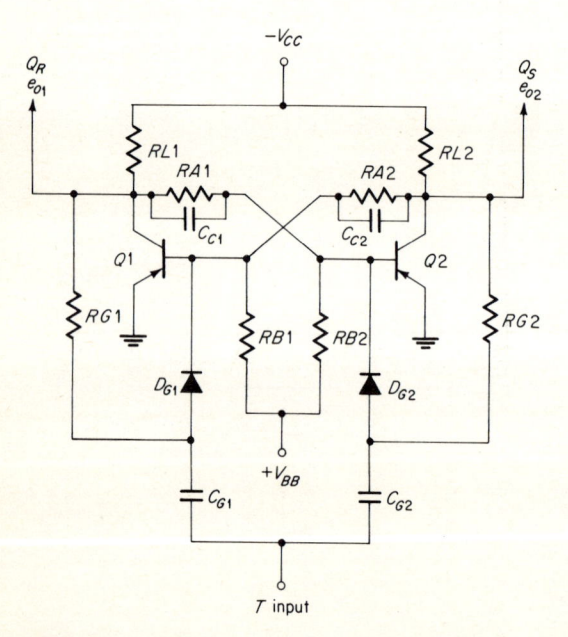

Fig. 12-24. A circuit to be used for waveform analysis.

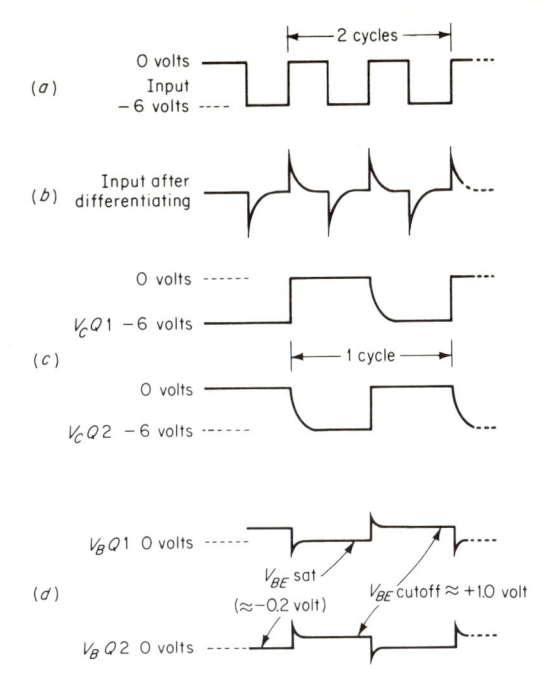

Fig. 12-25. Typical bistable-multivibrator waveforms: (*a*) input square wave; (*b*) input after differentiating; (*c*) collector waveforms yielded by the input shown; (*d*) base waveforms showing the undershoots and overshoots.

pulse effects the transition of the flip-flop. On the assumption that $Q1$ was initially off and $Q2$ was on, the collector waveforms show the transition clearly. $Q1$ goes from cutoff to saturation while $Q2$ goes from saturation to cutoff. At the same time, the base waveforms indicate the same action.

Note a significant factor: Two input cycles are required to produce a single cycle at the output. *The bistable multivibrator evidently is dividing the number of input pulses by* 2. This fact points to a basic property of the flip-flop, called *counting*. Such a circuit is capable of counting in a very real sense, and in Chap. 13 this aspect of circuit action will be thoroughly investigated. Careful inspection of input-versus-output waveforms reveals that if four cycles are presented to the input, two will appear at the output; eight pulses in result in four out, and so on. Additionally, because of the bistable nature of the circuit, it is capable of "remembering" the last state in which it was placed. The circuit is therefore used in applications where it can provide a memory function. Again, this aspect of circuit behavior will be dealt with later.

npn **Flip-flop**

Figure 12-26 illustrates a complementing flip-flop using silicon *npn* transistors. From the waveforms given in Fig. 12-27, the only significant difference can be seen to be the somewhat larger $V_{BE,\text{sat}}$ for these transistors. Of course, the supply voltage is positive with respect to ground and so the

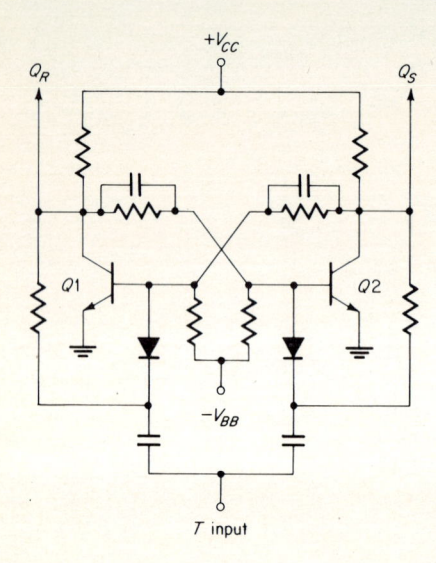

Fig. 12-26. *npn* version of standard flip-flop.

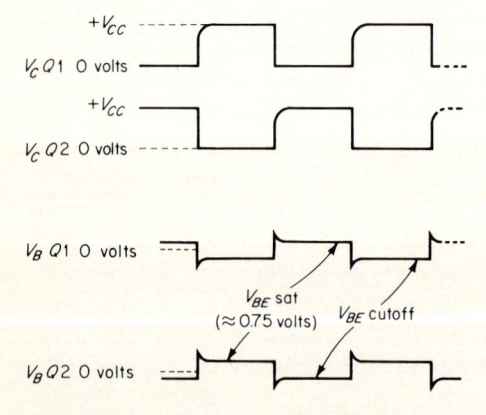

Fig. 12-27. Waveforms for the *npn* flip-flop.

collector waveforms vary between $+V_{CC}$ and ground, but otherwise the circuits will function identically.

12-4 MONOSTABLE MULTIVIBRATORS

As its name implies, the monostable multivibrator, or *one-shot*, has one stable state and therefore one quasi-stable state. Such a circuit is shown in Fig. 12-28, where close inspection reveals that only $Q1$ can be in its on

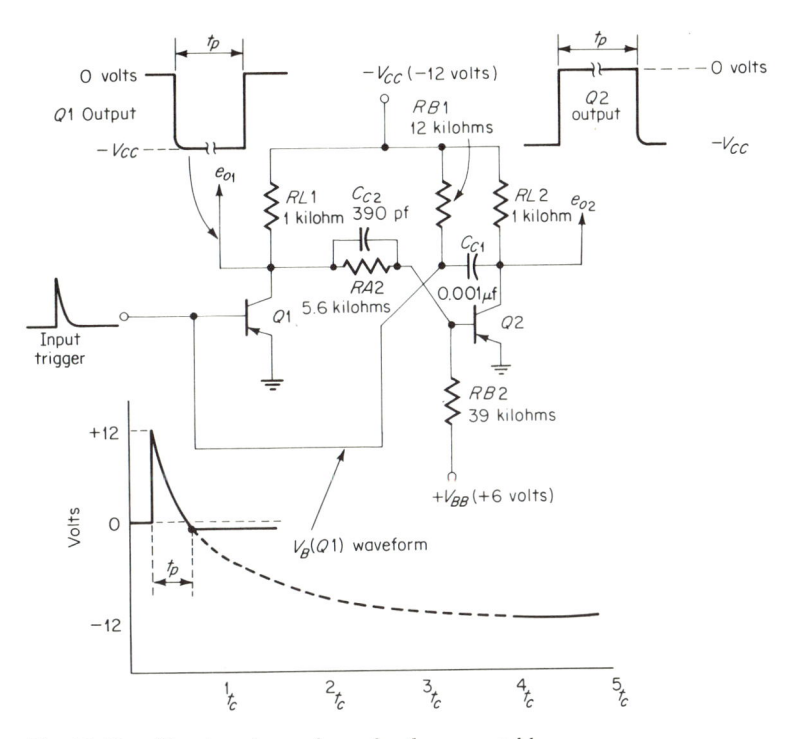

Fig. 12-28. Circuit and waveforms for the monostable multivibrator.

state quiescently. $Q2$, on the other hand, may be in either state because of the resistive voltage divider in the base circuit. If $Q1$ is on, the base of $Q2$ is more positive than its emitter and hence $Q2$ is off. If $Q1$ is off, the base of $Q2$ is returned to $-V_{CC}$ and $Q2$ is in saturation. Thus the quiescent state of the circuit is with $Q1$ saturated and $Q2$ off.

To yield an output the base of $Q1$ is presented with a short differentiated pulse (positive with respect to ground) and $Q1$ is thereby turned off.

As its collector rises toward $-V_{CC}$, the base of $Q2$ goes progressively more negative and $Q2$ is turned on. At this point, with the loop gain greater than 1, the input pulse may disappear, since the regenerative action will now take over and complete the cycle.

As $Q2$ comes on, its collector falls rapidly to ground, as indicated on the waveform at e_{o2}. Since C_{C1} cannot change its state of charge rapidly, the positive-going excursion of the $Q2$ collector is felt by the base of $Q1$. Since the action of C_{C1} is to cause the base of $Q1$ to go more positive (as will subsequently be proved), $Q1$ now goes off. The length of time $Q1$ is held off is a function of the time constant of C_{C1} and $RB1$. This, of course, determines the length of time of the quasi-stable state.

As time progresses, the base of $Q1$ drops from a positive voltage with respect to ground, and as this voltage begins to go slightly negative, the base of $Q1$ becomes forward-biased and $Q1$ comes on, terminating the quasi-stable state. The circuit then remains in its stable state, awaiting the next input pulse.

Basic Circuit Action

The analysis of such a circuit is relatively simple. As just described, the quiescent state of the circuit is the condition with $Q1$ on and $Q2$ off since there is no provision for a positive dc voltage to be applied to the base of $Q1$.

To verify that $Q2$ is firmly off during the stable state, the quiescent base voltage may be found. The voltage divider $RA2$ and $RB2$, as well as the voltage to which the divider is returned, determine this. Since if $Q1$ is firmly saturated its collector is essentially at ground potential, the divider has 6 volts across it:

$$V_{B2} = \frac{RA2}{RA2 + RB2} \times V_{BB} = \frac{5.6 \text{ kilohms}}{44.6 \text{ kilohms}} \times 6 = 0.1256 \times 6 \cong 0.75 \text{ volt}$$

This, of course, places $+0.75$ volt on the base of $Q1$, relative to ground, and $Q2$ is indeed off.

That $Q1$ is firmly saturated can be easily determined. Assuming that h_{FE} is no less than 40, the required base current to just saturate the transistor is

$$I_{B,\text{sat}} = \frac{I_{C,\max}}{h_{FE}} = \frac{|V_{CC}|/RL}{h_{FE}} = \frac{12/1{,}000}{h_{FE}} = \frac{0.012}{40} = 300 \ \mu\text{a}$$

Thus, if the base current is equal to or greater than 300 μa, $Q1$ must be in saturation.

$$I_{B,Q1} \cong \frac{|V_{CC}|}{R_{B1}} = \frac{12}{12 \text{ kilohms}} = 1 \text{ ma}$$

Therefore, more than three times the required base current is flowing and $Q1$ is firmly saturated.

When an input arrives with sufficient amplitude to raise the loop gain to more than unity, a transition occurs, with $Q2$ going rapidly to saturation. Its collector goes from -12 volts to nearly ground, a 12-volt positive-going excursion. Because C_{C1} has a relatively long time constant, this 12-volt change is seen by the base of $Q1$. But, since the base of $Q1$ was essentially at ground potential, it must rise to a voltage that is 12 volts more positive than ground. With $+12$ volts at its base, $Q1$ is, of course, turned off and will remain off until the charge on C_{C1} can leak off. This period of time (the quasi-stable state) is determined by the size of C_{C1} and $RB1$ and is a function of the time constant.

$$t_c = RC = (1.2 \times 10^4)(0.001 \times 10^{-6}) = 0.0012 \times 10^{-2} = 12 \ \mu sec$$

The length of time $Q1$ is held off, then, is determined by the following relation.

$$t_p = RC \ln \frac{E}{E - e_c} = 12 \ \mu sec \ \ln \frac{24}{24 - 12.1} \cong 12 \ \mu sec \ \ln 2$$

$$= 12 \ \mu sec \times 0.694 \cong 8.33 \ \mu sec$$

Thus, $Q1$ remains off for slightly more than 8 μsec. Note that this is determined by the same relationship used for a simple astable multivibrator since the base waveform is the same.

Waveform Analysis

Figure 12-29 shows typical one-shot waveforms on a common time base. The input-trigger pulses arrive at 13-μsec intervals, and, using the previous circuit values, the one-shot produces output pulses of 8-μsec duration.

Noting first the collector waveform of $Q1$, it is observed that since the transistor is quiescently on, its collector is at ground potential. While not clearly evident in the figure, this voltage is actually about -0.1 or -0.15 volt above ground. As a trigger pulse arrives, a transition occurs, with $Q1$ turning off and $Q2$ turning on. Because of minority-carrier storage, storage-time delay prevents the collector from rising rapidly to $-V_{CC}$, and hence the exponential curve as $Q1$ turns off. Also, any capacitive loading in the external circuitry (such as speedup capacitors, for example) will cause further rounding of the waveforms.

Eventually, however, the collector rises to $-V_{CC}$ and remains there for the duration of the quasi-stable state. Then, when the multivibrator *times-out*, or terminates, $Q1$ comes back on. Note that there is virtually no visible delay when turning on since storage time is not a problem at this

292 PULSE AND SWITCHING CIRCUIT ACTION

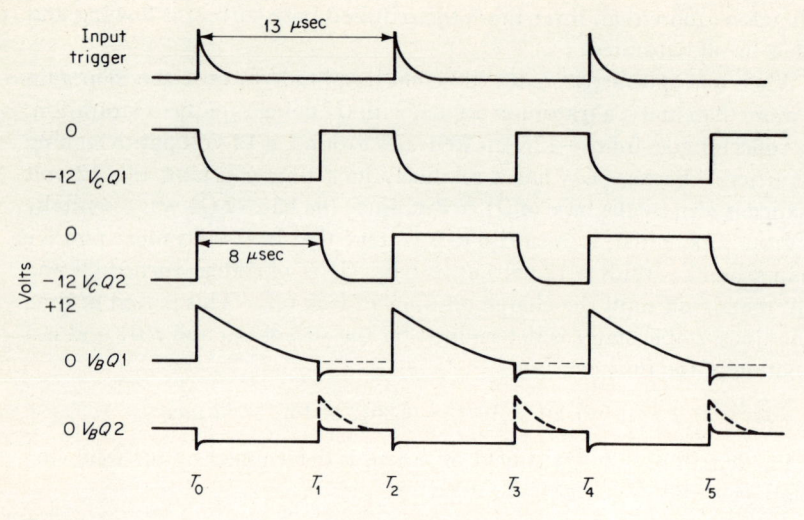

Fig. 12-29. Input, base, and collector waveforms for the monostable multivibrator on a common time base.

time; also, the transistor drives a capacitive load much better during turn-on time. The waveform, then, exhibits a very rapid and square-shaped excursion.

The waveform at the collector of $Q2$ is seen to be similar to that of $Q1$. It is, however, 180° out of phase, as, of course, it must be. The sharp positive-going edge and the exponential negative-going rise toward $-V_{CC}$ are explained in the same way as the waveform for $Q1$. The base waveform of $Q1$ shows the initial very sharp rise from about ground potential to $+12$ volts as $Q2$ comes rapidly on. This action is caused by the presence of C_{C1} and is quite similar to that of the coupling capacitor's action in the astable multivibrator circuit.

Once $Q1$ is turned off, the base-emitter junction of $Q1$ is reverse-biased and hence the only possible charge or discharge path for C_{C1} is by way of $RB1$. Thus, the rate of the slow exponential fall of V_B back toward ground is a function of the time constant of $RB1$ and C_{C1}, as explained earlier. As the base voltage crosses 0 volts and just begins to go more negative than ground, the base-emitter junction becomes forward-biased and $Q1$ again conducts. This, of course, causes $Q2$ to be cut off, and its collector now rises toward $-V_{CC}$. Since now $Q2$'s collector rises in a negative direction, the base of $Q1$ feels this change through C_{C1}. This is the cause of the small undershoot slightly evident on the waveform.

The base waveform of $Q2$ is somewhat different than that of $Q1$. This is to be expected in view of the different configuration of the base circuitry. Quiescently, the base is at a potential of +0.75 volt. As Q1 is turned off by the input trigger, its collector rises toward $-V_{CC}$; this change is first seen at the base of Q2 by the action of the speedup capacitor C_{C2}. After some time has elapsed, the voltage divider itself provides base drive for Q2. As long as V_{CQ1} is at $-V_{CC}$, the Q2 base receives a very large current—more than enough to saturate the transistor. The value of this base current can be estimated very closely by Thevenizing the base circuit. With Q1 off, the base of Q2 looks into an impedance, or, in this case, resistance, of

$$R_{th} = \frac{R_T \times RB2}{R_T + RB2} = \frac{6600 \times 39{,}000}{6600 + 39{,}000} = 5.64 \text{ kilohms}$$

where $R_T = RL1 + RA2$.

The Thevenin-equivalent voltage V_{th} is the open-circuit voltage at the junction of R_{A2} and R_{B2}, with the base disconnected:

$$V_{th} = \left(\frac{R_T}{R_T + RB2} \times 18\right) - 12 = (0.1447 \times 18) - 12$$
$$= -9.395 \text{ volts} \cong -9.4 \text{ volts}$$

The base-emitter saturation resistance $R_{BE,\text{sat}}$ of Q2 might reasonably be expected to be on the order of 200 ohms, as suggested by the Thevenin-equivalent circuit of Fig. 12-30. The current in this circuit, which is,

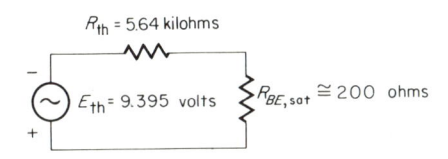

Fig. 12-30. Thevenin-equivalent circuit used to determine base current.

of course, base current, is determined as shown:

$$I_{B,\text{sat}} = \frac{|V_{th}|}{R_{th} + R_{BE,\text{sat}}} = \frac{9.4}{5640 + 200} \cong 1.61 \text{ ma}$$

Since collector current at saturation is the same as in the case of Q1 (12 ma), this is more than enough to just cause saturation. Note that the inclusion of $R_{BE,\text{sat}}$ hardly influences the value of $I_{BE,\text{sat}}$ at all since it is very much smaller than R_{th}. It could, then, be ignored without greatly decreasing accuracy.

It has been determined that $Q2$ is indeed in saturation when $Q1$ is off, and the waveform between T_0 and T_1 (VB_{Q2}) verifies this. The small spike at T_1 is again due to C_{C2}, the speedup capacitor. If this capacitor is too large, the positive spike may be similarly large, as indicated by the dotted line on the VB_{Q2} waveform. If the capacitor is too large, the exponential waveform falls back to $+0.75$ volt too slowly and $Q2$ can be held off for an excessively long time, which could cause the circuit to fail to perform properly.

It should be noted that the length of the output waveform T_0 to T_1, for example, is determined by the internal time constant of $RB1$ and C_{C1}, while the repetition rate ($1/T_0$ to T_2) is determined solely by the input trigger. The longer the period between input pulses, the longer the time between T_0 and T_2, T_2 and T_4, etc., but the period T_0 to T_1 remains the same, as does T_2 to T_3. Also, if an input trigger occurs during the quasi-stable state, the circuit will simply ignore it, providing the total energy of the trigger is relatively small and dissipates relatively fast. Often, a low-value resistor (≈ 1.0 kilohm) will be placed at the input terminal to provide a path to ground for rapidly dissipating this energy. The high end of this resistor is isolated from the base by a suitable diode, of course, to prevent altering the time constant of C_{C1}.

Triggering Requirements

The circuit just described, Fig. 12-28, assumed a trigger waveform that had already been developed into a spike. Usually, the differentiating and clipping circuits are incorporated as an integral part of the one-shot. Such a circuit is shown in Fig. 12-31. It is similar to the circuit given previously, except that it has a typical waveshaping network at the input. The pulse input for such a circuit, then, is a simple square wave, with the components $R6$, $D1$, $D2$, and $C3$ producing the short, sharp spike required at the base of $Q2$.

Typical waveforms for this circuit are shown in Fig. 12-32. The square-wave input is seen to be a 17-μsec pulse with a period of 34 μsec. If $D1$ and $D2$ were not included in the circuit, the base voltage would be a simple differentiated waveform. The purpose of $D1$, however, is to clip the negative spike. It does this since, when its cathode is driven more negative than ground, its internal resistance becomes very low because it is at this time forward-biased. Essentially it shorts the negative spike to ground, bypassing the resistor $R6$.

While this is occurring, any small negative voltage still left, owing to the finite resistance of $D1$, back-biases $D2$, and this prevents unwanted coupling to the base of $Q2$. The base of $Q2$, then, sees only a positive spike of relatively short duration. The time constant of $R6$ and $C3$ is 0.56 μsec. The full width of the spike, then, is five times this figure, or 2.8 μsec, which is sufficient to ensure a transition but short enough to disappear early in the quasi-stable period.

One further point must be made regarding the input possibilities of the one-shot. Occasionally, it is desirable to direct-couple a signal to one of the bases. This can have rather surprising effects under some conditions. If, through a suitable current-limiting resistor, a negative signal is applied to the base of $Q2$ that is of *longer duration* than the quasi-stable period, $Q2$ is, of course, turned on and a normal transition occurs, with $Q1$ being turned off by the action of C_1. When $Q1$ is timed-out by the decreasing base voltage, $Q1$ again comes on, as might be expected. However, since the input still appears at the base of $Q2$, it, too, is on and until the signal disappears *both* transistors remain on. $Q2$ only turns off when the input is removed; this causes the output waveforms to be altered from their usual appearance.

Additionally, if a direct-coupled input more positive than ground is applied to the base of $Q1$, it is turned off and will remain off for the duration of the input. Because of the direct coupling from the collector of $Q1$ to

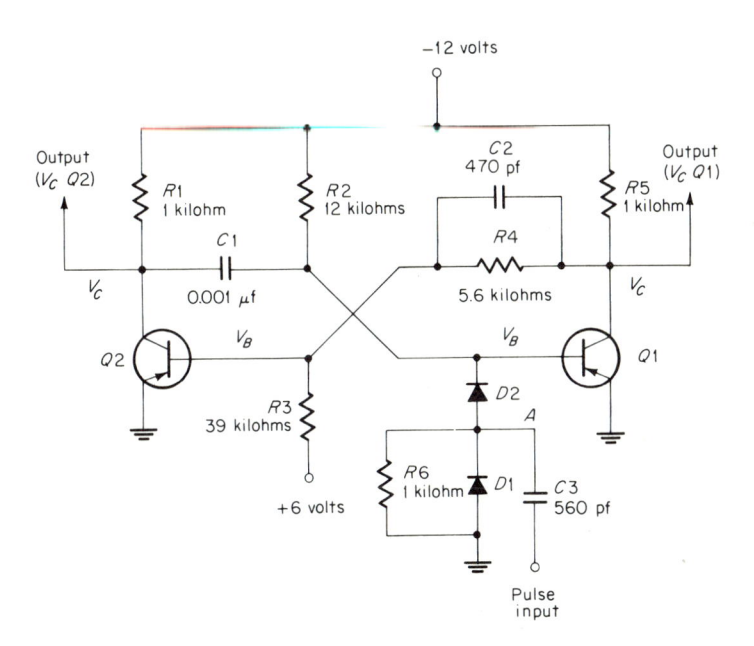

Fig. 12-31. A monostable circuit with differentiating input circuitry.

296 PULSE AND SWITCHING CIRCUIT ACTION

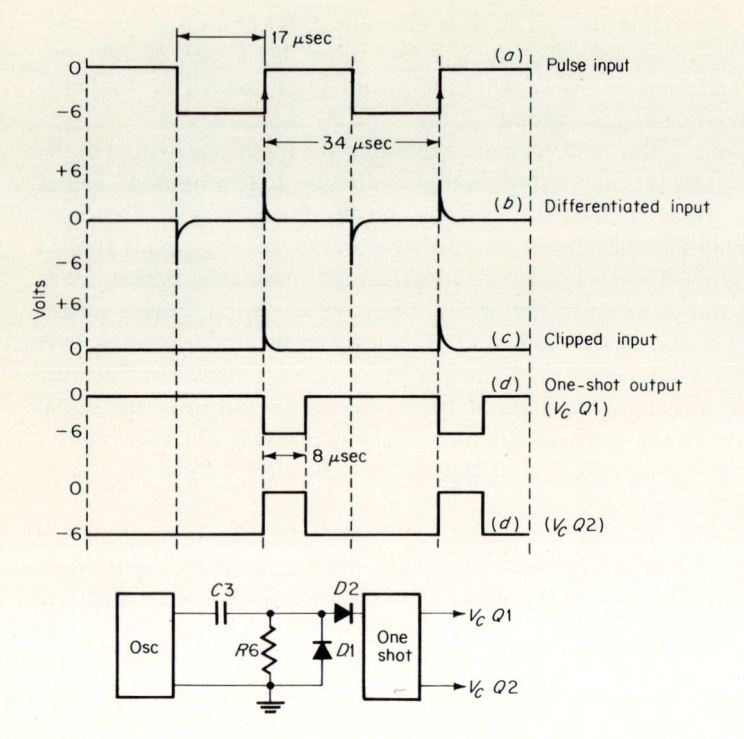

Fig. 12-32. Input and output waveforms for the circuit in Fig. 12-31.

the base of $Q2$, $Q2$ will remain on also for the duration of the input. In this instance the capacitor $C1$ has no effect on the circuit at all, and in reality the transistors are acting as cascaded inverters.

12-5 THE SCHMITT - TRIGGER CIRCUIT

The Schmitt-trigger circuit is a bistable multivibrator of a special sort. The conventional circuit has dc coupling from each collector to the opposite base, and in that case, the regenerative feedback is quite evident. In the case of the Schmitt circuit, one of these feedback paths does not exist as such, and another means is used to obtain regenerative feedback. In Fig. 12-33, $Q1$ is conventionally coupled via $R2$ and $R3$ to the base of $Q2$, as in any bistable circuit. The coupling from $Q2$ back to $Q1$, however, is via the emitter resistor $R5$ which is common to both transistors. By doing this, the circuit is given characteristics quite different from the usual bistable circuit.

The Schmitt-trigger circuit is unlike its more conventional counterpart in the fact that it is sensitive to changes in the *level* of the input voltage. Its main use, then, is converting a slowly changing waveform applied at the input to a square-wave output with very fast rise and fall times. As might be expected, it is stable in one state, and when triggered by a slowly changing level at the input, it makes a very fast transition to the alternate state. Then, when the input returns to its original value, the circuit returns to its stable state, again making a very fast transition. The width of the output pulse is therefore determined by the shape and width of the input waveform rather than any internal circuit constants.

Circuit Description

In the circuit of Fig. 12-33, $Q2$ is normally on. This is true because only the $Q2$ base is returned to $-V_{CC}$. With $Q2$ firmly on, $Q1$ is off. In the circuit shown, when $Q2$ is on, it is saturated and the emitter is at a level of 0 volts with respect to ground. This must be true since the internal resistance of the transistor $Q2$ is essentially zero. Thus each load resistor $R4$ and $R5$ must drop 6 volts, leaving 0 volts at the emitter, measured to ground. With zero on the emitter and the base of $Q1$, it is off. If now a negative-going signal of sufficient amplitude is applied to the coupling capacitor $C1$, $Q1$ will turn on. When this happens, its collector will fall toward 0 volts, and the base of $Q2$ will see a more positive voltage. $Q2$,

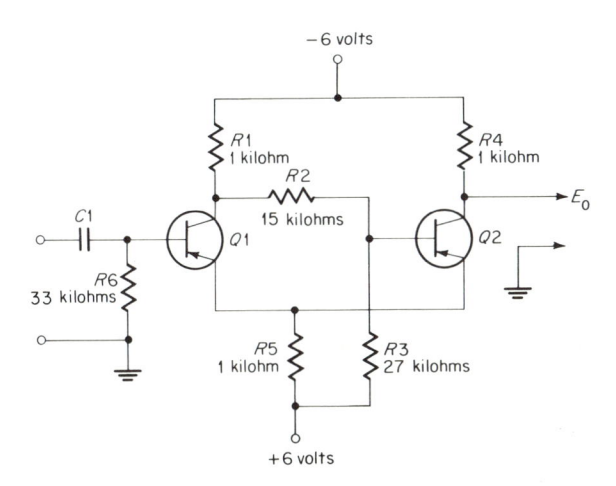

Fig. 12-33. The basic Schmitt-trigger circuit.

therefore, will turn off, and the circuit has completed the first half of its complete transition: Q1 is on and Q2 is off. The output E_0 rises from 0 to −6 volts.

As long as the input holds the base of Q1 in the condition of forward bias, Q1 will stay on and this will keep Q2 off. When the input begins to go positive again, at one point Q1 will turn off and this will bring Q2 back into conduction. The circuit has completed a full transition.

A typical set of waveforms for this circuit is quite enlightening, and in Fig. 12-34 these are shown. The waveform labeled "input" is shown as a slowly changing positive voltage; that is, it never goes more negative than ground. The waveform labeled V_B is viewed across the 33-kilohm resistor, and it does go somewhat below ground owing to the fact that the capacitor will not pass dc and the waveform tries to center around the base line.

As the base of Q1 begins to go very slightly negative, Q1 starts to turn on. It is, however, only very slightly in the active region at this first instant (slightly beyond T_0). Q2, of course, is firmly in saturation, and because of the equal resistors in its collector and emitter circuits, the emitter voltage is 0 volts with respect to ground. Hence, the emitter of Q1 is also at ground potential, held there by the large current through Q2.

Now, as Q1 starts to go on slightly, its emitter current adds to that of Q2 and the emitter goes very slightly more negative. At the same time, the collector of Q1 begins to fall toward ground, and this voltage change is larger than that at its emitter, indicating that the voltage gain of Q1 at this point (between T_0 and T_1) is considerably in excess of unity. This is true since the clamping action of Q2 on the emitters causes the emitter of Q2 to be essentially grounded. Thus, the voltage gain of Q1 is relatively large.

As the collector of Q1 goes more positive, its emitter is going more negative. Thus, the base of Q2 is driven more positive, while the emitter of Q2 is driven more negative. Either of these actions would eventually cause Q2 to turn off. At some point, then, Q2 begins to turn off. This helps drive the emitters more positive, which turns Q1 on harder. As Q1 goes further toward saturation, its collector goes more rapidly toward ground and Q2 is turned further off. Hence, each transistor helps the other, and at this point the regenerative transition occurs, shown as time 1 on the waveform. In other words, the loop gain is now greater than 1. Note the very sharp edges. This is indicative of the fact that the circuit produces regeneration since each stage is capable of amplification, and the circuit literally *snaps* from one state to the other.

Now, when the input begins to rise in the positive direction, a point is reached that causes Q1 to begin to turn off. Q1 starts into the active region, and its collector begins to rise from ground toward −6 volts. As

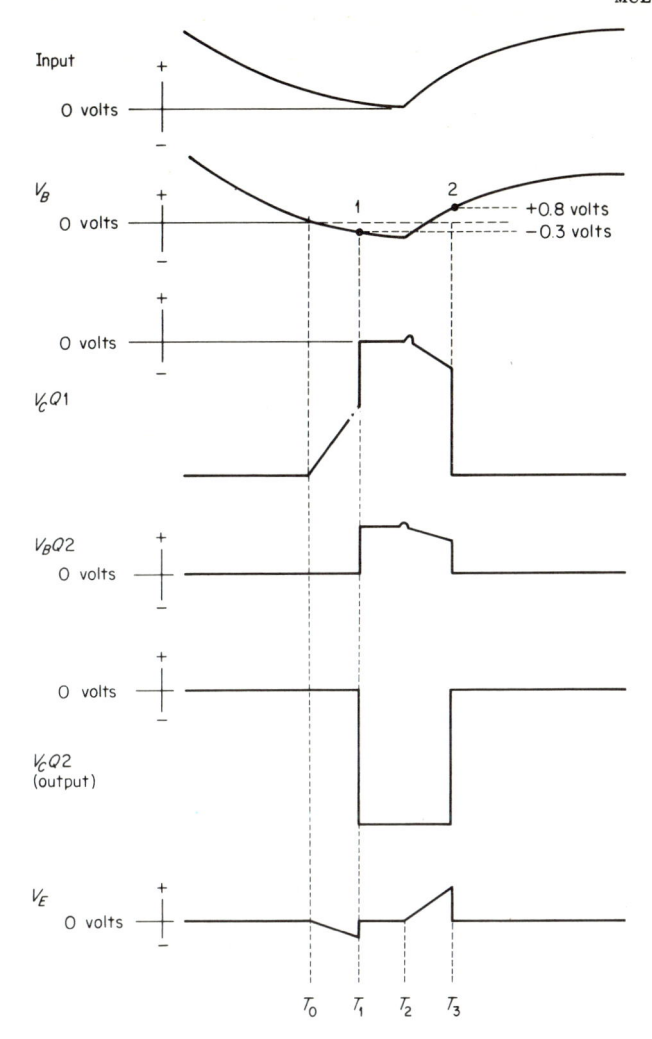

Fig. 12-34. Waveforms for the Schmitt-trigger circuit.

the collector reaches some negative value that is sufficient to begin to affect $Q2$, $Q2$ starts to turn on. Again, regeneration plays an important part in returning the circuit to its original condition, and the circuit snaps back.

Note that the point where $Q1$ turns on is about -0.3 volt but the point where it turns back off on the reverse transition is about $+0.8$ volt rather than slightly more positive than -0.3 volt. This peculiarity is typical of the Schmitt circuit and is called the *hysteresis*, or *backlash*, of the circuit.

The hysteresis is caused by several factors. One of these is related to the fact that a transition cannot occur until *both* transistors are driven into the active region. With the values shown, the base of Q2 must rise only 1.5 volts to begin to turn off. However, when it begins to turn back on, it must fall 2.25 volts from the full off condition before it comes back on. If the values of R2 and R3 were changed, this effect could be minimized, but another factor must also be considered.

During the quiescent state of the circuit, Q2 is on and Q1 is off. The emitter of Q1 is held to 0 volts by Q2 being in saturation. Q1, then, for all practical purposes, has its emitter tied to ground. This point will remain at ground until Q2 is turned off far enough to enter the active region. By then, Q1 will be far on. While Q1 is turning on, the voltage gain of Q1 is quite high, so that a very small change at the base will result in a large change at the collector. By the time the base of Q1 is only about -0.2 volt, Q2 is beginning to turn off and a transition occurs.

Now, when the first transition is completed, Q1 is on while Q2 is off. With Q2 off, it cannot affect the operation of Q1. As the input voltage begins to rise toward a more positive value, eventually causing a reverse transition, Q1 reenters the active region from saturation. But now, Q1 has a 1-kilohm load in the collector and a 1-kilohm load in the emitter. Hence, the voltage gain is unity (1). Now, in order that the collector can make a 1-volt excursion, the base will necessarily have to go much farther than before in the positive direction to begin a transition. On the waveform shown in Fig. 12-34, point 2 is noticeably more positive than point 1. The difference between these two values is the hysteresis.

With the output pulse produced by a negative-going input, the resultant pulse is wider than might be expected. On the other hand, if Q1 is held on quiescently and then turned off to produce the output, the result is a narrower pulse. Thus, the hysteresis must be considered as an important part of the circuit operation.

A very similar circuit is shown in Fig. 12-35, sketched in two ways. Each of these drawings is electrically identical to the other. Note that the simple voltage divider is replaced by an emitter follower, and this is used because it reduces the loading on the collector of Q1. Other than this change, there is no significant difference between the two circuits. The normal input to the circuit is as shown, and it is seen that the input holds Q1 in the on condition. With Q1 on and in saturation, its internal resistance is very low and the drop across the collector and emitter resistors is limited by the resistors: $I = E/R = 12/2000$, or 6 ma. Now, with 6 ma through each 1000-ohm resistor, there must be 6 volts dropped across each and the emitter of Q1 tries to be at 0 volts, or ground. However, in this circuit, the input line holds the base at about -1 volt, and with the transistor in saturation, the emitter and collector are forced to be at -1 volt, too. When the capacitor in the base lead of $Q1$ begins to charge to a more positive value (determined by something outside of the circuit shown), the input to $Q1$ goes positive. As the base, emitter, and collector go more positive, a point is reached where they go slightly

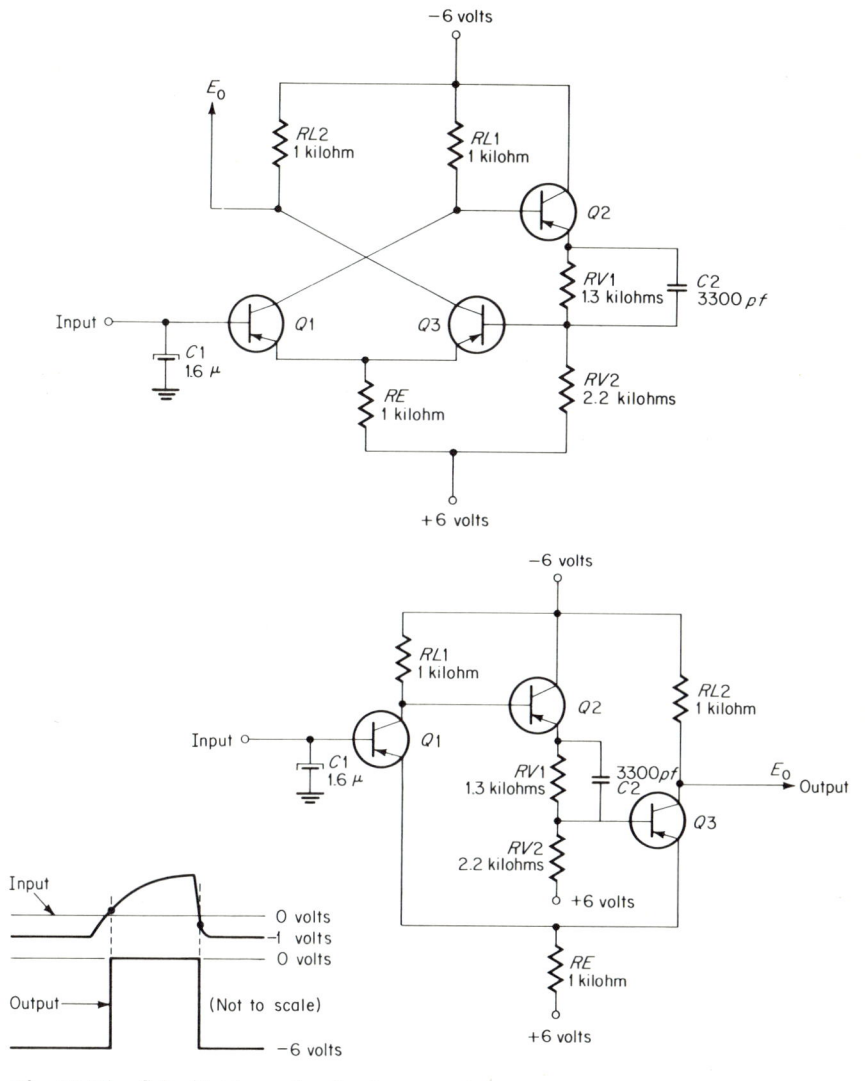

Fig. 12-35. Schmitt-trigger circuit using an emitter follower to isolate $Q1$ and $Q2$.

beyond the 0-volt point and $Q1$ begins to go into the active region. When this happens, the collector of $Q1$ begins to go in the negative direction. Through the emitter follower, this will, of course, drive the base of $Q3$ more negative and turn $Q3$ on. The base of $Q1$ was going positive and so was the emitter; this is the same as the base of $Q3$ going more negative, and so $Q3$ is helped on by $Q1$. As $Q3$ is driven to saturation, its emitter is driven back to zero and $Q1$ is very rapidly turned off. As $Q1$ turns off very rapidly, its collector goes toward -6 volts and $Q3$ is turned fully on very fast. The output makes a rapid transition from -6 volts to 0 volts (ground), as shown on the waveforms, and will stay there until the input goes back toward ground again, when a reverse transition occurs. Note the hysteresis that is evident in the waveforms, where the first transition occurs at a point on the curve that is more positive than ground while the opposite transition occurs when the input goes somewhat more negative than ground. In this instance, the length of time that the pulse is at ground is the pulse length, and so the hysteresis causes the pulse to be somewhat shorter than one might expect.

The emitter follower serves to isolate the collector of $Q1$ from the base of $Q3$ to reduce the loading on the collector circuit of $Q1$. This has the effect of speeding up the transitions, and capacitor $C2$ helps in this respect. It couples the fast changes from the collector of $Q1$ to the base of $Q3$, as well as helping to rid $Q3$ of its stored base charge as it goes out of saturation.

Q-Point Determination

Often, the quiescent operating condition of a Schmitt circuit is in doubt; that is, which of the two transistors is on and which is off is not always immediately apparent. Such a circuit is shown in Fig. 12-36. It operates in the same manner as those previously described, and the only functional difference is the steady-state condition, or rather the determination of this condition.

The most straightforward approach to the quiescent analysis of this circuit is to assume that one or the other transistor is on, regardless of whether or not this is actually the case. As an example, assume $Q1$ is on in this circuit. The emitter voltage V_E, with respect to ground, is a function of the voltage divider $R3$ and $R4$ and the total applied voltage.

$$E_{RE} = \frac{R4}{R4 + R3} \times V_{total} = \frac{500}{1500} \times 12 = 4 \text{ volts}$$
$$V_E = V_{EE} - E_{RE} = 6 - 4 = 2 \text{ volts}$$

With $Q1$ in saturation, then, the emitter is 2 volts more positive than

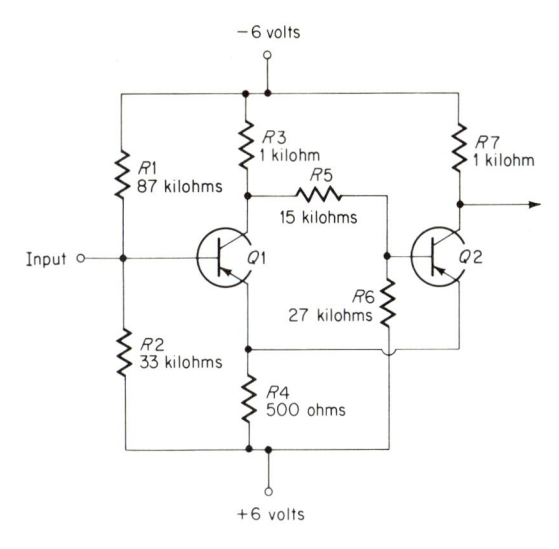

Fig. 12-36. Alternate circuit with Q1 biased by a voltage divider.

ground. The voltage divider in the base circuit of $Q1$ will determine the base voltage V_B.

$$E_{R2} = \frac{R2}{R1 + R2} \times V_{total} = \frac{33,000}{120,000} \times 12 = 3.3 \text{ volts}$$
$$V_B = V_{EE} - E_{R2} = 6 - 3.3 = 2.7 \text{ volts}$$

The base, then, would try to be 0.7 volt *more positive* than the emitter; hence $Q1$ cannot be in saturation.

To verify that $Q2$ is the transistor that is quiescently on, apply the same reasoning to it:

$$E_{RE} = \frac{R4}{R7 + R4} \times V_{total} = \frac{500}{1500} \times 12 = 4 \text{ volts}$$
$$V_E = V_{EE} - E_{RE} = 6 - 4 = 2 \text{ volts}$$

With $Q2$ on, $Q1$ must be off and the base of $Q2$ returns to $-V_{CC}$ through $R5$ and $R3$. The base voltage is therefore determined by the supply voltage and the divider, consisting of $R3$, $R5$, and $R6$.

$$E_{R6} = \frac{R6}{R6 + R3 + R5} \times V_{total} = \frac{27,000}{43,000} \times 12 = 7.5 \text{ volts}$$
$$V_B = V_{EE} - E_{R6} = 6 - 7.5 = -1.5 \text{ volts}$$

304 PULSE AND SWITCHING CIRCUIT ACTION

This indicates that the base is trying to be -3.5 volts with respect to the emitter (the difference between $+2$ and -1.5 volts), and $Q2$ is therefore far into saturation. Of course, the base cannot actually be -1.5 volts with respect to ground since the forward-biased emitter-base junction clamps the base to approximately -1.8 volts, allowing a 0.2-volt drop across the junction. Thus, $Q2$ can be quiescently on and $Q1$ cannot, so that it is quite evident that $Q2$ is on while $Q1$ is off.

12-6 THE BLOCKING OSCILLATOR

The blocking oscillator is not a true multivibrator but resembles the Armstrong oscillator in some respects. The energy from the collector circuit is inductively coupled to the base to produce regeneration. However, there is no resonant tank, and the circuit does not produce a sinusoidal output waveform. Its output is a short, narrow pulse of relatively large power. It can be designed to be either astable (free-running) or monostable (one-shot). The frequency stability of the astable version is poor, but the circuit can easily be synchronized by an external signal.

A typical astable blocking oscillator is shown in Fig. 12-37. The pulse

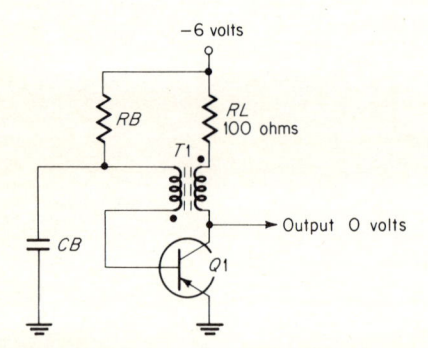

Fig. 12-37. Astable blocking oscillator.

transformer is connected in such a way as to yield regeneration and often has 1:1 turns ratio. The resistor labeled RL is actually a current-limiting device rather than a true load. In many circuits, the dc resistance of the transformer windings is negligible, and thus the need for some current limiting is apparent. For the circuit shown, maximum collector current is set at 60 ma for the given values:

$$I_{C,\max} = \frac{|V_{CC}|}{RL} = \frac{6}{100} = 0.06 \text{ amp} = 60 \text{ ma}$$

RB is simply the base-biasing resistor, allowing a dc return path for base current. Finally, CB is the timing capacitor that allows the repetition rate, or frequency, to be set to some particular value. The larger the value of CB, the lower the repetition rate.

Fundamentally, the circuit action is not unlike that of any regenerative oscillator. When power is first supplied, base and collector current increase from zero, and as collector current rises, a counter-emf is induced in the primary of the transformer such that the upper end (black dot) is negative.

In the secondary (base) circuit, the dotted end also becomes negative, turning the transistor on more fully. This in turn drives the base still more negative. The regenerative action very quickly drives the transistor to saturation.

When collector current reaches its maximum value, there is no longer a changing condition, and the induced voltage is zero. During the time collector current is increasing, the base is driven more and more negative. At this same time, the upper plate of CB is driven positive from the undotted end of the secondary winding. During this time, this positive voltage does not affect the transistor since the induced secondary voltage isolates the upper plate of CB from the base.

When the collector current has stopped increasing, the secondary voltage is zero. Now the capacitor voltage *does* influence the transistor, and $Q1$ begins to turn off. Current does not, of course, decrease instantly because of both minority-carrier storage and the induced voltage of the transformer, as well as the series resistance ($t_c = L/R$).

When the transistor reaches cutoff, the remaining charge on CB must dissipate by way of RB and the transistor is held off for a period of time that is a function of the time constant of RB and CB. As the upper plate of CB begins to go to perhaps -0.1 volt, $Q1$ again turns on and the entire sequence begins again.

A monostable blocking oscillator is shown in Fig. 12-38. This circuit is not free-running and yields an output only when triggered by an outside source. Normally, $Q1$ is held off by the voltage drop across $R1$, which places a small positive voltage on the base. A short negative spike at the input causes $Q1$ to conduct momentarily. A regenerative action now occurs, driving $Q1$ into saturation.

As soon as collector current stops increasing, the secondary voltage drops to zero. The voltage divider again allows the base to rise to a small positive voltage, and $Q1$ turns off, where it remains until the next trigger pulse. The output, then, is a positive excursion from $-V_{CC}$ to ground and back to $-V_{CC}$ again.

The purpose of $D1$ is to reduce the inductive *backswing* or overshoot,

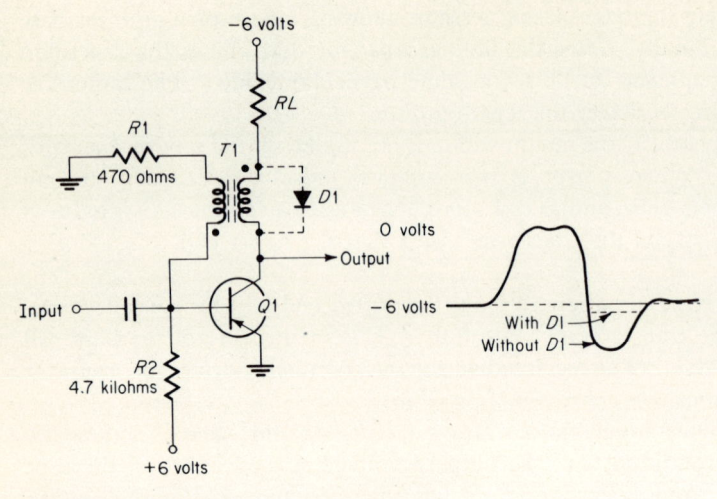

Fig. 12-38. Monostable blocking oscillator.

that occurs as the transistor is turning off. When the collector of Q1 is driven more negative than −6 volts, the diode conducts and clamps the collector to perhaps −6.5 volts. This, of course, prevents transistor damage, which might occur if the overshoot exceeded the BV_{CEO} rating of the transistor.

The transformers used in such a circuit range from the tiny "gumdrop" pulse transformers used for pulses in the range of microseconds to large iron-core transformers for repetition rates in the 60-cycle range. In general terms, the wider the pulse, the greater the necessary inductance of the transformer.

QUESTIONS AND PROBLEMS

12-1 Briefly define the word *regeneration*.

12-2 Briefly define the word *positive-feedback*.

12-3 Refer to Fig. 12-5. In this circuit, if $\beta \geq 33$ for each transistor, the circuit will not reliably self-start. Briefly describe why, if $\beta < 33$, the circuit *will* self-start.

12-4 Refer to Fig. 12-5. The load resistors are to be changed from 1 to 2 kilohms. To make the circuit self-starting, the transistors must be hand-picked to have a beta less than a specified amount. Determine this amount.

12-5 Refer to Fig. 12-5. *RB2* is to be changed from a 33- to a 66-kilohm value, while CC_2 is to be changed to a 0.002-µf value. Using these new components, is the circuit symmetrical or unsymmetrical?

12-6 Refer to Fig. 12-5. One collector-load resistor is to be changed from 1000 to 2000 ohms. Will this have any significant effect on the circuit symmetry?

12-7 Refer to Fig. 12-5. Compute the length of time that *Q*2 is held off if CC_2 is changed to a 0.002-µf value.

12-8 Refer to Fig. 12-5. Compute the length of time *Q*2 is held in the off condition if CC_2 is changed to a 1.0-µf value.

12-9 Refer to Fig. 12-10. Briefly describe the effect of halving the value of the two 27-kilohm resistors.

12-10 Refer to Fig. 12-10. The emitter resistor is to be changed to a 2200-ohm value. Briefly describe the effect on the circuit.

12-11 Refer to Fig. 12-15. Assume that *Q*1 is on and *Q*2 is off and that *SW*1 and *SW*2 are continuously open. Briefly and in general terms describe the circuit action.

12-12 Refer to Fig. 12-15. With the circuit as drawn, briefly describe the conditions necessary to cause a transition.

12-13 Refer to Fig. 12-18. Briefly describe the purpose of the *T* input.

12-14 Refer to Fig. 12-18. Assume that *RG*1 and *RG*2 are returned to ground rather than to the collectors.
(a) Would this be beneficial?
(b) Briefly describe the resulting circuit action.

12-15 Refer to Fig. 12-28. Changing only the value of CC_1 from 0.001 to 0.1 µf, determine the length of the quasi-stable state.

12-16 Refer to Fig. 12-28. Changing only the value of CC_1 from 0.001 µf to 100 pf, determine the length of the quasi-stable state.

12-17 Refer to Fig. 12-36. Changing only *R*1 from 87 to 100 kilohms, determine which transistor is quiescently on.

12-18 Refer to Fig. 12-36. Change only *R*1 from 87 to 62 kilohms. Determine which transistor is on.

13
COUNTERS

The subject of electronic counters is a large one. A counter can be described as a tallying device that tallies, or counts, some number of events. Electronic counters are somewhat analogous to mechanical counters, an example of which is the odometer of an automobile. The odometer counts the number of miles traveled by the car by means of a mechanical linkage between the several counter wheels, upon which the digits 0 to 9 are printed. Eventually, as many miles are put on the car, the capacity of the counter is reached (usually 99,999.9 miles), and with further rotation of the least significant wheel, the counter then resets to zero, ready to start all over again.

Electronic counters function in much the same general way. Pulses are introduced at the input, and the counter tallies each and every one, keeping a cumulative total that can be retained indefinitely when the counting stops. If the ultimate capacity of the counter is reached and if pulses are still present to be counted, the counter resets to zero and starts again to count from the beginning.

One advantage of the electronic counter is the tremendous speed with which it can count. Pulse-repetition rates well in excess of one million per second (1 MHz) can be easily handled by modern counters. A mechanical device, on the other hand, is hard put to exceed 30 operations per second.

An electronic counter can process information originating anywhere, providing only that the data can be converted into electric pulses, each of which represents one bit of information, or one *happening*.

A counter, then, has the following general characteristics. It accepts pulses and counts them in a quantity up to its inherent capacity. Then it automatically resets and continues counting if pulses are still presented to the input. The counter may have read-out facilities that allow the instantaneous state of the counter to be known or sensed. When it fills up and resets (or overflows), an output can be developed to indicate that the overflow existed and, perhaps, to further tally the number of overflows.

As an example, assume a certain counter has an inherent capacity of 10. Thus, each overflow that exists indicates that the counter has tallied 10 more counts. The first overflow, therefore, indicates that a total of 10 counts has been received, the second overflow that 20 counts have been received, etc. When the pulses stop arriving (or the counter is stopped), a residual count may exist. By suitable design, this can be made to be exhibited for easy determination. If the total input pulses numbered 35 in the foregoing example, the overflow counter would indicate three overflows (30 counts) and the residual-count mechanism would show five counts still in the counter.

The residual count can be made visible by several means. Neon bulbs (NE-2 and similar types) activated by the counter can be caused to light behind an opaque cutout for each digit to be represented, thus making the above five counts visible. Another method is to "paint" the digits on the face of a cathode-ray tube. Alternatively, special neon bulbs are available with the digits formed from fine wire, each of which is brought out to an external connection. As each wire is energized, that particular digit is caused to be surrounded with a luminous sheath and thus to become visible.

Most counters have additional features, among which are a reset bus to clear the counter prior to a new counting sequence and a means to *set* into the counter a particular set of conditions without the necessity to count up to this point. Also, feedback loops are often included as a part of the counter to alter the counting sequence and thereby change its inherent count capacity. Examples of many of these special attributes will be given in this chapter.

13-1 BINARY UP - COUNTERS

When several flip-flops are connected together in a certain way, they form a counting chain that is capable of actually counting or tallying events. The event can be anything, provided only that its existence can be converted into an electric pulse. Such a counter is shown in Fig. 13-1, which represents a scale-of-16 binary counter consisting of four flip-flops properly connected together.

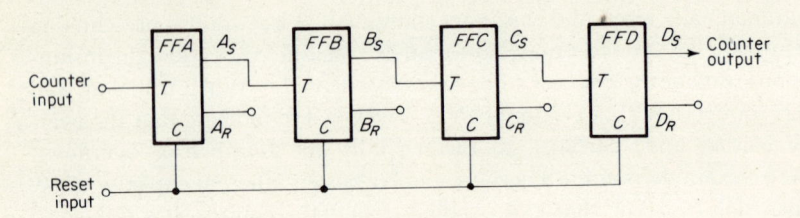

Fig. 13-1. Logic diagram of a scale-of-16 binary up-counter.

As shown, the pulses to be counted are applied to the T input of the A flip-flop. Succeeding flip-flops are connected to the set-side output of the preceding circuit through their T-input connections. The counter output is taken from the set-side D flip-flop. The clear (reset) inputs connect to the common reset inputs.

For purposes of explanation, the circuit will be assumed to be constructed with *pnp* transistors using a negative supply and negative logic. This will allow the circuit voltages, as well as the logic, to be explained in terms of the waveforms generated at each collector. These waveforms are shown in Fig. 13-2.

Initially, before a meaningful count can begin, a reset pulse must appear to cause a clear action. The state of the counter must be known before a count is started. When reset, the outputs of each flip-flop are such that the set sides are at logic 0 (ground), while the reset sides are at logic 1 (−6 volts). The logical equation for the reset condition is $\bar{A}\bar{B}\bar{C}\bar{D}$, and this simply indicates in Boolean language that logic 1 is at the reset-output wires.

The pulses that will be counted are represented in Fig. 13-2 at the top. This is recognized as a continuous train of negative-going pulses. As the square-wave pulses begin, the first excursion does nothing because it is negative-going and is blocked by the diodes in the base circuit of the A flip-flop. (The circuit of a flip-flop that might be used in this application is given in Fig. 12-18.) Now, when the input waveform makes its positive-going excursion, it will cause a transition in the A flip-flop, as shown by the collector waveform.

Note the arrowheads on the flank of certain parts of the waveforms. These arrowheads indicate that this rapid change in voltage causes a

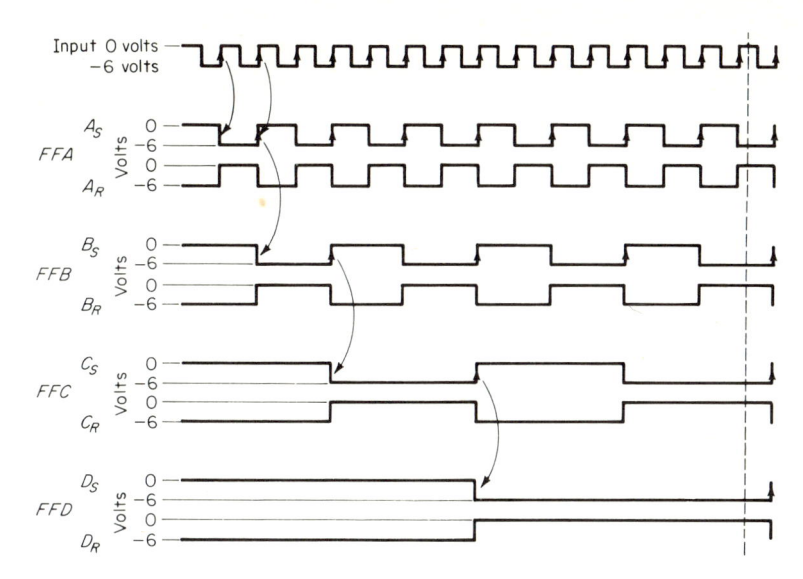

Fig. 13-2. Collector waveforms for the scale-of-16 counter.

transition in the following stage. The long curved lines show that, for example, the first positive-going excursion of the input causes a transition in the A flip-flop, which is evident on the A_S wire and is a negative-going excursion. This, of course, implies that the B flip-flop cannot transfer. When the A flip-flop again transfers on the next pulse, its collector now goes in the positive direction and this does now transfer the B flip-flop. However, because the B_S wire goes negative, the C flip-flop does not change its state.

This action continues, with the A flip-flop transferring at every pulse, the B flip-flop every other pulse, the C flip-flop every fourth pulse, and the D flip-flop every eighth pulse. When the input pulse has made 15 positive-going excursions (dotted line), the collectors are *all* in the opposite condition from reset for the first time. Now, -6 volts is at A_S, B_S, C_S, and D_S, while A_R, B_R, C_R, and D_R are all at 0 volts. The very next input pulse will cause A_S to go positive, which transfers the B flip-flop. B_S is going positive and so will transfer the C flip-flop. Because C_S is positive-going, the D flip-flop is transferred.

After 16 total input pulses, an output (the positive-going excursion) is produced from the D flip-flop. The circuit has returned to the reset condition and is ready to continue counting if more pulses are provided at the input. It is only necessary to have a mechanical device (or a separate counter) to tally each output from the D flip-flop to record the total number of input pulses. Each output of the D flip-flop (a positive-going excursion) indicates that there have been 16 input pulses; for example, if there are exactly 10 pulses from the D flip-flop, there must have been 160 input pulses.

Close inspection of the waveforms will reveal some interesting and enlightening facts. Note that flip-flop A makes twice as many transitions as flip-flop D. Flip-flop B, by the same token, makes twice as many transitions as flip-flop C, etc. This is a most significant observation since it indicates that *each flip-flop is dividing the output of the previous stage by two.* In other words, with 16 input pulses, flip-flop A makes only $16/2$ or eight complete transitions, flip-flop B makes $8/2$ or four transitions, flip-flop C makes $4/2$ or two complete transitions, and flip-flop D makes $2/2$ or one complete transition.

The circuit is counting by twos, and so each flip-flop can be weighted according to the way it responds to the counting sequence. The input-circuit flip-flop A is weighted 1, flip-flop B is weighted 2, flip-flop C is weighted 4, and flip-flop D is weighted 8. (This idea is similar to the column weighting of 10 in our decimal system. See Chap. 11 for more complete information on binary arithmetic.) The arithmetic weighting of the flip-flops simply determines what the worth is of any residual count left in the counter after the counting is finished. The weighting also indicates how many input pulses are required at the input to the entire circuit to transfer the flip-flop from the reset to the set (or transferred) condition. For instance, flip-flop C is weighted 4, and it requires four input pulses into flip-flop A to cause flip-flop C to transfer from reset to set. By the same token, flip-flop D is weighted 8, and from the waveforms in Fig. 13-2 it can be seen that it requires eight pulses into flip-flop A to transfer flip-flop D from reset to set.

The main advantage in using an electronic counter to actuate a mechanical counter is that it is capable of operating with tremendous speed. The input pulse can arrive much faster than a mechanical counter can follow—in this instance, 16 times faster. The electronic counter can be made to count several million pulses per second if needed, and this, of course, is far beyond the capability of a mechanical device.

Another advantage is that the circuit has a built-in memory for any remainder that should occur. For instance, suppose 23 pulses appeared at the input, but no more. At the sixteenth pulse, the D flip-flop will produce an output but the counter goes on through 17, 18, 19, 20, 21, 22, and 23; it will stop counting after the twenty-third pulse since no more input pulses are available.

If it is necessary to determine the residual count that remains in the counters, the state of each flip-flop will depend upon the relative voltages at the collectors, which reflect the logic condition. These voltages can be measured to determine if the flip-flop is set (transferred) or reset. If it is reset, the binary weight of the flip-flop is zero, but if transferred, the weight of the flip-flop as previously determined is added to the total remainder. In the above instance, with 23 pulses applied, there would be a residual count, or remainder, left in the counter. This would be indicated by the fact that certain flip-flops would be left in the set condition rather than reset.

Since $16 + 7 = 23$, the residual count must be 7; therefore the A, B, and C flip-flops would be in the set condition, with D remaining in the reset condition left by the sixteenth count. Thus, according to the previously discussed weight of each flip-flop, with flip-flops A, B and C in the set condition, $A_S = 1$, $B_S = 2$, $C_S = 4$, and $D_S = 0$; the residual count is then $1 + 2 + 4 + 0 = 7$.

Now, using negative logic and with $V_{CC} = -6$ volts; if the A flip-flop is in the set state, $A_S = -6$ volts and $A_R = 0$ volts, but if it is reset, $A_S = 0$ volts, and $A_R = -6$ volts. The state of the flip-flop can be measured with a voltmeter to determine whether it is set or reset, but a better way is to use a light bulb at each collector to show if a transistor is on or off. Such a scheme is shown in Fig. 13-3a.

If the flip-flop is reset, $Q1$ is off and the light bulb is connected between -6 and -6 volts and so cannot light. With the light out, then, the flip-flop is reset. If the flip-flop is caused to be set, $Q1$ is on and its collector is firmly at ground. The bulb is now connected between -6 volts and ground and so will light, thus indicating the set condition. If the counter is rapidly scaling through a series of counts, the bulb will probably not light up, but when the counting sequence is finished, any flip-flop containing a residual count will indicate its set condition by the fact that its light will then come on. ($Q1$, of course, must be able to adequately handle the extra current drawn by the bulb. In some circuits, an emitter follower is used between $Q1$ and the bulb to more adequately supply bulb current.) The lights are labeled according to the weights of the flip-flops to which they are connected. An *npn* version is given in Fig. 13-3b.

In the foregoing example of the scale-of-16 counter with a residual count of seven, such an array of lights would appear as shown in Fig. 13-4. Many other systems of residual read-out have been devised using neon bulbs, digital read-out tubes, cathode-ray tubes, as well as others.

A scale-of-16 counter or any other counter is of no use unless it can perform meaningful work. As mentioned earlier, one advantage of an electronic counter is the tremendous speed with which it is capable of operating, as opposed to a mechanical system. Also, a given counter can

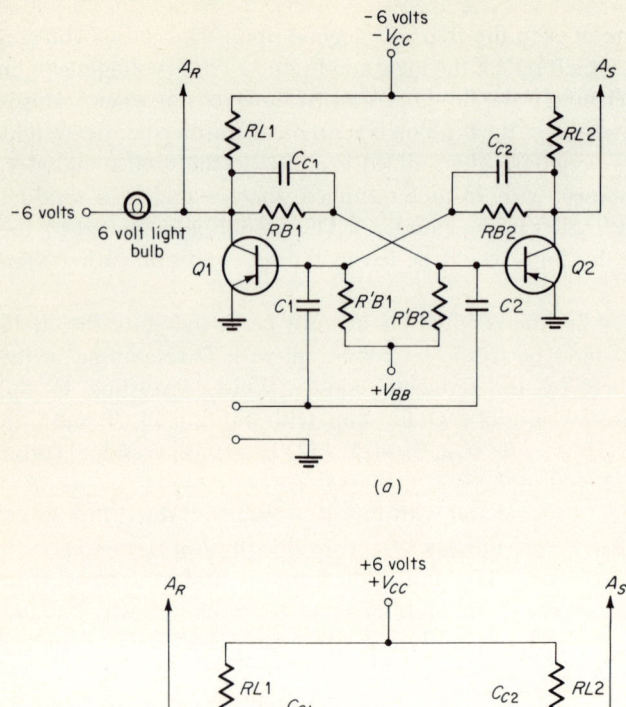

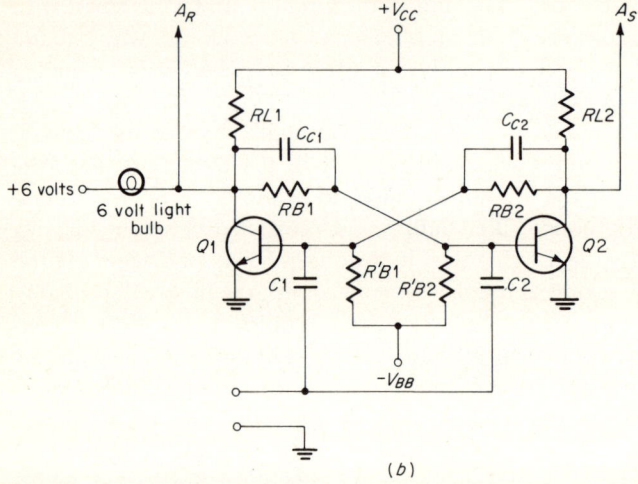

Fig. 13-3. (a) *pnp* circuit using a lamp to indicate whether the flip-flop is set or reset; (b) *npn* circuit with an indicator lamp.

be constructed so as to have a total count capacity of any reasonable number, limited only by the available space to put the many bistable circuits. For instance, if 20 bistable circuits are connected as a counter, it requires 1,048,576 input pulses to yield one output pulse. A counter with so large a capacity is not uncommon in certain fields.

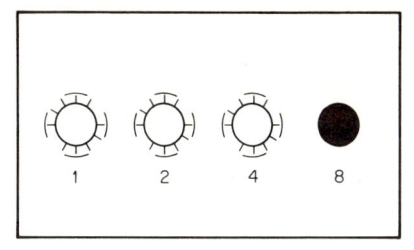

Fig. 13-4. A visual indication of the set-reset condition of a scale-of-16 counter having a residual count of seven.

The number of flip-flops, then, determines the total count capacity, according to the following relationship:

Capacity = 2^n

where n = number of stages (flip-flops).

Examples:

Capacity = 2^4 = 16
Capacity = 2^{20} = 1,048,576

An example of a typical counter and how it might be used is shown in Fig. 13-5. Without going into any detail of the matrix itself at this time (see Chap. 14), its purpose is to convert the binary conditions of the collectors (counting by twos) to decimal conditions on the wires labeled

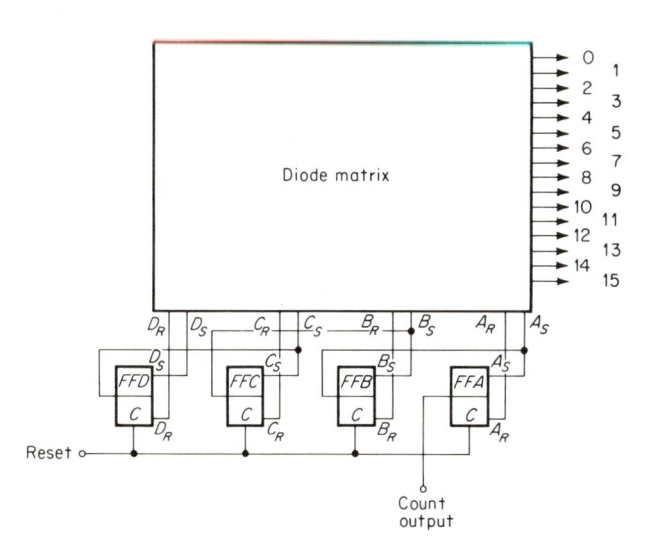

Fig. 13-5. A counter with a decimal decoding matrix and decimal output lines.

0, 1, 2, 3, etc. Thus, with the counter at reset, the line labeled 0 will be energized (−6 volts, perhaps) and all others will be nonenergized (0 volts).

The first input pulse will cause line 1 to be energized; the second pulse will energize line 2; etc. Therefore, the circuit could be classed as a *decimal-to-binary-to-decimal* converter since this is its main function. The pulses arriving at the pulse input are decimal by nature, and the output of the counter is coded in binary. The matrix converts the binary representation back to decimal. At this point, however, the wire will stay energized for any length of time necessary, whereas the input may have been a pulse of only a few microseconds duration.

There are several ways in which a counter can accept an input. One of these is via the normal counting T input, and in this case the counter will count, or scale, in the usual manner just described. But it is possible for the counter to accept a signal via other inputs, one of which is called the *preset input*. Using these inputs, we can literally force a binary number into the counter without waiting for it to count up. Figure 13-6 shows a

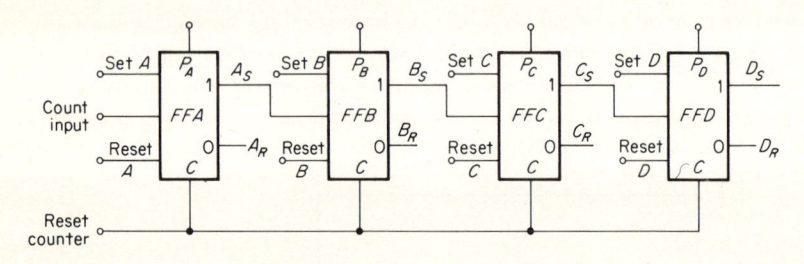

Fig. 13-6. A counter with various inputs to increase the inherent flexibility.

typical circuit in a logic block diagram, clearly indicating the preset P inputs. These are directly connected to the base of the transistor, usually through a small current-limiting resistor, and if the flip-flop is initially reset, will cause a transition by turning on the appropriate transistor when the input swings negative for *pnp* transistors or positive for *npn* transistors. This causes these flip-flops to become set.

As an example, suppose the counter is initially reset (logic 0 at A_S, B_S, C_S, and D_S) and it is necessary to instantly put a binary-coded 7 in the counter. By providing a logic 1 (dc level) at P_A, P_B, and P_C but a logic 0 at P_D, a binary 7 will be set into the counter in a length of time approximately equal to the pulse rise time. Again, if flip-flops A, B, and C are set while flip-flop D is reset, the counter state represents a 7. Note that the flip-flop action occurs on the *leading* edge of the inputs as the circuit is usually designed.

Other inputs shown are the set and reset inputs drawn parallel with the T input. These inputs require a transition (pulse) input and will cause a transition on the *trailing* edge of the pulse. This is very similar to the action of the T input.

It should be noted that the use of individual reset inputs should be effected with caution. Under certain conditions, unexpected counter settings can occur. For example, suppose flip-flop A is set while all others are reset. If the reset A input is used, the A flip-flop will certainly reset *but the B flip-flop will become set*. This occurs since, when a flip-flop makes a transition from set to reset, it toggles the next stage. If, as usually would be the case, this is an undesired result, other means must be used. The use of the reset-counter input will cause all stages to become reset; or the reset A input can be used if additional gating is provided at the T input to the B, C, and D flip-flops to inhibit unwanted toggling of B, C, or D.

13-2 PERMUTED COUNTERS

One definition of permutation is: ". . . *the act or process of changing the lineal order of an ordered set of objects*." This is exactly the meaning we ascribe to the word when we speak of a permuted counter. If the natural order of a counter is changed, or altered, the counter is said to be permuted. The natural order of a scale-of-16 (*hexidecimal*) counter is as described previously, with each flip-flop dividing its number of input pulses by two and producing the collector waveforms of Fig. 13-2. It is often necessary to have a counter that will count to some other base than the normal binary progression. A very common base is that used in the decimal system, although many others have been used, as will become evident.

By properly changing the circuitry to perform the correct permutation, it is possible to change the natural order of a counter to nearly any other base. We shall show several examples of permutation as used in typical equipment.

In one instance, an otherwise normal scale-of-16 counter is called a *modulus counter*, and it is a permuted counter. Without going into the philosophy of the equipment and why the counter needs to be permuted, the circuit itself can be discussed as an excellent example of a permuted counter. A block diagram of this counter is shown in Fig. 13-7. The individual flip-flops are labeled A, B, C, and D and are interconnected to perform normal counting. An additional bit of circuitry is shown that takes an output from the D flip-flop, inverts it, and presents it back to the

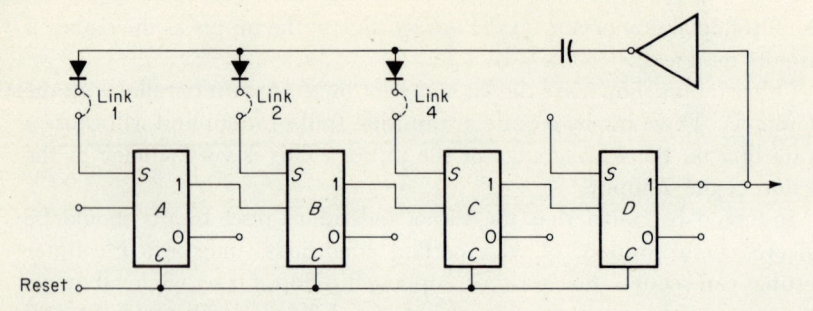

Fig. 13-7. A permuted counter with feedback paths that can be changed at will.

bases of the A, B, and C flip-flops. This is the means whereby the permutation is accomplished.

The circuit counts normally up to, and including, the seventh pulse. The eighth input pulse will cause the first transition of the D flip-flop, and at the same time the A, B, and C flip-flops will be reset. This can be verified from the collector waveforms in Fig. 13-8. At the instant the D flipflop makes its transition, the collector of $Q2$ goes to V_{CC}, which is inverted and reapplied to the $Q2$ bases of the A, B, and C flip-flops. The zero-going pulse will set these three flip-flops even though they have just been reset an instant before. With the A, B, C, and D flip-flops all in the

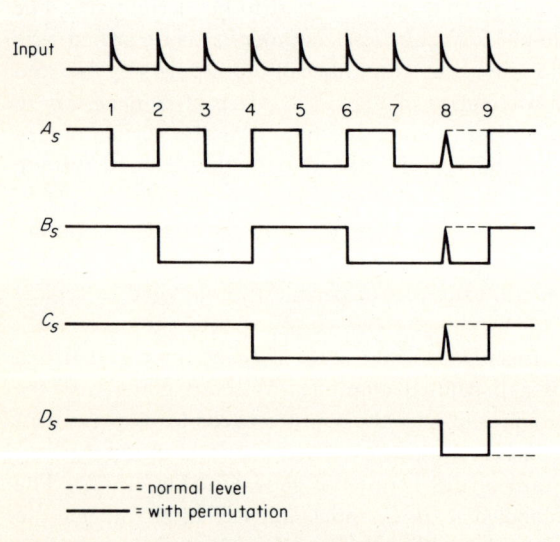

------ = normal level
———— = with permutation

Fig. 13-8. Waveforms of a counter permuted for scale-of-nine operation, using negative logic.

set condition, it will take only one more pulse (the ninth) to return the counter to its reset condition. Rather than requiring a total of 16 input pulses to scale clear around, the counter requires only 9 pulses to return to "home." It has been changed (permuted) from a scale-of-16 to a scale-of-9 counter.

What has really been accomplished is to set (not count) an extra seven counts into the counter during the counting process. The sequence of counting is as follows:

Input pulses: 0   1 2 3 4 5 6 7 8   9
State of counter: Reset 1 2 3 4 5 6 7 8 → 15 (16 or reset)

It is easily seen that the counter now requires only nine input pulses to return to reset, and this is obviously a scale-of-nine counter.

In this particular circuit the number of counts that are fed back to be set into the counter can be changed. In other words, any combination of the binary bits 1, 2, or 4 can be applied as feedback by removing the appropriate jumpers or links, labeled 1, 2, and 4 on the block diagram. For example, if the No. 2 jumper is removed, the feedback pulse can only get to the A and C flip-flops, and so rather than seven counts set in, there will only be five. In this case, the counting sequence would be as follows:

Input pulses: 0   1 2 3 4 5 6 7 8   9 10   11
State of counter: Reset 1 2 3 4 5 6 7 8 → 13 14 15 (16 or reset)

Having removed the No. 2 jumper, the counter becomes a scale-of-11 counter and at the eleventh pulse will return to reset.

To cause such a counter to scale decimally, the links are installed so that six counts are fed back. Installing links Nos. 2 and 4 but removing No. 1 will cause this action. As before, the counting sequence can be specified in tabular form:

Input pulses: 0   1 2 3 4 5 6 7 8   9   10
State of counter: Reset 1 2 3 4 5 6 7 8 → 14 15 (16 or reset)

The waveforms for a scale-of-ten counter are given in Fig. 13-9 for both the negative- and positive-logic cases.

13-3 THE SHIFT REGISTER

A normal binary counter, such as a scale-of-16 counter, accepts its information in serial form; that is, the counts come into the input one after the other, spread out in time. Often, it is desirable to put data into a counter in such a manner that data is not counted in, but rather *shifted* in, with all

320   PULSE AND SWITCHING CIRCUIT ACTION

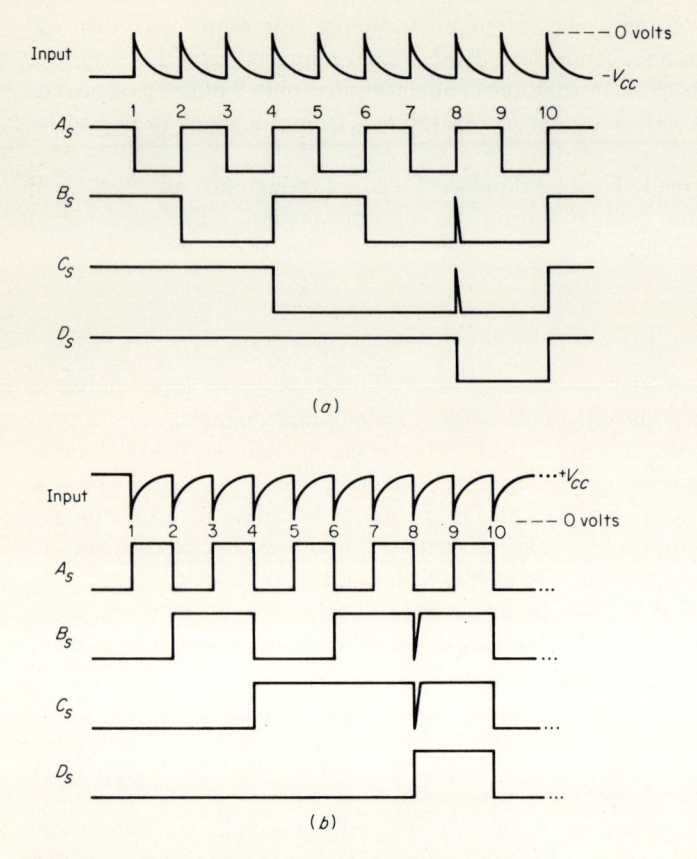

Fig. 13-9. (a) Waveforms for a scale-of 10 (decade) counter, using negative logic; (b) waveforms for a scale-of-10 (decade) counter using positive logic.

binary bits moving into the counter at the same time. In this instance, the individual flip-flops together form a *register*. A register is usually used as temporary storage for a group of binary bits. It has the advantage that bits of information can be accepted from another circuit while, at the same time, its own information is delivered to still another circuit. This operation is called *parallel shifting* of data because all bits are transferred at the same time.

Often, registers are organized to use the binary-coded decimal representation of the decimal digit. To code a decimal digit in the binary-coded decimal system, as previously mentioned, the decimal digits are expressed as binary bits, but only those combinations that agree with

digits 0 to 9 are allowed. Thus, a decimal 1 is expressed as 0001; 5 is expressed as 0101; 9 is expressed as 1001. If a number consists of *more* than one digit, each digit must be given separately but still coded in binary. For example, 13 in pure binary is 1101, but in BCD (binary-coded decimal) it must be written as two separate digits: 0001 0011.
$$\quad\quad\quad\quad\quad\quad\quad\quad\quad\quad\quad\quad\quad\quad\quad\quad\quad\quad\quad\quad\quad\quad\quad\quad\quad\quad\quad\quad\quad\quad\quad\quad\quad\quad\quad\quad\quad\quad\quad\quad\quad\quad\quad\quad\quad\quad\quad\quad\quad\quad\quad\quad\quad\quad\quad\;(1)\quad\;(3)$$
The column position of individual groups of digits determines the decimal weight as in the decimal digits of an ordinary decimal number.

In order to code a decimal digit in BCD, we note that the largest digit is 9 and that the minimum of binary bits to represent the 9 is 4; that is, a four-stage register must be used to hold a count equal to 9. Each decimal digit, then, that is to be stored in the register requires a four-stage set of flip-flops. Thus, to contain a three-digit decimal number, a register will necessarily consist of three sets of groups of four, or 12 individual flip-flops.

Now, it is true that a counter or register consisting of four flip-flops is capable of holding a count of 0 to 15 (or 1 to 16), but the conditions beyond 9 are invalid when using BCD because there are no digits beyond 9 in our decimal system and so these combinations are not used.

A typical *accumulator* might be a register that contains 64 flip-flops organized (wired) in groups of four so as to use the BCD system. It can therefore accumulate up to 16 decimal digits that can be shifted either left or right. A simple example of a shift register is given in block-diagram form in Fig. 13-10. To keep the drawing simple, only three decimal columns are shown, along with the shift-left mechanism, although

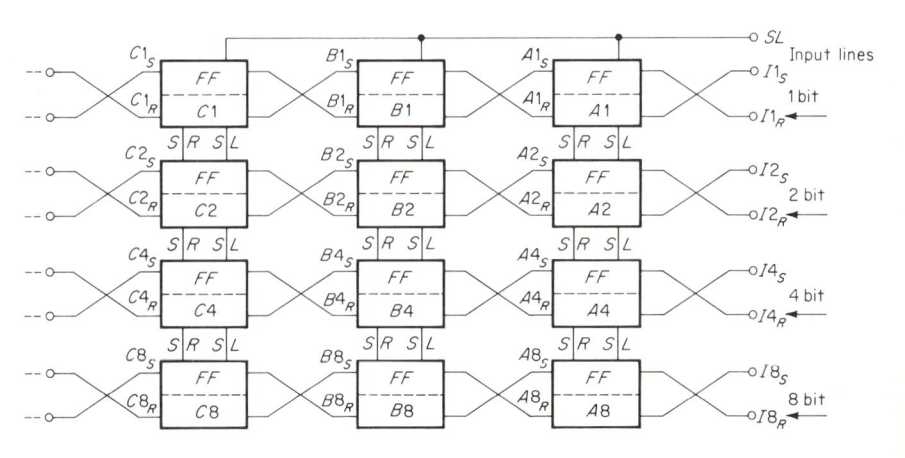

Fig. 13-10. Four-bit, three-stage, shift-left register. Input lines are on the right side to allow the vertical columns to represent decimal digits in proper relation to the decimal point.

in reality such a register can usually shift right also, as a later example will show.

The inputs are labeled $I1_S$ (for set input-1 bit), $I1_R$ (for reset input-1 bit), etc. A shift register requires that signal lines be in pairs rather than a single wire. For example, $I8_S$ and $I8_R$ originate at the collectors of a flip-flop so that if one is at ground, the other is at, say, -6 volts. For purposes of explanation, let us say that logic 1 is -6 volts and logic 0 is ground. If $I8_S$ is at ground and $I8_R$ is at -6 volts, this represents a logic 0 at the 8-bit input since the $I8_S$ input is the one that determines the logic at this point. When it is desired to shift a digit into the register, the input lines must first assume the voltage levels that represent the decimal digit, and these levels must remain there for some period of time, usually from a minimum of several microseconds up to a considerably longer time. Thus, they are called *dc signals*, although they are not dc in the same sense that a battery produces direct current. However, since a pulse may last for only a microsecond or so, this nomenclature helps to distinguish the dc level from a pulse since the level will remain static for a much longer time than the pulse.

Assume that the decimal digit is to be shifted into the register and that this digit is 5. In binary notation this is 0101. Now, a logic 0 is the condition where a 0-volt level exists at $I1_S$, $I2_S$, $I4_S$, and $I8_S$. This also implies that -6 volts is at the reset lines. To set up the lines to represent a 5 requires a logic 1 at the 1- and 4-bit lines. Therefore, -6 volts appears at $I1_S$ and $I4_S$ but 0 volts appears at $I2_S$ and $I8_S$. At the same time, 0 volts appears at $I1_R$ and $I4_R$ and -6 volts appears at $I2_R$ and $I8_R$. These levels appear simultaneously at all eight input lines, but they in no way affect the flip-flops in the A column of the register at this time.

Only when the proper shift pulse is caused to appear at the terminal labeled "shift left" (*SL*) will these logic levels be shifted into the input flip-flops. Note that the shift-left line connects to *all* flip-flops, and this suggests that information is shifted into and out of all flip-flops at the same instant. That is, any levels appearing at the logic lines are shifted into the A column at the same time that the information in the A column is transferred to the B column and information in the B column is transferred to the C column. All information moves one decimal column to the left at the same instant.

Although we have shown and described only a shift-left operation, it is just as possible to produce a shift-right transfer of data. This would require simply a separate line connecting to all flip-flops labeled "shift right" and the appropriate input-output connections between flip-flops. A pulse at this line would then move data one decimal column to the right for each *SR* pulse.

Circuit Description

The circuit diagram for typical shift-register flip-flops is shown in Fig. 13-11. Except for the input gates, the circuit is not unlike a conventional

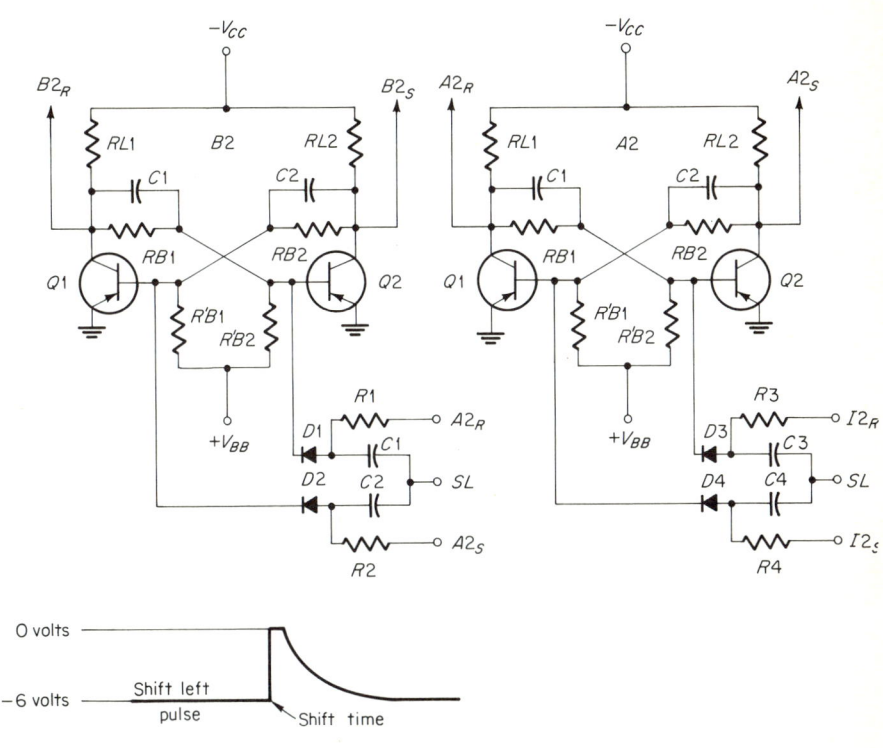

Fig. 13-11. Two stages of the shift register.

flip-flop. Keep in mind the fact that while the circuit is called a *shift register* it still possesses all of the attributes of a bistable multivibrator. If $Q1$ is off, $Q2$ is on and the circuit will remain in this state until instructed to change. Also, if $Q1$ is on, $Q2$ is off and again the circuit will remain in this state until told to make a transition. In the circuit shown, we have chosen to illustrate the $A2$ and $B2$ flip-flops of the block diagram. This will allow a more simple description of a shift-left operation from the input lines to the A flip-flop and from the A flip-flop to the B flip-flop.

The group of six components, composed of two diodes, two capacitors, and two resistors, form what is known as the *input gate* to the flip-flop. This is a +ac gate, as discussed in Chap. 11. The input gate to the $A2$

flip-flop has three signals at its input. These are $I2_S$, $I2_R$, and shift left. The three signals at the input to the B2 flip-flop are $A2_S$, $A2_R$, and shift left.

The *SL* lines are all common; that is, they are all connected together. But to keep the drawing uncluttered, the line is not drawn connected from one to the other. It is the function of the input gate to transfer the information from the input lines to the *A* flip-flop itself. This is accomplished by causing the coupling capacitors to temporarily store the information from the input lines. In the case of the *A*2 flip-flop, *C*3 and *C*4 store the information that exists at the $I2_S$ and $I2_R$ lines; at the instant the positive-going transition of the *SL* pulse occurs, the state of charge on these two capacitors will produce an appropriate change in the *A*2 flip-flop if a change is called for.

As an example of how the circuit works, assume that the two flip-flops have been reset and that logic $1 = -V_{CC} = -6$ volts and logic $0 = 0$ volts. Both *Q*2 sides are conducting, and so $A2_S$ and $B2_S$ are at ground while $A2_R$ and $B2_R$ are at -6 volts; this constitutes a logic 0 at each stage. If now the input lines are made to have a logic 1 on them, $I2_S$ will be at -6 volts and $I2_R$ will be at ground. This will in no way affect the flip-flop, and $A2_S$ will still be in the logic 0 state. At this point in the analysis it becomes necessary to determine the state of charge on the two coupling capacitors *C*3 and *C*4. The voltage to which each plate is returned is known, and it is a simple task to ascertain the charge attained by both capacitors.

In the case of both *C*3 and *C*4, the right-hand plates are returned to the shift-left line, and, according to the waveform shown, this line is normally at -6 volts. *C*3 is connected to $I2_R$ through *R*3 and so returns to ground. *C*4 returns to $I2_S$ and so is connected to -6 volts through *R*4. The right plate of *C*3 is at -6 volts, and its left plate is at ground; thus the capacitor is charged to 6 volts. *C*4, on the other hand, has -6 volts on both plates, and since there is no potential difference across it, there is no charge. The anode of *D*4 is at -6 volts.

This is the condition that exists at the time just prior to the occurrence of the shift pulse. Remembering that *Q*2 is on, when the shift pulse occurs, the *SL* line instantly goes to ground and carries the right-hand plates of the capacitors along with it. Since the capacitors cannot change the state of their charge in so short a time, the left plates must do the same thing; that is, they go 6 volts in the positive direction *from their original value*. The anode of *D*4 will be driven from -6 volts to a value that is 6 volts more positive than this, or 0 volts (ground). The anode of *D*3, on the other hand, will be driven from 0 to 6 *volts more positive than this*, or $+6$ volts on the base of *Q*2, which will turn it off. When *Q*1 turns on, the flip-flop has made a transition. With $Q2$ off, flip-flop $A2$ represents a logic 1; the logic 1 at the input has been transferred to the flip-flop. At the same time, the logic 0 that was in the $A2$ flip-flop was transferred to the $B2$ flip-flop, etc., and all bits of information have been shifted one column to the left.

As mentioned previously, a shift right is just as feasible, and Fig. 13-12

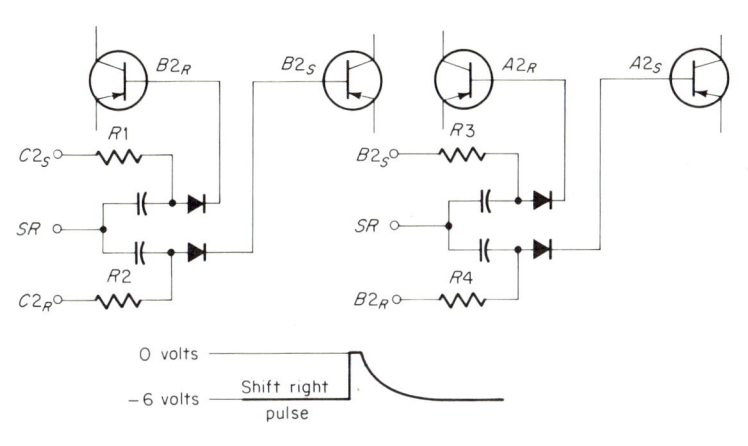

Fig. 13-12. Conditions necessary for a shift right.

shows the circuitry to perform this function. The only difference between this circuit and the shift-left circuit is that the qualifying resistors $R1$, $R2$, etc., are connected from the capacitors to the collectors of the flip-flops to the *left* of the flip-flop into which the information is being shifted. Also, a separate pulse is used to perform the shift, and this pulse is called the *shift-right pulse*. The information is shifted from $C2$ (not shown) to $B2$ and from $B2$ to $A2$, and this is a shift right, or a shift from one decimal column to the adjacent next least significant decimal column. Again, the coupling capacitors will assume a state of charge that depends upon the state of the collectors of the stage immediately to the left. Then, when the SR pulse occurs, the information from the $C2$ flip-flop will be transferred to the $B2$ flip-flop and the information in the $B2$ flip-flop will be transferred to the $A2$ flip-flop.

The overall operation of a small shift register will now be described in terms of the decimal-number processing. Figure 13-13 shows the register that will be used as an example. This is a block diagram of a three-column, 4-bit register, and it is similar in some respects to that of Fig. 13-10. With the register reset, all set sides of the flip-flops are at logic 0.

326   PULSE AND SWITCHING CIRCUIT ACTION

$A1_S$ is at logic 0, and $A1_R$ is at logic 1; $A2_S$ is at logic 0 and $A2_R$ is at logic 1, etc. (The criss-crossing that is necessary at the collectors of the transistors to cause the proper transfer of logic is not shown.)

To visualize the shifting operation in full detail, a series of numbers will be shifted into the register of Fig. 13-13. First assume that all registers

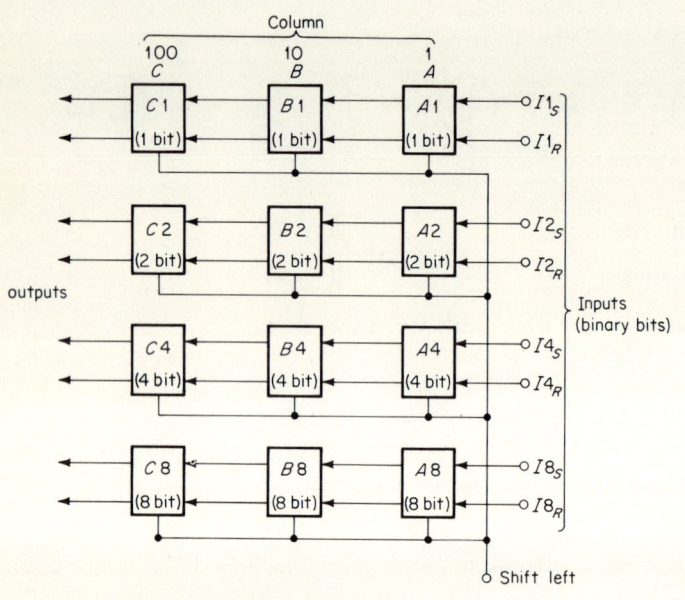

Fig. 13-13. Simplified shift-left register showing decimal-column weighting (1s, 10s, 100s).

are reset. The first digit will be a decimal three (3). The input lines must therefore be in the following states: a logic 1 at the $I1$ lines; a logic 1 at the $I2$ lines; a logic 0 at the $I4$ lines; and a logic 0 at the $I8$ lines. This is to be a shift-left operation, and when the SL signal occurs, flip-flops $A1$ and $A2$ will transfer while flip-flops $A4$ and $A8$ will stay in the zero state. Now, the state of the A flip-flop is 0011, and this is a decimal 3. The decimal number now in the register is 003. If the second digit is a 4, the the input lines will now change their state and a logic 1 will be at the $I4$ line while a logic 0 will be at all others. The next shift-left signal will transfer the 3 in the A flip-flops to the B flip-flops at the same time that the 4 is transferred to the A flip-flops. The decimal number in the register is now 034. Finally, assume that the last number to be shifted in is a 9. In this case, the input lines will be in the condition where logic 1 will be at the 1- and 8-bit lines, but logic 0 will be at the other two. The shift-left signal will now move the digits already in the register to the left and will shift the 9 standing at the input to the *A* flip-flop. The content of the register is now 349, and on three successive shifts, the three digits have been shifted into the register.

13-4 RING COUNTERS

The counters that have been described up to this point have been, for the most part, straight binary counters, or permuted binary counters. That is, they count in pure binary fashion, using 1, 2, 4, and 8 bits in a normal binary progression. There are times when certain counters use different methods of counting, and one of these is known as the *creeping*, or *Johnson*, code. The circuit to implement this code is often called the *switch-tail ring counter*. In this method of counting, the flip-flops progress as shown in Fig. 13-14.

A counter using the creeping code is very useful for easily producing

a

Johnson, or Creeping, Code						Count
5	4	3	2	1	–	Count
						0
0	0	0	0	0		reset (zero)
0	0	0	0	1	–	1
0	0	0	1	1	–	2
0	0	1	1	1	–	3
0	1	1	1	1	–	4
1	1	1	1	1	–	5
1	1	1	1	0	–	6
1	1	1	0	0	–	7
1	1	0	0	0	–	8
1	0	0	0	0	–	9
0	0	0	0	0	–	0 (reset)

b

Binary Code					Count
8	4	2	1	–	Count
0	0	0	0	–	0
0	0	0	1	–	1
0	0	1	0	–	2
0	0	1	1	–	3
0	1	0	0	–	4
0	1	0	1	–	5
0	1	1	0	–	6
0	1	1	1	–	7
1	0	0	0	–	8
1	0	0	1	–	9
1	0	1	0	–	10
1	0	1	1	–	11
1	1	0	0	–	12
1	1	0	1	–	13
1	1	1	0	–	14
1	1	1	1	–	15
0	0	0	0	–	16 or 0

Fig. 13-14. Comparison of the "creeping" code (*a*), to the binary code (*b*).

328   PULSE AND SWITCHING CIRCUIT ACTION

both up-counters and down-counters. An up-counter is one that counts from 0, 1, 2, 3, etc., to 9, while a down-counter counts from 9, 8, etc., back down to 0. In the following sections, both up- and down-counters using the creeping code will be described.

Up-counters

Figure 13-15 shows a typical up-counter using the creeping code. In this counter, known as a *switch-tail ring counter*, the flip-flop 1 is the least

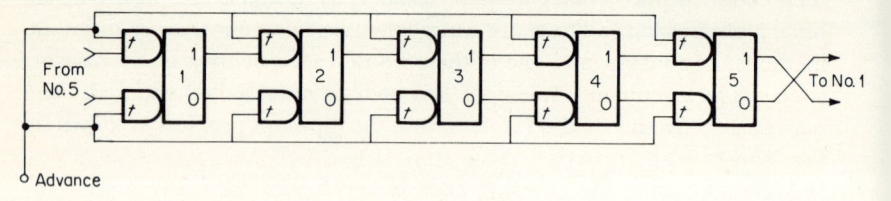

Fig. 13-15. Logic diagram of the switch-tail ring counter generating the so-called "creeping code."

significant of the five shown while the flip-flop 5 is the most significant. The counter is reset when all flip-flops contain a logic 0, and the signal to advance the counter to the first count is applied to all counters at the same time via the advance bus. In a sense, the counter does not count in the same manner that a straight binary counter does; rather it shifts information from one flip-flop to the next adjacent flip-flop, and the manner of connecting each flip-flop to its neighbor produces the effect of counting. The advance input, then, is really a shift pulse, much like the shift pulse described earlier.

If there is a logic 0 in each register, it would seem at first glance that nothing but zeros could be shifted from any flip-flop to its neighbor. Such is not quite the case, however. Notice the transposition of collector leads out of the flip-flop 5. A logic 0 at flip-flop 5 is changed to a logic 1 at flip-flop 1 by this transposition of wires since the output of flip-flop 5 is the input to flip-flop 1.

The first advance pulse will shift a logic 1 into the first flip-flop, as shown in the truth table of Fig. 13-14. Also, another zero is shifted into flip-flop 2 and all other flip-flops. After the first input, the counter is in the following state, which represents the condition for digit 1 in the counter: 0 0 0 0 1.

The next advance pulse will shift another 1 into the flip-flop 1 from flip-flop 5 since No. 5 is still reset. But No. 2 now contains a logic 1 since the 1 from flip-flop 1 has been shifted into it. Now the state of the counter is: 0 0 0 1 1. As further advance pulses come in, the counter advances until all flip-flops contain a logic 1. This corresponds to a count of five, and will, of course, require five input (advance) pulses. Note how the logic 1 creeps across the truth table.

The fifth input pulse shifted a logic 1 into flip-flop 5, and now for the first time its output is different from the original condition. The sixth input pulse must shift a logic 0 into flip-flop 1, again because of the transposition of the collector leads from flip-flop 5. After the sixth input, then, the state of the counter is 1 1 1 1 0, and as additional input signals arrive at the input, the logic 0 creeps across the truth table, eventually returning the counter to reset, where all flip-flops contain a logic 0. It requires ten input pulses to cause the counter to scale around once, and such a counter is obviously a scale-of-ten counter. Figure 13-16 shows very graphically why this circuit is called a *ring counter*.

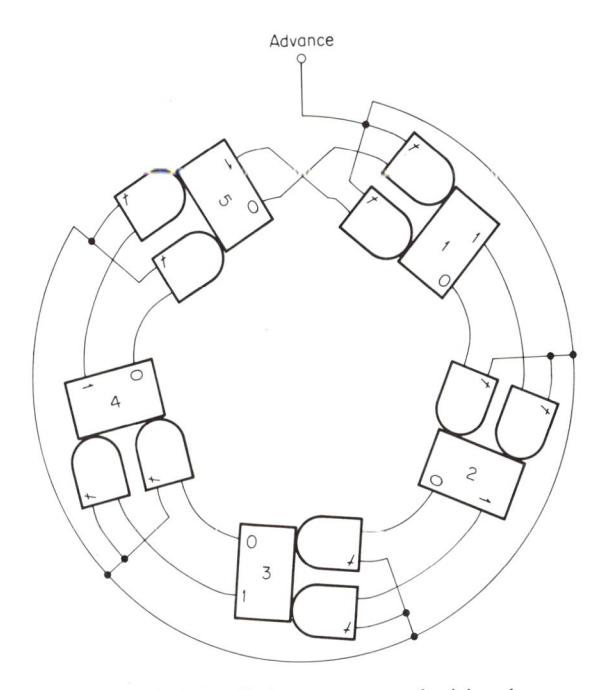

Fig. 13-16. Switch-tail ring counter emphasizing the ring-type connections.

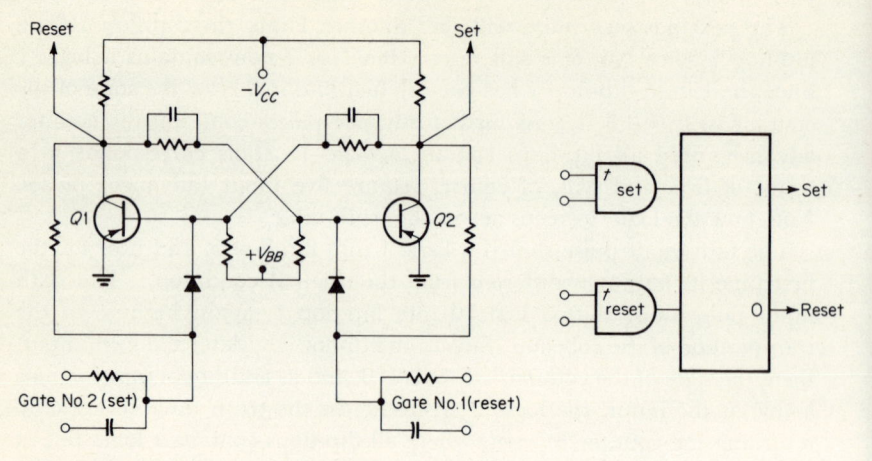

Fig. 13-17. Flip-flop used in a switch-tail ring counter. The gates are labeled for positive logic.

The circuitry of the flip-flop in this counter is shown in Fig. 13-17, along with the appropriate logic diagram. Note that in the block representation the gates are not considered an integral part of the flip-flop and are shown as a separate entity by themselves. Each gate is a simple circuit, consisting of a resistor, a capacitor, and a diode. The normal T input to the flip-flop is not used in this counter. The out-of-the-ordinary fact of the block diagram that must be kept in mind is that the input to the flip-flop shown on the set side actually connects to the base of the opposite transistor; that is, a logic 1 into the gate on the set side of the flip-flop results in a logic 1 out of the set side. To accomplish this, the actual connection must be to the opposite base. The same idea holds true for the reset side. This is borne out on the schematic diagram, Fig. 13-17, where gate 1 connects to $Q2$ (set side) and gate 2 connects to the base of $Q1$ (reset side). In this system, using *pnp* transistors, a 0-volt level is logic 1 while about -8 volts is logic 0. Thus, when the positive-going transition occurs, a positive pulse may be applied to the base of the transistor if the other leg of the gate is properly qualified. Since the positive pulse will turn off the transistor, we must look to the opposite transistor to see a valid, or ground, signal. The reader will recognize that these gates are simply the $+$ac gates previously described.

Down-counters

Figure 13-18 shows the circuit for a down-counter, and it can be seen that it is not much different from the up-counter, except for the fact that the signals travel from left to right in the up-counter but from right to left in the down-counter, as drawn. (An alternate way of looking at this is to simply redefine the flip-flops of the up-counter, so that the flip-flop 1 becomes No. 5, and the flip-flop 2 becomes No. 4, etc.) The truth table shown with Fig. 13-18 indicates the manner in which the counter functions.

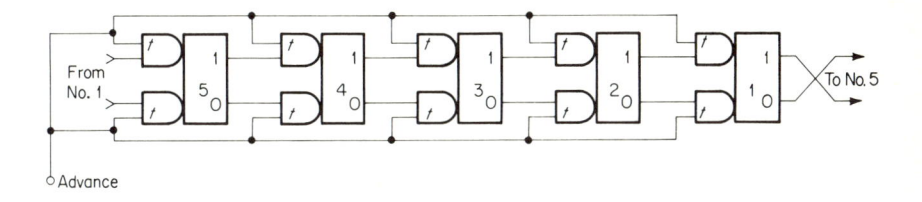

Counter	Input	State of counter (complement)
0 0 0 0 0 —	0	(10 − 0 = 10)
1 0 0 0 0 —	1	(10 − 1 = 9)
1 1 0 0 0 —	2	(10 − 2 = 8)
1 1 1 0 0 —	3	(10 − 3 = 7)
1 1 1 1 0 —	4	(10 − 4 = 6)
1 1 1 1 1 —	5	(10 − 5 = 5)
0 1 1 1 1 —	6	(10 − 6 = 4)
0 0 1 1 1 —	7	(10 − 7 = 3)
0 0 0 1 1 —	8	(10 − 8 = 2)
0 0 0 0 1 —	9	(10 − 9 = 1)

Table of 10's complements

Fig. 13-18. Logic diagram of the switch-tail ring counter, operating in the down-count mode.

The first pulse at the advance line causes a logic 1 to be shifted into flip-flop 5 because of the transposition of wires from the collectors of the flip-flop 1. Since the counter condition is now 1 0 0 0 0, it is in the state corresponding to the count of nine. The second pulse transfers the flip-flop 4 to make the condition 1 1 0 0 0, and this is equivalent to the count of eight. Thus the counter is progressing in a backward, or downward, direction.

One of the uses of the down-counter is to provide the complement of a number. The complement of a number is that number subtracted from the radix, or base, of the number system. (The complement of a number may also be the difference between a number and an arbitrary number.) The 10s complement of 2 is (10 − 2) = 8. The 9s complement of 2 is (9 − 2) = 7. A table of the 10s complements is shown and clearly indicates that the complement of the input is given by the down-counter.

13-5 BINARY UP - DOWN–COUNTERS

In the foregoing section one kind of ring counter that uses the creeping, or Johnson, code was described. This counter was found to possess the ability of counting either up or down. It is just as possible to arrange the circuitry of a binary counter so that it will count either up or down.

Consider Fig. 13-19, which shows a binary counter arranged for con-

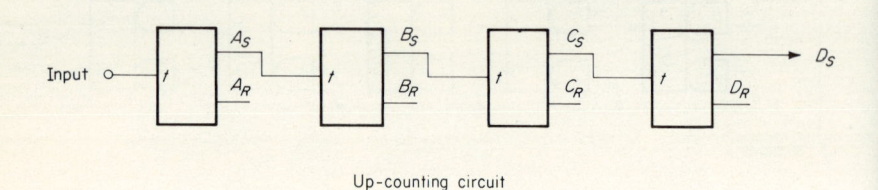

Up-counting circuit

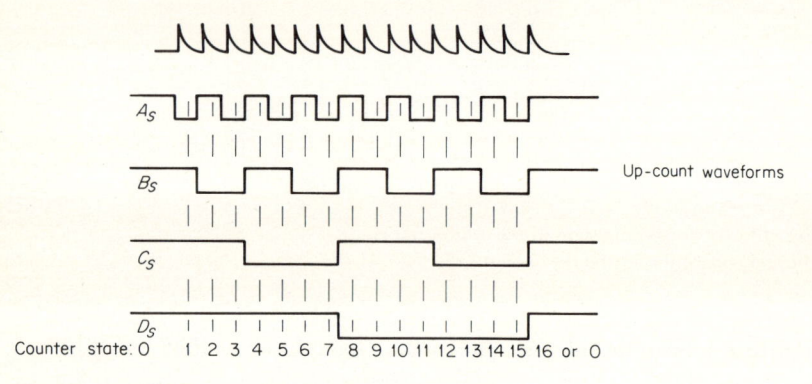

Up-count waveforms

Fig. 13-19. Binary up-counter and corresponding waveforms.

ventional or up-counting. This circuit configuration is quite familiar, being identical to the binary counter discussed earlier. To briefly recap the circuit operation, the first input pulse transfers (sets) flip-flop A. The wire labeled A_S makes a negative-going transition, and since the T input of B requires a positive-going transition, the B flip-flop does not transfer (*pnp* transistors). The second input pulse again transfers the A flip-flop, and now the A_S wire goes positive when the A flip-flop resets. This sets the B flip-flop, as the waveform shows.

Because the B_S wire is now negative-going the C flip-flop does not set.

The third input pulse again transfers the *A* flip-flop to the set condition, but *B* does not transfer. The counting sequence continues until, after the fifteenth input pulse, all four flip-flops are set, as indicated by the waveforms. Finally, at the sixteenth pulse, all flip-flops become reset.

The foregoing counting sequence represents the normal or up-counting ability of the counter. The waveforms help to determine the count that remains after some number of input pulses, and they also serve to identify the fact that this is an up-counter. A counter is said to be counting up when the first pulse transfers only the input flip-flop; the second pulse then transfers, or resets, the first flip-flop and sets the second flip-flop, while the third pulse sets the input flip-flop again, etc. As will soon be explained, a down-counter has much different counting characteristics.

The very same counter can be reconnected to provide down-counting, as indicated in Fig. 13-20. Also shown are the waveforms, and they can be seen to be different in their relation to each other when compared with the up-counter waveforms. One quite noticeable difference is that the very first input pulse *transfers all flip-flops*. In terms of the up-counter, this

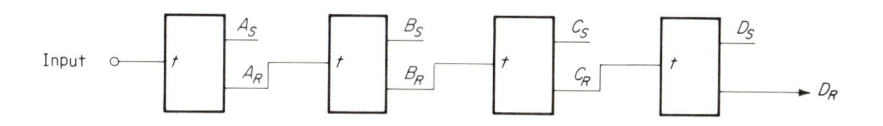

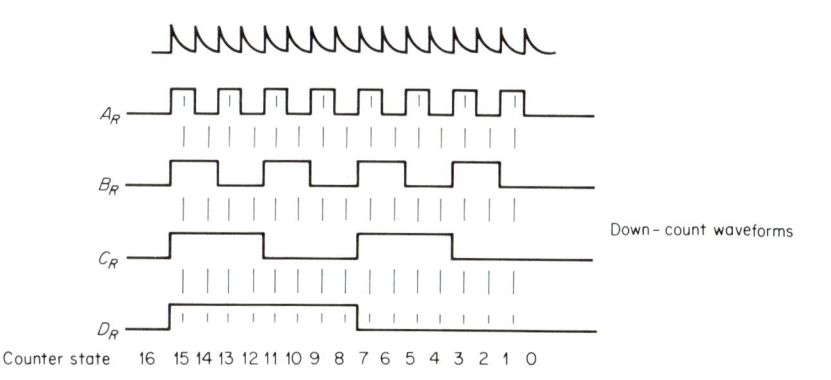

Fig. 13-20. Binary down-counter and corresponding waveforms.

corresponds to the fifteenth condition. Hence, the first input pulse causes the counter to advance from zero to the fifteenth condition.

The second input pulse retransfers the A flip-flop, but as the waveforms show, the A_R wire is negative-going, and so the B flip-flop will not be affected. Comparing these waveforms with those of the up-counter, the present state of the counter corresponds to the fourteenth state. Obviously, the circuit is counting down.

Note that the only difference between the two circuits lies in which of the two available output wires from each flip-flop is used. If the output is taken from the set side of each flip-flop (A_S, B_S, C_S, D_S), the circuit counts up. On the other hand, if the output is taken from the reset side (A_R, B_R, C_R, D_R), the circuit counts down.

A truth table can be constructed that shows the progression of counts in the counter, as shown in Fig. 13-21. The Boolean expression for each condition of the counter is shown, where the output wire of each flip-flop that has a logic 1 on it is called out. For example, the expression $\bar{A} \cdot \bar{B} \cdot \bar{C} \cdot \bar{D}$ tells us that logic 1 is on the reset wires, and so the wires A_S, B_S, C_S, D_S must be at logic 0. This, of course, is the *reset* condition. Each of these expressions signifies a unique state of the counter for a given number of input pulses.

Up-counter State	Up	Input Pulses	Down	Down-counter State
0	$\bar{A}\ \bar{B}\ \bar{C}\ \bar{D}$	0	$\bar{A}\ \bar{B}\ \bar{C}\ \bar{D}$	0
1	$A\ \bar{B}\ \bar{C}\ \bar{D}$	1	$A\ B\ C\ D$	15
2	$\bar{A}\ B\ \bar{C}\ \bar{D}$	2	$\bar{A}\ B\ C\ D$	14
3	$A\ B\ \bar{C}\ \bar{D}$	3	$A\ \bar{B}\ C\ D$	13
4	$\bar{A}\ \bar{B}\ C\ \bar{D}$	4	$\bar{A}\ \bar{B}\ C\ D$	12
5	$A\ \bar{B}\ C\ \bar{D}$	5	$A\ B\ \bar{C}\ D$	11
6	$\bar{A}\ B\ C\ \bar{D}$	6	$\bar{A}\ B\ \bar{C}\ D$	10
7	$A\ B\ C\ \bar{D}$	7	$A\ \bar{B}\ \bar{C}\ D$	9
8	$\bar{A}\ \bar{B}\ \bar{C}\ D$	8	$\bar{A}\ \bar{B}\ \bar{C}\ D$	8
9	$A\ \bar{B}\ \bar{C}\ D$	9	$A\ B\ C\ \bar{D}$	7
10	$\bar{A}\ B\ \bar{C}\ D$	10	$\bar{A}\ B\ C\ \bar{D}$	6
11	$A\ B\ \bar{C}\ D$	11	$A\ \bar{B}\ C\ \bar{D}$	5
12	$\bar{A}\ \bar{B}\ C\ D$	12	$\bar{A}\ \bar{B}\ C\ \bar{D}$	4
13	$A\ \bar{B}\ C\ D$	13	$A\ B\ \bar{C}\ \bar{D}$	3
14	$\bar{A}\ B\ C\ D$	14	$\bar{A}\ B\ \bar{C}\ \bar{D}$	2
15	$A\ B\ C\ D$	15	$A\ \bar{B}\ \bar{C}\ \bar{D}$	1
0	$\bar{A}\ \bar{B}\ \bar{C}\ \bar{D}$	16 (or 0)	$\bar{A}\ \bar{B}\ \bar{C}\ \bar{D}$	0

Fig. 13-21. Counter states for both up- and down-counting.

Now, to provide a single counter that is capable of both up- and down-counting, the circuits just shown must be somewhat modified. Obviously, a way must be provided to allow a choice of inputs—either from A_S, B_S, C_S, and D_S to effect the up-counting, or from A_R, B_R, C_R, and D_R, to effect the down-counting.

A very simple and straightforward method actually in use is illustrated in Fig. 13-22. This is a reliable way of accomplishing the objective but

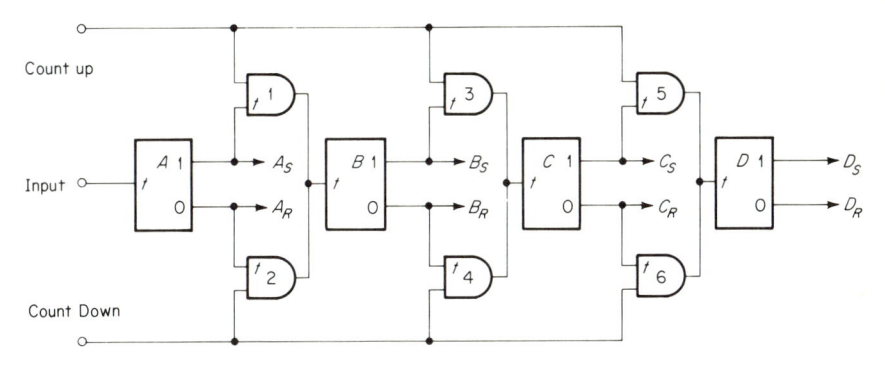

Fig. 13-22. Binary up-down–counter.

requires a relatively large number of components. The four flip-flops are connected together by special ac gates (resistor-capacitor gates) that are somewhat different from the usual T inputs. Normally, the T input is considered an integral part of the flip-flop, as shown for the input of flip-flop A. Flip-flops B, C, and D, however, have *dual-T* inputs that are not completely self-qualified; that is, the dc or enable, inputs of these gates are not only qualified by the collectors of the flip-flops, but also can be qualified externally.

If the line labeled "count-up" is enabled (and, of course, the count-down line is inhibited), the counter will count up. On the other hand, if the count-down line is enabled, the counter will count down. That this is true is evident when the signals used in each case are considered.

With ac gates 1, 3, and 5 enabled, the output of each flip-flop is taken from the set side, and this will cause up-counting. With the count-down line enabled, the signals are taken from the reset sides of the flip-flops, causing down-counting of any input pulses. Hence from a block-diagram viewpoint, the up-down–counter of Fig. 13-22 is not difficult to understand.

The circuitry of the B, C, and D flip-flops is also not difficult to understand if approached properly. First, it is desirable to review briefly the action of a conventional T input. Once this is thoroughly understood, the modified dual-T input will be seen to be very similar, differing only in relatively minor ways.

Figure 13-23 shows a conventional flip-flop with a self-gated T input.

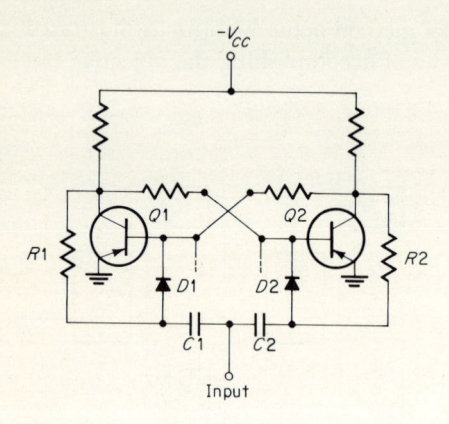

Fig. 13-23. Flip-flop with the regular T input.

Assume that at first $Q1$ is on while $Q2$ is off. The collector of $Q1$ is at ground potential, while that of $Q2$ is at $-V_{CC}$, or nearly so. Because the collectors of $Q1$ and $Q2$ are resting at different potentials, the two input-coupling capacitors are returned to different voltages. The output (left) plate of $C1$ is returned to ground, while the output (right) plate of $C2$ is returned to nearly $-V_{CC}$. Therefore, when a positive-going pulse arrives at the input, the output, or left plate of $C1$, is driven more positive than ground and $Q1$ is turned off through diode $D1$. At the same time, the output, or right plate of $C2$, is simply driven from $-V_{CC}$ to ground. Since the anode of $D2$ is not driven positive enough to become forward-biased, the base of $Q2$ does not see any part of the incoming pulse.

The pulse is therefore steered to only the base of the on-transistor and is prevented from appearing at the base of the off unit. After the flip-flop has made the transition, the voltage conditions at the collectors reverse, and now $C2$ is enabled while $C1$ is inhibited. The next pulse, then, will find its way to the base of $Q2$, now on, while it will be blocked from the base of $Q1$. The normal operation of the T input is seen to result in the application of the input pulse to only the on-transistor, thus ensuring reliable triggering.

The dual-T input used in one form of a binary up-down–counter is shown in Fig. 13-24. The operation of this circuit is not difficult to under-

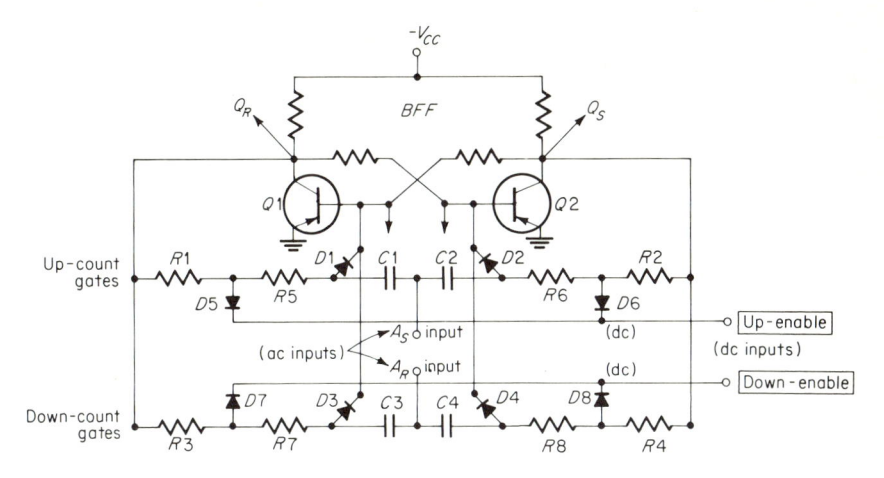

Fig. 13-24. Flip-flop with a dual-T input to provide for both up- and down-count capability.

stand if the block diagram of Fig. 13-22 is kept in mind. The circuit shown is the B flip-flop of Fig. 13-22. (Some of the essential parts of the flip-flop itself are not shown to simplify the drawing.) The input gates are divided into two sections. One of these, labeled the "up-count gate," has as its ac, or pulse, input the set-side output of the preceding flip-flop, the A_S signal. This gate is enabled by the dc input to the gate labeled "up-enable." The other gate is the down-count gate, and its ac input comes from the reset side of the preceding flip-flop, A_R. It is enabled by the down-enable line.

The function of each of the components will now be discussed. Suppose the counter is to be enabled for up-counting. The up-enable line will therefore be at ground potential, while the down-enable line is at $-V_{CC}$. Furthermore, assume that $Q1$ is on and $Q2$ is off. The collector of $Q1$ is at ground, while the collector of $Q2$ is at nearly $-V_{CC}$.

The voltage to which the output plates of $C1$ and $C2$ will charge is a function of the conditions at the collectors *and* the dc enabling input. Both the collector of $Q1$ and the cathode of $D5$ are at ground potential; the output plate of $C1$ is therefore at 0 volts. If the A_S input happens to be negative at this time, the input side of $C1$ is at $-V_{CC}$, as, of course, is the input side of $C2$. However, the output plate of $C2$ is presented with a different set of conditions. The collector of $Q2$ is at $-V_{CC}$, essentially, which places a negative voltage at the anode of $D6$, reverse-biasing it. Thus, the output plate of $C2$ is at $-V_{CC}$.

When the A flip-flop makes a transition (from $-V_{CC}$ to ground), $C1$ is enabled while $C2$ is inhibited, and the base of $Q1$ is presented a positive pulse, turning $Q1$ off and effecting a transition at the B flip-flop. Since when A_S goes positive, A_R goes negative, the condition of the down-count gate is unimportant at this time because the negative transition cannot affect the flip-flop.

However, consider the case where the signal A_S goes negative; in this case the signal A_R goes positive. If the circuit is enabled for up-counting, this positive-going transition at A_R must be prevented from appearing at either base. Of course, this is accomplished by virtue of the inhibiting voltage at the down-enable input.

With $-V_{CC}$ at the cathodes of $D7$ and $D8$, the diodes are forward-biased and will clamp both of the junctions of the two groups of resistors ($R3$, $R7$, and $R4$, $R8$) to $-V_{CC}$. Now, for example, if $Q1$ is on, its collector is at ground, but this is prevented from being applied to the output plate of $C3$ by the clamp voltage at the dc input (down-enable). Thus, the full V_{CC} drops across $R3$. Now, both $C3$ and $C4$ are inhibited, with each output plate at $-V_{CC}$, and as the signal A_R goes positive, neither capacitor can deliver a positive voltage to either base.

When the circuit is enabled for down-counting, the up-count gate is inhibited and the down-count gate is allowed to function as a normal T input.

QUESTIONS AND PROBLEMS

13-1 A binary counter has a total of three flip-flops. Determine the count capacity.

13-2 A binary counter has a total of six flip-flops. Determine the count capacity.

13-3 A binary counter has a total of seven flip-flops. Determine the count capacity.

13-4 A binary counter has a total of 10 flip-flops. Determine the count capacity.

13-5 A binary counter has a total of 12 flip-flops. Determine the count capacity.

13-6 A binary counter has a total of 13 flip-flops. Determine the count capacity.

13-7 Refer to Fig. 13-1. The counter is initially reset. Determine the number of input pulses necessary to cause C_R to go from logic 1 to 0.

13-8 Refer to Fig. 13-1. The counter is initially reset. Determine the number of input pulses necessary to cause D_R to go from logic 1 to 0.

13-9 Refer to Fig. 13-4. Lamps 1 and 4 are lighted, while lamps 2 and 8 are out. Determine the residual count.

13-10 Refer to Fig. 13-4. Lamps 2 and 8 are lighted, while lamps 1 and 4 are out. Determine the residual count.

13-11 Refer to Fig. 13-5. Flip-flops B and C are set, while the others are reset. Determine the output line that is energized.

13-12 Refer to Fig. 13-5. Flip-flops A and D are set, while the others are reset. Determine the output line that is energized.

13-13 Refer to Fig. 13-5. Flip-flops A, B, and D are set. Determine the output line that is energized.

13-14 Refer to Fig. 13-5. All flip-flops are set. Determine the output line that is energized.

13-15 Refer to Fig. 13-7. Links 1, 2, and 4 are open. Determine the total count capacity.

13-16 Refer to Fig. 13-7. Links 1 and 4 are open, but link 2 is installed. Determine the total count capacity.

13-17 Refer to Fig. 13-11. $-V_{CC} = -18$ volts, and $V_{BB} = +9$ volts. Determine the maximum voltage that appears on the appropriate base to turn the on-transistor off. Ignore small semiconductor voltage drops.

13-18 Refer to Fig. 13-11. $-V_{CC} = -12$ volts, and $V_{BB} = +6$ volts. Determine the maximum voltage that appears on the appropriate base to turn the on-transistor off. Ignore small semiconductor voltage drops.

13-19 A Johnson-code counter has a total of three flip-flops. Determine the total count capacity.

13-20 A Johnson-code counter has a total of 10 flip-flops. Determine the total count capacity.

14
MATRICES

A matrix is a device that is capable of extracting count information from a binary counter, decade counter, or other counter or register. As normally used, the matrix is a group of logical gates arranged in a specific fashion, with the output of each gate representing the residual count in the counter or register at that moment. Such action is termed *decoding*.

A matrix may be used in reverse to provide a binary, or BCD, representation of a decimal number. Any other code may also be used, and to accomplish this it is usually necessary to rearrange the gating elements only slightly. However, to keep the following discussions simple, only the binary, BCD, Johnson, and decimal codes will be used since the application principles are approximately the same for all codes.

14-1 DECODING MATRICES

Figure 14-1 illustrates a typical three-stage binary counter. The logic condition of the collector wires determines the state of the counter. Recalling from Chap. 13 the sequence of the counting process, illustrated in Table 14-1, it can be seen that the collector wires represent the instan-

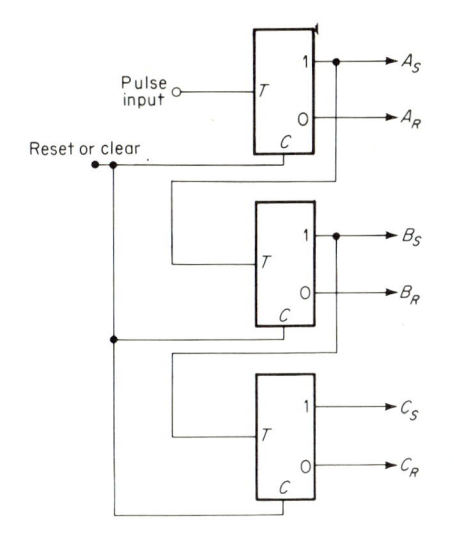

Fig. 14-1. Three-stage binary counter used as a source of waveforms to be decoded.

taneous state of the counter. When reset, the set-side wires are at logic 0, while the reset-side wires are at logic 1.

As pulses are applied to the count input (T input of flip-flop A), the counter scales upward, according to the truth table, in binary fashion. The condition of the counter at any residual count may be determined by measuring the collectors with a voltmeter to determine whether the individual flip-flops are set or reset.

For example, if the meter indicates that flip-flops A and B are set while C is reset, the counter has a residual count of 3 (0011). Of course, such a crude method of determining the residual count is not practical. A much

Table 14-1
Logic Description of the State of the
Flip-flops in a Three-stage Binary Counter

Count	C_S	B_S	A_S
0	0	0	0 reset
1	0	0	1
2	0	1	0
3	0	1	1
4	1	0	0
5	1	0	1
6	1	1	0
7	1	1	1
0 or 8	0	0	0

more efficient method of determining the state of the counter, doing it virtually instantaneously, is by the use of a diode matrix. This allows the information residing in the counter to be *read out* continuously, so that the data is available for use whenever needed. The diode matrix consists of, usually, very simple AND or OR gates. The simplicity of such a read-out system is its greatest attribute.

To begin to appreciate the general subject of matrices, consider the AND gates illustrated in Fig. 14-2. The logic diagram indicates a general case,

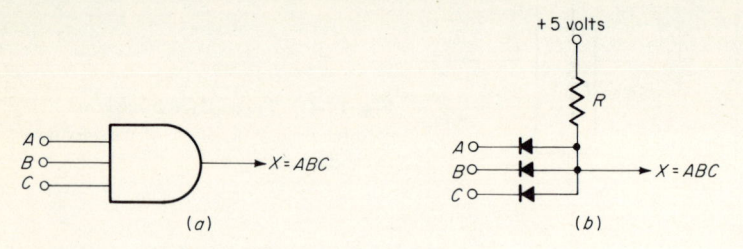

Fig. 14-2. (*a*) Logic diagram of the AND function (logic 1 may be either positive or negative); (*b*) schematic diagram of a positive AND diode gate (*Y* gate).

which may be used for either positive or negative logic. The schematic diagram, on the other hand, is more specific, being in this case a negative OR, positive AND gate (*Y* gate). Also used is the *X* gate, used as the negative AND, positive OR gate (not illustrated). The foregoing circuits, then, are the basic building blocks of a matrix.

In Fig. 14-3 is shown a scale-of-eight binary counter with eight AND gates connected in a manner to allow decoding all possible states of the counter. The eight gates comprise the matrix, and the matrix action will now be described. Assume first that the counter is reset; hence the set-side outputs are at logic 0, which implies that logic 1 appears at the reset-side outputs. Inspecting the various gates shows that gate A_0 has as its inputs $\bar{A}\bar{B}\bar{C}$. Since all inputs are at logic 1, this gate yields an output that gives an indication that the counter is reset. Note carefully that *no other gate* shown, A_1 to A_7, can have an output since one or more inputs are at logic 0. Therefore, one and only one gate can yield an output at any one time.

Suppose that now the counter is caused to advance by one count by the application of an input pulse at flip-flop *A*. After the required transitions have occurred, the counter is now in the state where the *A* flip-flop is set but the *B* and *C* flip-flops are reset. This corresponds to the Boolean

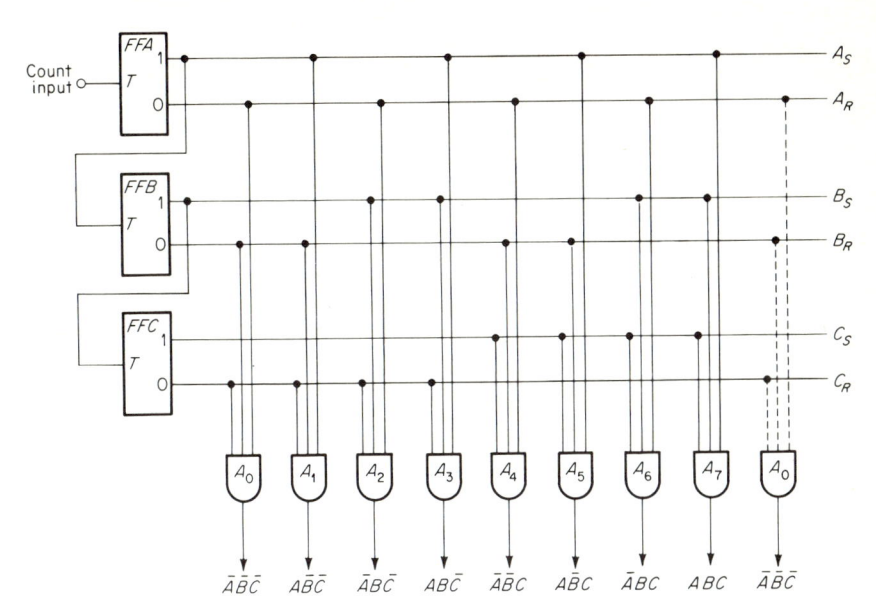

Fig. 14-3. Logic diagram of a three-stage counter (binary) and decoding network, or matrix, producing binary-to-decimal conversion.

equation $A\bar{B}\bar{C}$. Again inspecting the various gates, note that gate A_1 now has at its inputs the required conditions to yield an output. This gate now has a logic 1 at its output, but all others have a logic 0 since one or more of their inputs are at logic 0.

It now becomes apparent that as the counter advances, count by count, successive gates are energized, one at a time, and each one corresponds to one unique state of the counter. After two input pulses have arrived, which will place the counter in the state described by the Boolean equation $\bar{A}B\bar{C}$, gate A_2 yields an output that is at logic 1. As before, all other gates yield logic 0.

As a point of interest, note the symmetrical appearance of the connection points on the collector wires. On the A_S–A_R lines, the connection points alternate at each gate. On the B_S–B_R wires the connection points alternate each two gates, while on the C lines this occurs every four gates. This, of course, is related to the action of the corresponding flip-flop itself. The A flip-flop transfers at every input pulse, while the B flip-flop transfers at every second input pulse. Hence, the diodes of the individual gates must be connected in this way to yield the desired result.

Very often it is necessary to decode for a variety of conditions, as

344 PULSE AND SWITCHING CIRCUIT ACTION

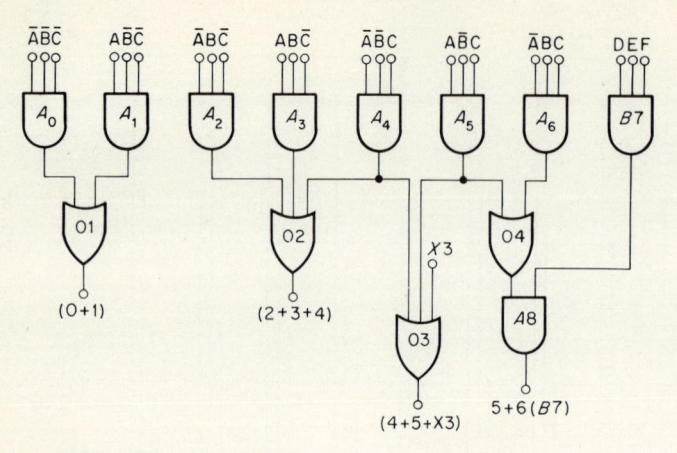

Fig. 14-4. Matrix used for developing combination signals to be used as instructions in other parts of the equipment. The inputs of the AND gates originate at scale-of-16 counters.

illustrated in Fig. 14-4. A network such as this often forms signals that become the *instructions*, which are conditional signals that occur only when the input requirements are met.

If it is necessary to know, for one reason or another, whether a counter is either reset or at the one-count condition, gates A_0 and A_1 will yield such a signal. When the counter to which this matrix is connected is at either its 0 or its 1 state, the O_1 gate will yield a logic 1 output. By inspecting the input requirements for these two AND gates, this output can be verified. With the counter in any other state, the output from the OR gate will be at logic 0.

When the counter advances to the 2d, 3d, or 4th state, gate O_2 will yield an output. By the same token, gate O_3 yields an output when the counter advances to either the 4th or 5th state. Also, note that a signal that originates elsewhere, X_3, will also produce an output. Such mixing of signals from separate sources is quite common to produce the desired circuit action.

Another output, $(5 + 6) \cdot (B7)$ results when the second counter (the B counter) is in its 7th state (DEF) and the first counter is in either its 5th $(A\bar{B}C)$ or 6th $(\bar{A}BC)$ state. In a similar manner, signals from many different sources may be intermixed and subsequently decoded for many different circuit conditions.

The matrices illustrated thus far have been drawn using standard logic symbols. This is perfectly legitimate and straightforward but often requires more space for drawing than is available. Also, it is sometimes necessary to show such a circuit in schematic form rather than in logic form. A somewhat more compact way of drawing a matrix, from which its name is derived, is illustrated in Fig. 14-5. The components labeled

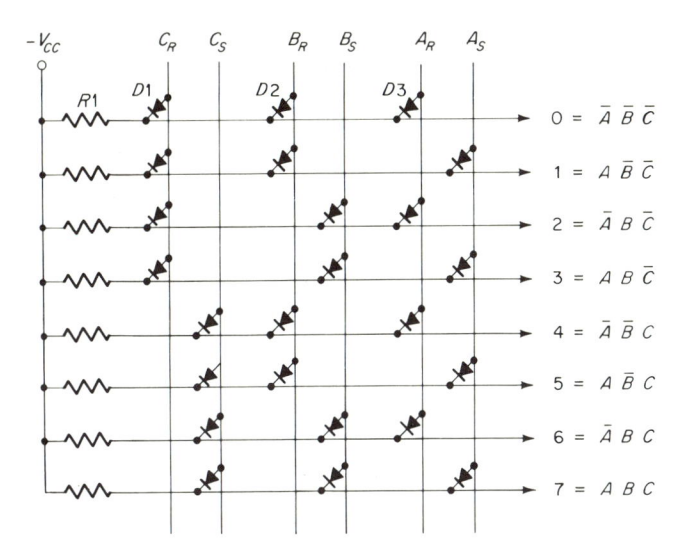

$$0 = \bar{A}\,\bar{B}\,\bar{C}$$
$$1 = A\,\bar{B}\,\bar{C}$$
$$2 = \bar{A}\,B\,\bar{C}$$
$$3 = A\,B\,\bar{C}$$
$$4 = \bar{A}\,\bar{B}\,C$$
$$5 = A\,\bar{B}\,C$$
$$6 = \bar{A}\,B\,C$$
$$7 = A\,B\,C$$

Fig. 14-5. Diode matrix for use with *pnp* transistors and negative logic.

$R1$, $D1$, $D2$, and $D3$ comprise a diode gate that corresponds to the A_0 gate of Fig. 14-3. Here, the actual circuitry is shown, and hence the proper signal levels must be specified in terms of voltage. The circuit shown requires a logic 1 that is more negative than logic 0 and would ordinarily be used with *pnp* transistors.

Such a circuit is often drawn showing only one diode, with the remainder given as only a diagonal line properly placed, which, of course, saves much drafting time. Note also that the diodes are simply placed according to the Boolean expression for that particular decoding situation. For example, the expression for the 4th state of the counter is $\bar{A}\bar{B}C$. Note that diodes are connected to A_R, B_R, and C_S. Thus, with flip-flops A and B reset but C set, an output will be evident at the line labeled 4.

Figure 14-6 illustrates a diode matrix for use with positive logic, which would ordinarily be used with *npn* transistors. These individual gates comprise positive AND gates since it is normally convenient to use positive logic with *npn* units.

If it should be necessary to decode a decimal counter, this is accomplished as easily as before. A decade counter is shown in Fig. 14-7, with a suitable

346　PULSE AND SWITCHING CIRCUIT ACTION

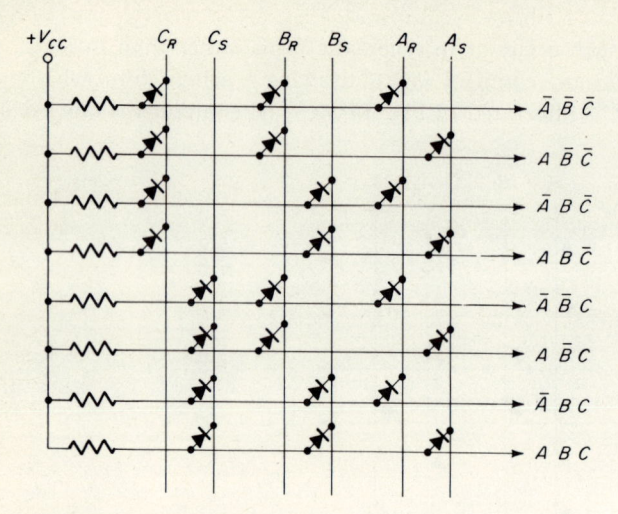

Fig. 14-6. Diode matrix for use with *npn* transistors and positive logic.

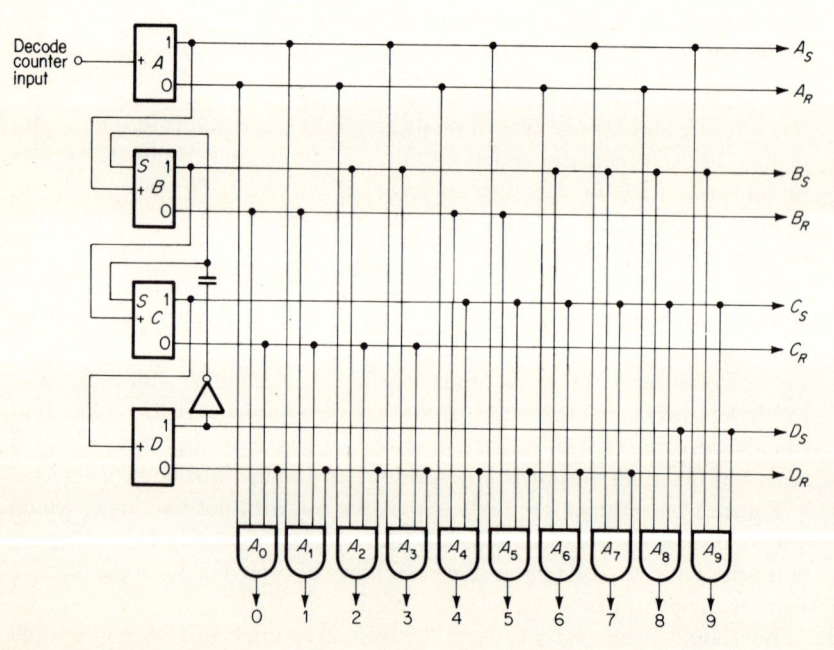

Fig. 14-7 Matrix for a scale-of-10 (decade) counter.

matrix to decode the 10 separate states of the counter. Recall from Chap. 13 the operation of a typical decade counter. Using the waveforms given in Fig. 13-9, the eighth count causes the counter to automatically advance to the 14th state. Thus, the normal counter conditions for the 8th to the 13th states are simply not decoded; note that these do not appear as input conditions for the 10 gates shown.

Gates 1 to 7 are perfectly straightforward and are no different than those employed with any conventional counter using the binary code. However, the eighth input pulse causes the counter to instantly advance to the 14th state, and hence the gate labeled $A8$ has as its inputs the 14th state of the counter. This condition is specified by $\bar{A}BCD$, and these are the counter lines to which the four inputs of $A8$ are attached. By the same token, gate $A9$ must have as its inputs the 15th state, or $ABCD$. This, too, is seen to be the case in Fig. 14-7. Thus, decoding the outputs of a decade counter is no more difficult than is the case with a simple binary unit.

Rather than use simple AND gates, it is possible and often desirable to construct a decoding matrix using NAND/NOR gates. This is especially appropriate if integrated circuits are being used. Figure 14-8 illustrates

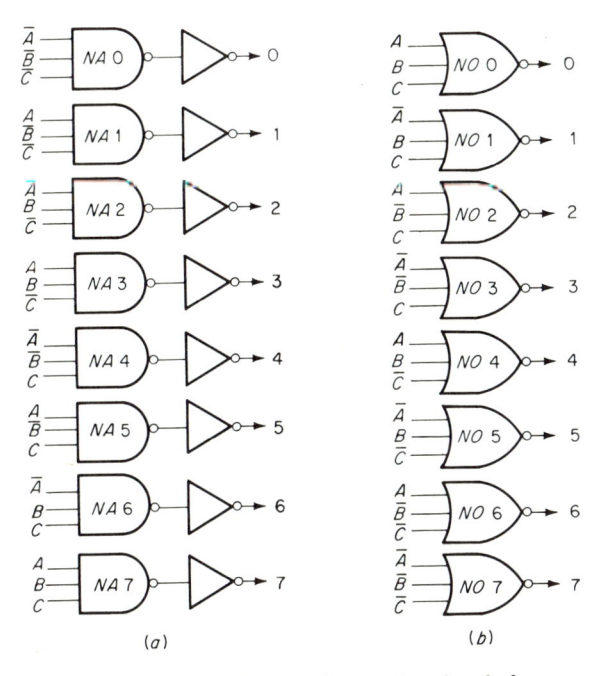

Fig. 14-8. NAND/NOR matrices used to decode from binary to decimal. These would, in all probability, be implemented with integrated circuits.

both NAND and NOR gates used in a decoding matrix. Note that because the NAND gate actually provides an inverted AND function, it is usually necessary to provide an additional stage of inversion. Alternatively, it is ordinarily possible with NOR gates to use as inputs the opposite to the conditions shown; that is, the 0-state gate would have ABC as its inputs rather than $\bar{A}\bar{B}\bar{C}$ etc. Such a solution is used with the NOR gates shown, where the additional inverter is not necessary. This solution cannot be used with NAND gates.

A final example of a decoding matrix is illustrated in Fig. 14-9. This

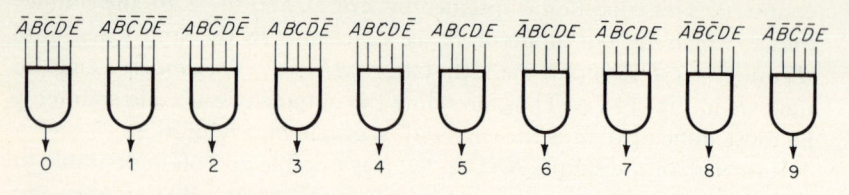

Fig. 14-9. Decoding matrix for a Johnson-code counter (switch-tail ring counter).

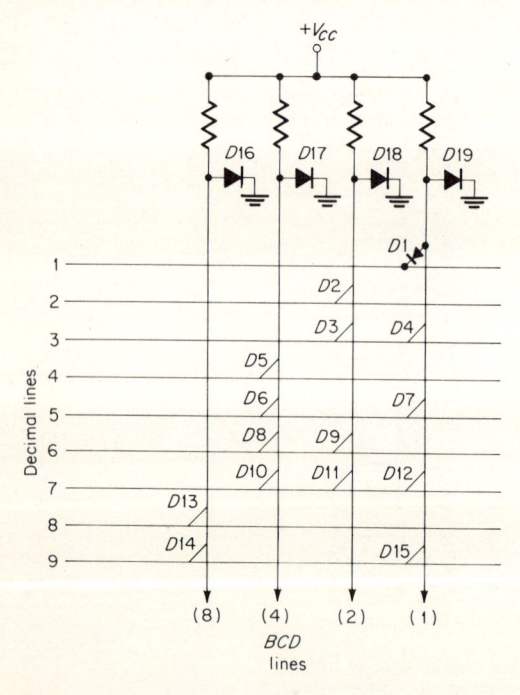

Fig. 14-10. Decimal to BCD encoder.

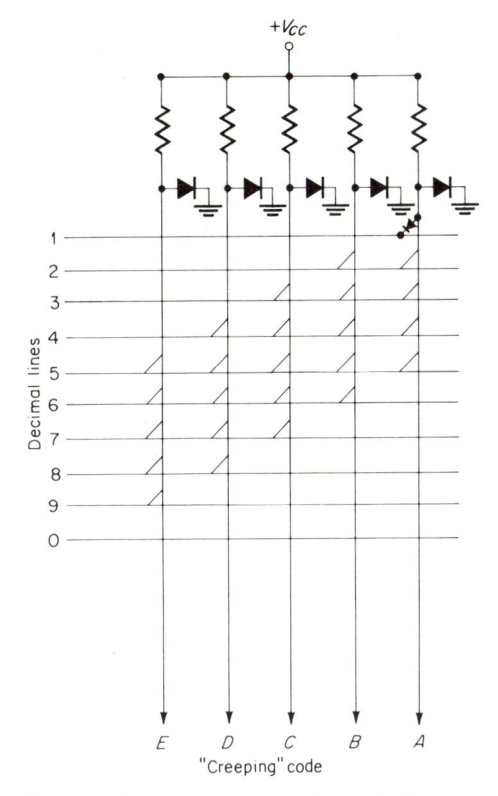

Fig. 14-11. Decimal to creeping-code (Johnson-code) encoder.

matrix allows a switch-tail ring counter to be decoded for each of its 10 states. As before, only one gate at a time will yield an output, depending upon the state of the counter. Because this counter requires five flip-flops, the gates must have five inputs.

14-2 ENCODING MATRICES

A simple circuit to convert the logic 1 on a single wire representing a decimal digit to an equivalent binary-coded-decimal digit is shown in Fig. 14-10. The diodes are installed so that the appropriate binary line will be clamped to logic 1 (a negative voltage in the circuit shown) for each of the binary bits necessary. If no diode is connected, for some given condition, the binary line will remain at ground potential.

350 PULSE AND SWITCHING CIRCUIT ACTION

To illustrate circuit action, assume that the decimal line 3 is at some negative voltage while all others are ground. Diodes $D3$ and $D4$ will become forward-biased and will conduct, clamping the 1- and 2-bit binary lines to the negative voltage. Output lines (BCD) 4 and 8 will still be at ground, and hence the condition of the output wires is 0011, or a 3 in BCD. If decimal line 6 is at logic 1, diodes $D8$ and $D9$ will cause the 2- and 4-bit lines to be at logic 1 (0110). Thus, any one decimal line can cause the corresponding binary code to be formed by clamping these binary lines to logic 1. The purpose of $D16$ to $D19$ is to clamp the binary lines to ground when not otherwise energized. They become back-biased when logic 1 appears on a binary wire.

Figure 14-11 illustrates an encoding matrix for a situation requiring the Johnson code. The placement of the diodes determines the code that is created.

QUESTIONS AND PROBLEMS

14-1 Refer to Fig. 14-1. Write the Boolean equation to decode the 4th counter state (three input pulses).

14-2 Refer to Fig. 14-1. Write the Boolean equation to decode the 6th counter state (five input pulses).

14-3 Refer to Fig. 14-3. Of the six binary wires shown, which will be at logic 1 if four input pulses are presented at the counter input (counter initially reset)?

14-4 Refer to Fig. 14-3. Of the six binary wires shown, which will be at logic 1 if six input pulses are presented at the counter input (counter initially reset).

14-5 Refer to Fig. 14-8a and b. Determine the proper configuration to decode a scale-of-16 counter using NAND logic for its 14th state (13 input pulses).

14-6 Refer to Fig. 14-8a and b. Determine the proper configuration to decode a scale-of-16 counter using NAND logic for its 12th state (11 input pulses).

14-7 Refer to Fig. 14-8a and b. Determine the proper configuration to decode a scale-of-16 counter using NOR logic for its 14th state (13 input pulses).

14-8 Refer to Fig. 14-8*a* and *b*. Determine the proper configuration to decode a scale-of-16 counter using NOR logic for its 12th state (11 input pulses).

14-9 A Johnson-code counter has seven stages. Determine the code for the counter state after 13 input pulses (the counter is initially reset).

14-10 A Johnson-code counter has seven input stages. Determine the counter state after 13 input pulses (the counter is initially reset).

15
SPECIAL CIRCUITS AND DEVICES

In this chapter we shall investigate a variety of circuits and devices. Pulse amplifiers of several configurations, sawtooth and staircase generators, and inductively coupled circuits will be dealt with, among others. Also, devices classed as other than junction transistors, such as the SCR, the FET, tunnel diodes, and many others, will be described.

The pulse amplifiers in this section will be recognized as true linear amplifiers for the most part. Nearly all equipment using memory devices (a magnetic-core memory, for example) requires a linear amplifier to amplify the weak pulses derived from the memory and to provide a substantial pulse for the logic circuitry. The waveforms used with these circuits qualify them for inclusion herein since they are used with, for the most part, square waves. Hence, their primary purpose is to convert and amplify a signal that is usually anything but a good square wave and to yield as an output a waveform that is acceptable to the logic circuitry.

15-1 SINGLE - ENDED SMALL - SIGNAL LINEAR PULSE AMPLIFIERS

The purpose of an amplifier such as that shown in Fig. 15-1 is to amplify a weak signal originating at a memory device such as an acoustic delay line. The circuit illustrated performs this function admirably, amplifying

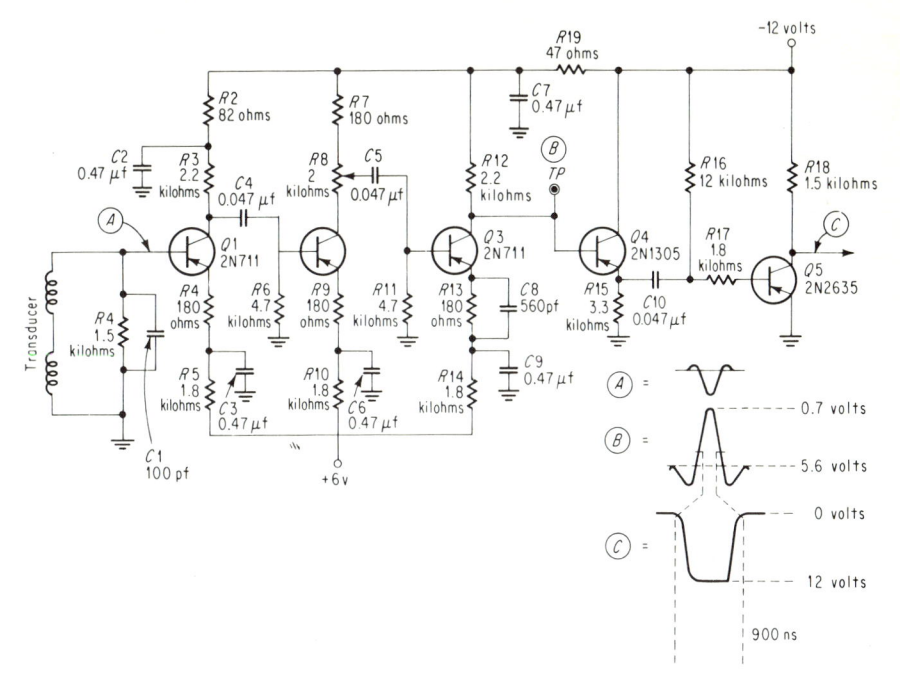

Fig. 15-1. Small-signal pulse amplifier.

a 25-mv signal about 200 times as measured at the collector of $Q3$. The output, measured at test point TP, is therefore on the order of 5 volts.

$Q1$, $Q2$, and $Q3$ are biased as class A linear amplifiers and provide most of the gain in the overall amplifier. $Q4$ is an emitter follower and serves to isolate $Q3$ from $Q5$. The output stage is an overdriven amplifier, as can be seen on the waveforms. Its output is a 12-volt pulse.

To begin the circuit description, the method of biasing used for $Q1$ to $Q3$ will be briefly discussed; a simplified circuit, shown in Fig. 15-2, will serve

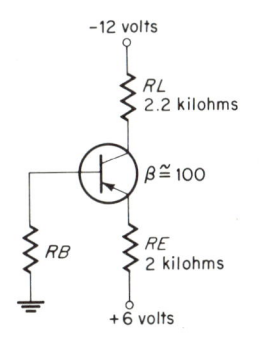

Fig. 15-2. Approximate dc equivalent circuit for $Q1$.

354 PULSE AND SWITCHING CIRCUIT ACTION

to illustrate this point. This simplified drawing represents the dc state of any of the three input transistors.

First, it is necessary to insure that the transistor is biased in, or near, the center of the transistor's active region; that is, the transistor must be turned on to some degree, but not fully on. In terms of the circuit limits, V_{CE} must be greater than zero but smaller than V_{CC}. To accomplish this, the base-emitter junction must be forward-biased but base current must be limited to some intermediate value.[1]

Figure 15-3 illustrates the fact that the base is returned to ground, which

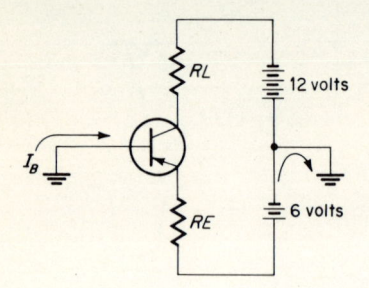

Fig. 15-3. Base-current path.

in this circuit is 6 volts more negative than the supply voltage to which the emitter is returned. Hence, base current flows in some amount, and thus the emitter junction is forward-biased.

Referring again to Fig. 15-2, the dc voltages that exist in the circuit with no signal impressed are easily computed. For example, if this circuit represents the circuit of $Q1$, RB is essentially 0 ohms since the transducer coils have virtually no resistance. The base voltage, then, must be essentially ground (0 volts). On the assumptions that the base-emitter voltage is negligible and that the base-emitter junction is forward-biased, then the emitter is also at ground potential. If the emitter is very nearly at ground potential, the current through the emitter resistor must be a function of this voltage drop and the value of RE.

$$I_E = \frac{E_{RE}}{RE} = \frac{6}{2000} = 3 \text{ ma}$$

The base current, then, is determined in this circuit by the transistor itself. If $\beta = 100$,

$$I_B = \frac{I_E}{\beta + 1} \cong 30 \text{ }\mu a$$

[1] For a complete discussion of linear amplifier biasing, see Henry C. Veatch, "Transistor Circuit Action," McGraw-Hill Book Company, New York, 1968.

On the assumption that the transistor's h_{FE} (beta) is 100 and therefore alpha is 0.99, the collector current is determined as follows:

$$I_C = \alpha I_E = 0.99 \times 0.003 \cong 0.00297$$

Since collector current is so nearly the same value as emitter current, it can be assumed that each is equal to 3 ma.

Once the transistor's currents have been given a value, the voltage drops around the circuit can be evaluated. First, the drop across the collector resistor RL can be found:

$$E_{RL} = I_C \times RL = 0.003 \times 2200 = 6.6 \text{ volts}$$

The collector voltage in reference to ground is just as easily determined:

$$V_C = -V_{CC} + E_{RL} = -12 + 6.6 = -5.4 \text{ volts}$$

Referring to Fig. 15-1, the actual circuit conditions in the original circuit will now be correlated with the foregoing figures. First, note the emitter resistors of $Q1$, Fig. 15-1. From a dc standpoint, the 180- and 1800-ohm resistors appear in series, and hence their sum is 1980 ohms. This is so nearly 2000 ohms that the two circuits are essentially identical.

The two resistors in the collector lead of $Q1$, $R2$, and $R19$ were ignored in the simplified version. In the actual circuit the 82-ohm resistor drops 0.25 volt, while the 47-ohm resistor drops about 0.4 volt. For this reason, the actual collector-supply voltage for $Q1$ is on the order of 11.35 volts rather than 12 volts. Also, $Q2$ and $Q3$ have about 11.6 volts for their supply voltage.

The bases of $Q2$ and $Q3$ are returned to ground through the 4700-ohm resistors, which raises the possibility of the bases being at something other than ground potential. If it is assumed that the transistors have an h_{FE} of 100, the actual base voltage may be determined. It was determined earlier that the base current is 30 μa. The drop across the base resistor, then, is easily found:

$$I_B = \frac{I_E}{\beta + 1} = \frac{0.003}{101} \cong 30 \text{ μa}$$

$$V_B = I_B \times RB = 0.00003 \times 4700 = 0.14 \text{ volt}$$

In most cases this small voltage could be ignored in the calculations. When analyzing such a circuit, it is always wise to double-check for the base voltage, for if RB were very much greater in value, the voltage drop would then be appreciable. A rule of thumb states that the percentage error may be up to, but no greater than, the tolerance of the components used in the circuit. If the resistors used in this circuit are $\pm 5\%$, then the same degree of error can be allowed in simplifying the circuit. In the present case, 0.14 volt is slightly greater than 2% of the collector-supply voltage and, for most purposes, could be ignored since the resistors themselves will yield greater deviation from the calculated values.

To continue the circuit description, note that $Q3$ is direct-coupled to $Q4$ and that the steady-state condition of $Q4$ depends on the collector voltage of $Q3$, which is -5 volts. Since the collector resistor of $Q3$ is also the base resistor of $Q4$, a large base current flows in $Q4$. Because $Q4$ is an emitter follower, the emitter is also nearly at -5 volts with no signal applied.

The steady-state condition of $Q5$ is determined by the circuitry in the base lead. The base is returned to the negative supply through a total of 13,800 ohms. $Q5$, then, is biased quiescently on, and since base current must be large, it can be assumed that the transistor is saturated. Therefore, the steady-state condition of the entire circuit is not difficult to determine.

To describe the signal response of this circuit, the waveforms given on Fig. 15-1 will be used. Two factors stand out as being the most important of all signal conditions. These are the *overall voltage gain* and the *upper-frequency limit*. The voltage gain of the circuit determines whether or not the output transistor will provide a full 12-volt pulse with a typical input. The frequency response determines whether or not the rise time of the output pulse is fast enough to properly actuate the logic circuitry.

For midband conditions, the voltage gain for any amplifier stage is a function of the total signal load upon the transistor and any unbypassed resistance in the emitter lead. For example, the total signal load seen by $Q1$ is the parallel resistance of $R3$, $R6$ and the input resistance of $Q2$. This can be determined as follows (assume beta is 25):

$$\text{load}_{ac} = \frac{1}{1/R3 + 1/R6 + 1/r_{i,Q2}} = \frac{1}{0.000881} = 1140 \text{ ohms}$$

$$A_v = \frac{\text{load}_{ac}}{RE} = \frac{1140}{180} = 6.3$$

where $r_{i,Q2} = (\beta + 1)R9$. Hence, the voltage gain of $Q1$ is slightly greater than 6 for midband conditions. The voltage gains of $Q2$ and $Q3$ are determined in the same way. The maximum overall gain of $Q1$, $Q2$, and $Q3$ is in excess of 400, with the moveable arm of $R8$ at the collector of $Q2$. As used in the application for which the amplifier was designed, the potentiometer is set so that the voltage gain is on the order of 200 for best results.

The low-frequency amplification of this circuit is made deliberately poor at frequencies below 5000 Hz. Since the circuit should never have to operate at these frequencies, the bypass capacitors $C3$, $C6$, $C8$, and $C9$ are made small. The high-frequency response is accentuated to ensure a fast rise time, and this is accomplished by the addition of $C8$, placed in parallel with R13. Above 1.6 MHz, this capacitor acts to raise the voltage gain of Q3 by bypassing the resistor. At frequencies below 1.6 MHz, the reactance of the capacitor is quite high and the resistor provides degeneration, thus reducing the gain somewhat. However, at values above the critical one, the capacitor appears as a lower impedance than the resistor and the effects of degeneration are minimized. This compensates, at least in part, for the normal decrease in gain as the frequency increases.

The collector of Q1 is decoupled from the power supply by the decoupling filter, consisting of R2-C2. This prevents unwanted feedback from the other transistors by way of the internal resistance of the power supply. The filter R19-C7 also serves to further decouple the three-stage amplifier from variations in the supply line.

The purpose of Q4 and Q5 is to further amplify and shape the output of Q3 and to insure a full 12-volt output. Q4, of course, is simply an emitter follower. Q5, however, performs a function that is not immediately apparent, in addition to its normal functions of amplifying and inverting.

The input circuit to Q5, R16, R17, and C10 serves to provide a *threshold* voltage that prevents a full output with minor disturbances at the input to the amplifier. That is, very small voltages at point A will be amplified, along with the desired signals, and can produce very significant output-voltage swings. The threshold voltage prevents any unwanted outputs from the amplifier unless they are of sufficient amplitude to overcome the threshold.

To explain the circuit action of these components, consider the state of the circuit just prior to the arrival of a pulse. Because of the voltage at the collector of Q3, the emitter of Q4 is at a level of about -5.5 volts with respect to ground. The voltage at the right plate of this capacitor is set by the voltage divider of R16 and R17.

$$E_C = \frac{1800}{12{,}000 + 1800}(-12) \cong -1.6 \text{ volts}$$

Note that Q5 is normally on; the voltage at its base must rise to approximately ground before it can be turned off. The emitter of Q4, then, must go more positive than 1.6 volts before Q5 can turn off. Variations smaller than this will not drive Q5 off and hence are said to be *discriminated* against. Since the valid inputs will produce much larger swings, only they will produce a valid output signal.

15-2 PUSH - PULL LINEAR PULSE AMPLIFIERS

The circuit shown in Fig. 15-4 is a linear push-pull amplifier, used as the input circuit in a *read amplifier* that extracts data from a magnetic-disc

358 PULSE AND SWITCHING CIRCUIT ACTION

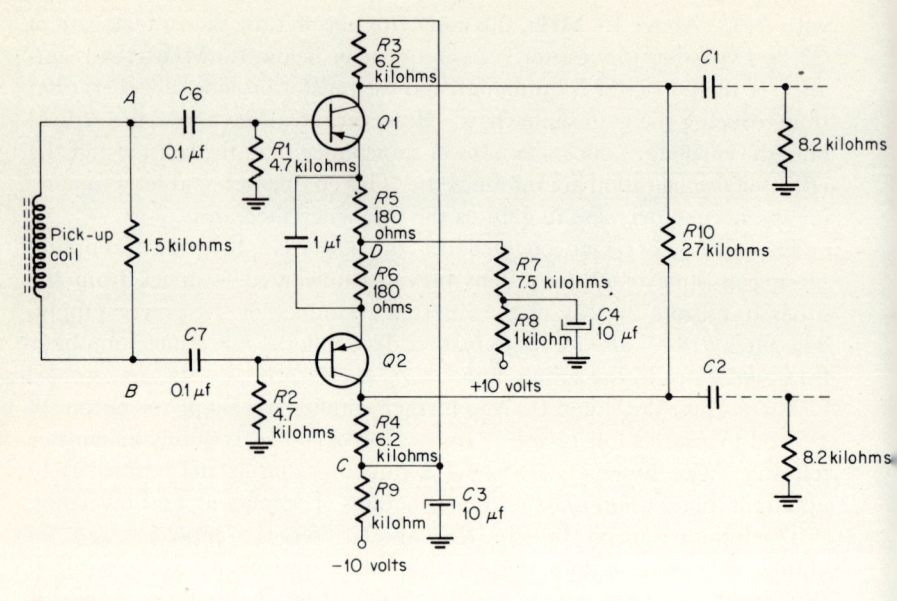

Fig. 15-4. Differential amplifier.

storage medium. This type of circuit is often called a *differential amplifier* since it will respond only to differential signals. An ideal differential amplifier is one in which amplification occurs for only the instantaneous *difference* between two signals. Such a circuit is said to have *common-mode rejection* since the inverse of a differential signal is a common-mode signal. In theory, an ideal circuit will reject all common-mode signals and readily amplify differential signals. In practice, some common-mode signal is usually apparent in the output.

The first step in describing the circuit action is to specify the dc, or quiescent, voltages and currents. To facilitate the dc analysis, the circuit is redrawn in Fig. 15-5. This will provide a better mental picture of the purpose of each component.

The circuit is seen to be symmetrical in that the load resistors, emitter resistors, etc., are the same in each transistor circuit. A small voltage is induced in the pickup coil by a rapid change in the magnetic flux, and the desired signal will drive one base positive while driving the other negative. If some 60 Hz energy is induced in the base circuit, both inputs will be driven in the same direction at the same time and the amplifier will reject this unwanted signal. The output, taken from each collector, will be further amplified and shaped before being used by the logic circuitry.

The quiescent base voltage at either base must be very nearly ground.

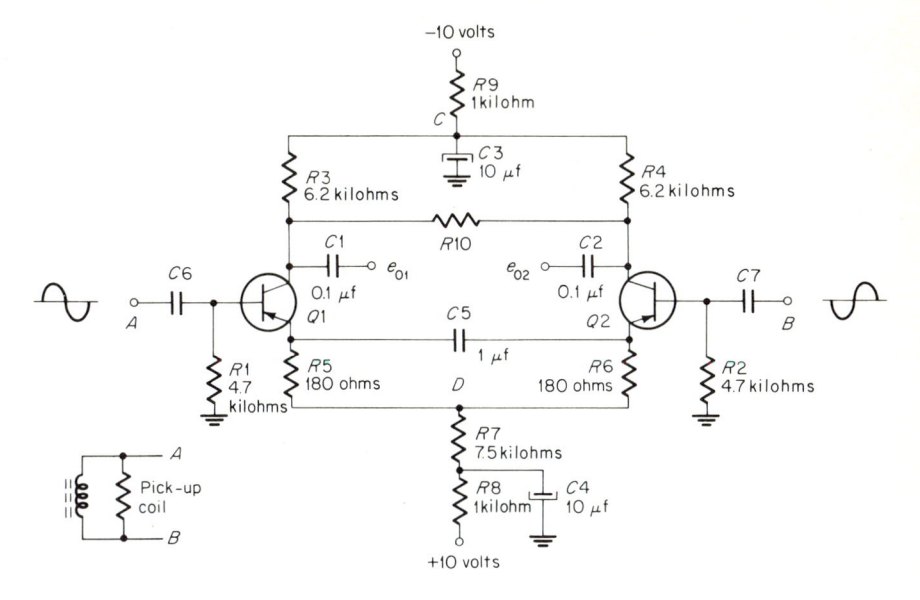

Fig. 15-5. Differential amplifier redrawn.

While base current does flow through $R1$ and $R2$, the amount is quite small and the voltage drop across either resistor is correspondingly small. On the assumption that the drop from base to emitter is also small, the emitters of both transistors can be assumed to be approximately at ground potential. One might approximate this as a total of 0.2 volt, dropped across the base resistors and V_{BE} of the transistors. Hence, the emitters are very nearly at $+0.2$ volt above ground.

Noting that the 180-ohm resistors are only on the order of 2% of $R7$ and $R8$, the drop across these can be ignored for the present. Since the voltage at point D is known to some degree of accuracy, the current through $R7$ and $R8$ can easily be found:

$$I = \frac{E}{R7 + R8} = \frac{9.8}{8500} = 1.15 \text{ ma}$$

Inspecting the circuit closely reveals the fact that the same current flows through $R9$. The voltage drop across $R9$, then, is

$$E_{R9} = I \times R9 = (0.00115)(1000) = 1.15 \text{ volts}$$

The voltage at point C, then, is $-V_{CC}$ less the drop across $R9$.

$$E_C = -V_{CC} + E_{R9} = -10 + 1.15 = -8.85 \text{ volts}$$

Now, because the circuit is symmetrical, the current through $R9$ must divide equally between $Q1$ and $Q2$. The collector current for each transistor is therefore one-half the above figure:

$$I_C = \frac{I_{R9}}{2} = \frac{1.15}{2} = 0.575 \text{ ma}$$

Knowing the voltage drop across $R3$ and $R4$ will allow the collector-to-emitter voltage to be determined:

$$E_{RL} = I_C \times RL = 0.000575 \times 6200 = 3.57 \text{ volts}$$
$$V_C = -V_{CC} + E_{R9} + E_{RL} = -10 + 1.15 + 3.75 = -5.1 \text{ volts}$$

Because the emitter of either transistor is assumed to be at $+0.2$ volt with respect to ground, the collector-to-emitter voltage must be 5.3 volts. Thus, with approximately 5 volts across the transistor and nearly 4 volts across the load resistor, the transistors are biased well within the active region. Note that the foregoing figures are slightly in error since, among other things, the transistor alpha was ignored, but they are nevertheless surprisingly close to actual measured values.

The signal conditions can be described briefly in terms of the waveforms at both input and output, shown in Fig. 15-6. As the read head passes a change in flux, it generates a voltage that is fed to the transistor bases A and B. As A goes more positive, B goes more negative and $Q2$ is

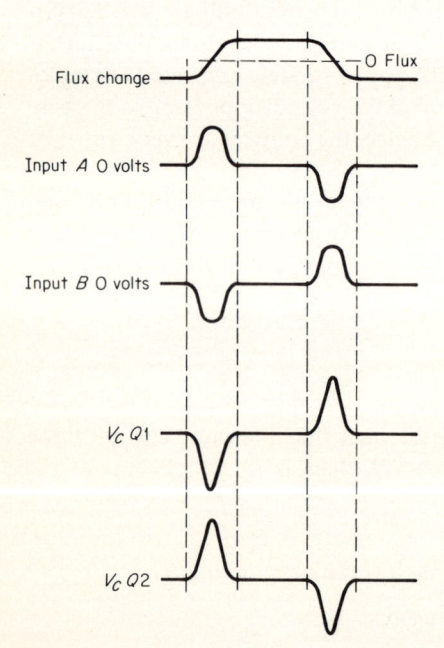

Fig. 15-6. Differential amplifier waveforms (not to scale).

SPECIAL CIRCUITS AND DEVICES   361

turned on somewhat while Q1 is turned off somewhat. Since this is a linear amplifier, the transistors can be expected to remain continually in the active region, never reaching either circuit limit (cutoff or saturation). The input signal is perhaps 100 mv or less, and the output signal is considerably larger than this. The actual amount of gain is determined by the total signal load (not shown) and cannot be calculated with the information given.

15-3   BIASED - UP DIFFERENTIAL AMPLIFIERS

A slightly different circuit is shown in Fig. 15-7. This is also a differential amplifier but is used in an application where a separate preamplifier has

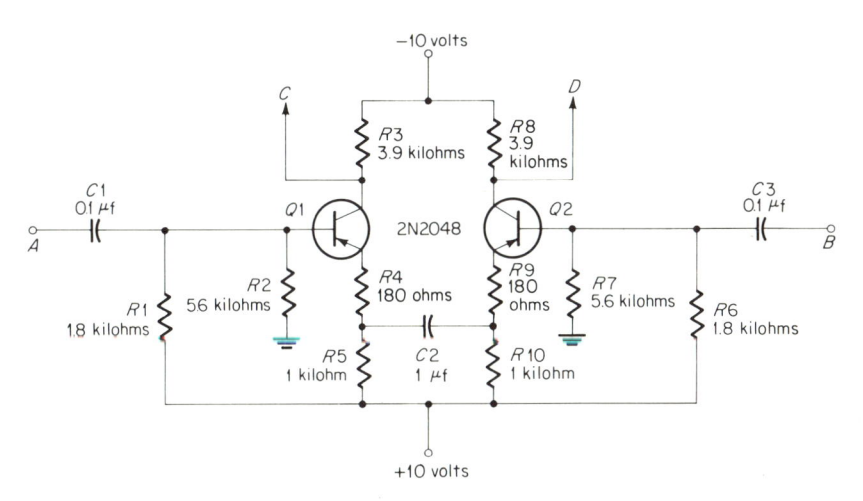

Fig. 15-7.   Biased-up differential amplifier.

already done a considerable amount of amplifying. Hence, the output of this circuit may exceed 10 volts, and since the collector power supply is only 10 volts, measures must be taken to prevent clipping of the signal. To accomplish this, the transistors are biased so that a greater portion of the available 20 volts is usable by the circuit. If the bases of Q1 and Q2 were simply returned to ground, the maximum output of either transistor would be limited to no more than 10 volts.

However, the collectors can make an excursion of perhaps 18 volts, peak to peak, because of the biased-up feature. The resistors R1, R2 and R6, R7 perform this function of lifting the bases above ground by their voltage-divider action. Because the signal activity of this circuit is much like that of the previous circuit, except for the greater signal amplitude, the following discussion concerns the calculation of the dc state of the circuit.

The dc voltage at either base, measured with respect to ground, is determined by the voltage divider mentioned above. Considering the circuit of $Q1$, this voltage divider consists of an 1800- and a 5600-ohm resistor. The base voltage of $Q1$, then, is V_{B1}:

$$V_{B1} = \frac{R2}{R1 + R2} \times V_{EE} = \frac{5600}{1800 + 5600} \times 10 = 7.6 \text{ volts}$$

Because the circuit is symmetrical, the same voltage is evident at the base of $Q2$ also. Now, allowing a 0.1-volt drop across the base-emitter junction of the transistor, the emitter must be at a potential of 7.7 volts, referred to ground.

Emitter current is a function of the voltage across the emitter resistors.

$$I_E = \frac{V_{EE} - V_E}{R4 + R5} = \frac{10 - 7.7}{1180} = 1.95 \text{ ma}$$

Again, making the assumption that I_C is nearly equal to I_E, the drop across the load resistors can be found.

$$E_{RL} = I_C R_L = 0.00195 \times 3900 = 7.6 \text{ volts}$$

The collector voltage, referred to ground, can now be determined.

$$V_C = -V_{CC} + E_{RL} = -10 + 7.6 = -2.4 \text{ volts}$$

The collector-to-emitter voltage is the difference between V_C and V_E, both referred to ground.

$$V_{CE} = (-2.4) - (7.7) = 10.1 \text{ volts}$$

As shown, the circuit allows the collectors to make maximum excursions from -10 volts to approximately $+8$ volts, a nearly 18-volt peak-to-peak excursion.

The voltage gain of the amplifier is on the order of 22. One factor that determines this is the unbypassed resistor in the emitter of each transistor. The lower end of each 180-ohm resistor is at a *virtual ground*, produced by the bypass capacitor $C2$. This results in the same action that would occur if each resistor were bypassed separately to ground by individual capacitors.

To verify this statement, Fig. 15-8 is offered. In Fig. 15-8a the original circuit is shown, using two bypass capacitors, each of which is 2 μf. This version works quite well, but, as will be seen, one of the capacitors is

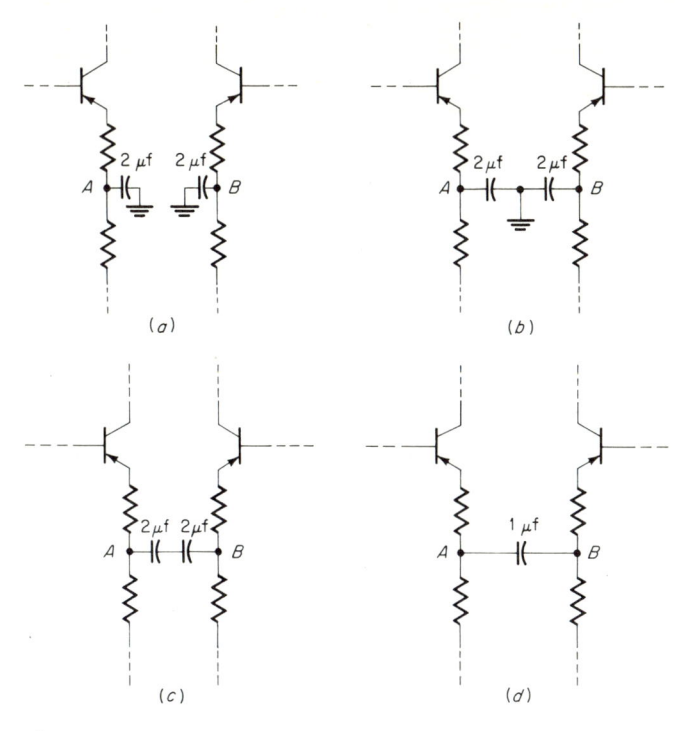

Fig. 15-8. Developing a *virtual ground* at A and B.

unnecessary while the other is twice as large as required. The bypassed currents flow in each of these capacitors from points A and B out to ground. Each of these currents will be equal to the other since the two transistor circuits are symmetrical. However, they are flowing in opposite directions at any one instant because of the push-pull action of the transistors.

In Fig. 15-8b the circuit is slightly modified by causing the connection to ground to be effected through a single wire. Note that the common wire can now have *no current through it* since the currents are of equal amplitude but opposite direction. A wire through which no current ever flows may be removed, as it accomplishes nothing, and this is shown in Fig. 15-8c. The two series capacitors now have a total effective value of 1 μf since they are in series, and thus they can be removed and replaced with a single unit having a capacitance of 1 μf. Thus, the circuit is simplified with no sacrifice in performance. Indeed, performance is improved somewhat over the circuit in Fig. 15-8a because should an unequal input occur (perhaps one base going more positive than the other one goes negative), the circuit will tend to compensate by providing inverse feedback and correct for the unbalance in input signals.

15-4 CRT DEFLECTION AMPLIFIERS

An amplifier designed to drive the deflection plates of a CRT (cathode-ray tube) is shown in Fig. 15-9. This circuit is a large-signal linear class A

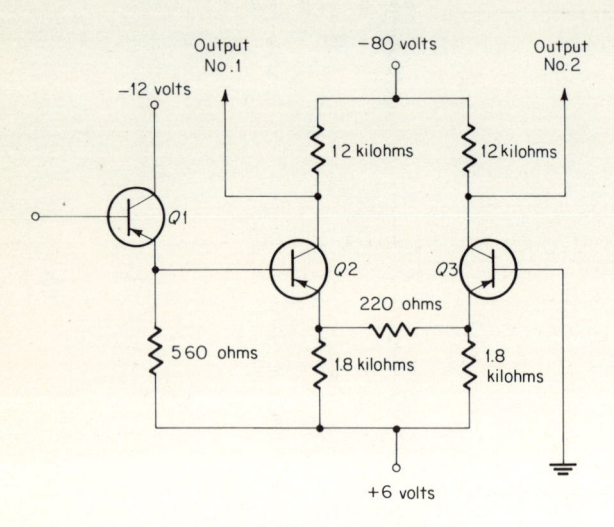

Fig. 15-9. Large-signal differential amplifier.

differential amplifier. It differs from the preceding examples in several ways. One of these is the considerably larger collector-supply voltage, which is necessary to allow a 150- to 160-volt collector-to-collector voltage swing. The circuit, then, truly classifies as a large-signal amplifier. Another difference is that it accepts at its input a single-ended signal while producing a double-ended, or push-pull, output signal. It thus provides phase inversion in addition to the usual amplification. Additionally, the two transistors $Q2$ and $Q3$ exhibit all three basic circuit configurations, with $Q2$ performing as both a common-emitter and a common-collector circuit. $Q3$, on the other hand, performs as a common-base amplifier. This circuit, then, possesses many interesting features not usually found in a single circuit.

The circuit is designed to operate in the center of the dc load line. Because the outputs connect directly to the deflection plates of a CRT, there is no signal load as such. In this special case, the dc and signal load lines coincide with each other. Thus, for maximum symmetrical output, the proper bias point for the circuit is at the center of the dc load line. The peak-to-peak voltage at the collectors is sufficient to drive the spot on the CRT a total of perhaps 3 to 4 in. The circuit is used for decimal-character display on the tube, which is more than enough for this purpose.

This circuit has certain advantages over other ways of implementing for the same result, among which are simplicity, common-mode rejection, excellent frequency response, and good balance between outputs. In order to make the circuit description more clear, it will first be described in general terms, without proof. Then, the description will be repeated in an attempt to prove certain salient points.

The signals applied to the base of $Q2$ vary between ground and about -0.8 volt. The waveshape, which is a very complex staircase, need not concern us to any extent. As the base of $Q2$ is driven more negative by the emitter follower $Q1$, $Q2$'s emitter must follow its base. $Q2$ is therefore turned on more fully, and more collector current flows, causing the collector to fall toward ground. Thus, as the base of $Q2$ goes more negative, its collector goes more positive, as is expected from a common-emitter circuit.

Now, as the emitter of $Q2$ is driven more negative by its base, this negative-going waveform is coupled to the emitter of $Q3$ by way of the coupling resistor connected between the emitters. As collector current increases through $Q2$, the current flowing in the coupling resistor causes collector current in $Q3$ to decrease by the *same amount* as the increase in $Q2$. The collector voltage of $Q3$, then, rises toward $-V_{CC}$ by the same amount that the collector voltage of $Q2$ falls toward ground. The two outputs, therefore, are 180° out of phase with each other.

To verify the foregoing description, the circuit will be investigated more thoroughly. To simplify the description of the dc state of the circuit, it will be assumed that the base of $Q1$ is connected to a potential that will place the base of $Q2$ at essentially ground. The dc state of $Q2$ is therefore as shown in Fig. 15-10, where it is seen that the base is connected to ground.

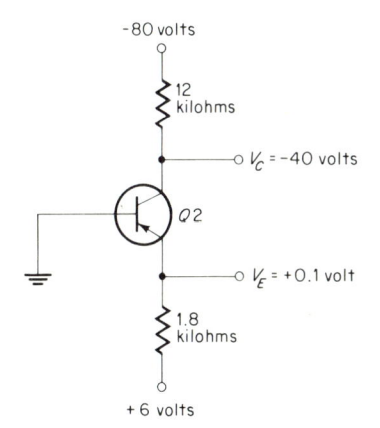

Fig. 15-10. The dc condition of $Q2$.

To determine the dc voltages and currents that specify the bias conditions, the first step is to determine the emitter current.

Because the base is at ground potential, the emitter is also near ground, differing only by the drop across the base-emitter junction V_{BE}. Assuming this to be a germanium transistor, this might be expected to be on the order of 0.1 volt. The emitter, then, is at a potential of +0.1 volt. The 1800-ohm resistor therefore has a voltage across itself of +6 − 0.1 = 5.9 volts. The current through this resistor, which is also the emitter current, is easily found:

$$I_E = \frac{E_{RE}}{RE} = \frac{5.9}{1800} = 3.28 \cong 3.3 \text{ ma}$$

On the assumption that collector current is very nearly equal to emitter current, the drop across the load resistor can also be found:

$$E_{RL} = I_C \times RL = 0.0033 \times 12{,}000 = 39.6 \text{ volts}$$

This allows collector voltage, referred to ground, to be found:

$$V_C = -V_{CC} + E_{RL} = -80 + 39.6 = -40.4 \text{ volts}$$

Q2 is quite obviously biased at the center of the dc load line since the drops across the load resistor and the transistor are very nearly equal.

Q3 is in the same equivalent circuit as Q2, as evidenced by the circuit of Fig. 15-11. Because of the similarity between the two equivalent circuits,

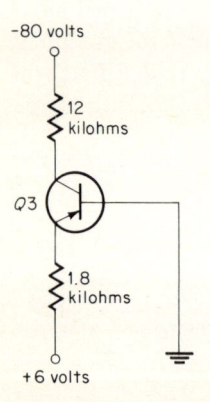

Fig. 15-11. The dc circuit of Q3.

one might expect that the dc voltages and currents for the quiescent condition are similar, which is, of course, the case. The base voltage is 0 volts (ground), and the collector voltage in reference to ground is, perhaps, −40 volts, while the emitter is 0.1 volt more positive than ground.

At this point, one must consider the 220-ohm resistor that is connected from emitter to emitter, for this may upset the previous calculations. However, note that *each* emitter is at a potential of approximately 0.1 volt, and hence there can be no current through this resistor. It cannot, therefore, change any of the previously determined values.

Now that the quiescent state of the amplifier has been specified, the manner in which it responds to a signal can be investigated. Assume that the signal makes a rapid transition from ground (0 volts) to -0.5 volt. The emitter of $Q2$ will follow this change and will make a change from $+0.1$ to -0.4 volt. When the emitter arrives at a potential of -0.4 volt, there is a difference of potential across the coupling resistor of 0.5 volt. Current now flows in this resistor, and this current must be considered a signal current. However, note that the 1800-ohm resistor, also in the emitter circuit, has a different voltage across it and so has a different current through it.

Signal current flows from the emitter of $Q2$ and into the coupling resistor. Figure 15-12 illustrates this action. Figure 15-12*b* shows the signal current,

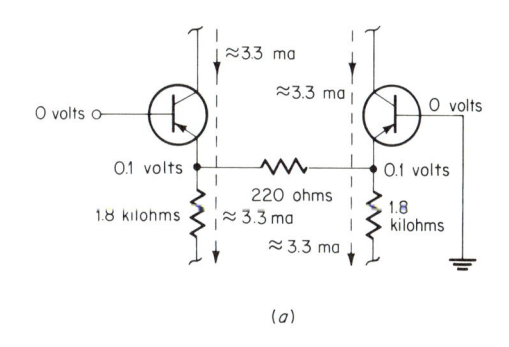

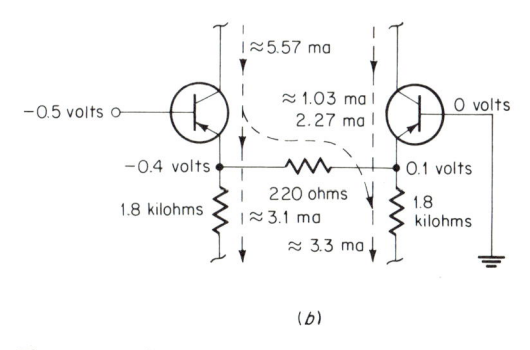

Fig. 15-12. (*a*) Quiescent emitter current; (*b*) emitter current with signal.

while Fig. 15-12a depicts the quiescent currents. Note that each transistor has approximately 3.3 ma flowing in the collector circuits and that quiescently there is no current in the coupling resistor. As the input transistor is turned on more fully, its collector current is seen to increase to 5.35 ma but its emitter current changes hardly at all. Instead, the increase in current flows into the coupling resistor and the 1800-ohm resistor *in the emitter circuit of Q3*. Note very carefully that because the voltage across this resistor is still the same as before, the current through it *must remain the same*. This can only occur if the collector current of Q3 decreases. As it happens, the collector current of Q3 will decrease in the same amount that the collector current of Q2 increases.

The magnitude of these currents can easily be found. The quiescent current in the collector of each transistor is stipulated as 3.3 ma. The current through the coupling resistor is a function of the voltage across it and the value of the resistor.

$$I = \frac{E}{R} = \frac{0.5}{220} = 2.27 \text{ ma}$$

The total collector current of Q2 is therefore equal to the sum of these two currents:

$$I_{C,\max} = 0.0033 + 0.00227 = 0.00557 = 5.57 \text{ ma}$$

Because the 1800-ohm resistor in the emitter circuit of Q3 must still have 3.3 ma through it, and if 2.27 ma is contributed by Q2, then the collector current of Q3 must have decreased by 2.27 ma.

$$I_{C,\min} = 0.0033 - 0.00227 = 1.03 \text{ ma}$$

These currents give rise to the desired circuit action at the collectors, with the collector of Q2 going in the positive direction and the collector of

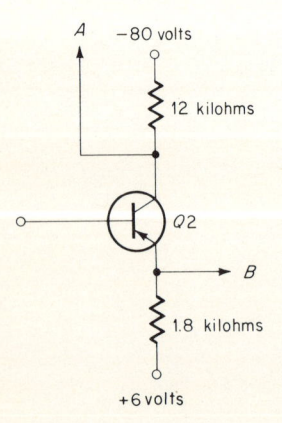

Fig. 15-13. Q2 has two outputs: *A* is a CE output, while *B* is a CC output.

SPECIAL CIRCUITS AND DEVICES   369

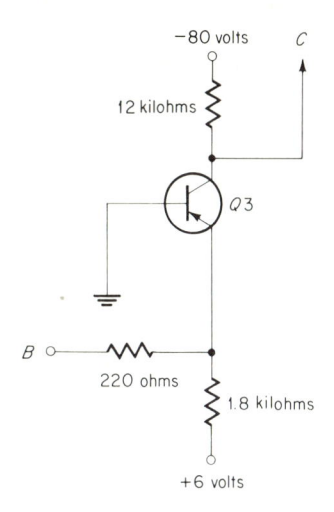

Fig. 15-14. *Q*3 is a common-base amplifier.

*Q*3 going in the negative direction. Because the two collector currents are of equal amplitude, the voltage changes at the two collectors are equal but of opposite sign. The circuit is providing its own phase inversion without benefit of additional circuitry.

To verify that *Q*2 acts as both a common-emitter amplifier *and* a common-collector amplifier, consider Fig. 15-13. Point *A* is the common-emitter output, while point *B* is the common-collector output. Figure 15-14 verifies that *Q*3 is a true common-base circuit, with the input applied at the emitter and the output taken from the collector, the base being at signal ground.

To determine the amplification of the circuit, the drop that occurred across the load resistors must first be found:

$$\Delta E_{RL} = \Delta I_C \times RL = 0.00227 \times 18,000 = 27.2 \text{ volts}$$

$$A_V = \frac{e_0}{e_i} = \frac{27.2}{0.5} = 54.4$$

Thus, the circuit amplifies an input signal by the factor of 54.4 times. It should be noted that if the circuit is measured from one collector to the other, the output-voltage swing will be two times the foregoing value; that is, the collector-to-collector–voltage swing, with a 0.5-volt input, will be 108.8 volts.

15-5   NONLINEAR WRITE AMPLIFIERS

The amplifier circuit shown in Fig. 15-15 is used to provide a short, sharp pulse across the load resistor. This pulse becomes data that is written into

370 PULSE AND SWITCHING CIRCUIT ACTION

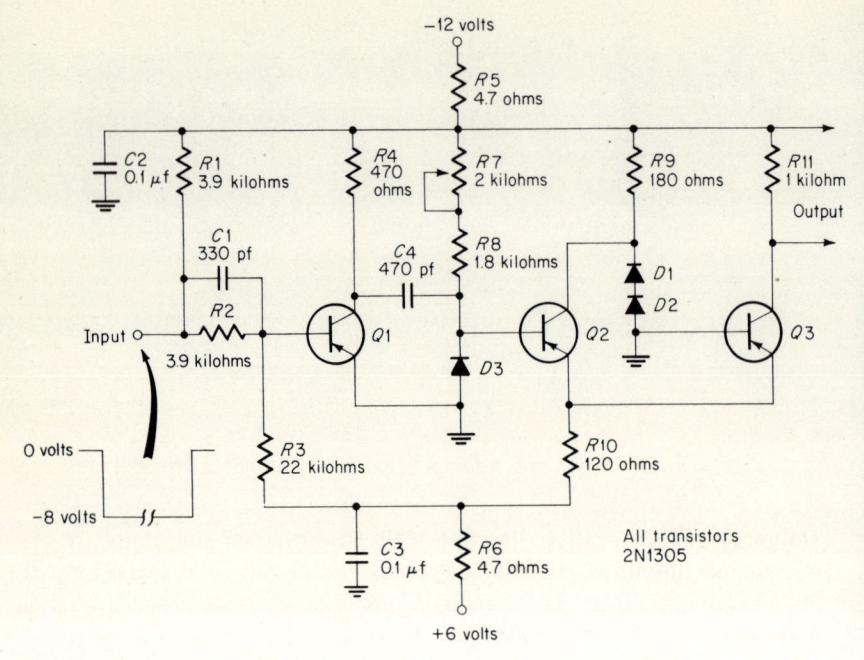

Fig. 15-15. The circuit diagram for the write amplifier.

a memory device, thus the name *write amplifier*. To make a sensible description of the circuit, the first determination is the static operating condition.

The input terminal is clamped to ground by the input signal, and so the base of $Q1$ sees a voltage that is determined by the voltage divider $R2$ and $R3$:

$$V_B = \frac{R2}{R2 + R3} \times V_{\text{total}} = \frac{3900}{25{,}900} \times 6 = +0.9 \text{ volt}$$

Thus, with 0 volts at the input, $Q1$ is quiescently off since its base is more positive than its emitter. The collector of $Q1$ is therefore at -12 volts since there is no current through the load resistor. During static input conditions, $C4$ can in no way influence $Q2$, and so the state of $Q1$ has no immediate bearing on the state of $Q2$. It does, however, help determine the state of charge of $C4$, as will be seen.

The state of $Q2$ depends upon the values of the resistors in the base circuit as well as the load resistor in the emitter. Diode $D3$ is normally conducting, and so the base voltage of $Q2$ is on the order of 0.5 volt.

The emitter of $Q2$ is at a potential that is approximately 0.1 volt more positive than its base, or -0.4 volt. Because the emitters of both Q2 and Q3 are tied together, the emitter voltage of Q3 must also be -0.4 volt. Note that the base of Q3 is directly grounded; it must therefore be normally off since the base is more positive than the emitter. Thus, the normal condition of Q3 is off, and there is no voltage drop across its load resistor. Before the input pulse arrives, both Q1 and Q3 are off while Q2 (an emitter follower) is on, with its emitter very nearly at ground potential.

When the input pulse arrives, the input is driven to about -8 volts. Now, the voltage divider is across a total of $-8 - (+6) = -14$ volts. The base of Q1 is therefore driven to a negative value. If the voltage divider tries to place a voltage $V_{B'}$ at the base that is more negative than V_{BE} by any significant amount, the transistor will be in saturation:

$$V_{B'} = \frac{R2}{R2 + R3} \times V_{\text{total}} = \frac{3900}{25,900} \times -14 = -5.0 \text{ volts}$$

Q1, then, is driven firmly into saturation, with V_B clamped to a value that might be approximated as -0.2 or -0.3 volt.

When Q1 turns on, its collector falls to ground and the left plate of C4 must also fall to ground. Since this capacitor plate changes from -12 to 0 volts very rapidly, the other plate must also go 12 volts in the positive direction from the level it had prior to the change. It was at -0.5 volt owing to the drop across the diode D3; thus a 12-volt change from this level is $-0.5 + 12 = 11.5$ volts positive with respect to ground.

At this instant, the base of Q2 is driven to $+11.5$ volts, and its emitter tries to follow this change. However, as soon as its emitter goes perhaps to $+0.2$ volt above ground, the emitter-base junction of Q3 becomes forward-biased and Q3 turns on. This, of course, clamps the emitter of Q2 to $+0.2$ volt. Q2 therefore cuts off and remains cut off for a period determined by the length of time that its base is more positive than the emitter. Q2 is, by definition, an emitter follower, but note that it can only act as a true emitter follower when the emitter is between the limits of -0.4 and $+0.2$ volt. Q2, then, performs simply as a noninverting switch, providing a low-impedance, high-valued current source for Q3. This helps Q3 turn on and off in a minimum of time. As long as Q3 is on, the output pulse exists across its load resistor.

The length of time that Q2 remains off is determined by the time constant of C4 and the associated resistance. As long as C4 can keep the base of Q2 more positive than the emitter, it remains off. The resistor R7 allows this period to be adjusted. The time constant of C4, and R7 plus R8, is about 1.8 μsec, with R7 set at the midpoint. C4 will keep Q2 off for approximately 980 nsec (0.98 μsec), and Q2 will return to the on condition in about 70% of one time constant.

As $Q2$ comes back on, its emitter returns to -0.4 volt and $Q3$ turns off, terminating the output pulse even though the input may still be at -8 volts. As the input returns to ground, $Q1$ turns off and $C4$ recharges to its original condition rapidly. The primary purpose of $D3$ is to allow very rapid recharge of $C4$ through a low-impedance path. If the input pulses, for example, were arriving every 3 μsec, the capacitor would never fully recharge, for without $D3$ the normal time for recharge is about 20 μsec. With the diode, however, $C4$ can recharge in 1.1 μsec, which allows ample time before the next input pulse.

To complete the circuit discussion, a few obscure but important points will be brought out. Note the two diodes $D1$ and $D2$ in the collector circuit of $Q2$. These play an important part in the operation of $Q2$. They are placed in the circuit as they are to provide a local constant-voltage source that differs in value from any available power-supply value. The drop across a forward-biased diode is essentially constant for reasonable changes in current through it. In this instance, the need is for a power supply of about 1 volt, which is of course not available from the regular supply.

The reason for this requirement is easily explained. The collector power dissipation for $Q2$ is

$$P_C = I_C \times V_{CE} = 0.047 \times 12 = 0.564 \text{ watt}$$

A transistor of this type seldom is rated for a power dissipation of greater than 150 mw. With the two diodes in the circuit, the dissipation is reduced to 47 mw, which is well within the capability of the transistor.

15-6 SAWTOOTH GENERATORS

Waveforms such as those that appear in Fig. 15-16 are called *sawtooth* waveforms for obvious reasons. A sawtooth generator is a circuit that produces just such a repetitive waveform. There are many ways to do this, the simplest being to repetitively charge and discharge a capacitor. In

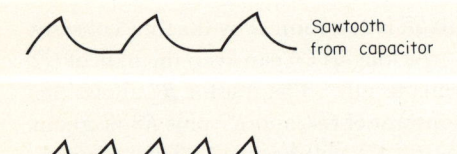

Fig. 15-16. Sawtooth waveforms.

many cases, the curvature caused by the capacitor charging produces nonlinearities that are objectionable. Still, this is without doubt the easiest way to obtain a sawtooth waveform.

A very simple circuit that produces a sawtooth waveform is shown in Fig. 15-17. This is a relaxation oscillator, which uses a tiny neon lamp

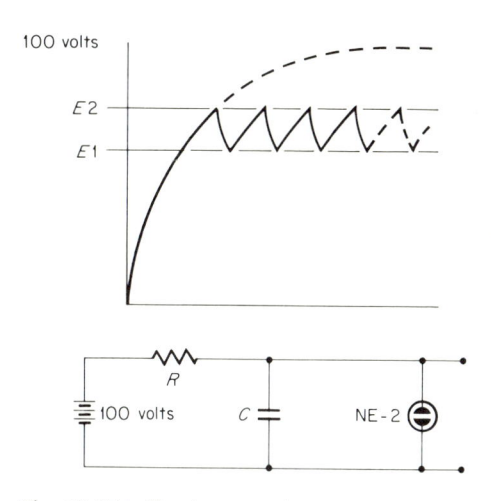

Fig. 15-17. Simple sawtooth generator.

(NE-2) and a resistor-capacitor network to produce the waveform shown. If C is a 1-μf capacitor and if R is between 1 and 5 megohms, the light will flash repetitively several times per second. After the initial charging of the capacitor, the value $E2$ is reached, which is the ignition potential of the lamp. As the lamp ignites, its internal resistance becomes very low and it conducts quite heavily, discharging the capacitor. $E1$ is often on the order of 70 volts. As the capacitor discharges, the voltage from plate to plate decreases and finally reaches the value of $E1$, the extinguishing potential of the lamp. Because this is measurably lower than $E1$, the waveshape shown is produced. When the lamp goes out, its resistance is virtually infinite and the capacitor again begins to charge. This action continues until power is removed from the circuit, and the result is the sawtooth waveform shown in the figure.

A close inspection of one cycle of this waveform will reveal its shortcomings. Figure 15-18 illustrates an amplified portion. Note the departure from a straight line that exists. Ideally, the sawtooth edges would be perfectly straight, but with such a simple circuit this is not possible.

374 PULSE AND SWITCHING CIRCUIT ACTION

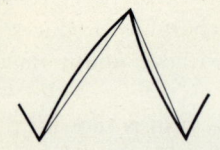

Fig. 15-18. Deviation from a perfect sawtooth.

There is, however, a way of making a sawtooth voltage more linear that is relatively simple and requires very little in the way of more circuitry. To understand how this is possible, consider Fig. 15-19. The first waveform (Fig. 15-19a) shows nearly a complete charge curve for a capacitor. By using only a small part of this curve, as shown in Fig. 15-19b, the curvature is less severe and the line appears to be somewhat straighter. By the same reasoning, Fig. 15-19c exhibits even less curvature and Fig. 15-19d is virtually straight.

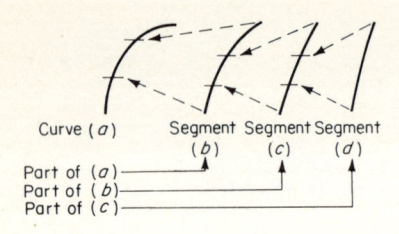

Fig. 15-19. Using smaller and smaller segments to approach linearity.

The segment of the curve in Fig. 15-19d occupies but a very small part of the overall curve, as shown in Fig. 15-20. If the total amplitude of the

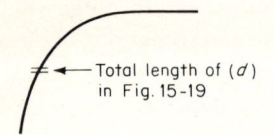

Fig. 15-20. Final segment used to achieve linearity.

curve in Fig. 15-19a were 1 volt, the straight-line segment might have an amplitude of only 0.01 volt. It is conceivable that this would be too small to be of use.

Another way of approaching the problem of obtaining a straight-line segment is to charge a capacitor to a much larger voltage, for example, 100 volts. Then, the part representing the same relative part of the curve would yield a 1-volt segment. This would, in all probability, be sufficient for most purposes.

If a very straight line segment is necessary, an alternate system must be used. The major reason for the curve of the capacitor charge is that the capacitor voltage increases as it accepts and stores the charges. The effective circuit voltage, therefore, continually decreases until, when the capacitor becomes fully charged, no further current flows. To compensate for this, a scheme such as the one illustrated in Fig. 15-21 is often used.

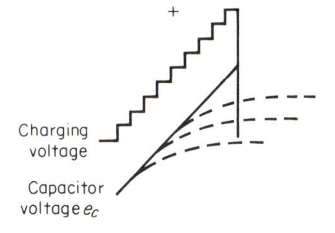

Fig. 15-21. Increasing the charging voltage as the capacitor charges yields good linearity.

If the charging voltage is increased at the same rate that the capacitor charges, the rate of charge of the capacitor must be linear. The very beginning part of the charging curve is the most linear. By constantly changing the applied voltage, the capacitor is constantly recharging from the beginning of its charge curve in its effort to catch up to the applied voltage. A nearly perfect linear rise in voltage can be obtained by this method. An example of such a charging circuit for the capacitor is a constant-current generator that yields a constant current into the capacitor by, perhaps, the use of a constant-current diode (see Chap. 15, Sec. 15-9).

Sawtooth-generator Example

A simple sawtooth generator that illustrates many of the foregoing principles is shown in Fig. 15-22. $Q1$ is an inverter amplifier that is used as a high-speed switch. When the gating signal goes positive, it turns off $Q1$ and lifts ground from the capacitor. This allows $C1$ to begin to charge toward 100 volts. Twenty μsec later $Q1$ turns back on, and now $C1$ loses its

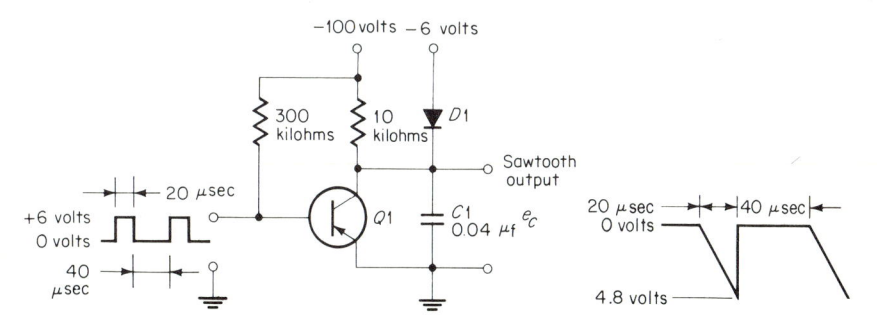

Fig. 15-22. Basic sawtooth generator.

376 PULSE AND SWITCHING CIRCUIT ACTION

charge through the nearly zero resistance of $Q1$. Each time the input signal turns $Q1$ off, $C1$ begins to charge and the voltage across it becomes the sawtooth, as shown.

A closer look into the circuit reveals some interesting items. Note the diode connected from the -6-volt supply to the collector of $Q1$. This is a protection diode, used to protect the transistor in the event the input voltage remains positive for a longer-than-normal time. Should this occur, $Q1$ would be damaged since the average transistor used in such a circuit is seldom rated for more than 30 volts. Without the diode, the collector of $Q1$ could attain a voltage of 100 volts if the input to $Q1$ would allow it. With the diode, the collector can never go more negative than about -6.5 volts since the collector is clamped to the -6-volt supply by the diode. Because the maximum voltage at the collector is -4.8 volts, this diode normally never comes into conduction.

The linearity of this circuit is reasonably good, as can be verified by determining the time constant of $C1$ and the load resistor in relation to the ramp time. As shown, the ramp (sawtooth segment) exists for 20 μsec. This is determined by the external circuitry that produces the input signal. The time constant is simply the product of $C1$ and the 10,000-ohm load resistor, or 400 μsec. Five times this value is the total charge time to bring $C1$ up to the supply voltage. Thus, 5×400 μsec = 2 msec. Note that the design of the circuit allows exactly one-hundredth of the full charge time for the ramp time. The part of the curve being used, then, is one-hundredth of the total. This gives the circuit a reasonable linearity, with the curvature held to a minimum for the purposes for which the circuit is used.

A slight variation of the preceding circuit is given in Fig. 15-23. Certain

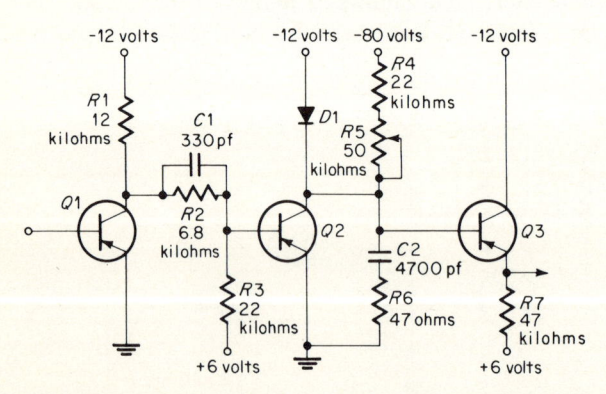

Fig. 15-23. Sawtooth generator with variable charge time.

SPECIAL CIRCUITS AND DEVICES   377

refinements are added to make the circuit more usable from a practical standpoint. The inverter $Q1$ is the means used to turn $Q2$ on or off. If $Q1$ is on, $Q2$ is off; if $Q1$ is off, $Q2$ is on. $Q2$ is the shunt switch that determines whether or not $C2$, the charging capacitor, will be able to charge.

The series-charging resistance for $C2$ is variable to allow for setting the time constant to suit necessary external circuit needs. The output transistor $Q3$ is an emitter follower, used to isolate the charging capacitor from the external load. Note the rather large emitter resistor used for $Q3$. This imposes a very small load on the charging capacitor, which is necessary so that the voltage across $C2$ will not be interfered with and thus upset the charging curve.

The 47-ohm resistor in series with $C2$ limits the current through $Q2$ at the instant $Q2$ comes on. Without this resistor, the transistor might be damaged by excessive current. If $C2$ should be charged to its maximum voltage of 12 volts, $R6$ will limit the current through the transistor to 0.255 amp. Since many switching transistors are rated for 300 ma, this is well within a safe value. Except for these small differences, the circuit works much like the previous one.

15-7 STAIRCASE GENERATORS

A staircase (or stairstep) waveform can be produced by using one of several techniques. Such a waveform is shown in Fig. 15-24, and the reason for its name is evident.

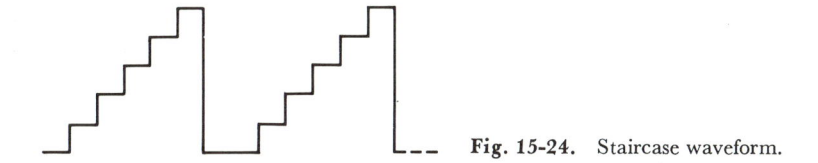

Fig. 15-24. Staircase waveform.

One of the simplest ways of generating this waveform is to utilize the charging characteristics of a capacitor in a suitable circuit. The charging curve of a capacitor is shown in Fig. 15-25a, and if the capacitor is allowed to charge uninterruptedly, it will follow the curve exactly. But suppose the charging source is intermittently opened while at the same time no discharge path is provided. Then the waveform in Fig. 15-25b must be evident across the capacitor. During the short time the capacitor is allowed to charge, it builds up a small voltage. Then, when the switch

378   PULSE AND SWITCHING CIRCUIT ACTION

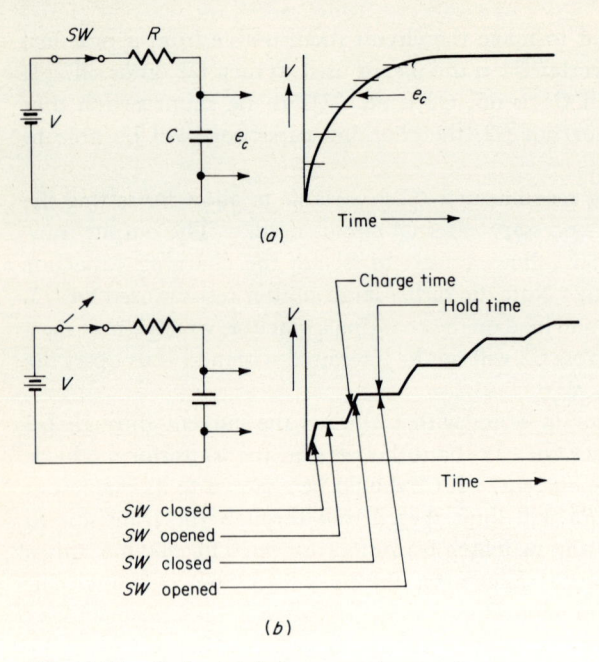

Fig. 15-25. Basic circuit for generating a staircase waveform.

is opened, the voltage across the capacitor must remain at the last value since no discharge path is provided. When the switch is closed again, the capacitor begins to charge and accumulates a greater voltage, the amount depending upon the time allowed for charge and the charging time constant. This action will continue until the voltage on the capacitor equals the supply voltage. As the capacitor begins to reach full charge, the step becomes less steep because the capacitor is still charging according to its charge curve. This is illustrated in Fig. 15-26.

Because of the changing slope as the capacitor reaches its full charge, it is desirable to operate on the lower part of the curve only. This will insure that each change from one level to another will occur in a reasonably short time. In other words, if it is necessary to produce several steps, each should be as small a part of the overall charging curve as possible. Thus, each step will be very nearly like any other, assuming equal charging times. Note that in Fig. 15-26 the total step period is longer for step e than it is for step a. The flat portion is the same in each case, and since each step requires a longer time to rise to the same incremental voltage, the total step time increases for each successive step.

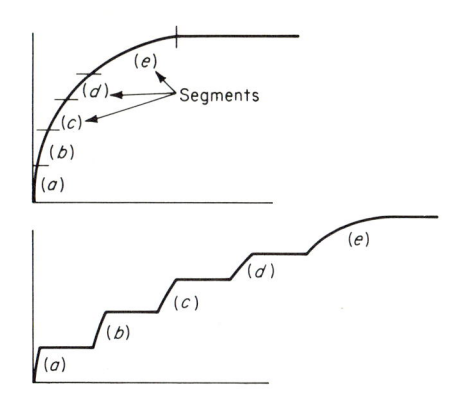

Fig. 15-26. Varying-sized steps produced as the capacitor charges.

By using just the segment a of the curve of Fig. 15-26 as the entire allowed charging curve, the incremental steps can be made quite similar, as illustrated in Fig. 15-27. If the time between the charge periods is very long relative to the time required to charge the capacitor to the new level, the appearance of the waveform will be nearly like the ideal staircase illustrated in Fig. 15-24. In this case, the time allowed to charge is perhaps on the order of microseconds. Viewed on an appropriate time base, the rise time, or charge time, appears to occur almost instantly.

Any loading upon the capacitor will necessarily create a discharge path. If the capacitor discharges appreciably during the hold time, the voltage will sag and the waveform will not consist of flat-topped steps.

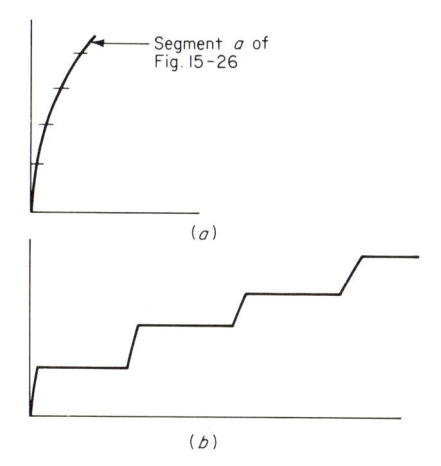

Fig. 15-27. Nearly equal steps produced by using small segment of full charge curve.

380   PULSE AND SWITCHING CIRCUIT ACTION

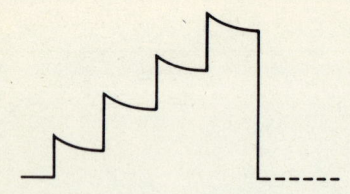

Fig. 15-28. Effect of capacitor discharge during hold time.

The waveform of Fig. 15-28 will then result. That this might occur must be considered when the equipment is first designed.

A practical circuit for producing a staircase waveform is shown in Fig. 15-29. In this circuit, the staircase voltage is produced across $C2$, the staircase-producing capacitor. Since $C2$ charges toward -80 volts through $R3$, the waveform starts at ground and proceeds in a negative-going direction. $Q1$ is the switch that allows or disallows the charging of $C2$.

Note that $Q1$ is shunted around $C2$ rather than being in series with it. Except for this minor change and the addition of other components that help $Q1$ in doing its job, the circuit works very much like the elementary circuit of Fig. 15-25. To describe the circuit action, first assume that the reset switch has momentarily connected the upper plate of $C2$ to ground. This simply removes any residual charge and shorts it to ground. After this, the wire marked "reset switch" is open-circuited and so cannot further affect the circuit. $Q1$ is normally on, and so its collector is at ground. Diode $D1$ is reverse-biased, while $D2$ has simply 0 volts across it. With 0 volts at the base of the transistor $Q2$, 0 volts exists at the output. A series of input pulses will alternately turn $Q1$ off and on.

As $Q1$ turns off, $C2$ is free to accept some amount of charge and its

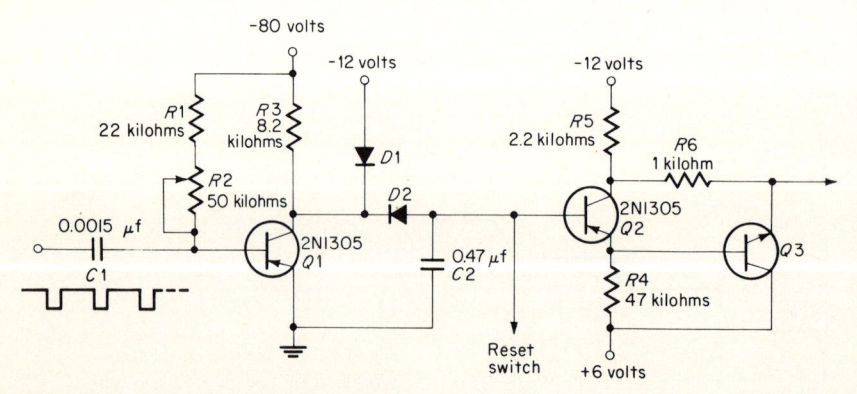

Fig. 15-29. Practical staircase generator.

upper plate swings in the negative direction to produce the first step. After a certain time has elasped, $Q1$ comes back on and at first glance it would seem that it would short out the capacitor. But $D2$ is now backbiased and thus prevents the discharge of $C2$. The capacitor then holds its charge for a considerable time, which is limited only by the leakage currents of $D2$ and $Q2$.

When the next input arrives, $Q1$ again turns off and again $C2$ begins to charge to a greater voltage than before. The action continues until the proper number of steps have been produced, which is controlled by the number of pulses available at the input. This, of course, is a function of circuitry external to that shown. When the required number of steps have been produced, the reset line is suddenly connected to ground, relieving the capacitor of its charge and terminating the staircase. This is illustrated in Fig. 15-30, which shows typical output waveforms for this circuit.

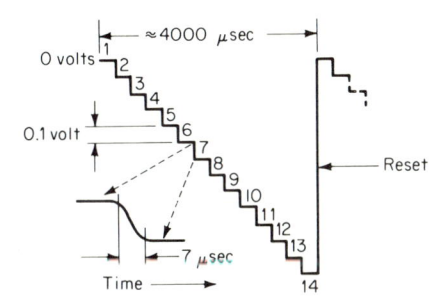

Fig. 15-30. Staircase waveform.

The waveform shows 14 steps in a negative-going direction. In the particular application for which this circuit was designed, each step duration is of the order of 288 μsec, and it requires about 7 μsec for the waveform to change from one level to another.

A detailed examination of the circuit shows that the charge time is determined by the length of time that $Q1$ is held off by the coupling capacitor $C1$. When the input swings in a positive-going direction, $C1$ drives the base of $Q1$ more positive than ground, and the length of time that $Q1$ is held off will depend upon the time constant of $C1$, $R1$, and $R2$. As long as the right-hand plate of $C1$ is more positive than ground, $Q1$ is held off. When $C1$ begins to recharge toward the -80-volt supply, $Q1$ is turned on again and that particular charge period is terminated. The purpose of $R2$ is to provide some adjustment for the discharge time of $C2$ and thus also the time that $Q1$ is held off. As before, $D1$ clamps the collector of $Q1$ to no more than about -12 volts should the transistor be held off for a longer-than-normal time.

The circuits of $Q2$ and $Q3$ are interesting because of the action upon the capacitor $C2$. During the hold time for this capacitor, any load will drain energy from it and the plate-to-plate voltage will sag. Alternatively, if there is a source that has a path to $C2$, the voltage will tend to rise, again altering the desirable waveform. The rather unconventional circuitry of these two transistors is designed to minimize these effects as far as possible.

The base of $Q2$ presents a possible path of discharge for $C2$, and hence the emitter resistor is a large value. This reduces the base-current drive and so reduces the loading on the capacitor. $R5$ acts to reduce the leakage current of $Q2$ (which would tend to charge $C2$) by reducing the collector-emitter voltage. Additionally, the leakage current from $Q3$ flows in the $Q2$ circuit in such a way as to tend to reduce $Q2$'s leakage. This further reduces the tendency to recharge $C2$.

The voltage-divider method of producing a staircase voltage is useful for generating a small number of steps. Such a circuit appears in Fig. 15-31. The output will appear as a series of steps, providing the inputs to A and B are correct. At the start, A and B are clamped to ground, and hence the output is also at ground since there is no current through the 1-kilohm resistor. As illustrated in Fig. 15-32, the inputs are one factor in determining the output. The other factor is the circuit values, where one resistor string has a value that is twice the value of the other.

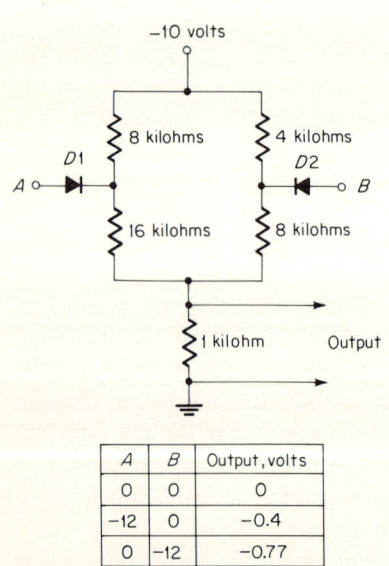

A	B	Output, volts
0	0	0
−12	0	−0.4
0	−12	−0.77
−12	−12	−1.11

Fig. 15-31. Simple staircase generator.

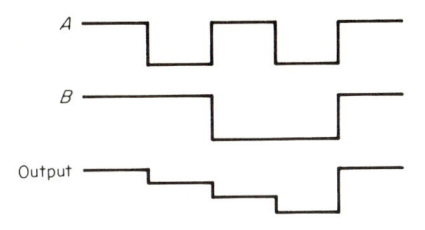

Fig. 15-32. Waveforms of Fig. 15-31.

To make the first step, input A is caused to go to -12 volts. This back-biases $D1$ and removes the ground clamp from that side of the circuit. Current now flows in the three resistors in the amount of 0.4 ma. The drop across the 1-kilohm resistor is therefore 0.4 volt, and this is the first step. After a period of time, A returns to 0 volts, but at this same time point B goes to -12 volts. Current now flows in the amount of 0.77 ma, and the output is now -0.77 volt. Finally, with both inputs at -12 volts, the output rises further to 1.11 volts. Note that the steps are not exactly twice the value of the preceding step. This occurs because of the constant value of the 1-kilohm resistor. Nevertheless, for many applications this is a perfectly usable staircase waveform.

This circuit configuration is particularly suitable for use with a pair of flip-flops, each driving one of the inputs and connected as a two-stage counter. The truth table in Fig. 15-31 will verify this.

15-8 INDUCTIVELY COUPLED CIRCUITS

In the study of digital circuitry, one encounters circuits coupled by inductive means more or less infrequently. Enough of them exist, however, to make the study of them well worth the time expended. The methods used are somewhat different from those used in linear circuits, as will be seen in the following material.

A typical transformer used in such a circuit is illustrated in Fig. 15-33. This is a so-called "pot-core" transformer, constructed of a ferrite material, which allows use at rather high frequencies without undue core loss. The primary and secondary coils are simply laid in the recess and the unit assembled, with the leads extending out the wire exit.

Either step-up or step-down of voltage is possible by setting the turns ratio to a suitable figure. In one instance, the primary has 8 turns and the secondary has 200 turns, center-tapped. A circuit to use such a transformer is shown in Fig. 15-34; this is a *read* amplifier, operating from a core memory. The action of the core itself will not be discussed in this book. Suffice it to say that under certain conditions the core is capable of generat-

384 PULSE AND SWITCHING CIRCUIT ACTION

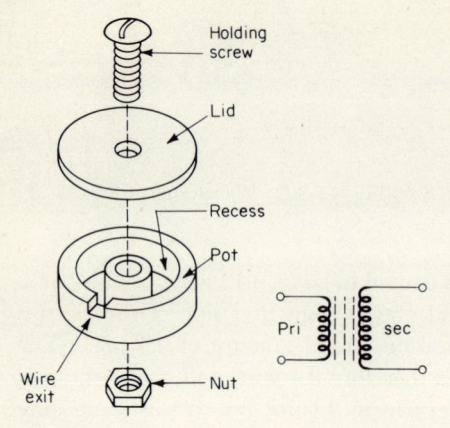

Fig. 15-33. Ferrite pot-core transformer.

ing a small voltage in the wire that threads through it. This voltage is on the order of 25 mv. The pulse transformer, then, serves to couple this small voltage to the base of $Q1$ and, at the same time, to step up the small voltage to a value suitable for the transistor.

Assuming no voltage drop across the diodes, about 0.3 volt is presented to the base when the core generates a signal. The voltage across the secondary appears as a damped oscillation, as shown, and, after rectification by the diodes, becomes a series of negative swings below ground. In the application illustrated, the input pulse (rectified damped oscillation) at the base has a duration of nearly 5 μsec.

Note that the polarity of the signal at the transformer is of no consequence since the diodes dictate the polarity at the base. $Q1$ provides an output waveform that is essentially a square wave and that has an amplitude of

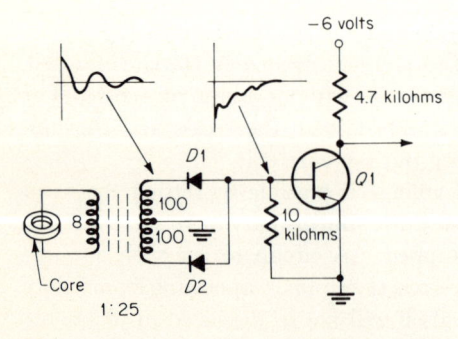

Fig. 15-34. Simple read-amplifier circuit.

nearly 6 volts. This is further amplified and shaped in circuitry that is not shown.

In the preceding circuit, the polarity of the input signal is of no importance. In Fig. 15-35, a circuit is shown that requires an input voltage of a

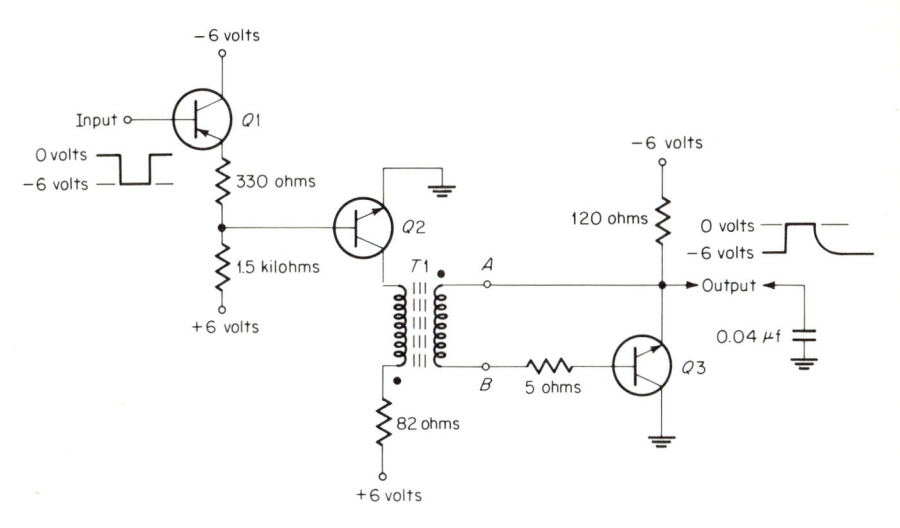

Fig. 15-35. Inductively coupled pulse circuit driving a capacitive load.

certain polarity to function properly. This circuit has the function of acting as a high-current switch since the load is highly capacitive. Q3 acts as a shunt switch for the 0.04-μf capacitor, allowing a very fast discharge with attendant short time constant. Of course, Q3 must be capable of handling the maximum discharge current and, as such, is a power-type transistor.

The circuit operation is relatively simple. The input waveform applied to the base of Q1 is as shown and will cause the emitter follower to duplicate this waveform at its emitter. The voltage at the base of Q2, however, is quiescently at a slightly positive value owing to the voltage divider in the emitter of Q1. Hence, Q2 is quiescently on, whereas Q3 is off since its base and emitter are connected together through the secondary of the transformer. The output wire is therefore at −6 volts, and the capacitor is charged to 6 volts.

As the input pulse goes to −6 volts, the emitter of Q1 goes quite negative and Q2 is turned off. The quiescent current of Q2 is on the order of 75 ma, and this now decreases toward zero. Figure 15-36 depicts the circuit

386 PULSE AND SWITCHING CIRCUIT ACTION

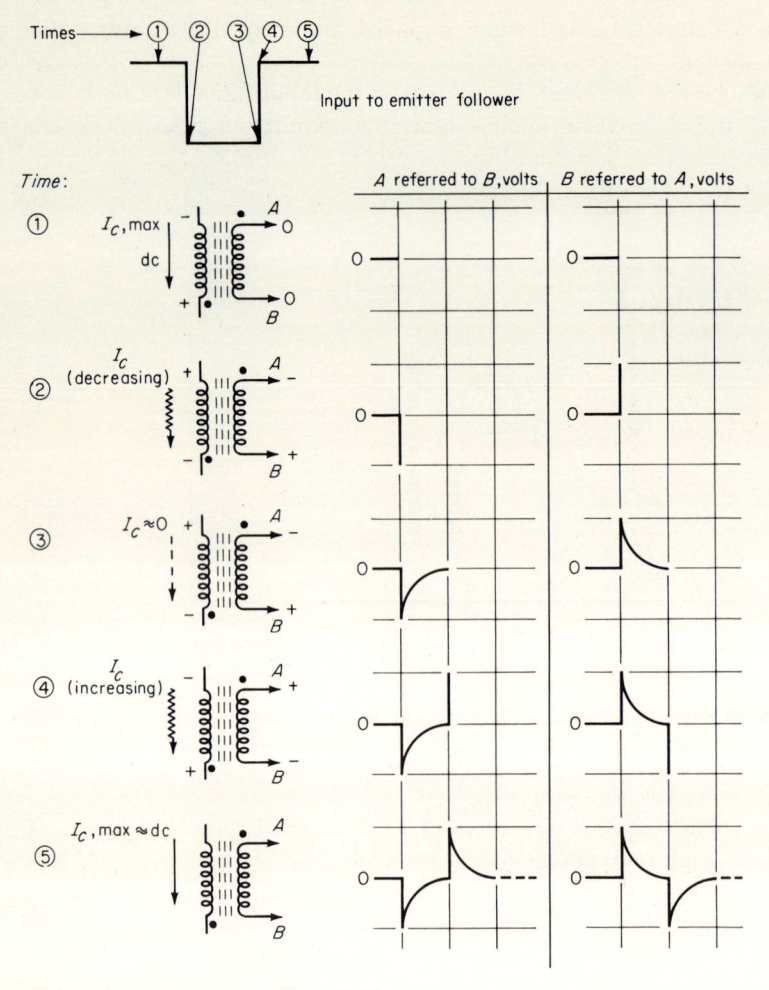

Fig. 15-36. Waveforms of Fig. 15-35.

action during this period of time. The induced voltage at the base of $Q3$ is positive, as shown, and this turns on the transistor. Because base current is hardly limited to any extent, base current is very large and $Q3$ is turned on very hard. The capacitor may now discharge rapidly, and the output wire falls to ground very fast.

The length of the output pulse is dependent upon which of two things happens first. If the input pulse terminates quickly, the base of $Q3$ sees a negative spike, as shown, and it turns off. The output wire, and so also the capacitor, rise toward the supply with a time constant RC, R being the

120-ohm resistor and C the 0.04-μf capacitor. If the input pulse remains at -6 volts for any great length of time, the output terminates because the transformer current has dropped finally to zero and hence cannot generate any secondary voltage. In this instance, the pulse length (output pulse) is a function of the total energy-storage capability of the transformer.

A final example of an inductively coupled circuit is shown in Fig. 15-37.

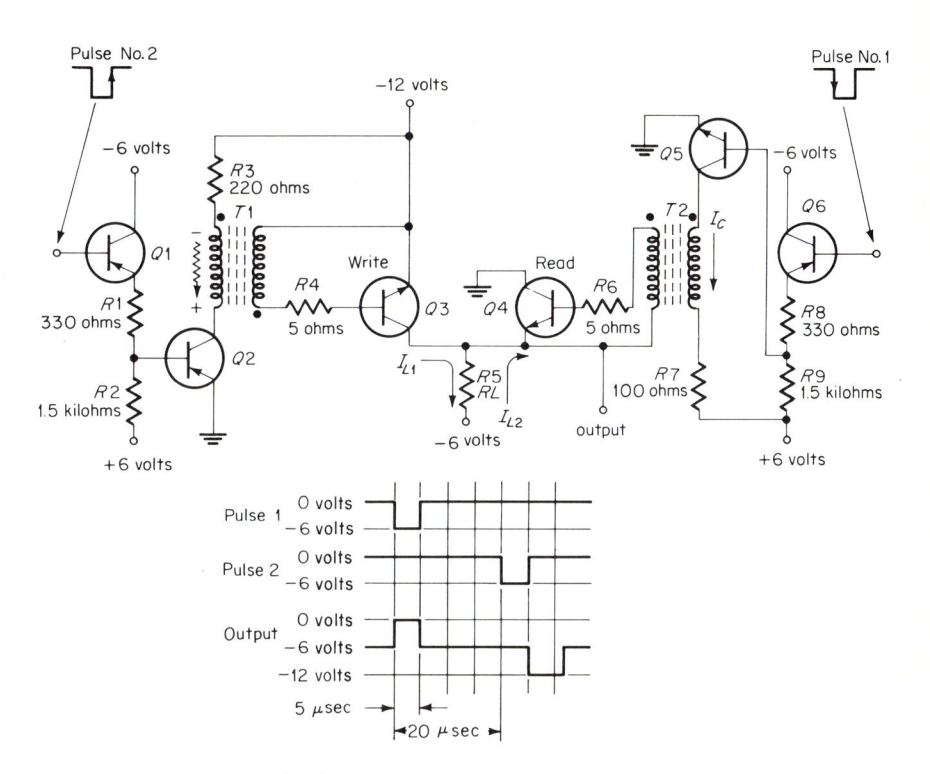

Fig. 15-37. Read-write circuitry.

The circuit performs not unlike those described earlier, and therefore the following description is brief. The primary purpose of the circuit is to afford the possibility of current through the load $R5$ in either of two directions. If $Q3$ is on, current flows from the -12-volt supply out to the -6-volt supply. If $Q4$ is on, current flows from the -6-volt supply out to ground. The two input pulses are spaced in time so that they can never occur together. This circuit is a portion of the *read-write* circuitry used in conjunction with a magnetic-core memory. With $Q4$ on, information is read out of memory, while with $Q3$ on, information is written into memory. The resistor labeled $R5$ is actually the resistance represented by the wire that threads through the many cores in a typical memory system. The waveforms depict how the circuit functions.

15-9 SPECIAL DEVICES

The varied kinds of devices to be described in this section range from special-purpose diodes, to complex control units, to special types of active units. Each is basically a semiconductor of one sort or another and is constructed so as to take advantage of one particular characteristic. This characteristic is therefore emphasized and gives the device its own peculiar identity and special application to digital circuitry.

Because these devices are many in number, each having a complex function, the following descriptions are necessarily brief. The basic characteristics are given, and, where possible, a brief and simple application is described to illustrate typical usage. Further information can be obtained by referring to the bibliography at the end of this book or by consulting the various manufacturers of these devices.

The Unijunction Transistor

The unijunction transistor is a three-terminal device that behaves very differently from a regular junction transistor. As the name implies, it has but one junction, with two base leads and one emitter lead. This is shown in Fig. 15-38 both schematically and diagrammatically.

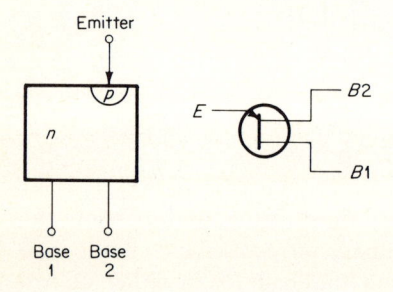

Fig. 15-38. The unijunction transistor.

A chip of n-type silicon is used as the base material, and the p-type emitter is formed on it, creating a pn junction. One of the base leads ($B2$) is closer to the emitter than the other. The resistance between the two base leads is typically 6 to 8 kilohms. In a normal circuit, the resistance from the emitter to either base is variable, depending upon the circuit and the applied voltages.

Base 1 is usually placed at circuit ground, and the circuit power supply V_{BB} is applied to base 2. The emitter, then, becomes the input connection, while base 2 provides the output. Other connections are also possible. However, the unijunction transistor is not capable of linearly amplifying a signal. Its main uses depend upon its *negative-resistance* characteristic. Some of these uses are

1. SCR triggering
2. Oscillator
3. Timing
4. Pulse generator
5. Bistable circuits
6. Sensing circuits

In all these instances, use is made of the fact that as a signal is injected, the transistor goes from the off to the on condition or vice versa. There is no useful in-between area of stable operation.

A characteristic curve for a typical unijunction transistor is shown in Fig. 15-39. The ordinate of the graph is labeled V_E (emitter voltage) and

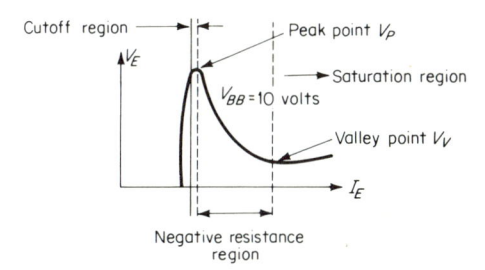

Fig. 15-39. Typical unijunction-transistor characteristics.

increases in the upward direction. The abscissa is emitter current and increases to the right. Note that in the negative-resistance region, as emitter current increases, emitter voltage *decreases*. This is exactly the opposite of what we might expect. Any device that has smaller voltage drop as the current is increased is said to possess negative resistance.

Any device exhibiting negative resistance is capable of regeneration and thus capable of oscillating.

The region to the left of the peak point is called *cutoff*, while the region to the right is called *saturation*. Some of the more important terms used in conjunction with unijunction transistors are listed and explained below.

1. *Interbase resistance R_{BB}:* The interbase resistance is the ohmic resistance measured between base 1 and base 2 with the emitter open.

2. *Intrinsic stand-off ratio η:* The intrinsic standoff is a number less than 1 that represents the amount of applied voltage necessary to fire (turn on) the transistor. Mathematically,

$$\eta = \frac{V_P - V_D}{V_{BB}}$$

Typical values of η are from 0.4 to 0.8.

3. *Peak-point current I_P:* The peak-point current is the emitter current at the peak point. This is the minimum current needed to turn on the transistor.

4. *Peak-point emitter voltage V_P:* This is simply the emitter voltage at the peak point.

5. *Emitter reverse current I_{eo}:* This is equivalent to I_{cbo} in a conventional transistor. It is measured between base 2 and the emitter, with base 1 open.

6. *Valley voltage V_V:* This is the emitter voltage at the valley point.

7. *Valley current I_V:* This is the emitter current at the valley point.

8. *Diode voltage V_D:* This is the drop across the *pn* junction, equal to

$$V_D = V_P - \eta V_{BB}$$

Unijunction-transistor-circuit Example

A typical oscillator circuit will serve to show how the device can be effectively used. Such a circuit is shown in Fig. 15-40. This is a basic relaxation oscillator.

At the beginning of a cycle of operation, say, point A on the waveforms, the capacitor begins to charge toward $+V_{BB}$ through $R3$. At this time the transistor is off and does not influence circuit action. This is shown at B. When the emitter voltage reaches the peak-point value V_P, the unijunction transistor turns on and the capacitor discharges into $R1$

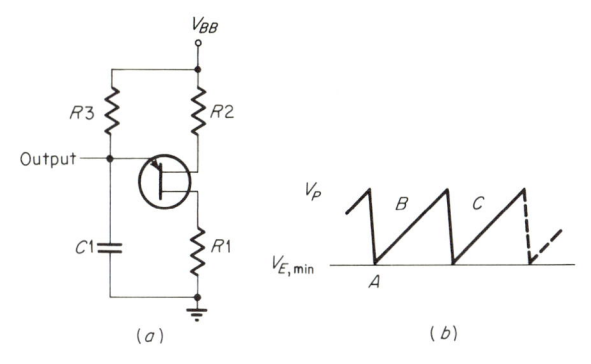

Fig. 15-40. (a) Unijunction-transistor circuit; (b) output waveform.

and the emitter, shown at point C. When the capacitor voltage drops very low, the transistor turns off, and the cycle begins over again.

This circuit could be used for timing purposes, as a pulse generator, as a trigger circuit, or as a sawtooth-wave generator.

The Field-effect Transistor (FET)

The field-effect transistor (FET) is a relatively recent development as used today. Only since about 1960 has it occupied a place of prominence in the industry. Its usage is increasing of late, and it is reasonable to assume that one might encounter such a device in various circuits. Therefore the well-informed technician should be familiar with it. The following material relates to the FET as used in a general application and is more concerned with basic operation than with application. The digital application of FET's is discussed in Chap. 16.

There are two major types of field-effect transistors, the *junction field-effect transistor* and the *insulated-*, or *isolated-*, *gate field-effect transistor* (abbreviated JFET and IGFET, respectively). To begin to understand how these devices work, consider Fig. 15-41. Figure 15-41a represents a wafer of silicon that acts much like an ordinary resistor. The two connections are labeled the "source" and "drain." The device as shown would conduct current if a voltage were impressed upon the terminals, the amount of which would be proportional to the value of voltage. In Fig. 15-41b two additional areas have been added, gates 1 and 2. These areas have been doped with p material, while the main bar of silicon is n type; thus each causes the formation of a junction. As with all pn junctions, a depletion region is formed at the junction, as in Fig. 15-41c. The region between

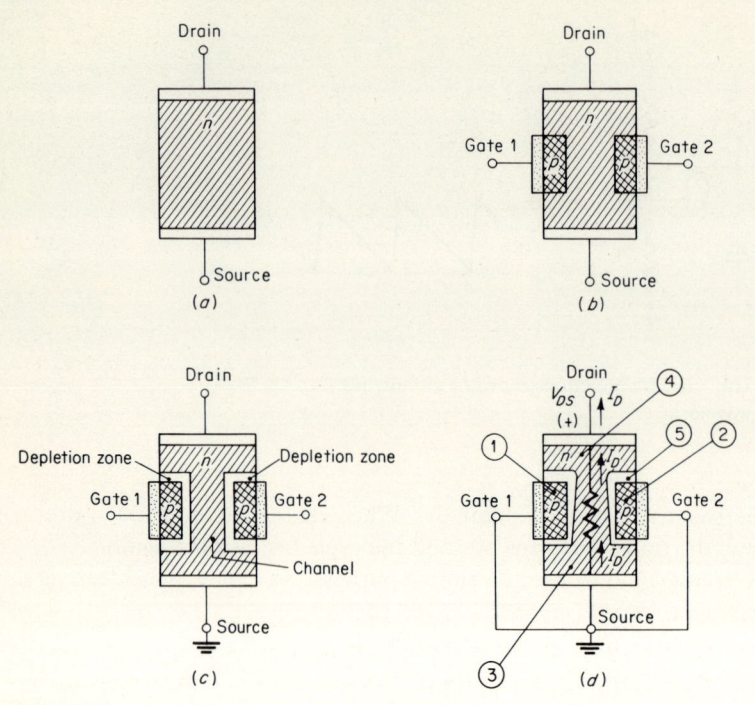

Fig. 15-41. Development of junction field-effect transistors. (*From Application Note AN 211A, Motorola Semiconductor Products, Inc.*)

the gates is called the *channel*, which is a part of the main bar, or wafer, of silicon, known as the *substrate*. Any current flowing through the device must pass through the channel.

If current is caused to flow, as shown in Fig. 15-41*d*, the shape of the depletion region is altered, and as current is increased to a large value, the depletion regions meet and restrict any further increase in current. The applied voltage that just causes this effect is called the *pinch-off voltage* V_P.

Since it is difficult to produce the gates on both sides of a bar as shown in Fig. 15-41, a single-ended method is used to produce the practical field-effect transistor. Figure 15-42 shows such a structure, and the various parts are numbered to correspond with the numbers in Fig. 15-41*d*. Area 2 is the substrate, which forms gate 2 and is the main part of the structure. Area 5 is the *n*-type channel that is diffused into the surface of the block of silicon. Finally, area 1 is formed by diffusing a *p*-type material to form gate 1, as shown. Electron flow, from source to

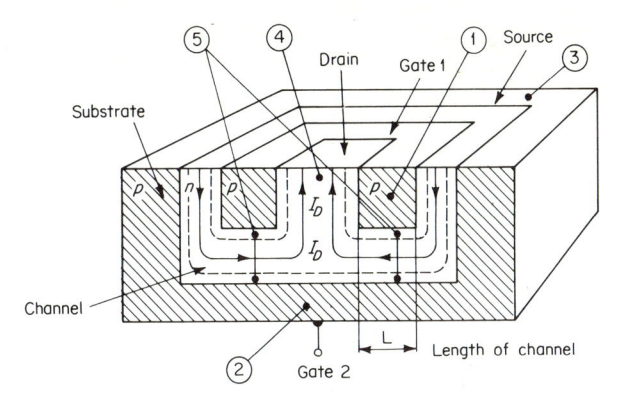

Fig. 15-42. Single-ended configuration. (*From Application Note AN 211A, Motorola Semiconductor Products, Inc.*)

drain, is also shown, and note that it must pass through the channel. A depletion region surrounds the channel at every *pn* junction, and this is the mechanism by which the drain current can be made to vary according to a signal. If gate 1 is caused to vary its voltage, the depletion regions will vary and the current flow will also vary in step with the input at gate 1.

The electrical symbol for the JFET is shown in Fig. 15-43. As can be

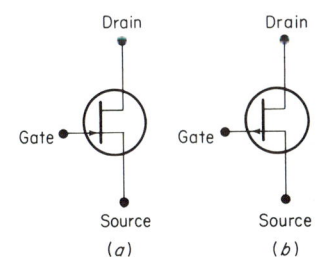

Fig. 15-43. Junction FET's: (*a*) *n*-channel device; (*b*) *p*-channel device.

seen the gate connection is used as an input, with the drain as the output. The source and the substrate connections are usually tied to signal ground.

The insulated-gate field-effect transistor is constructed somewhat differently from the junction counterpart. Figure 15-44 shows how the IGFET is developed. In this case, two separate *n*-type regions are diffused into the substrate to form the insulated source and drain. Then the entire surface is covered with silicon dioxide, which is an excellent insulator. Openings are made in the oxide layer to allow contact to be made between the source and drain connections and the *n* channels.

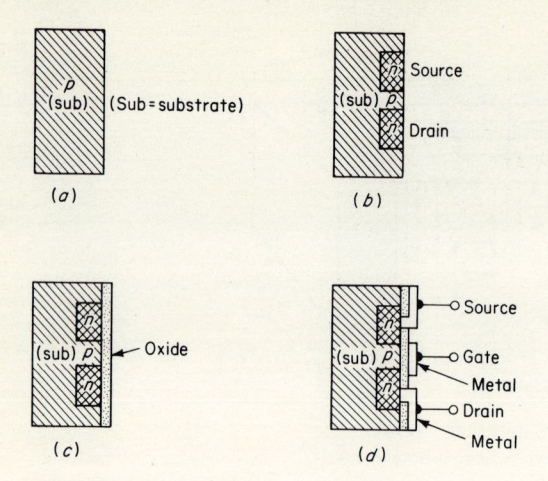

Fig. 15-44. Development of insulated-gate FET. (*From Application Note AN 211A, Motorola Semiconductor Products, Inc.*)

Next a metal covering is laid over the center section, which becomes gate 1. There is no physical connection from gate 1 to the semiconductor proper. Thus the metal connector, the oxide insulator, and the channel just beneath form a capacitor. This is a most important idea because any voltage impressed upon gate 1 can affect the transistor *only* through this capacitance.

Because the source and the drain are isolated by the substrate, the drain current I_D is essentially zero with zero gate voltage. This is true because the internal junctions between the source and the drain act like back-to-back diodes, and with any applied voltage, one of them will be back-biased. In order to cause current to flow through the device, it must be turned on. This is accomplished by applying a positive voltage to the gate with a normal positive voltage applied to the drain. As gate 1 is made positive, an *induced* channel is formed, as shown in Fig. 15-45. As the upper plate of the effective capacitor becomes more positive, the lower plate, just below the oxide layer, becomes more negative. The accumulated electrons become carriers and will now allow current to flow from the source to the drain. The number of electrons available to carry current is a function of the gate voltage, and so the drain current can be made to vary in proportion to the applied signal voltage at the gate. Increasing the gate voltage is said to *enhance* the drain current.

One advantage of this kind of transistor is that its input impedance is

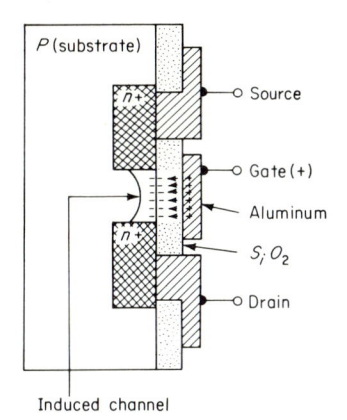

Fig. 15-45. Channel enhancement. (*From Application Note AN 211A, Motorola Semiconductor Products, Inc.*)

very high since the gate acts much like a capacitor. The electrical symbol for the IGFET is shown in Fig. 15-46.

Comparing the two types of FET's just described, we find that the first of these, the JFET, operates fully on with zero gate-1 voltage and can only be turned *more off*. This is called the *depletion* mode of operation. On the other hand, the IGFET is normally off and can only be turned *more on* by the application of gate voltage. This is called the *enhancement* mode of operation.

The depletion mode refers to the decrease of carriers in the channel due to an increase in positive gate voltage. Enhancement refers to the increase of carriers in the channel due to the increase of positive voltage on the gate.

A third type of field-effect transistor is also possible. This is known as the *depletion-enhancement* type. In this case, the zero-gate-voltage drain current is intermediate between full on and full off. Hence both depletion and enhancement of the carriers are normal for this device. This result is accomplished by diffusing a thin n channel between the source and the drain of a regular IGFET just below the oxide layer. This yields a layer that is conductive, and the carriers in the conductive layer can be augmented or decreased by the application of gate voltage.

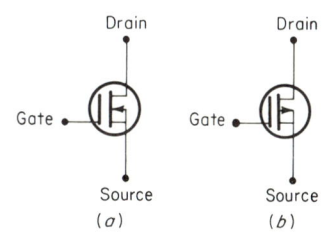

Fig. 15-46. Insulated-gate FET's: (*a*) n-channel IGFET; (*b*) p-channel IGFET.

The Silicon-controlled Rectifier (SCR), or Thyrite

The silicon-controlled rectifier is a special case where the properties of semiconductors are used to produce a special effect. The SCR is a solid-state device having certain properties quite similar to a thyratron gas tube. That is, the control element, called the *gate*, can maintain sufficient control to keep the device off (open-circuited) for any period of time provided the device is already off. By injecting a current into the gate, the device is turned on and will conduct current. Once the conducting state is reached, the gate loses its ability to control current flow and can neither turn the main device off nor further on. If now the current path is interrupted, the gate can once again regain its ability to keep the device off, even though an otherwise complete circuit is reestablished.

An SCR is made of four alternate layers of p- and n-type silicon. Its physical construction is symbolically illustrated in Fig. 15-47. The SCR

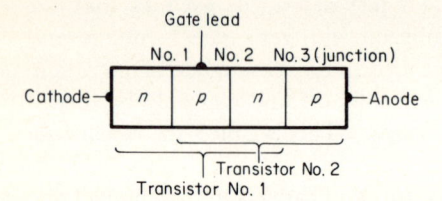

Fig. 15-47. SCR ($pnpn$ switch).

is seen to be the equivalent of two separate transistors, an npn unit (1) and a pnp (2). The way it operates can best be appreciated if we split the two center sections in two pieces, as shown in Fig. 15-48b and connect them together with ordinary wire.

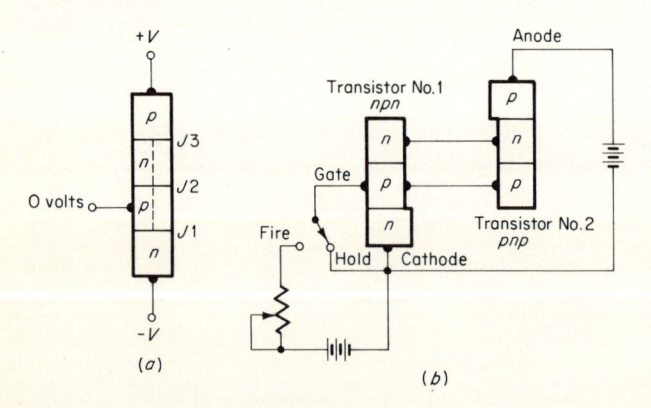

Fig. 15-48. Equivalent circuit for an SCR.

With the voltages applied as shown in Fig. 15-48b and with the switch in the gate lead as shown, little or no current flows in the anode circuit. Inspection of the three junctions reveals the fact that junction 2 ($J2$) is reverse-biased; thus no appreciable current can flow since junction 1 is not forward-biased.

The base of the *npn* transistor is not forward-biased because of the switch position; thus the *npn* unit is nonconducting. Any current flow in the anode circuit (*pnp* unit) must flow through the *npn* part, and so the anode current is essentially zero. (A very small temperature-dependent current similar to the I_{cbo} in a transistor will exist.)

If now the switch is transferred with a large resistance in series, a small base current will flow in the *npn* unit. (The circuit is redrawn in Fig. 15-49a.) The collector current will then be beta times the *npn* base

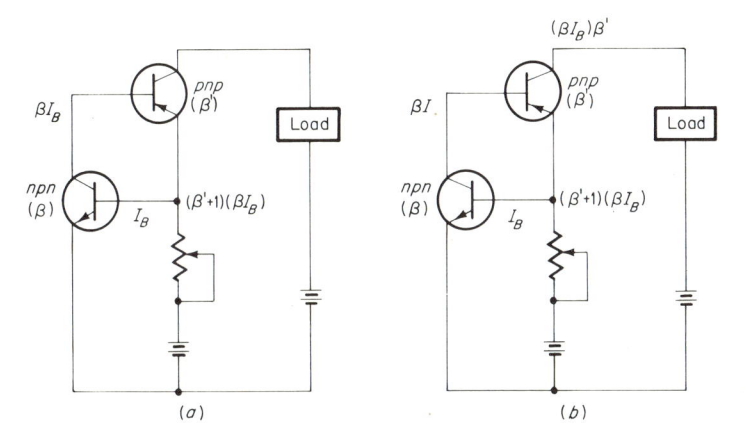

Fig. 15-49. SCR relationships.

current. This current is injected into the base of the *pnp* unit and will appear at the emitter of the *pnp* unit with an amplitude of $(\beta' + 1)(\beta)$ times the original base current (β = *npn*, β' = *pnp*). If we gradually reduce the resistance in the *npn* base, more and more current will flow in the base circuit. When the initial base current is made large enough so that the current in the emitter of the *pnp* part is exactly the same value as the original base current, the loop current gain is 1; and now if the base of the *npn* part is disconnected, the device will *continue to provide its own base current*. Both the *npn* and *pnp* parts will go into saturation, and a large current will flow in the anode circuit, the value of which is dependent upon the load resistance and the applied voltage.

Since the device is now supplying its own internal base currents, what is done to the external *npn* base circuit (gate) can in no way affect the conductivity, and the gate is said to have lost control. Anode current can be stopped only by either opening the anode circuit or by reducing the anode current below the point of regeneration. Once anode current stops, we can again regain control by stopping gate current, and if anode voltage is again applied, no anode current will flow until the gate current is again made large enough to produce regeneration.

Some of the terms used in describing SCR's are listed below for handy reference.

I_f Forward anode current. The value of anode current through the device when on.

$I_{f,off}$ Forward off current. The value of anode current through the device when in the off condition.

I_g Gate current.

I_h Holding current. The minimum anode current required to sustain the on condition.

I_r Reverse current. Any current through the device when negative voltage is applied to the anode.

V_{bf} Forward breakover voltage. Anode voltage that will cause the rectifier to switch to the on state, with no gate current.

V_f Forward voltage. The voltage drop between the anode and cathode at any specified forward current, when the device is on.

$V_{gr,rated}$ Gate voltage, reverse. Maximum allowable reverse voltage applied to the gate junction.

V_r Reverse anode voltage. Any negative value of voltage applied to the anode.

$V_{r,rated}$ Maximum inverse voltage allowed. To exceed $V_{r,rated}$ would cause entry into the avalanche region, and the device would go on, even though no I_g were present.

The schematic diagram of an SCR is shown in Fig. 15-50, with an

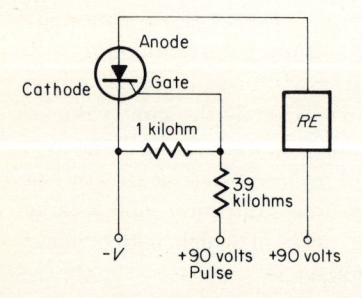

Fig. 15-50. SCR schematic circuit.

appropriate circuit that might be used to, say, energize the relay *RE*. The purpose of the voltage divider is to reduce the 90-volt pulse to a value suitable for application to the gate. With a 90-volt pulse at the input (1 msec duration or longer), slightly more than 2 volts will be applied to the gate, which will be sufficient to turn on the SCR.

The Silicon-controlled Switch (SCS)

The SCS is nearly identical to the SCR, with the primary difference being in the power-handling capability. The SCR can operate with currents in excess of 100 amp and operating voltages of 1000 or higher. The SCS, on the other hand, is intended for low-level logic and switching applications, where dissipation is typically a few watts or less. In other respects the SCS is identical in operation to the SCR.

Binistor, Trigistor, and Transwitch

These devices (the above are trade names) are four-layer silicon devices that are especially designed to be turned off, as well as on, by a gate signal (negative). Otherwise they are very similar to the SCS; they even share the same schematic symbol.

Thyristor

A thyristor is a three-leg, three-layer device having both the characteristics of the SCS *and* a junction transistor (*pnp*). At low values of current (about 10 ma or less), the unit performs as a conventional *pnp* germanium transistor. At higher currents, its characteristics are similar to a silicon-controlled switch, with the added advantage that a positive signal, applied to the base, will turn the device off.

The essential difference between a transistor and a thyristor lies in the collector. The collector contact consists of a tab of nickel, soldered to the germanium of the collector with an alloy of lead, tin, and indium. This results in the differing actions at low and high values of collector current.

At relatively low values of current, the nickel tab simply acts as a collector for the carriers (holes) in the collector region. But, as the current is increased, this contact region begins to inject electrons into the collector. Thus, in the vicinity of the collector, an additional *n*-type layer is produced and the device now behaves as a four-layer device. It therefore goes into avalanche conduction as the collector current rises to some critical point. The characteristic curves shown in Fig. 15-51 clearly indicate this action.

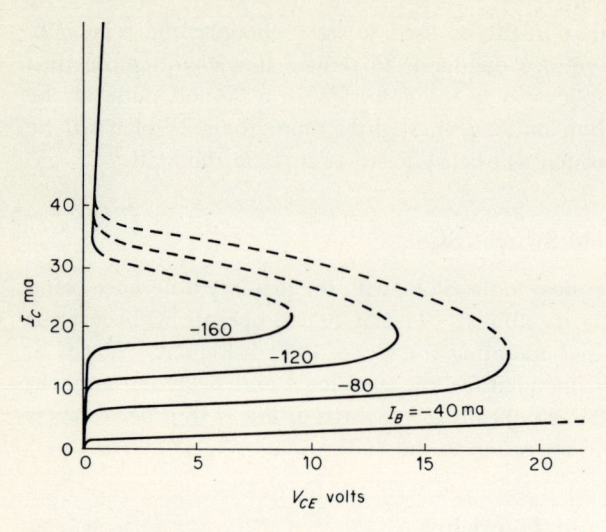

Fig. 15-51. A set of curves for a hypothetical thyristor.

Four-layer Diode

A four-layer diode is illustrated in Fig. 15-52, and its characteristics are illustrated in Fig. 15-53. Note that the device itself is identical to the SCR except that no connection is made to a gate. In use, the diode will not conduct if the cathode-to-anode voltage (forward voltage) is smaller than a critical value. If now the voltage is substantially increased, the diode is said to *break over*. The current is therefore limited by something outside the device, and it conducts heavily. The voltage drop from cathode to anode is typically 0.7 volt. The forward characteristics

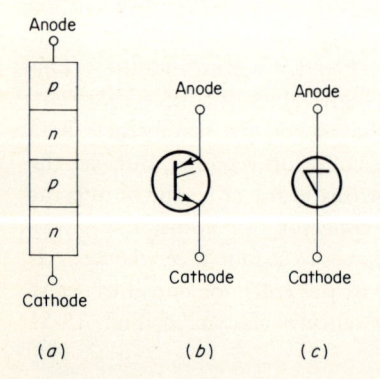

Fig. 15-52. (*a*) Four-layer diode; (*b*) and (*c*) circuit symbols.

SPECIAL CIRCUITS AND DEVICES   401

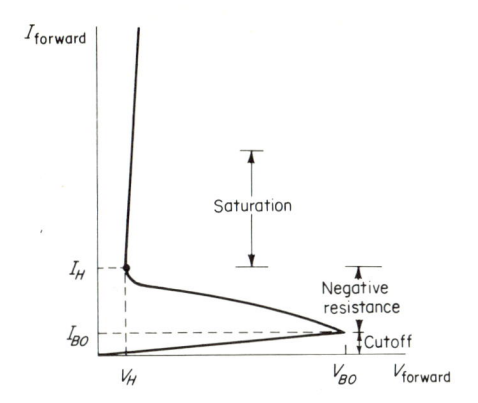

Fig. 15-53. Forward characteristics of a typical four-layer diode.

(Fig. 15-53) show that as forward voltage is increased, current increases very slowly and is essentially a low-value leakage current. At a value of V_{BO}, the breakover voltage, the device switches from its off to its on state, with current limited to a value greater than its holding value I_H.

Typical values of current and voltage are: V_{BO}, from ten to several hundred volts; I_{BO}, from ten to several hundred microamperes; I_N, from several to several hundred milliamperes; the holding voltage V_H ranges from perhaps 0.5 volt to 20 volts.

The Triac

The triac is a bidirectional semiconductor triode-type switch that may be triggered by a gate voltage from a nonconducting (blocking) state to a conducting state regardless of the polarity of applied voltage. In effect, it consists of two inversely connected SCR's operating with a single gate.

A typical triac circuit is illustrated in Fig. 15-54, along with the electrical and physical characteristics. The circuit action is straightforward. A suitable phase-shifting network is used to trigger the triac into conduction during some portion of the input cycle. The load, then, receives alternating current for some part of the total cycle, and the total energy delivered to the load can be made to be variable, as suggested by Fig. 15-55. Some examples of such usage are light dimmers, motor-speed controls, and variable-output power supplies. Typical characteristics for a hypothetical triac are given in Table 15-1.

402 PULSE AND SWITCHING CIRCUIT ACTION

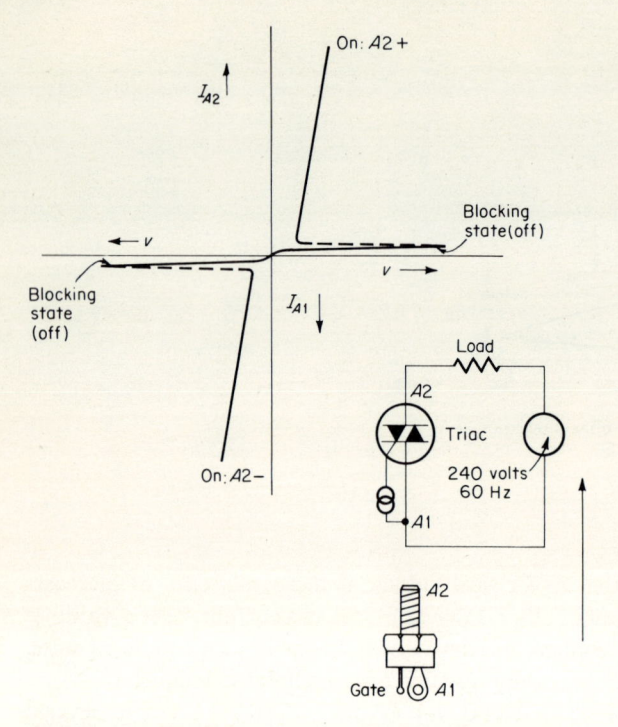

Fig. 15-54. Bidirectional switch (triac) used in ac circuits.

Table 15-1
Typical Characteristics of a Triac

Maximum Ratings
$V_{(BR)}$ = minimum breakdown voltage either direction   ±400 volts
I_{rms} = rms conduction-current rating   10 amp
$P_{G,(PK)}$ = peak gate-power rating   5.0 watts
P_G = average gate-power rating   0.6 watt
Peak 1-cycle surge   80 amp
T_S = storage temperature   −25 to 100°C

Characteristics
Peak blocking current at 400 volts either direction   5 ma
Maximum static gate current and voltage required to trigger
    A_2+, gate+   50 ma at 3 volts
    A_2+, gate−   75 ma at 3 volts
    A_2-, gate+   100 ma at 3 volts
    A_2-, gate−   50 ma at 3 volts
Conduction voltage drop   1.6 volts

(a) [waveform: Triac fires, Off, On] — Full delivery to load

(b) [waveform: Triac fires, Off, On] — Half delivery to load

(c) [waveform: Off] — No delivery to load

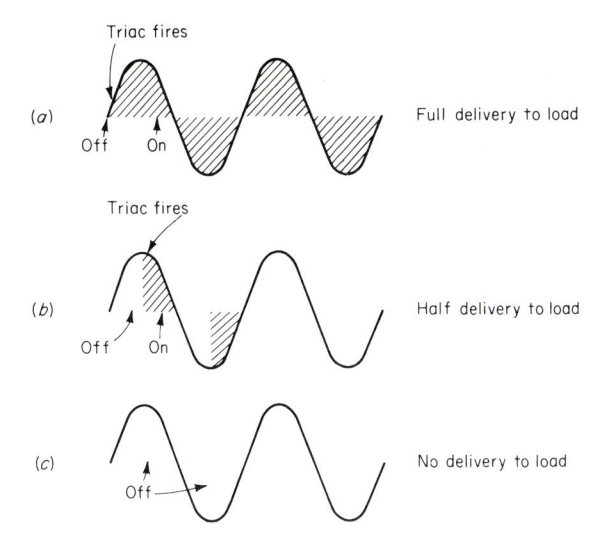

Fig. 15-55. Energy delivery to load.

The Tunnel Diode

The tunnel (or Esaki) diode is very similar in most respects to a conventional junction *pn* diode. In the *pn* diode, the impurity concentration is on the order of one part in 10^8. This results in a depletion-layer width of approximately 5×10^{-4} cm. The potential barrier thus formed prevents the majority carriers on one side of the junction from migrating to the other side unless a significant voltage is applied.

If, during manufacture, the impurity concentration is greatly increased to, for example, one part in 10^3, the electrical characteristics are completely changed. Now, the width of the depletion layer is on the order of 1×10^{-6} cm, which is considerably less than the wavelength of visible light. Normally, one would consider that a carrier must attain sufficient energy to go *over* the barrier to reach the other side. However, for ultra-thin barriers such as estimated for the tunnel diode, quantum mechanics indicate a very large probability that the carrier will be able to tunnel through. The net result of the tunneling effect is the unusually shaped curve of Fig. 15-56.

The diode is seen to be an excellent conductor in the reverse direction, where the *p* side is made negative and the *n* side is made positive. In the forward direction, the diode also conducts well up to the point where peak current exists. If the applied voltage increases beyond V_P, the current *decreases* until point V_V is reached. An increasing voltage with

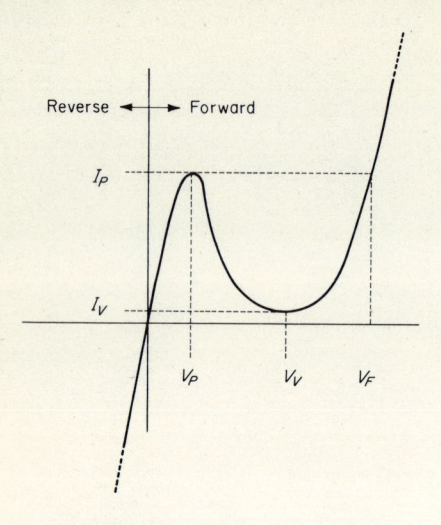

I_P = peak current
I_V = valley current
V_P = peak voltage
V_V = valley voltage
V_F = peak forward voltage

Fig. 15-56. Electrical characteristics of the tunnel diode.

attendant decreasing current indicates *negative resistance*. Beyond V_V the diode performs as one would expect. Operation between the point of origin and V_V is the normal operating area for the tunnel diode.

The diode is used primarily for the characteristics obtained by operation in the negative-resistance region. Because tunneling occurs at the speed of light, the diode is extremely useful as a very high speed switch. Switching times as short as 50 psec (50×10^{-12}) have been obtained.

The Backward Diode

Conventional silicon diodes used in rectifying or detecting applications have forward barrier voltages of, typically, 0.5 to 1.5 volts. To perform efficient rectification, the applied voltage must exceed this by a relatively large amount. Even a germanium diode requires, at the least, a few volts to rectify with any efficiency. It is possible, by carefully controlling the junction formation, to construct a tunnel diode that will rectify quite well voltages in the 100- to 200-mv region. Such a diode is called a *tunnel rectifier*, or *backward diode*.

To illustrate the general characteristics of such a device, consider Fig. 15-57. When the p side is made more positive than the n side by 100 or 200 mv, the diode conducts very little. A conventional diode would be considered to be forward-biased under these conditions. However, for small voltages, the backward diode is nonconducting with this applied voltage.

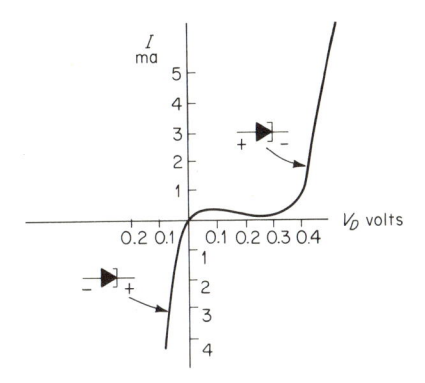

Fig. 15-57. Germanium-backward-diode characteristics (typical).

With reversed polarity, the diode conducts heavily, as is normal with a tunnel diode. Note that the significant difference between a conventional tunnel diode and the backward diode is that the value of peak current I_P is kept to a minimum in the latter case. Comparing the characteristic curves for the two devices, this difference is clearly seen. Other than this, the two devices are quite similar.

Constant-current Diodes

A zener diode provides a constant voltage by virtue of its reverse characteristics, which allow a large current change with attendant small voltage change. The constant-current diode has exactly the opposite effect. It provides essentially a constant current within some minimum-maximum voltage range; that is, the current through the diode remains constant regardless of external conditions within certain limits.

The diode is basically a field-effect transistor (FET) that has its gate lead connected directly to the source lead. A set of characteristic curves for a typical FET is shown in Fig. 15-58. Note that in the vicinity of the dotted line each characteristic curve is nearly horizontal, suggesting a nearly constant current delivery to the load circuits.

Figure 15-59 illustrates a FET with V_{GS} equal to 0 volts, along with the appropriate characteristics. Between the point on the curve labeled

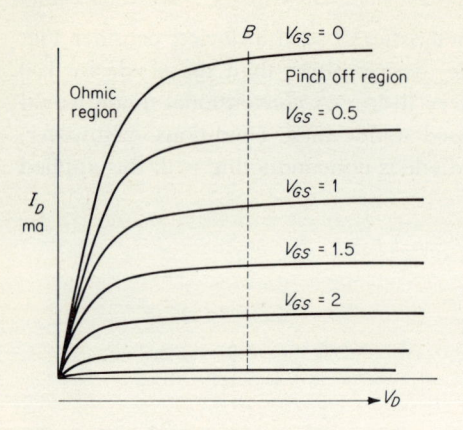

Fig. 15-58. FET characteristics.

"pinch-off" and the point labeled "breakdown" the curve is virtually flat. It is between these limits that constant-current operation is obtained. In practice, a constant-current diode is made by optimizing certain properties, namely, low pinch-off voltage, high breakdown voltage, and high dynamic impedance. As normally used, the quiescent operation is caused to be in the center of the flat portion of the curve, allowing operation on either side of the quiescent point.

Several symbols have been used to represent the constant-current diode; typical examples of these are given in Fig. 15-60. There are many applications for this device, one of which is shown in Fig. 15-61. A differential amplifier requires a constant-current source for emitter current, and such is the case in Fig. 15-61. Regardless of the transistor action, the current delivered to the emitters is constant, provided, of

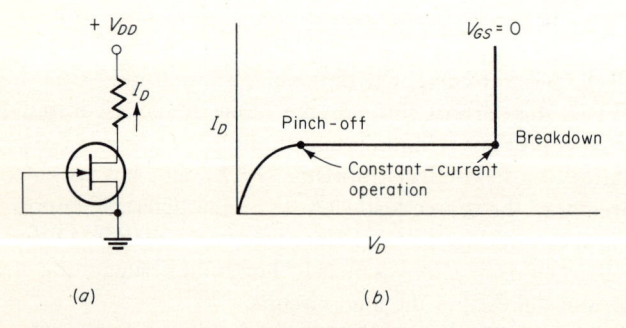

Fig. 15-59. Equivalent constant-current diode and characteristics.

Fig. 15-60. Symbols used for constant-current diodes.

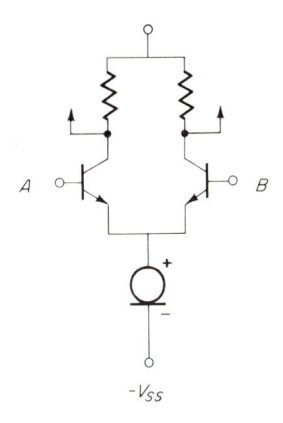

Fig. 15-61. Constant-current diode used as constant-current generator in the emitter-supply lead of a differential amplifier.

course, that the transistors do not cut off. Many other circuit configurations use the constant-current principle to advantage; space precludes comprehensive coverage here. Briefly, a few of these might be

1. Charging a capacitor from a constant-current source

2. Providing a constant low-voltage source through use of a fixed, accurate resistance

3. Noise, or spike, filter

4. Providing an accurate, constant-current source for measuring a transistor's beta

5. Used with a zener diode, providing much better regulation

QUESTIONS AND PROBLEMS

15-1 Refer to Fig. 15-1. Name the functions of $R2$ and $C2$.

15-2 Refer to Fig. 15-1. Name the functions of $R19$ and $C7$.

15-3 Refer to Fig. 15-1. Name the functions provided by $C10$, $R16$, $R17$, and $Q5$ other than amplification and inversion.

15-4 Refer to Fig. 15-1. Briefly describe the purpose and function of $C8$.

15-5 Refer to Fig. 15-7. Briefly describe the reason the circuit is biased-up.

15-6 Refer to Fig. 15-7. The lower ends of $R4$ and $R9$ are referred to as a point with a special name. What is this name?

15-7 Refer to Fig. 15-9. Name the purpose of the 220-ohm resistor.

15-8 Refer to Fig. 15-9. Name the basic transistor configurations represented by $Q2$ and $Q3$.

15-9 Refer to Fig. 15-15. Name the function provided by $D1$ and $D2$.

15-10 Refer to Fig. 15-15. Name the function of $R9$.

15-11 Name the most desirable characteristic of a perfect sawtooth voltage.

15-12 True or false? A perfect staircase waveform can be generated by utilizing the complete charge curve of a capacitor being charged by a constant-voltage source.

15-13 Refer to Fig. 15-29. Name the primary function of $D1$.

15-14 Refer to Fig. 15-29. Name the primary function of $R5$.

15-15 Refer to Fig. 15-35. Determine the circuit action if the input to $Q1$ consists of square waves operating between 0 and $+6$ volts.

15-16 Refer to Fig. 15-35. Determine the circuit action if the input to $Q1$ consists of square waves operating between -6 and -12 volts.

15-17 True or false? The unijunction transistor makes an excellent linear amplifier.

15-18 True or false? Referring to Fig. 15-40, slightly changing the value of $R2$ will cause the frequency of oscillation to change.

15-19 True or false? The tunnel diode is useful for rectifying small voltages.

15-20 Briefly describe the action and construction of a constant-current diode.

16
INTEGRATED CIRCUITS

Because integrated circuits are used so widely, it is felt that the entire subject is worthy of separate and more or less complete coverage. However, because of limited space, the discussion herein is, in certain areas, somewhat superficial. Nevertheless, it is hoped that the following information will allow the serious student to fully appreciate integrated circuits and to be able to work on and around them with greater confidence.

16-1 INTRODUCTION

Integrated circuits (IC's) are being used in an increasing number of applications, and it is expected that as time goes by, their use will become much more widespread. They are being used in nearly every type of commercial and military equipment where reduced weight, smaller size, and better reliability are required.

Microelectronics is the art of compressing more and more components in a smaller and smaller volume (package). In earlier days, using vacuum tubes and associated components, about 5000 components could be packed in a square foot of space. Later, when transistors became firmly established, on the order of 100,000 components could be mounted in the same space. Now, however, using integrated circuits, nearly 10,000,000 components might be found in a 1 cu ft volume. It becomes apparent, then, that such a saving in space, weight, and very often cost can be quite considerable.

Some of the most obviously useful applications for IC's are in space vehicles (or any extraterrestrial vehicle), computers, portable hand-carried equipment, etc. Many other applications come to mind with very little thought.

Integrated circuits have provided the means to reach a solution to one of the greatest problems ever encountered by the electronic design engineer. This problem is described as the *tyranny of numbers*, which refers to the ever-increasing complexity of modern-day equipment and the consequent multiplication of individual parts. As machines become ever more complex, the chance of machine failure due to component malfunction increases by leaps and bounds.

One of the advantages of integrated circuits lies in the lack of individual connections that in the past would have been installed by hand. A very large percentage of these connections are an integral part of the chip and as such are removed from the foibles of human frailty. By their very nature, integrated circuits are inherently reliable, consisting of a number of components, both passive and active, inseparably bonded together into an integral unit. This greatly increases total system reliability. Adding to this inherent reliability is the passivation process, which seals each part of the integrated circuit not only from all other parts, but from the outside world as well.

Another significant advantage is the extremely low cost per unit relative to the number of functions on a given unit. As will be seen, up to 1500 identical circuits are processed simultaneously on a single wafer, which is on the order of 1 to $1\frac{1}{2}$ in. in diameter. Because several hundred wafers can be processed at one time, the cost per circuit is very small. However, this does not necessarily mean that all integrated circuits are of inconsequential cost. After the chip is separated from its neighbors on the wafer, it must then be thoroughly tested, mounted in a suitable enclosure, and leads attached. These steps in the manufacturing process cannot be accomplished en masse, and from this point on, much hand labor is involved; it is here that most of the ultimate cost occurs.

Because of minor imperfections in the crystal structure, the possible infusion of dust and other contaminants, as well as other considerations, not every circuit measures up to standards. Some percentage of every batch must be discarded at some point in the overall process. This, of course, also has a large bearing on the ultimate cost. Nevertheless, disadvantages notwithstanding, the integrated circuit is here to stay, at least until an even more revolutionary process comes along to supplant it.

It must not be assumed that the integrated circuit has no disadvantages at all. For example, one limiting factor in the application of IC's is the relatively low power dissipation of the device. For many applications, this is not a severe drawback, but in others it is. Also, the integrated circuit finds greatest use in applications where there is considerable redundancy, that is, where the same circuit can be used over and over again, as in the case of a digital computer. In equipment that uses highly distinctive circuitry, IC's may not provide the most economical approach.

There are four basic techniques used in the manufacturing of IC's. These are the *hybrid, thin-film, monolythic,* and *compatible* circuits. The hybrid circuit is one in which separate component parts or dice (transistors, resistors, etc.) are attached to a ceramic base and interconnected to each other by means of either wire bonds or a metallization pattern. This construction is very similar to conventional means except that it is enclosed in a container perhaps $\frac{3}{8}$ in. in diameter and $\frac{1}{4}$ in. in height. A thin-film circuit consists of microscopically thin films of material deposited on a ceramic base to form the passive components. Then, active components must be added in discrete form to these thin-film networks. With the monolithic (single-stone) technique, all parts of the circuit, including transistors, diodes, resistors, and capacitors, are formed within the single wafer of silicon. Finally, the compatible circuits are those which have the active components formed within the chip of silicon but have the passive components deposited by thin-film techniques on top of the insulating layer covering the active components.

In the following discussion, we shall be concerned primarily with the monolithic integrated circuit since this is by far the most widely used method. (The bibliography lists a few good sources for more detailed information on the subject.) A later section will briefly describe the MOS integrated circuit, which operates upon somewhat different principles.

16-2 MANUFACTURING PROCESSES

While it is not necessary to know how IC's are made in order to apply them, certain advantages accrue when at least a cursory investigation is made. At the very least, some of the processes used are enlightening and quite interesting, being unique in the realm of electronics.

To introduce this subject, refer to Fig. 16-1. The object at the top of

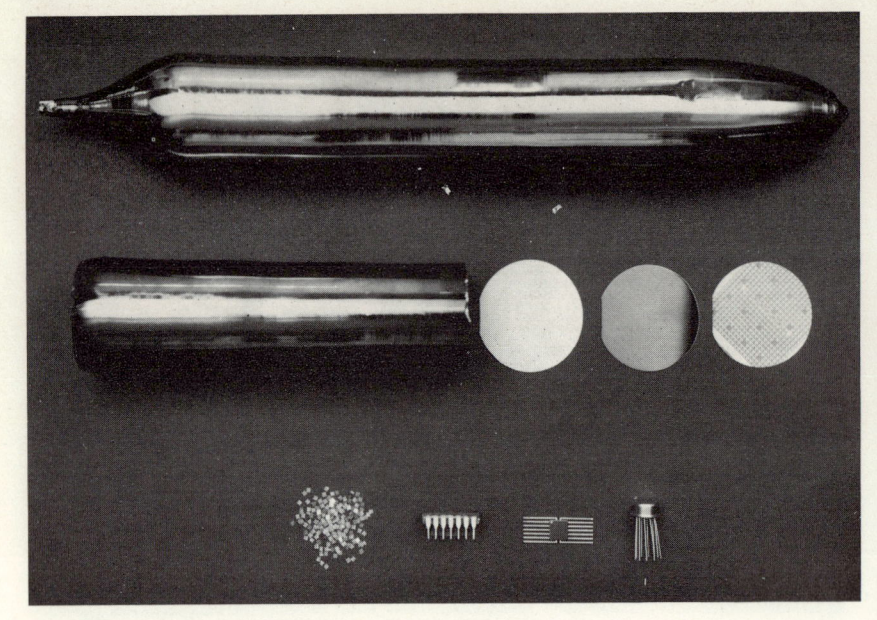

Fig. 16-1. Various steps in the manufacture of IC's.
(*Courtesy Fairchild Semiconductor.*)

the figure is a bar (ingot) of pure silicon that has been produced, or grown, by "pulling" a seed crystal from a mass of molten silicon under very controlled conditions. The ingot of pure silicon represents the first stage in *IC* production; it is about 6 to 8 in. in length and between 1 and 1½ in. in diameter. The next step in the manufacturing process is to cut the ingot into very thin slices, or wafers, that are on the order of 12 mils (0.012 in.) thick. Many, many wafers can be sawed from each ingot. The wafers are extremely brittle and will break quite readily, and hence they must be handled carefully. Each wafer is then lapped to about 6 mils, using a very fine grit abrasive, and chemically etched to form an extremely flat and smooth surface.

At this point in the overall process, the wafers are subjected to the epitaxial growth of a layer of silicon over the face of the wafer that will eventually hold the circuits. The need for the epitaxial growth lies in the fact that in the following processes, which are accomplished by diffusion, no more than three layers can be produced satisfactorily. The added layer produces an exact extension of the underlying crystal structure. That is, the exact structure of the atoms in the original crystal (the wafer) is duplicated in the epitaxial layer, which becomes simply an extension of the substrate about 10 μ in depth (10 μ = 10 $\times$ 10^{-6} m).

In practice, the epitaxial layer is a doped structure and can provide either a p- or an n-type layer. It is grown in a furnace, with a carefully controlled temperature, containing an atmosphere of silicon, chlorine, hydrogen, and a phosphorus compound for n-type doping or a boron compound for p-type doping. A typical structure is shown in Fig. 16-2.

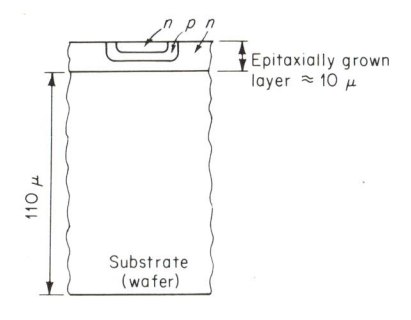

Fig. 16-2. The epitaxial layer.

The epitaxial layer is grown upon the substrate (body of the wafer), and the transistor junctions are formed in this layer by the process of diffusion, to be described subsequently. While transistors can be formed on the substrate with no epitaxial layer, the overall characteristics of the device (BV_{CBO} and $V_{CE,\text{sat}}$, in particular) are greatly improved. The formation of the junctions in the epitaxial layer affords a simple and effective way to produce a compromise between these conflicting requirements.

Next, the wafers are placed in a furnace containing an oxygen atmosphere at 1200°C. The oxygen penetrates the surface of the silicon and combines chemically with the atoms of the crystal lattice, forming silicon dioxide, a stable, inert glass. The dioxide layer envelopes the wafer and *passivates* the wafer surface, which greatly reduces the possibility of contamination. The wafer is now ready to begin the processes that will ultimately lead to the final integrated circuits, up to 1500 of which may be formed on the single wafer.

Before the next processes can be accomplished, several other steps must be taken in preparation for the actual formation of the IC's. First, the electrical design must be finalized so that when the IC's are finished, it is known for certain that the circuit will have the required electrical characteristics. Then, the basic breadboard is transformed into the required artwork drawings, one of which exists for each step of the process. Each drawing represents the areas on the chip that are to be operated upon during that particular step. The artwork is drawn first about 30 × 30 in. to insure the highest possible accuracy. Figure 16-3 illustrates a typical piece of artwork for a very simple device. In practice, these can become much more complex. Each panel is then reduced by photo-

Fig. 16-3. Typical IC artwork.

graphic means about five-hundred times. Then, a small glass plate, identical in size to the wafer, is treated with photosensitive material, and the image is exposed repeatedly across the face of the plate. Thus, as many as 1500 identical images may be produced on a single plate, which is used as a master to expose the silicon chip. With all master glass plates ready, the silicon wafers are ready to be transformed into completed integrated circuits.

Because of the complexity of a typical IC, it is not feasible to show the complete process. Figure 16-4, however, illustrates the portion of a typical circuit that we shall attempt to depict. Taking any of the complete circuits on the wafer and separating it from its neighbors, one single component (a transistor) from the circuit will be dealt with to illustrate in a very simplified manner how the IC is built up, step by step. Throughout the following discussion, keep in mind that the individual steps that are correlated with the various parts of the transistor are, *at the same time*, producing *all* collectors, *all* bases, and *all* emitters, as well as other components, such as resistors.

The passivated wafer begins to take the form of an integrated circuit when the collector cutout is made. The wafer is coated with a layer of photosensitive material and then is exposed to light through the glass mask that has been prepared to process all collector regions at once. The portions of the wafer that are exposed to light become hardened and in the subsequent rinse remain on the disc, with all other areas having a coating that is easily removed. Next, a hydrofluoric acid etch is used to remove the silicon dioxide from those areas not protected by the layer of photosensitive material (photoresist). In this manner, windows in the dioxide layer are produced to allow the next step, collector diffusion, to take place.

The wafer is then placed (along with many others) in a furnace whose atmosphere contains an *n*-type dopant. As the temperature is raised to

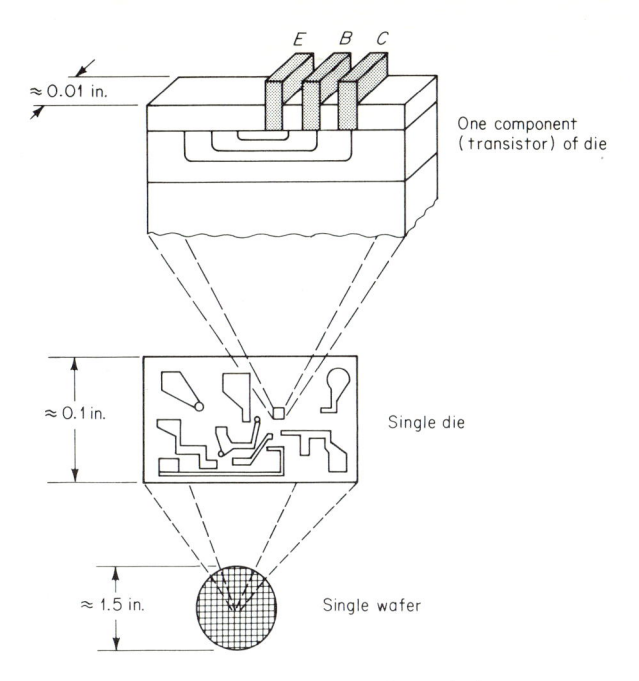

Fig. 16-4. Location of a single transistor relative to completed wafer.

about 1200°C, the dopant begins to diffuse into the surface of the wafer. That is, the atoms of the impurity are so violently agitated that they bombard the surface, and many of them penetrate into the inner regions and become a part of the silicon structure. This produces a highly doped n-type region that is the collector. The control of the depth of penetration is easily accomplished by accurate temperature control and timing of the diffusion process, and hence the collector can be constructed with exactly the desired characteristics. Figure 16-5 shows a portion of the wafer prior to the diffusion process, while Fig. 16-6 indicates the same section after the collector has been formed.

The next step is to form another layer of silicon dioxide over the entire layer (not illustrated) and a second window exposed and etched over the previous one. This window is smaller than before and is illustrated in Fig. 16-7. The wafer is again placed in the furnace, and this time a p-type dopant is used to diffuse the base region. This, of course, forms a pn junction that is to become the collector-base junction. The diffusion is carefully controlled to allow exactly the correct degree of penetration.

Once again, a layer of silicon dioxide is formed over the entire structure

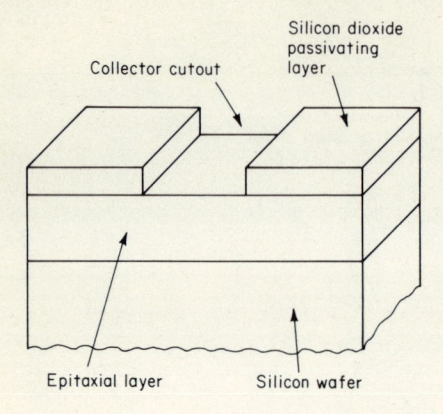

Fig. 16-5. Collector cutout prior to diffusion.

in preparation for the emitter diffusion. A still narrower window is etched, after which the wafer is again subjected to a furnace containing the n-type dopant. The formation of the emitter is shown in Fig. 16-8, where it is seen that the three areas are clearly defined and each extends up to the surface of the chip. A new silicon-dioxide layer is formed over the wafer to prepare it for the application of the leads, which must be formed so as to interconnect all necessary areas to each other. Now, many new windows are made, carefully aligned with the proper area, to allow the formation of the metallized leads. Typical windows for this stage are shown in Fig. 16-9. One method of doing this is the Metal-Over-Oxide process,[1] which is accomplished by evaporating metal onto the surface of the silicon wafer. Aluminum is literally boiled from a hot tungsten filament, depositing the metal in a thin, even coat over the entire wafer surface, as illustrated in Fig. 16-10. One final photoetching process is done at this stage to remove the aluminum over areas where it

[1] Metal-Over-Oxide is a patented Fairchild process.

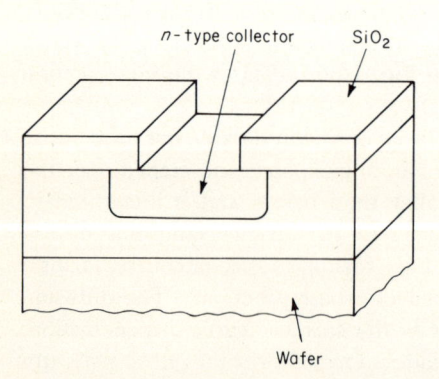

Fig. 16-6. Collector region after diffusion.

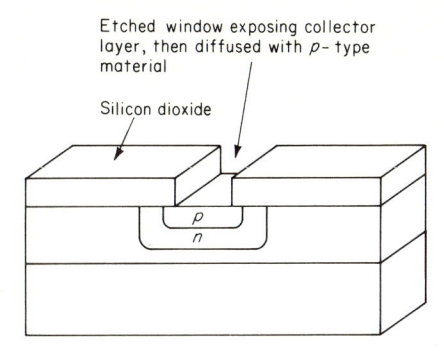

Fig. 16-7. Forming the base region.

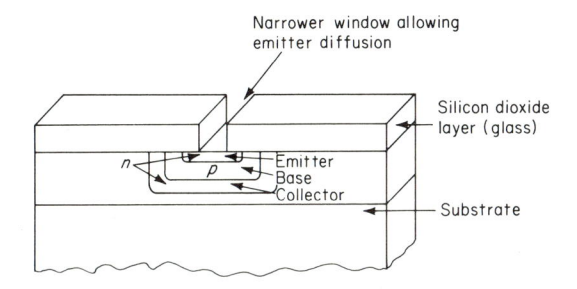

Fig. 16-8. Forming the emitter region.

is not desired. The net result is illustrated in Fig. 16-11, where it is clearly shown that the leads are firmly connected to the three parts of the transistor. The wafer on the right side of Fig. 16-1 shows a completed unit at this stage, with an enlargement of this wafer shown in Fig. 16-13.

The wafers are now ready for final processing. Each wafer consists of

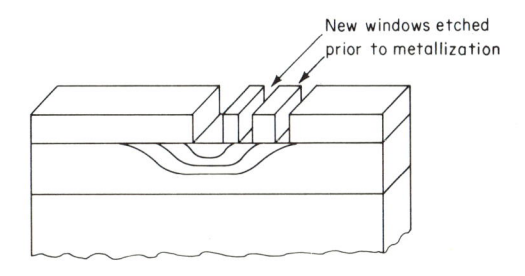

Fig. 16-9. Preparation of cutouts for metallization.

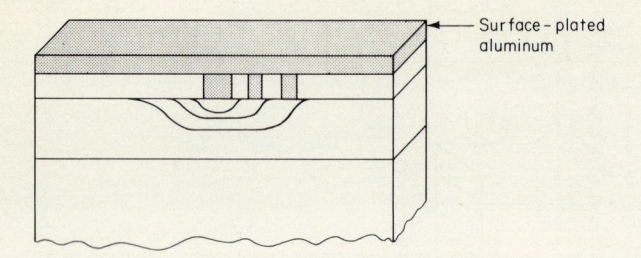

Fig. 16-10. Forming the interconnection paths.

up to 1500 individual IC's, each of which may contain several dozen (or possibly several hundred) individual components. By means of rather complex equipment, each circuit on an individual wafer is now tested automatically. Each wafer is inserted in a step-tester having microscopically fine probe tips prepositioned to contact the pads on the periphery of the individual circuits. The wafer is stepped to a position to allow the probe to be lowered, contacting the IC. The probes are connected to an automated tester that evaluates the proper electrical function of the circuit. If the circuit malfunctions, it is marked with dye and will later be destroyed. If it is satisfactory, the probes are lifted, the entire wafer is automatically moved so that the next circuit is in position, and the above process is repeated. Thus, every circuit is rapidly tested, Fig. 16-12, under identical conditions.

Next, a diamond-tipped tool scribes a fine line between each circuit, which will allow each to be separated from all others. Each circuit, now called a *die*, is a complete and functioning device. A complete wafer, containing several hundred circuits, is shown in Fig. 16-13 prior to dicing. Figure 16-14 illustrates the size of several typical chips relative to a common paperclip.

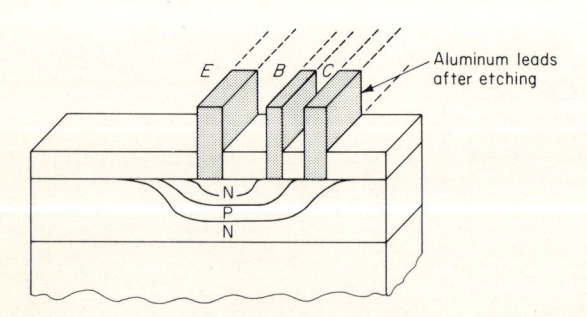

Fig. 16-11. The completed transistor.

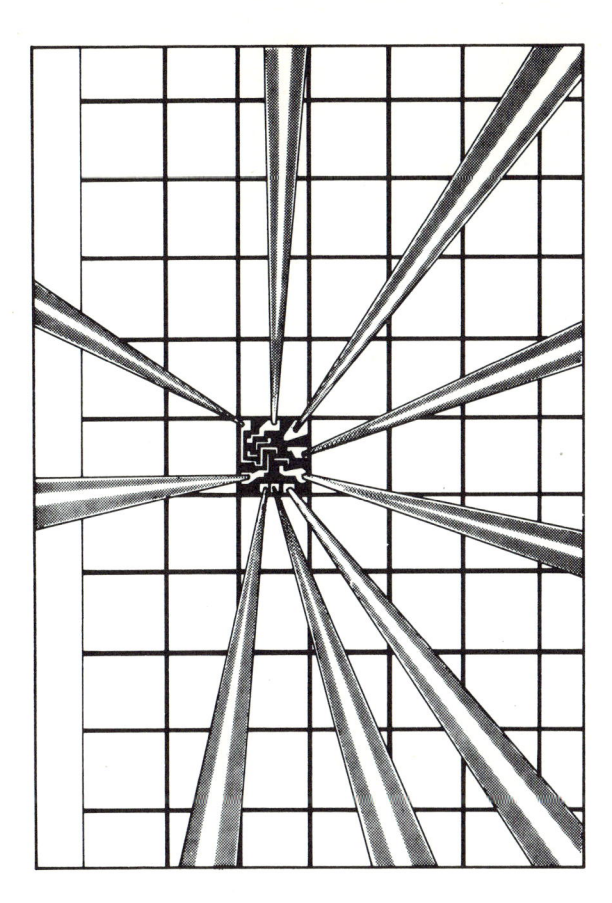

Fig. 16-12. Automated testing of the chip.

However, these dice cannot be used as shown, of course. In order to be able to be connected to the outside world, the unit must be mounted in a suitable container that will allow the connecting leads to be installed. Typical mounting arrangements are shown in Fig. 16-15. Also, Fig. 16-1 shows several possible arrangements. The pads on the chip that were formed during the final metallizing process are bonded to very fine gold wires that connect to the outer leads, by which the device will ultimately be connected in its permanent place. Completed IC's are shown in Fig. 16-16*a* and *b* in a closeup view. This particular unit (Fig. 16-16*b*) has 24 transistors, 6 diodes, and 36 resistors incorporated in its circuit, all on a chip about $\frac{1}{20}$ in. square.

One serious problem encountered in monolythic IC's is that of isolating

420   PULSE AND SWITCHING CIRCUIT ACTION

the various components from one another. The reader may have noticed that if two transistors are constructed side by side, the two collectors are, in effect, connected together through the rather low resistance of the epitaxial layer. This, of course, must be avoided at all costs. By doping the epitaxial layer with *p*-type material (for *npn* devices), this effect can be eliminated almost entirely. Figure 16-17a shows the effect with no provision for isolating the transistors on the same chip. The resistance from collector to collector may be very low in value and will certainly prevent the collectors from acting independently. In Fig. 16-17b the internal structure of the monolythic chip is shown, which indicates the manner of isolation. In this instance, the collectors and the epitaxial layer now form two diode junctions, which, if the proper connections are made, can be caused to be reverse-biased at all times, thus effectively isolating the two collectors. The epitaxial layer is connected to the most

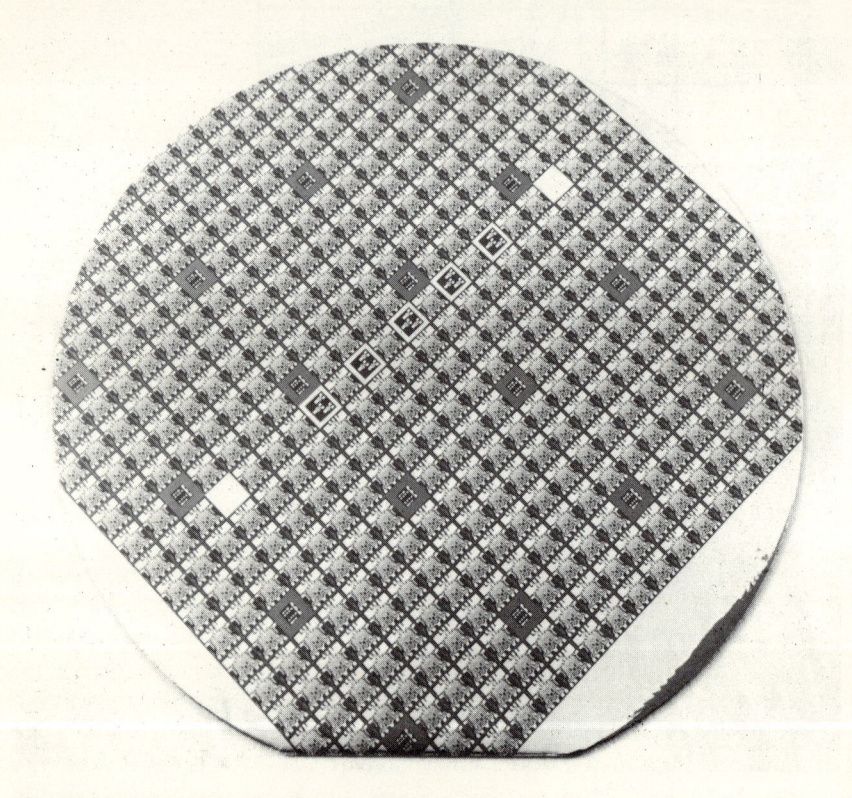

Fig. 16-13. A completed wafer containing many individual circuits before dicing. (*Courtesy Fairchild Semiconductor.*)

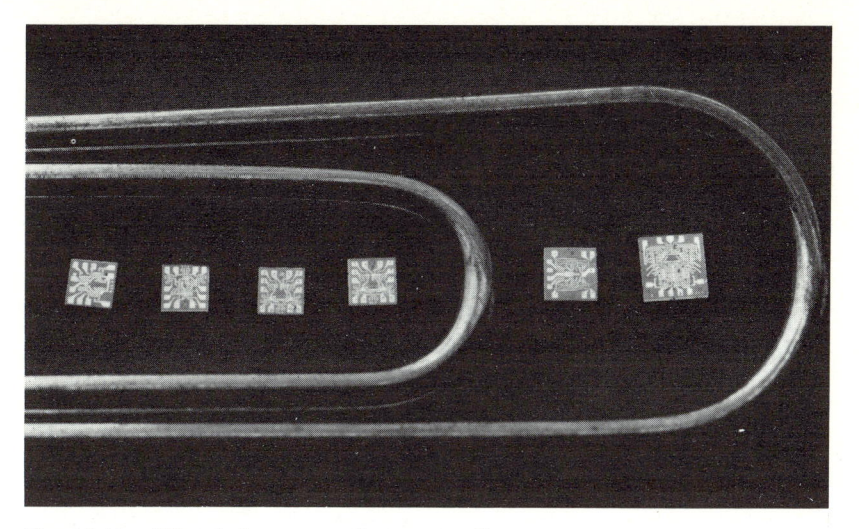

Fig. 16-14. IC's relative to an ordinary paperclip. (*Courtesy Fairchild Semiconductor.*)

negative point in the circuit (this is often ground), which will cause the diodes (Fig. 16-17c) to always be reverse-biased since all voltages in the circuit will be more positive than this.

One further problem is illustrated in Fig. 16-17d. The astute reader will have noticed that a second transistor exists at the location of each *npn* unit. The base of the *npn* unit, along with the collector and epitaxial layer, form a *pnp* transistor, and this so-called "parasitic" transistor can interfere with normal operation. As before, the connecting of the epitaxial layer to the most negative potential will never allow this parasitic transistor to become forward-biased since its base will always be more positive than the emitter. Certain manufacturing processes (notably gold doping) can reduce the beta of the *pnp* parasitic transistor to a value such that transistor action is negligible. The parasitic junctions, then, act simply as diodes, which can easily be made to be always reverse-biased.

The above method of component isolation is probably the most widely used, but other ways of accomplishing the same thing are possible. One such method is to layer the substrate with a thick coating of silicon dioxide, and then to etch tiny pockets in the insulating layer. Each active component is then constructed by the usual means completely within the confines of the pocket, thus effectively isolating each transistor, each diode, etc.

The formation of passive components (resistors and capacitors, primarily) is accomplished by diffusion techniques very similar to those used

Fig. 16-15. Integrated circuit in a TO-5 case, showing lead posts and lead bonds. (*Courtesy Fairchild Semiconductor.*)

in making transistors. Diodes, on the other hand, are formed by simply connecting the base to the collector of a transistor, as illustrated in Fig. 16-18. The emitter-base junction is used rather than the collector-base junction because the inherent capacity of this junction is less, and hence the circuit is usable at higher frequencies.

A typical silicon-dioxide capacitor is illustrated in Fig. 16-19. Such a device uses the insulating properties of silicon dioxide as the dielectric of the capacitor. One plate of the capacitor is the aluminum metallization, as shown. The other plate is formed from the heavily doped layer just beneath the oxide, labeled N^+. Values of capacitance up to 500 pf can be formed, with maximum voltages of about fifty. An alternate method uses tantalum oxide as the dielectric, producing values up to 5000 pf with working voltages of about twenty. Several other processes are used to limited degrees, each having certain advantages and disadvantages.

A resistor is formed by simply defining the dimensions of a certain volume of silicon, properly doped, and isolating it from the other com-

INTEGRATED CIRCUITS 423

Fig. 16-16. Completed IC's mounted in (*a*) plug-in receptacle and (*b*) 14-pin container.

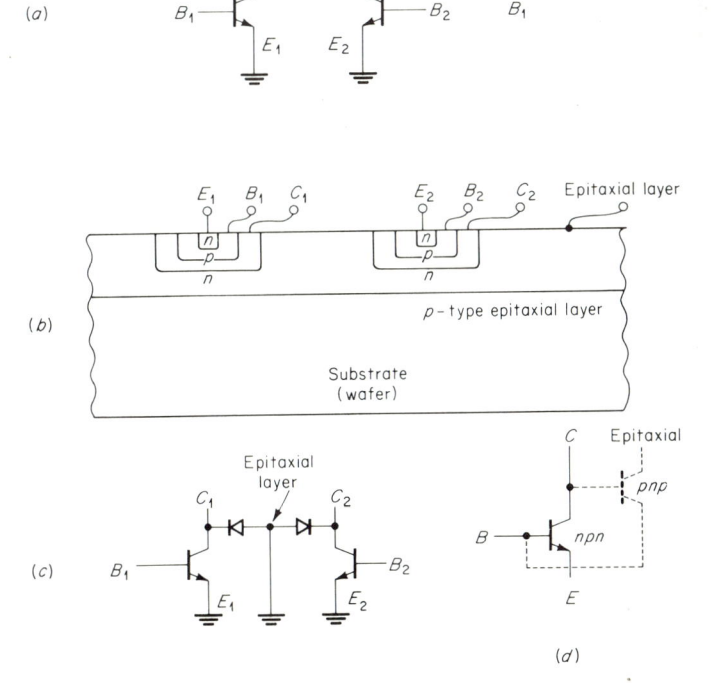

Fig. 16-17. One method of isolating active component parts in a monolithic IC.

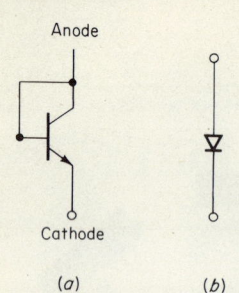

Fig. 16-18. (a) Diode formed by connecting the base to the collector; (b) the usual schematic representation.

ponents. The dimensions of a typical resistor are shown in Fig. 16-20a and the actual construction in Fig. 16-20b. Directly beneath the silicon dioxide is the p-type layer that is the resistor proper. Each end connects to the aluminum metallization pattern. Such a unit, with these dimensions and with typical diffusion densities, will have a resistance of approximately 4000 ohms. In considering the entire integrated chip, the resistors are usually formed at the same time as the transistor bases, and the value of each is determined by the dimensions of the pattern used. Note in Fig. 16-20 that there is a parasitic pnp transistor formed. For this reason, it is necessary to insure that the proper voltages are provided to keep the junctions reverse-biased at all times. This is accomplished by returning the n-type layer just beneath the p-type resistor to the most positive voltage available. Thus, the pnp transistor can never become turned on.

The foregoing description has been greatly simplified and should not be considered to be a complete dissertation on the subject. Various manu-

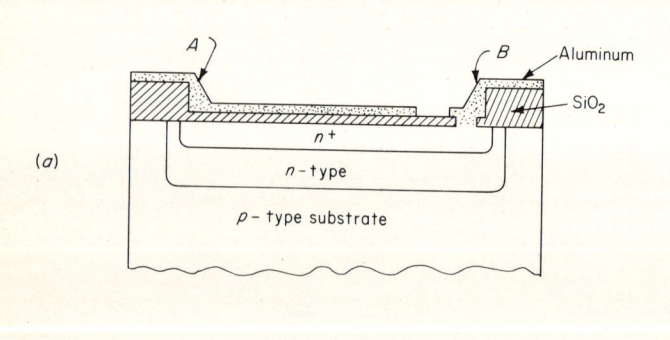

Fig. 16-19. (a) Silicon-dioxide capacitor used in monolithic IC's; (b) schematic representation.

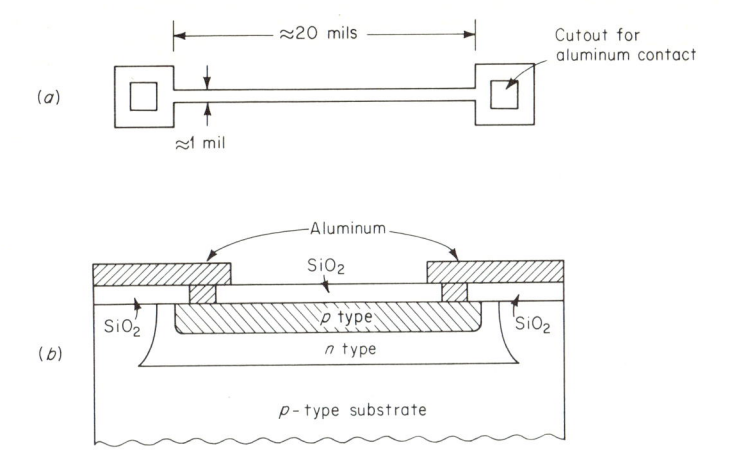

Fig. 16-20. (a) Approximate dimensions of a diffused resistor; (b) internal resistor structure.

facturers have differing processes, all of which have not been taken into account. Nevertheless, this discussion is adequate for allowing one to most efficaciously apply IC's.

16-3 SWITCHING AND GATING IC's

As of this writing, the number of applications for linear IC's is small relative to their use in digital equipment. Because the very nature of digital equipment implies redundancy, IC's find their greatest use in pulse and switching circuits. The circuit of a typical integrated circuit for such use is given in Fig. 16-21. Both the internal circuitry and the logic diagram are shown, as well as the implementation by normal discrete methods. This particular gating network is of the family called TTL, or *transistor-transistor logic*.

Note on the schematic diagram the input transistor $Q1$. Here is a technique that has never been used with discrete-component circuits. Forming multiple-emitter transistors is unique with IC's and greatly simplifies the circuitry. Each emitter forms a separate circuit, and its action is identical to that shown in the circuit of Fig. 16-21c; that is, each emitter acts the same as one of the diodes in the more conventional circuit. In the circuit of Fig. 16-21c, if either input is at ground, the emitter of $Q1$ is at ground. Therefore, the transistor is fully on and the collector is also at ground. However, when *both* inputs are at 5 volts, $Q1$ is off

426 PULSE AND SWITCHING CIRCUIT ACTION

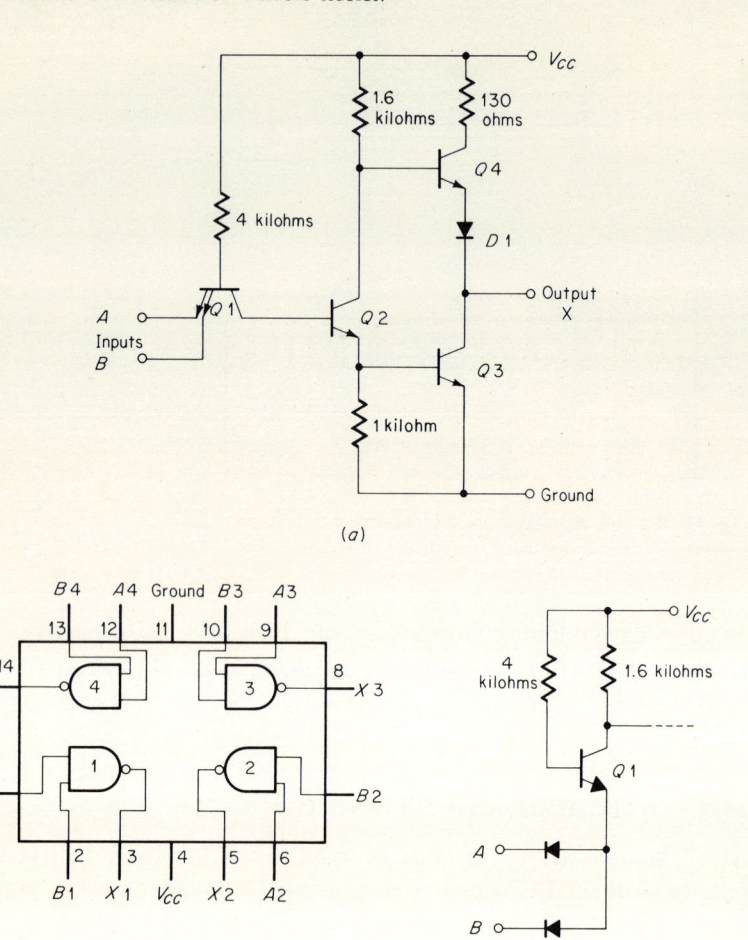

Fig. 16-21. SN7400 positive NAND gate (TTL). (*a*) Schematic; (*b*) logic diagram; (*c*) equivalent diode-transistor logical gate. (*Courtesy Texas Instruments.*)

since both the emitter and base are at the same voltage. The output, then, is also at 5 volts.

The input transistor of Fig. 16-21*a* functions in exactly the same way. If either input is at ground, the collector of $Q1$ is also at ground and $Q2$ and $Q3$ are off. Only when both inputs rise to 5 volts will $Q1$ turn off and in so doing turn on the other two transistors. The circuit of $Q1$ is often called a *collector follower* since the collector of the transistor follows the emitter and this is exactly descriptive of the circuit action. It can be appreciated that the emitter circuit of Q1 is acting as a gate, only allowing Q1 to perform some useful action when the input has the proper conditions. Also, the overall circuit is an inverter since Q3 is in the common-emitter configuration. A ground level at the base of Q3 is inverted to a 5-volt level at the output, while a 5-volt level at its base results in a 0-volt output (ground). The circuit, then, is a combined gate-inverter, and will perform as described to yield logical operations. In this particular unit, there are four separate circuits, having only ground and V_{CC} as common connections.

The circuitry of one of the gates will now be analyzed in some detail. While it is true that it is impossible to repair the circuit, nevertheless it is of some benefit to be aware of the details of circuit action. In the circuit of Fig. 16-21a, first assume that both inputs to Q1 are at approximately V_{CC}. Because this transistor is a collector follower, its collector is also at V_{CC}, or nearly so. With the base of Q2 at a high positive potential, Q2 is on and its emitter tries to follow the base in rising to the supply voltage. Since the emitter-base voltage of these silicon transistors is on the order of 0.7 volt, the emitter of Q2 cannot rise to more than the base-emitter voltage of Q3, or about 0.7 volt. Thus, Q3 is also on, and its collector is nearly at ground potential. Hence, with inputs A and B at positive 5 volts, the output is at ground.

At this same time, with the emitter of Q2 clamped to about 0.7 volt, its collector is also at roughly the same potential. Allowing 0.2 volt across the Q2 transistor, its collector can be no more positive than 0.9 volt. Therefore, the base of Q4 is also at 0.9 volt. To see whether Q4 is on or off, one simply determines the required voltage at the base to allow on-turning. The collector of Q3 is at about 0.2 volt, allowing for $V_{CE,\text{sat}}$, and D1 will require on the order of 0.7 volt to become forward-biased. Also, the base-emitter junction of Q4 will require 0.7 volt to be turned on, and the sum of these quantities is the value to which the base of Q4 must rise to allow Q4 to come on. Hence, the base of Q4 must become more positive than 1.6 volts before Q4 can be considered on.

Q4 is therefore off since the base is more negative than the emitter by nearly a volt. With a positive voltage at both inputs, Q1 and Q4 are off while Q2 and Q3 are on, and this details the circuit action for these input conditions.

Now, with either or both inputs at ground, the circuit action is somewhat different. Q1 is turned on very hard, and its collector falls to ground ($\cong 0.2$ volt). The base of Q2 must be more positive than 1.4 volts (2×0.7) to be on, and so it is therefore off. Its emitter falls to ground,

428 PULSE AND SWITCHING CIRCUIT ACTION

and hence $Q3$ is off also. However, the collector of $Q2$ rises toward V_{CC}, and $Q4$ is thereby turned on. This, of course, *actively lifts* the output wire toward V_{CC} and accomplishes what it is intended to do—it provides a low-impedance path to V_{CC} when $Q3$ is off. The output wire, then, rises toward V_{CC}, less the drop across the 130-ohm resistor and diode $D1$. If the ultimate load connected to the output is relatively high in value, the drop across the 130-ohm resistor and the diode is relatively small and the output can be considered to be at V_{CC}. $Q4$ is often known as a *pull-up* transistor since this is quite descriptive of its circuit action.

Only minor variations of TTL circuits are normally encountered, and one of these is illustrated in Fig. 16-22, where the input gate has eight separate inputs. Except for this difference, the circuit is nearly identical to the ones shown previously.

Diode-transistor logic (DTL) is also used in integrated circuits quite extensively, and a circuit such as this is shown in Fig. 16-23a. Rather

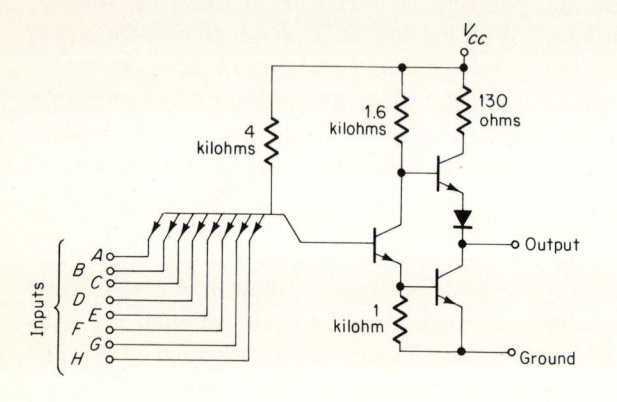

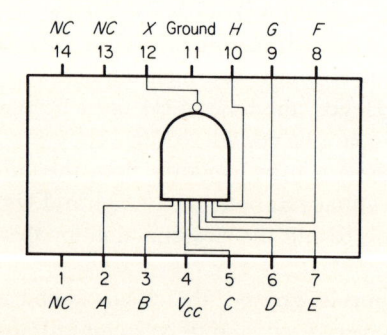

Fig. 16-22. Eight-input positive NAND gate, SN7430N (TTL). (*Courtesy Texas Instruments.*)

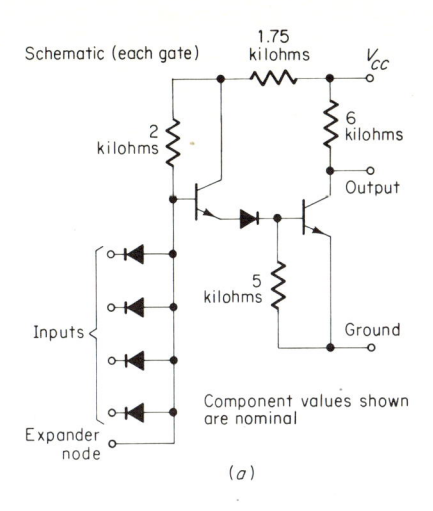

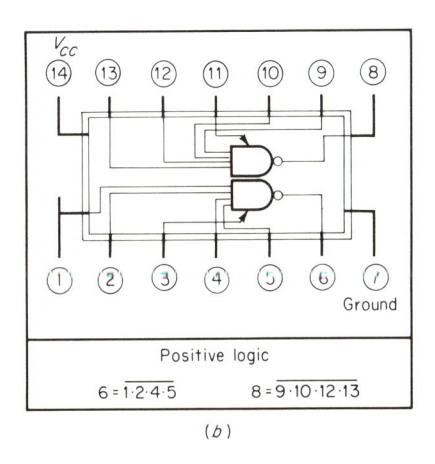

Fig. 16-23. Dual four-input positive NAND gate (DTL), SN15930. (*Courtesy Texas Instruments.*)

than have multiple-emitter transistors, separate diodes are used to perform the gating function. In this instance, note that there is no pull-up function connected with the output transistor and the output waveform may, under some circumstances, show a pronounced rounding on the positive-going slope of the pulse. The expander node is used to allow a greater number of inputs by adding additional diodes.

A slightly different approach is shown in Fig. 16-24, using DTL. This

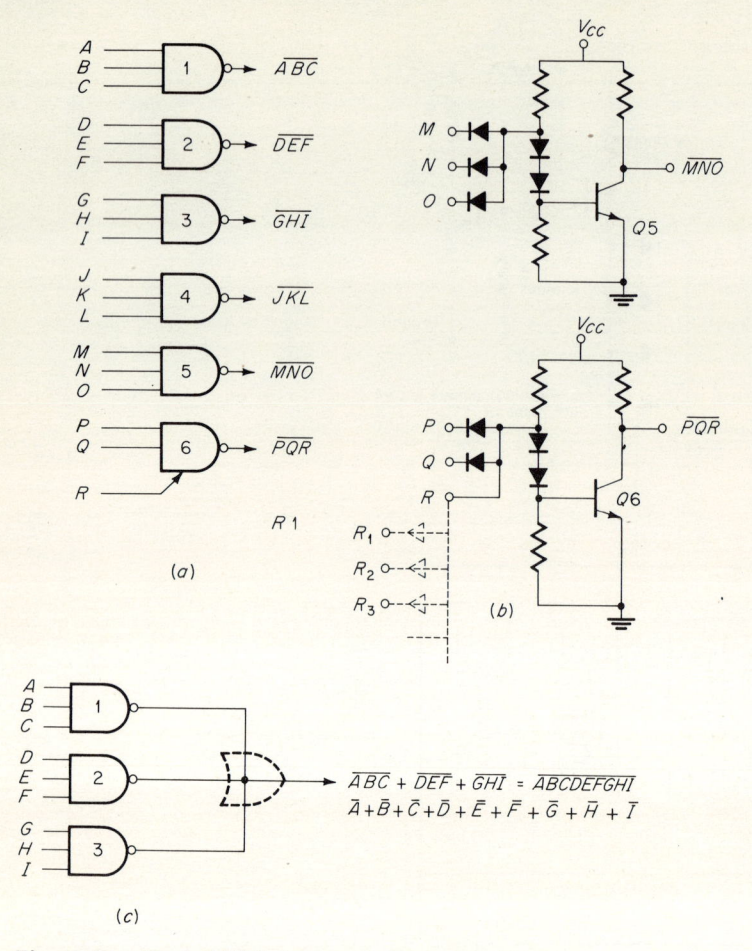

Fig. 16-24. Typical DTL integrated-circuit techniques:
(a) logic diagram of sextuple chip; (b) schematic; (c) "wired-OR" connection (for negative logic).

particular unit has five identical NAND gates and one with an expander input, which allows the input to have several additional diodes added to it, as before, so as to have more than three inputs at this gate. The two diodes in series with the base lead provide for positive turn-off. The input must be more positive than about 2.8 volts to turn the inverting transistor on, and hence any value very much less than this forces the inverter to be off. Again, there is no pull-up feature, and the output does not drive from low to high very well. In many applications this is no severe drawback, however.

An additional feature of this circuit, as well as others shown, is the so-called "wired-OR" connection, which can be formed by connecting several collectors together, as shown in Fig. 16-24c. The electrical connection formed by this performs a logical function, even though it is nothing more than a soldered connection. If the output is to be high, there must be a low at one or more of the inputs of *each* gate. Hence, such a connection performs an AND function for positive logic or an OR function for negative logic. Using positive logic, as is usually specified for such circuitry, the Boolean equation can be written as shown.

$$X = (\bar{A} + \bar{B} + \bar{C})(\bar{D} + \bar{E} + \bar{F})(\bar{G} + \bar{H} + \bar{I})$$

for the six-input gate shown. That is, if the inputs labeled A, D, and G are low, the output will be high. Of course, if more than one input is low at any one of the gates, this will not change the result.

Using negative logic, where the most negative voltage is logic 1, all inputs must be high at any one gate. In terms of an equation,

$$X = (\bar{A}\bar{B}\bar{C}) + (\bar{D}\bar{E}\bar{F}) + (\bar{G}\bar{H}\bar{I})$$

Many of the circuits shown thus far (those without the pull-up feature) have this attribute. Some IC's of this type have no collector resistor as a part of the chip, and they are tied together externally to a single common load resistor (discrete component). This, of course, provides a very low cost logic function where it can be used.

The foregoing examples are representative of currently available gating circuits utilizing IC construction. Most of the important points have been covered, and this discussion should enable the reader to apply the material to this kind of circuit.

16-4 IC MULTIVIBRATORS

The multivibrators to be discussed in this section will be found to be considerably different from those previously described. The circuitry itself will seem to be much more complex. One reason for this is the need for faster and faster response, and this can be accomplished by the various techniques of IC's, some of which have been covered in detail in preceding material. A typical example is the active pull-up circuit used in the NAND/NOR circuits. Another reason for increased complexity is the need to provide special attributes that were never available in discrete circuits, at least never available for a reasonable cost. Because extra transistors can be added with virtually no increase in costs, more elaborate circuit functions can now be made available as standard features.

The two multivibrators now to be discussed are representative of the many types available. Space limitations preclude a discussion of all available types, but these two will provide a firm foundation in the basic operation of such circuits.

JK Master-Slave Flip-flop

The master-slave principle as used in digital circuits is a unique application of the type of design principles that are possible with integrated circuits. While such a circuit would not be impossible using discrete components, it would be prohibitively expensive. IC's, however, allow the use of many more active components at no increase in cost.

Before the flip-flop itself is presented, several of the individual circuits used in this application must be described. One of these is given in Fig. 16-25, which is seen to be the multiple-emitter transistor that is already

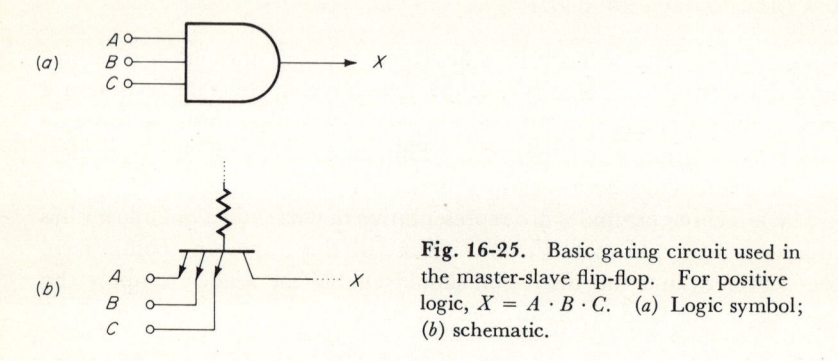

Fig. 16-25. Basic gating circuit used in the master-slave flip-flop. For positive logic, $X = A \cdot B \cdot C$. (a) Logic symbol; (b) schematic.

familiar. This circuit is acting, for positive logic, as a negative OR gate or a positive AND gate. That is, if any one input is low (ground) the output is low; if all inputs are high, the output is high.

A different circuit is shown in Fig. 16-26, which is a coupling pair of transistors in a somewhat unusual configuration. The emitters are tied together and comprise one of the circuit inputs. Each base, A and B, constitutes another input, and hence there are three inputs. The truth table shown in Fig. 16-26b gives the circuit action for all possible conditions of input. In effect, the circuit acts to allow the base voltages to be used as outputs only under certain input (emitter) conditions. Note that the last four conditions on the truth table indicate that the output is logic 1 for both wires when the emitters are at logic 1. Actually, this is

INTEGRATED CIRCUITS   433

A B C	A' B'
0 0 0	1 1
1 0 0	0 1
0 1 0	1 0
1 1 0	0 0
0 0 1	1 1
1 0 1	1 1
0 1 1	1 1
1 1 1	1 1

(b)

(a)

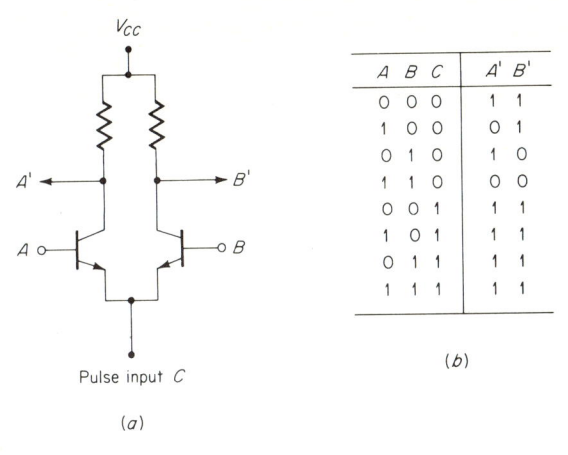

Fig. 16-26. (a) Transistor pair used to couple a pulse to the output for certain input conditions; (b) truth table.

not quite true since if one considers the voltage levels involved, the emitters and collectors are at V_{CC}. The transistors are cut off and hence are, in effect, out of the circuit. It is this condition that allows the circuit to act as it does. This must be explained in terms of the overall flip-flop circuit, and a further discussion will be deferred.

The final preliminary circuit to be investigated, shown in Fig. 16-27, has two inputs and two outputs. Again, the truth table gives the details of circuit operation, where it is seen that the X output yields the NOR function while the Y output yields the OR function. In terms of voltage levels, any high at any input gives a logic 0 (low) at X and a logic 1

A B	X Y
0 0	1 0
1 0	0 1
0 1	0 1
1 1	0 1

(b)

(a)

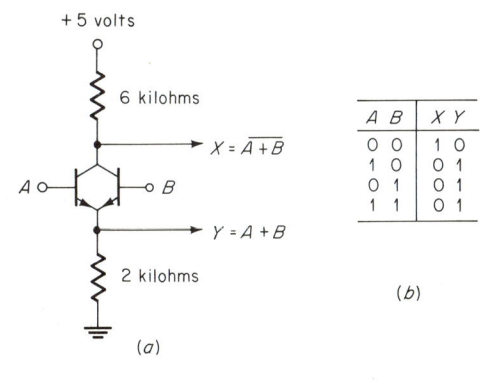

Fig. 16-27. (a) NOR/OR circuit; (b) truth table.

(high) at Y. If both inputs are low, the output at X is logic 1 (high), and at output Y there appears a logic 0 (low). A significant feature of this circuit is the actual voltage levels at either output in relation to V_{CC}. The voltage at X must operate between the levels of 5 and 1.25 volts. Also, at Y the levels must be ground and 1.25 volts, using the circuit values shown. That is, high for the Y output is 1.25 volts, not 5 volts, and low for the X output is 1.25 volts, not ground. These figures result from the voltage-divider action of the 6- and 2-kilohm resistors. In the actual flip-flop circuit these values are further modified somewhat by the action of other semiconductors.

The functional diagram of the entire flip-flop is given in Fig. 16-28a, while the internal action that occurs as a clock pulse arrives is detailed in Fig. 16-28b. First, note on the block diagram that the K input (clear) is internally gated with certain other signals generated within the flip-flop itself. This gives rise to some of the distinctive action of the circuit. Also, the J input (set) is similarly gated. The output of these gates feeds two cross-coupled NOR gates, which act as the master flip-flop. These, in turn, are connected through the coupling transistors to the

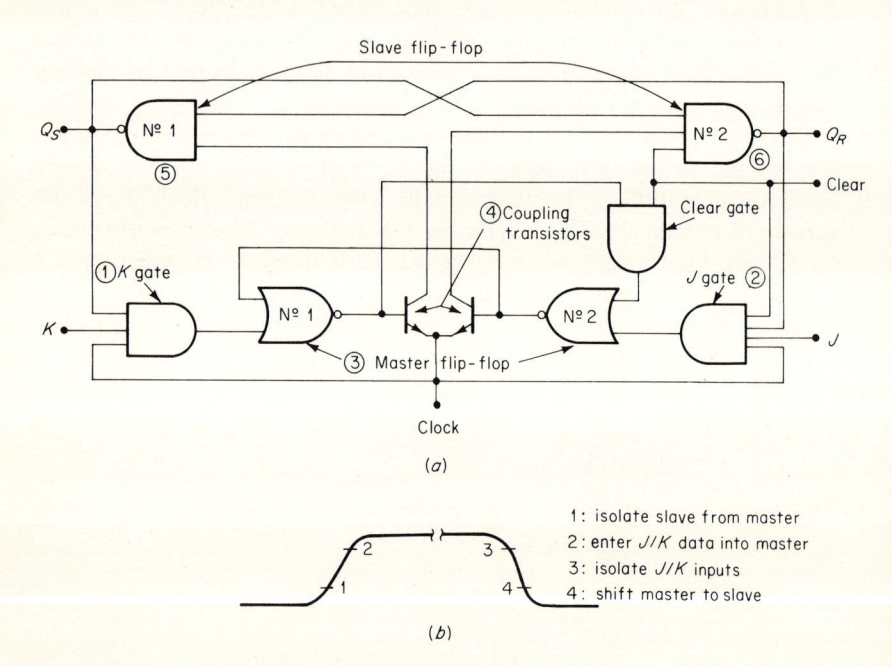

1: isolate slave from master
2: enter J/K data into master
3: isolate J/K inputs
4: shift master to slave

Fig. 16-28. (a) Functional block diagram of the master-slave JK flip-flop; (b) internal action when clock pulse goes high.

slave flip-flop, which, when set or reset, provides the output data on the Q_S and Q_R lines. One other input, labeled "clear," allows the circuit to be reset no matter what the condition at any other input.

Very basically, the circuit operates as follows. The inputs J and K are the logic inputs, and the information on these lines will ultimately be transferred to the output lines. If both J and K are low, the clock pulse will have no influence on the flip-flop and it will remain at its original state. If J is high and K is low, the unit will become set when the clock pulse is applied. If J is low and K is high, the flip-flop will become reset on the clock pulse. If either of these actions is to occur, the clear input must be high at all times. A low at this input will reset (clear) the flip-flop. One other mode of operation is possible, and this occurs when both J and K are high when the clock pulse arrives. In this case the flip-flop toggles and will reverse its state regardless of its original condition.

To investigate the circuit action a bit more closely, Fig. 16-28*b* illustrates how the clock pulse affects the various parts of the internal circuitry. Initially, the clock pulse is low. As it rises in the positive direction by about 1 volt, it disconnects the slave from the master. Then, as the input waveform rises further to about 4 volts, the master is allowed to accept data from either the J or K inputs. If at this time the J input is high (and K is low), the master flip-flop becomes set but cannot influence the slave as yet; the output remains as it was originally. If, on the other hand, the K input is high while J is low, the master becomes reset. In the period between points 2 and 3 nothing further happens.

At point 3 on the waveform the J and K inputs are disconnected from the master. From this point on, the J and K inputs can in no way influence the flip-flop. Finally, at point 4 the information (data) in the master is transferred, or shifted, into the slave unit. As soon as this occurs, the information is evident at the output of the circuit Q_S and Q_R. The clock pulse then goes to its most negative value (probably ground), and this concludes circuit action for this clock pulse.

The actual internal circuitry is shown in Fig. 16-29. Note the encircled areas; these represent the same numbered areas as in Fig. 16-28. Area 1 is the K gate, while area 2 is the J gate. The two groups labeled 3 comprise the master flip-flop, while 4 is the coupling unit. Number 5 is the set side of the slave flip-flop, while 6 is the reset side. Each function of the block diagram can similarly be correlated with the schematic diagram.

Proceeding now to the internal circuit action, note Fig. 16-30 (see color section). This is the functional diagram that shows the high-low states of all important connections in the flip-flop. Once this drawing is under-

436 PULSE AND SWITCHING CIRCUIT ACTION

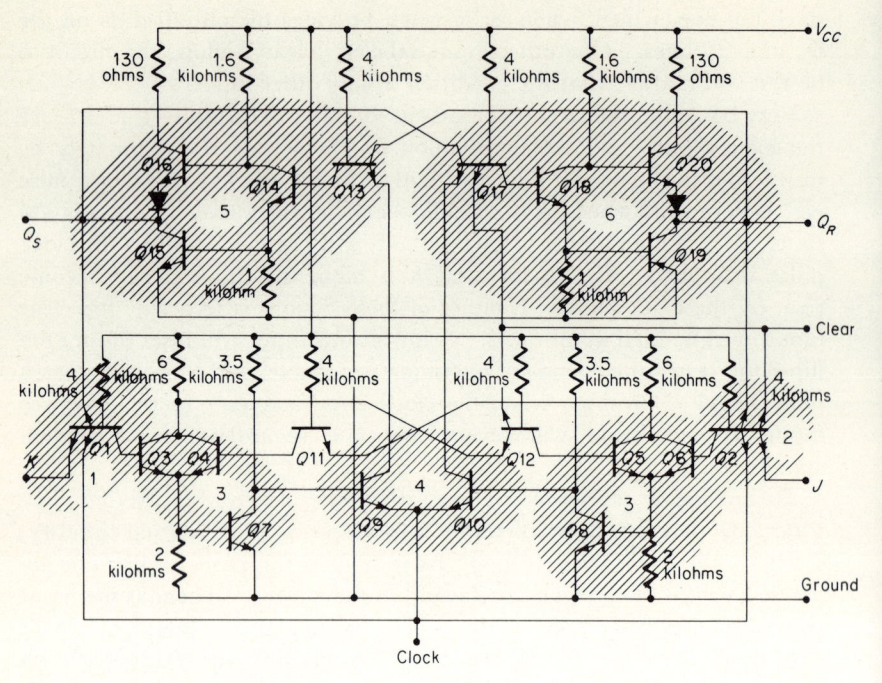

Fig. 16-29. Schematic diagram showing component functions of master-slave flip-flop.

stood, it is but one step further to understand the actual circuit functions, as given in Fig. 16-31 (see color section), the schematic diagram that also gives the logic states of the various parts. In these two figures (and several to follow), the colors used have great significance. The red lines represent the high state (more positive), while the green lines represent the low state. This approach precludes having to memorize the condition of each wire when analyzing the circuit action. It then becomes a simple matter to mentally make one change in color on the figure to see how the circuit reacts to a given set of input conditions. Figures 16-30 and 16-31 show the flip-flop in the reset condition.

Comparing Figs. 16-30 and 16-31, also note that the transistors are identified, so that each circuit may be compared with the other as to transistor location and function. First, note that the clock-input line is low in Fig. 16-31. This line connects to the emitters of $Q9$ and $Q10$, as well as one emitter of $Q1$ and one emitter of $Q2$. $Q2$, the J-input gate, has three of its four emitters at the high state, while only one is at low (the clock line). Since $Q2$ is effectively a negative OR gate, its collector must also be low, as shown by the green line. It will later be confirmed that the base of $Q5$ is low, and hence the bases of both $Q5$ and $Q6$ are low, as are the two emitters. The collectors, then, must be high. The line from these collectors feeds this high level through $Q11$ to the base of $Q4$, resulting in a high at the emitter of $Q4$. The collector of $Q7$ is therefore low. The collector of $Q9$ is high and that of $Q10$ is low.

These conditions, along with the Q_S and Q_R conditions shown, represent the static state of the flip-flop when reset. Now, to help visualize the circuit action as the clock pulse goes to logic 1, assume that the clock line is now colored red. At transistor $Q1$, nothing can occur since only one input is high. But at $Q2$, the clock pulse causes the final input to go high, and therefore the collector of $Q2$ also goes high. This causes the collectors of $Q5$ and $Q6$ to go low, which, in turn, causes the base of $Q4$ to go low and so also the base of $Q7$. At the same time, the collector of $Q4$ goes high. This is one collector of the master flip-flop, and hence the master flip-flop is now in the set condition. The collectors of $Q3$ and $Q4$ are high, while those of $Q5$ and $Q6$ are low. Note that this is the opposite condition to the one originally shown; the master was reset but is now set.

This action occurs at point 2 of the input-clock waveform. However, with the clock pulse high, transistors $Q9$ and $Q10$ are not conducting, and hence the state of the master cannot yet affect the slave unit. As the clock waveform begins to approach point 4, $Q9$ and $Q10$ begin to conduct, and since the master flip-flop is now set, the base voltages on $Q9$ and $Q10$ are reversed to that shown. Therefore, the collector of $Q9$ starts to go low along with the clock pulse. This causes the bases of $Q14$ and $Q15$ to go low, which starts the slave flip-flop into a transition. As the clock waveform passed point 3, the J input was disconnected since the clock voltage reverse-biases all other emitters at $Q2$. Q_S is now high and Q_R is now low, which represents the set condition for the entire flip-flop.

The set condition is shown in Figs. 16-32 and 16-33 (see color section). Since the K input is shown as being in the high state, the flip-flop will reset on the next clock pulse. It may be instructive for the reader to mentally apply a clock pulse and note the circuit action, thus becoming more familiar with circuit response.

At this point it is appropriate to mention the action that takes place when the clear input is caused to go low. As will be seen, the flip-flop will become reset as soon as the clear input goes low, regardless of the condition of the other inputs. Using Fig. 16-33 as an example, when clear goes low, $Q17$ has an emitter that is tied to the clear line and hence the output of this transistor goes low. This results in the output transistor

Q19 going high, which, of course, transfers the slave flip-flop immediately; the output wires now reflect the reset condition, to the exclusion of *any* other input.

A somewhat obscure point regarding this type of flip-flop is illustrated in Figs. 16-34 and 16-35 (see color section). In some applications, it is desirable to have the clock input remain at a high level. If, during this time, either the J or K input goes high, even for a brief period, the master flip-flop will become set or reset. Then, when the clock pulse goes toward ground, the master information is transferred to the slave unit and the entire flip-flop is in the state indicated by the input that went high.

The circuit action for this set of input conditions is easily described by referring to Fig. 16-35. As illustrated, the clock input is high and the J input goes from low to high to low. Since the master flip-flop is reset, this will cause a transition to the opposite state on the proper part of the J pulse. As the J input goes from low to high, the collector of $Q2$ goes high also, and this is the action that sets the master unit. Hence, in this mode, the J input is performing the same action that the clock pulse does, with the same effect.

Finally, it was mentioned that if both the J and K inputs are high at the same time and a clock pulse is applied, the flip-flop will toggle. The significant action in this case concerns $Q1$, $Q2$, $Q13$, and $Q17$, as well as the master flip-flop. If the master flip-flop is reset, the clock will cause it to become set, and on the trailing edge of the pulse this will be transferred to the slave unit. If the entire unit was originally set, the master unit will become reset, and either $Q1$ and $Q13$ or $Q2$ and $Q17$ will initiate the toggle on the trailing edge of the clock pulse. Thus, with J and K high, the flip-flop reverses its condition on the clock.

D-type Flip-flop

The logic diagram for the *D*-type flip-flop is shown in Fig. 16-36. The operation of this unit is somewhat different than that of a JK flip-flop in that it will not toggle in the usual sense. This particular unit is the SN7474N (Texas Instruments) and as produced has two separate flip-flops in each package. It is mounted in a 14-pin plastic package, similar to the left-hand unit shown in the bottom row in Fig. 16-1.

Operationally, the unit has the usual flip-flop characteristics, including complementary outputs Q_S and Q_R. The preset and clear inputs are independent of the others. A low signal (logic 0 for positive logic) at either input will cause the named action to occur. Both inputs must be high to allow the clock and D inputs to function. If the D input is low, the flip-flop will reset on the positive-going edge of the clock pulse. If, of

COLOR SECTION

Figures 16-30 through 16-35 and 16-38 through 16-41

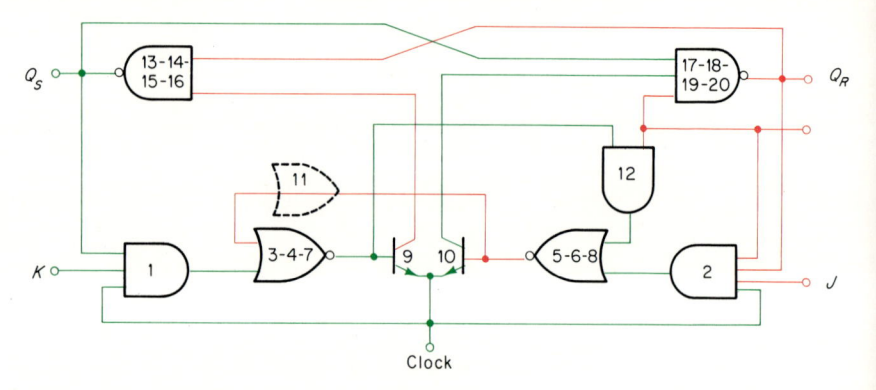

Fig. 16-30. Functional diagram showing the master-slave flip-flop in the reset condition. Red is high (logic 1), and green is low (logic 0).

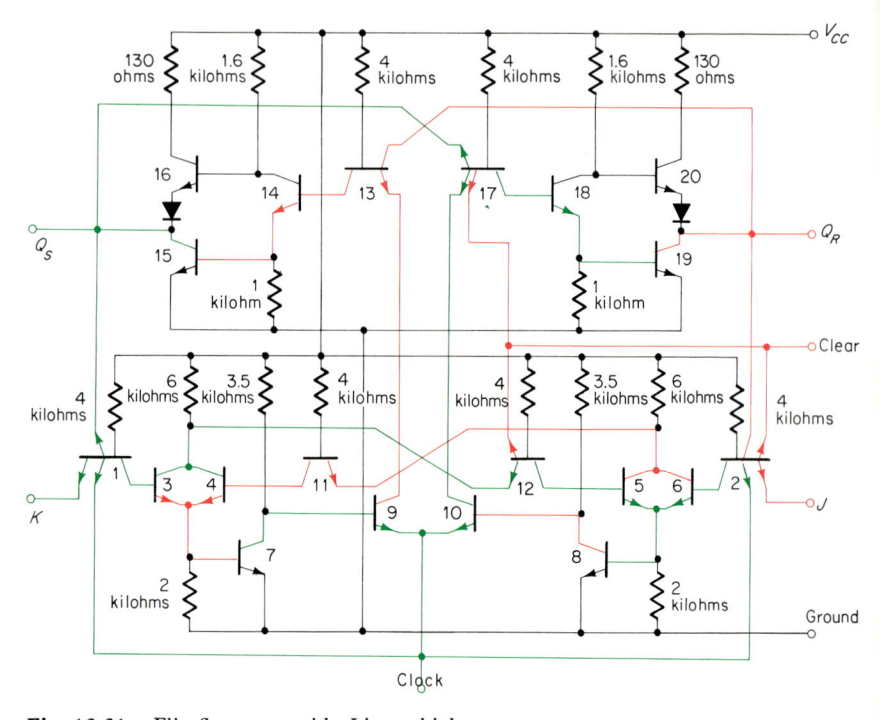

Fig. 16-31. Flip-flop reset, with J input high.

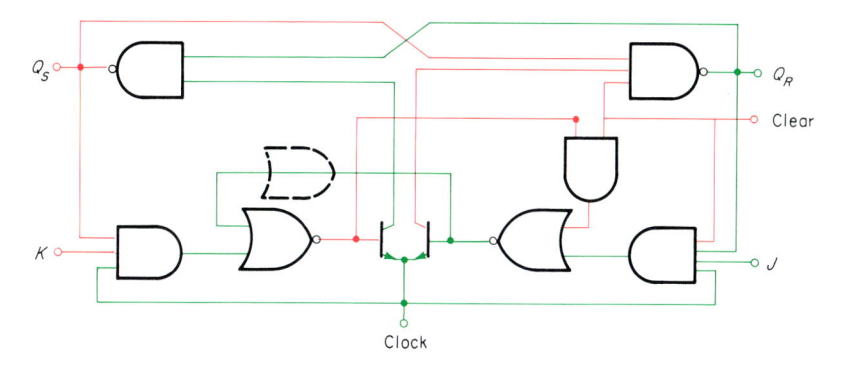

Fig. 16-32. Flip-flop set with K input high.

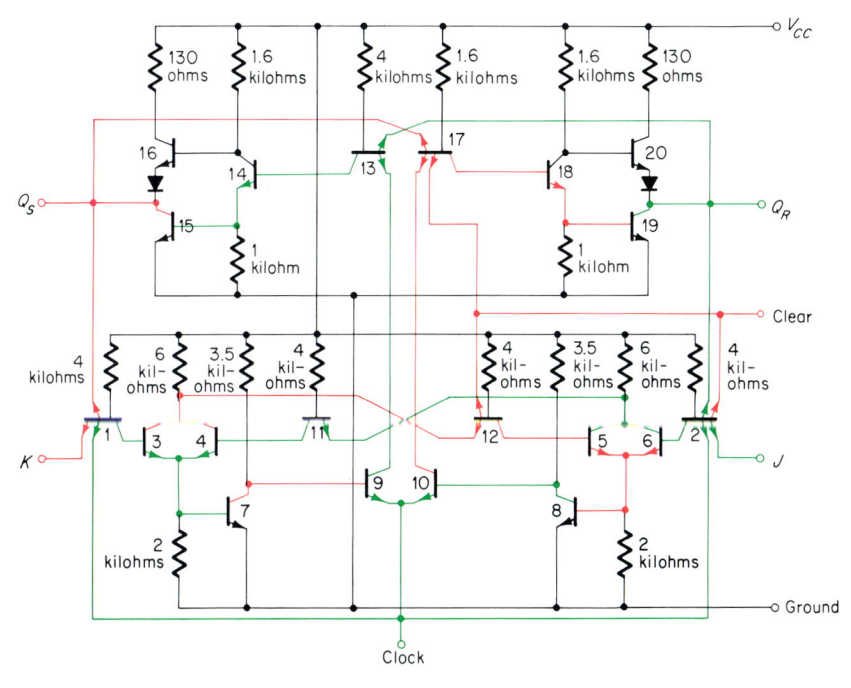

Fig. 16-33. Set condition illustrated by schematic diagram. (*Courtesy Texas Instruments.*)

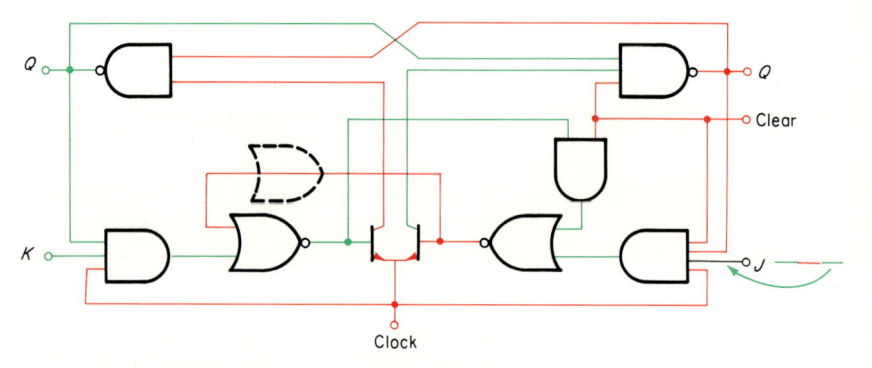

Fig. 16-34. Master section set by brief logic 1 at the J input while clock is high. (*Courtesy Texas Instruments.*)

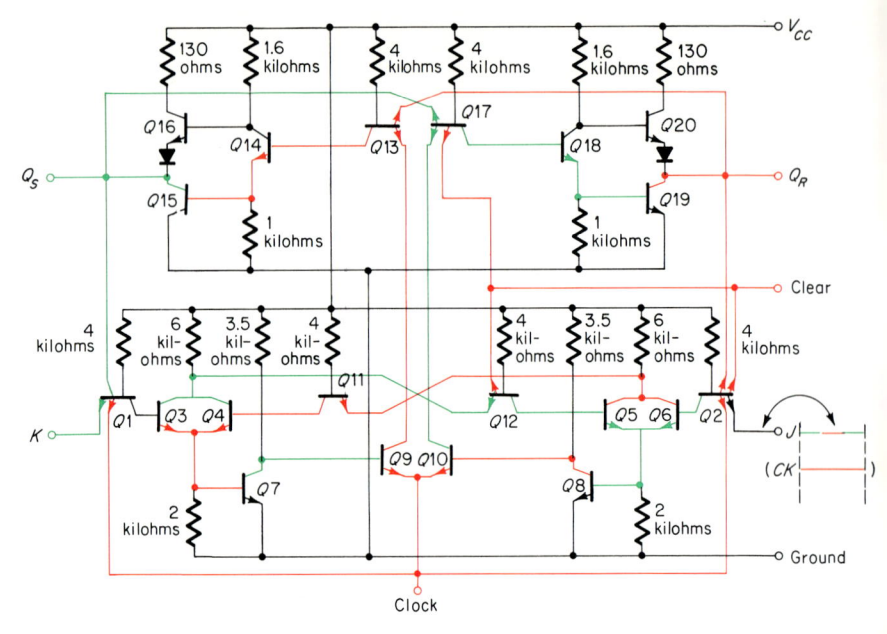

Fig. 16-35. Internal action if clock is high and J input goes briefly high. (*Courtesy Texas Instruments.*)

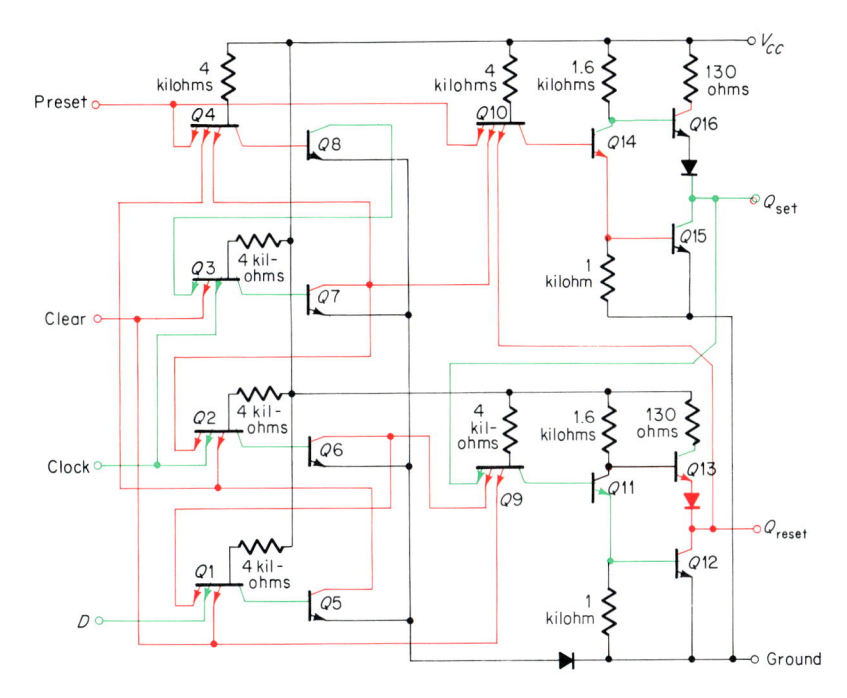

Fig. 16-38. *D*-type flip-flop showing static levels for the reset condition. *D* input is low, and when the clock goes high, nothing occurs since the unit is already reset (red is high, green is low). (*Courtesy Texas Instruments.*)

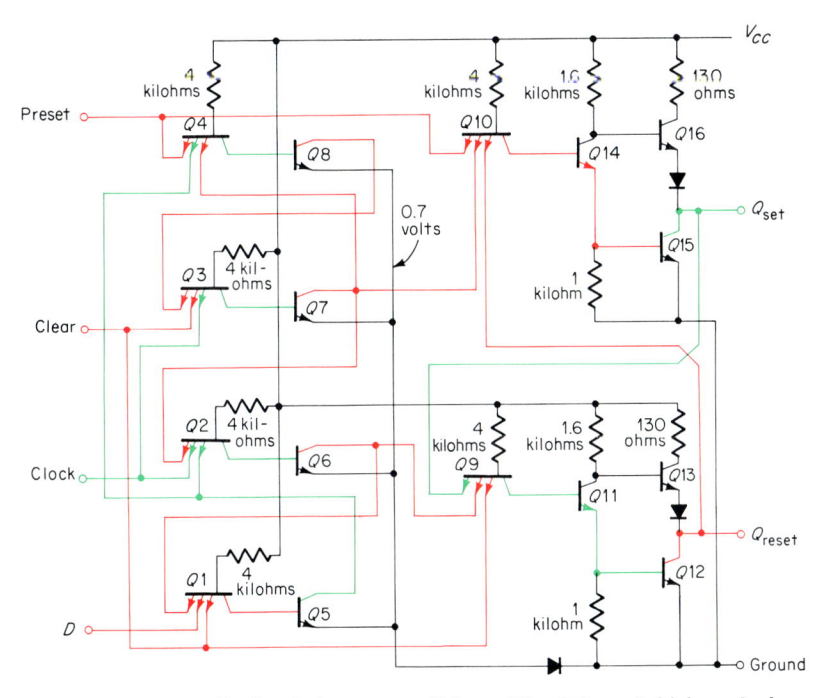

Fig. 16-39. *D*-type flip-flop in its reset condition. The *D* input is high, and when the clock goes high, the flip-flop will become set. (*Courtesy Texas Instruments.*)

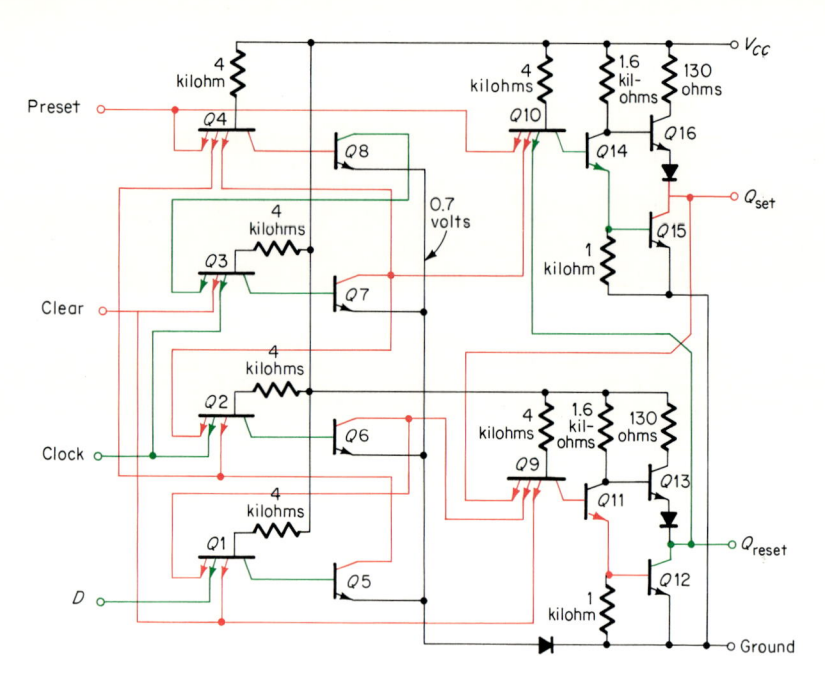

Fig. 16-40. Flip-flop is set and the *D* input is low. When the clock goes high, the flip-flop will be reset. (*Courtesy Texas Instruments.*)

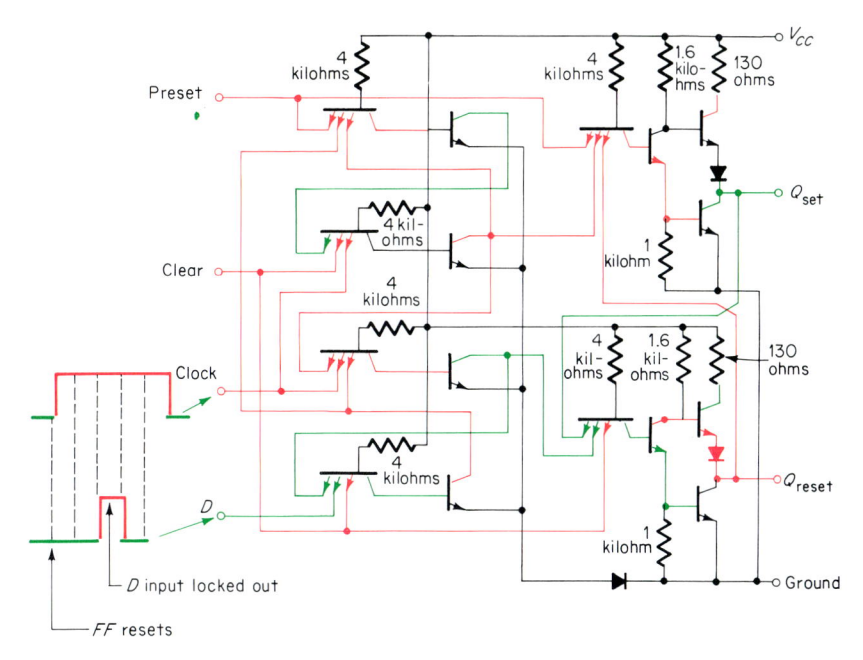

Fig. 16-41. When the D input is low and clock goes high, the flip-flop resets. If then the D input goes high for an instant, nothing occurs since the D input is then locked out.

Preset

Clock

D input

Clear

D-type flip-flop

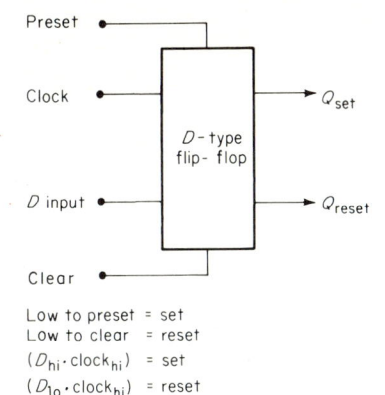

Q_{set}

Q_{reset}

Low to preset = set
Low to clear = reset
($D_{hi} \cdot clock_{hi}$) = set
($D_{lo} \cdot clock_{hi}$) = reset

Fig. 16-36. Logic diagram for the D-type flip-flop.

course, the flip-flop is already reset, no action occurs. If the D input is high, the flip-flop will become set on the clock pulse. This type of unit, then, is seen to be a direct-coupled flip-flop, where action occurs on the *leading-edge* of the clock pulse.

To describe the internal action in more detail, Figs. 16-37 and 16-38 (see Fig. 16-38 in color section) are offered. The first of these is a functional block diagram that delineates the various circuit functions. The

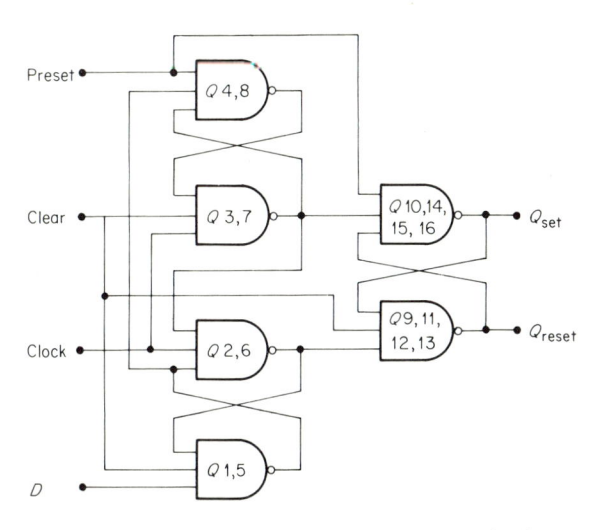

Fig. 16-37. Functional diagram of the D-type flip-flop, with Q designations correlating with the schematic diagram. (*Courtesy Texas Instruments.*)

gates are named according to the transistors that take part in the formation of the internal flip-flops. In effect, there are three flip-flops, formed by cross-coupling the NAND gates. Figure 16-38 shows the actual schematic diagram of one complete flip-flop circuit. Note that the logic condition of each connection is called out by the color of the line; as before, red denotes high ($\cong 5$ volts) and green denotes low ($\cong 0$ volts, or ground).

In this figure, the flip-flop is represented as being in the reset condition, with Q_S low and Q_R high. Preset and clear are both high to allow the logic inputs (clock and D) to function. With the flip-flop in this condition, a clock pulse at the input will cause no action since the unit is already reset. However, if the D input were to become high, shown in Fig. 16-39 (see color section), the emitters of $Q1$ would all be high, and therefore its collector would also be high. A high at the base of $Q5$ would yield a low at its collector, which is one of the inputs for $Q4$. Hence, the base of $Q8$ goes low and the collector goes high. This qualifies one leg of the $Q3$ unit, leaving only one leg still to be qualified.

Now, when the clock pulse goes high, note that the other leg of $Q3$ also goes high. For the first time, all emitters of $Q3$ are high, and thus its collector also goes high. This results in a low out of $Q7$. With one emitter of $Q10$ going low, its collector goes low, which drives the collector of $Q15$ high. $Q15$ is cross-coupled to $Q9$, and hence the flip-flop makes a transition to the opposite state and becomes set.

Another possible condition is with the flip-flop set and the D input low. With the D input low, the next clock pulse will cause the flip-flop to reset on the positive-going edge of the pulse. Figure 16-40 (see color section) illustrates the high and low levels set by this condition. The high transition at the clock input causes the emitters of $Q2$ to become high together, and hence the collector now goes high. This results in a low from $Q6$, and the emitter of $Q9$ to which it is connected goes low. This, of course, initiates a transition by driving $Q12$ high, which, in turn, drives $Q15$ low. This is the reset condition, which is specified by the D input being low.

Figure 16-41 (see color section) illustrates the case where the clock pulse remains at a high level for some period of time. During this time, the D input goes from low to high and back again. In this instance, the circuit action is first to cause the flip-flop to reset since when the clock pulse went high, the then-present condition of the D input was low. As soon as the clock goes high, the D input is locked out and can no longer influence the circuit. Hence, the positive-going pulse at D can in no way affect the flip-flop.

By carefully studying these diagrams, the reader can much more readily understand the circuit action of this and similar circuits.

16-5 LSI AND MSI

Large-scale integration (LSI) and medium-scale integration (MSI) are natural outgrowths of the IC as it is currently used. At present, LSI is encountered only occasionally since it has but recently been introduced. MSI, however, has been attained for some time, and is being used in commercial equipment very successfully. In the case of the former, it may be expected that before too long the logic of a complete small computer will be constructed on a single monolithic chip. Several companies are now working on such projects, with varying degrees of success.

By using conventional methods of silicon epitaxial monolithic construction, relatively large amounts of circuitry can be designed into a single chip. The present discussion, then, will be concerned with just such a device. Figure 16-42 illustrates a 5-bit Johnson-code counter that replaces a great amount of discrete-type circuitry. The entire function is mounted in a 24-pin container (Fig. 16-44c) that contains a total of five flip-flops, interconnected as a Johnson counter, plus many gates to perform a variety of functions. The circuit additionally has the ability to act as a shift register and will accept a 5-bit input on a shift pulse that will set in a coded number from another source. The dimensions of the overall package are only $1\frac{1}{4} \times 0.45$ in.

Recall the Johnson code for this type of counter, shown in Fig. 16-44b, which should be referred to as the discussion progresses. Again directing our attention to Fig. 16-42, note first the set and reset inputs for each flip-flop. The inputs for flip-flop 5 come from flip-flop 4; the inputs for flip-flop 4 come from flip-flop 3; etc. (To preserve clarity, the actual connections are not shown but are simply indicated by signal name.) This, of course, suggests that this is a ring counter since the output of each flip-flop is directed to the input of the next flip-flop.

Assume that the counter is reset and therefore the set sides of each flip-flop are at logic 0. The signal that will begin to cause counting originates at terminal 4, labeled $\overline{ADV\,A}$ (ADV = advance). A pulse at this terminal will cause the counter to advance according to the truth table for this code. The first action that occurs on the first pulse is that the $A1$ flip-flop goes from reset to set. The counter is now in the condition specified by the code 0 0 0 0 1 (LSD to the right). The next pulse causes the counter to advance one more step: 0 0 0 1 1. This condition of course represents the state of 2. The third input pulse at terminal 4 causes the counter to advance to 0 0 1 1 1, which represents the 3d state. This action continues until the counter has advanced through all 10 separate states, after which it repeats the above sequence if ADV pulses are still present.

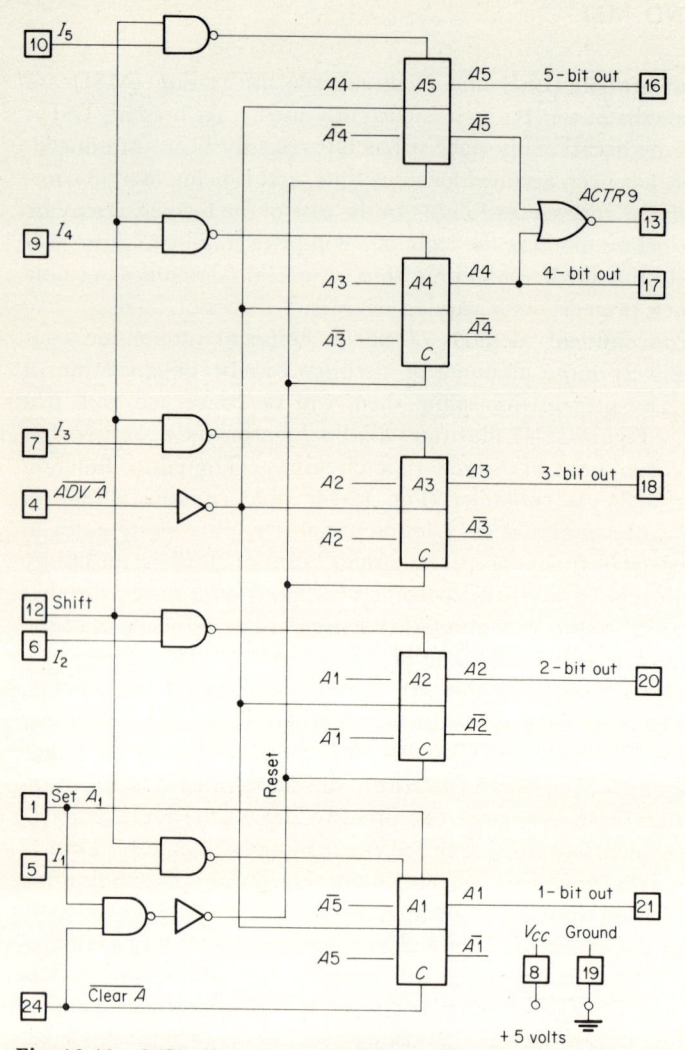

Fig. 16-42. MSI chip, where a multitude of functions appear on a single IC chip. (*Courtesy Friden Div., Singer.*)

The $\overline{ADV\ A}$ signal is applied to all flip-flops simultaneously, and it may not be immediately evident how the first flip-flop is caused to become set on the first input pulse and reset on the sixth pulse. Note flip-flops 5 to 2. The set-side inputs are all fed from the set-side outputs of the next-lower flip-flop. That is, the set-side input for $A5$ is the set-side output from $A4$, etc. The $A1$ flip-flop, however, has the reverse conditions evident at its

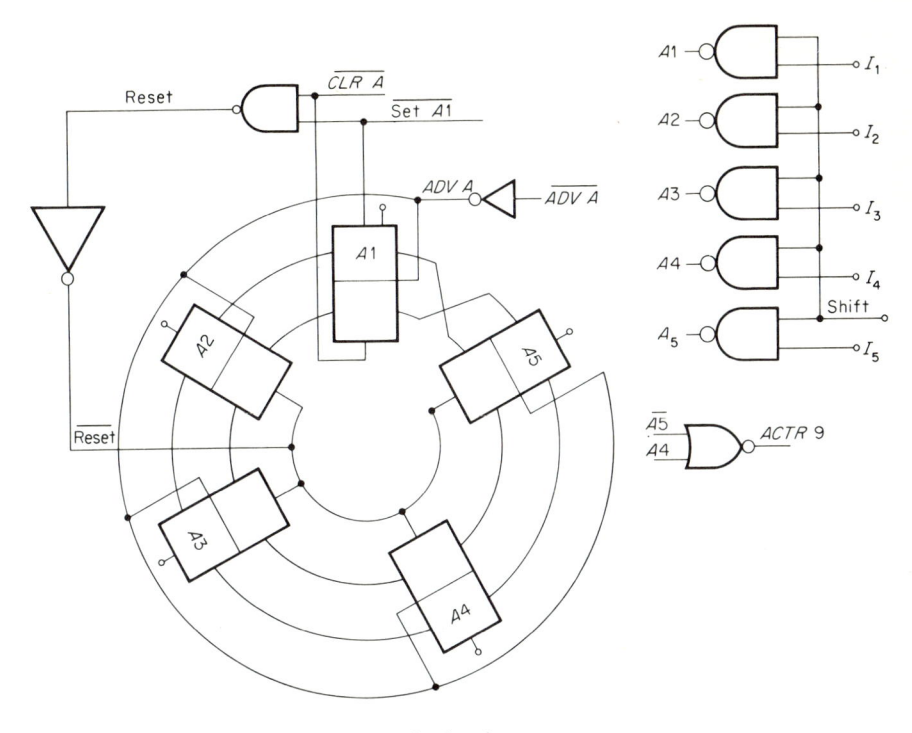

Fig. 16-43. MSI chip redrawn to emphasize the ring-counter function. (*Courtesy Friden Div., Singer.*)

input; the set-side input for $A1$ comes from $\overline{A5}$, and the reset-side input comes from the set-side output of $A5$. Recall an earlier discussion of a similar circuit where the output leads were transposed.

Note Fig. 16-43, which is another way of showing this counter. Here, the switching of the output leads of $A5$ is clearly evident, as is the fact that this is truly a ring counter. In terms of logic, if all flip-flops are initially reset (logic 0), then the input of the $A1$ flip-flop has dc levels upon it that represent a logic 1 because of the switched leads at this point. By themselves, these dc levels cannot influence the $A1$ flip-flop, but when the $\overline{ADV\ A}$ signal arrives, they will be shifted into the flip-flop. However, the $A1$ flip-flop was in the zero state, and so a 0 is shifted into the $A2$ flip-flop, and so on down the line. The next shift pulse ($\overline{ADV\ A}$) will shift another 1 into the $A1$ flip-flop and at the same time will shift the 1 from the $A1$ unit to $A2$. Now, flip-flops $A1$ and $A2$ have a 1 in them, while all others contain a 0. The logic 1 continues shifting down the line until all flip-flops have a logic 1 in them. This is the 5th state of the counter. Now, for the first time, flip-flop 5 is at logic 1, and hence its output wires reflect this new condition. The sixth shift pulse causes the logic 1 in the $A5$ flip-flop to appear as a logic 0 at the input to the $A1$ flip-flop, and hence $A1$ resets as shown by the truth table. Then, this logic 0 is shifted down the line by succeeding shift pulses, eventually returning the counter to its reset condition.

On this chip, provisions are made to clear the counter ($\overline{CLR\ A}$) and to set the $A1$ flip-flop ($\overline{SET\ A1}$). Also, it is desirable to know when the counter has attained the 9th state; the NOR gate, with $\overline{A5}$ and $A4$ as inputs, will signify this fact.

A simplified drawing of such a device is often used in large, complex logic schematics, shown in Fig. 16-44a. Of course, one must know the internal function of this kind of a device if such a simplified symbol is used, for none of the functions are readily discernible.

Truth table

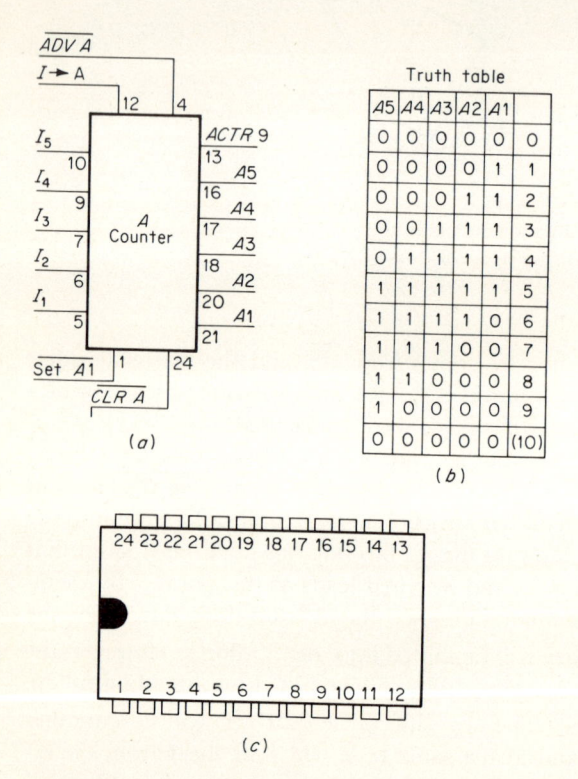

A5	A4	A3	A2	A1	
0	0	0	0	0	0
0	0	0	0	1	1
0	0	0	1	1	2
0	0	1	1	1	3
0	1	1	1	1	4
1	1	1	1	1	5
1	1	1	1	0	6
1	1	1	0	0	7
1	1	0	0	0	8
1	0	0	0	0	9
0	0	0	0	0	(10)

Fig. 16-44. (a) Logic diagram, (b) truth table, and (c) pin connection for the MSI chip.

Each of the flip-flops in this example of MSI is a complete JK master-slave flip-flop, and these, along with the associated gates, will allow one to appreciate the relative complexity of the overall unit. Approximately 135 transistors comprise the five flip-flops and the several gating networks.

16-6 MOS INTEGRATED CIRCUITS

The field-effect transistor is particularly adaptable to the techniques of IC's. There are two major types of FET's, the junction FET and the insulated- (isolated-) gate FET. The latter is variously known as the *metal-oxide semiconductor* (MOS), the *metal-oxide–silicon transistor* (MOST), or the *metal-oxide–silicon field-effect transistor* (MOSFET). Regardless of the title, these are all the same device. In order to understand how a MOS integrated circuit functions and how it is applied, one must first investigate the junction FET. While the FET was described in Chap. 15, the following brief discussion is given for review purposes, with especial emphasis upon the integrated-circuit construction.

The structure of a typical field-effect transistor is shown in Fig. 16-45a. The device operates upon a principle somewhat different than that of a bipolar transistor. The FET conducts electron current from the source to the drain by very conventional means. The region is doped as shown so as to exhibit n-type characteristics. Current must flow through the region called the *channel*, which is made to be very narrow (p-type channels are also used). Hence, with a voltage applied across the source and drain, a current will flow, limited only by the source-to-drain resistance of this region. That is, with the gate lead left open, the region between the source and drain acts like a very simple resistor, with a value from a few thousand to, perhaps, several hundred thousand ohms. Note that since there has been no mention of pn junctions, this current will flow as readily in one direction as the other. If the source is made negative and the drain is made positive, electrons will flow from the source to the drain, while if the reverse polarity is applied, current will flow in the opposite direction.

Now, to cause the device to behave like a transistor, a voltage must be applied to the gate as well as to the source-drain circuit. The device illustrated is an n-channel transistor and would normally be operated with the drain at the most positive voltage. The source, then, is grounded. To reduce the drain current to a reasonable value, the gate is operated at a value that will reverse-bias the gate junction. Thus, the gate is operated at a value of voltage more negative than ground. This causes the depletion regions that surround the junctions to increase in a direction such as

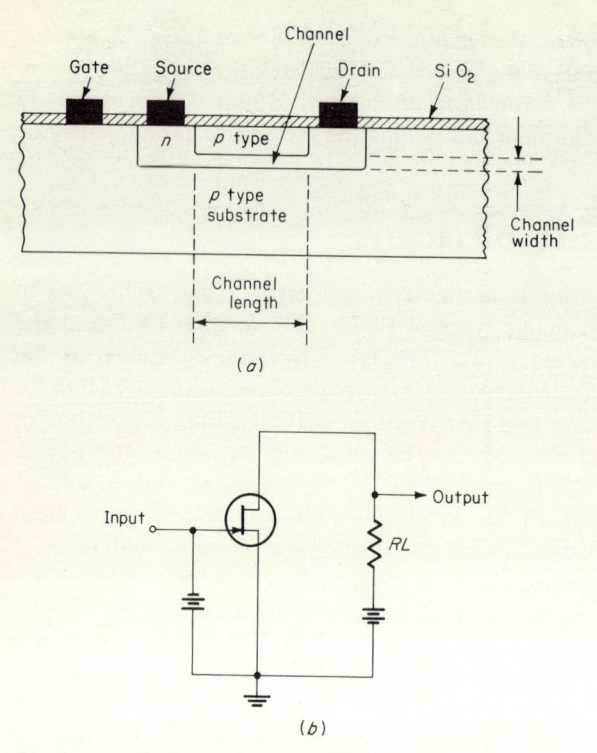

Fig. 16-45. (a) Structure of the depletion-mode junction FET; (b) basic amplifier circuit.

to reduce the channel area, thus restricting current flow. The more negative the gate is made, the smaller the drain current. Because the gate is reverse-biased, virtually no gate current flows; the voltage applied here looks into a very high resistance. Typical values of input resistance might be from 10^6 to 10^7 ohms.

Figure 16-45b also shows the basic circuit in which a typical n-channel device might appear. Shown are the bias voltages which must be applied for proper amplifier operation. The major characteristics of a circuit such as this are high input impedance, high voltage gain, virtually infinite current gain, and moderately high output impedance.

The insulated- (or isolated-) gate FET is illustrated in Fig. 16-46. In this device, the conduction is accomplished somewhat differently than before. Here, a conducting channel is not present until a voltage is presented to the gate of sufficient magnitude to induce a concentration of charge carriers between the two p regions. As before, the two electrode regions are called the *drain* and the *source* and they consist of *p*-type regions diffused into the *n*-type substrate. The two regions are separated by a distance of perhaps three-hundred millionths of an inch. The source and drain, together with the substrate, form two junctions (as shown) that produce the same effect between source and drain as two back-to-back diodes. Hence, with no gate voltage applied, no current can flow from source to drain, regardless of the applied polarity.

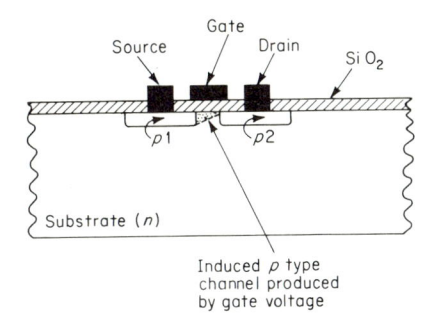

Fig. 16-46. Insulated-gate FET, enhancement-mode type.

Note that the gate electrode is insulated from the other elements on the wafer. There is no direct connection between the gate lead and any other part of the device. Hence, it would appear that the gate can in no way influence the action of the FET. Such, however, is not the case. If the gate is made sufficiently negative with respect to the source, holes are attracted to the surface of the *n*-type region under the gate electrode and cause it to change to *p*-type. A *p*-type region joins the other *p* regions, and current flows from source to drain, the magnitude of which is a direct function of the gate voltage. Operation of a device as described above is called the *enhancement-mode* since with no gate voltage applied no drain current flows; the forward bias applied to the base is said to *enhance* the concentration of carriers in the channel. For many such devices, the gate voltage must be at least 4 volts in amplitude to begin conduction, and they are therefore known as *normally-off* devices. Note that the forward bias applied to the gate of this type of transistor does not cause conduction in the gate circuit since the gate is insulated from the channel by the silicon dioxide.

A third type of FET, the *depletion-enhancement* type, is illustrated in Fig. 16-47. It can operate in either of the two modes of operation, and it differs from the other types in that there is a very thin channel diffused just beneath the dioxide layer. Hence, for the *p*-type device illustrated, a negative voltage at the gate will enhance the channel carriers while a

448 PULSE AND SWITCHING CIRCUIT ACTION

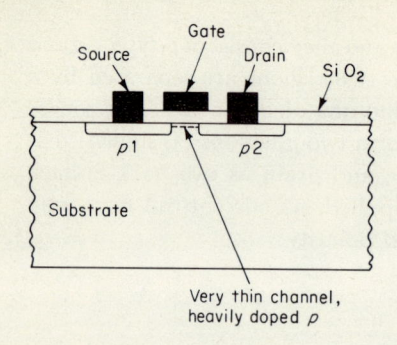

Fig. 16-47. Depletion-enhancement MOSFET.

positive voltage will deplete the carriers. Such a device is normally on with no applied gate bias.

The schematic symbols for the various kinds of FET's are shown in Fig. 16-48. A basic amplifier circuit for a junction FET is given in Fig. 16-49. The drain-supply voltage is applied through the load resistor to the transistor, as shown. The gate voltage is supplied as shown, and a signal at the input will be evident at the output, which is an amplified replica of the input but inverted. This circuit is shown in the common-source configuration, which is similar to the common-emitter configuration of the ordinary bipolar transistor. Also possible are the common-

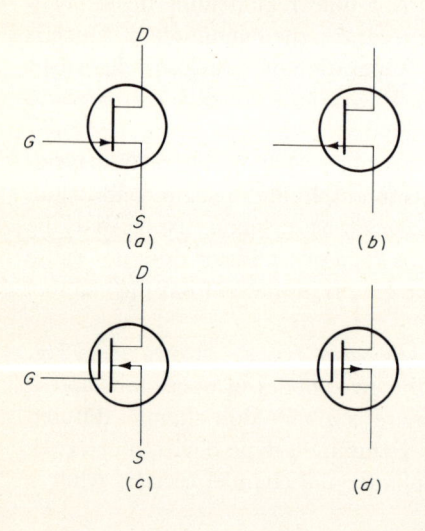

Fig. 16-48. Schematic symbols for FET's. (a) n-channel JFET; (b) p-channel JFET; (c) n-channel IGFET; (d) p-channel IGFET.

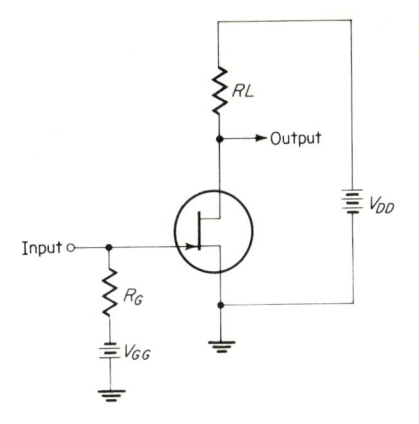

Fig. 16-49. Basic JFET amplifier circuit.

drain and common-gate circuits, which are approximately equivalent to the common-collector and common-base circuits, respectively.

A somewhat more complex circuit is shown in Fig. 16-50. This is a typical preamplifier application for audiofrequencies. Note that the transistors are biased by what might be called *vacuum-tube techniques* since these are normally-on devices. The drop across the source resistor R_S produces the bias for the device. Except for this difference, and the high input impedance, the circuit works much like any transistor amplifier for small-signal linear applications.

In the digital field, MOSFET's are used to a limited degree for nearly any logic application. Gating circuits, as well as flip-flops and counting circuits, are beginning to be used in several commercial applications. One advantage is the cost factor; MOS techniques cost measurably less

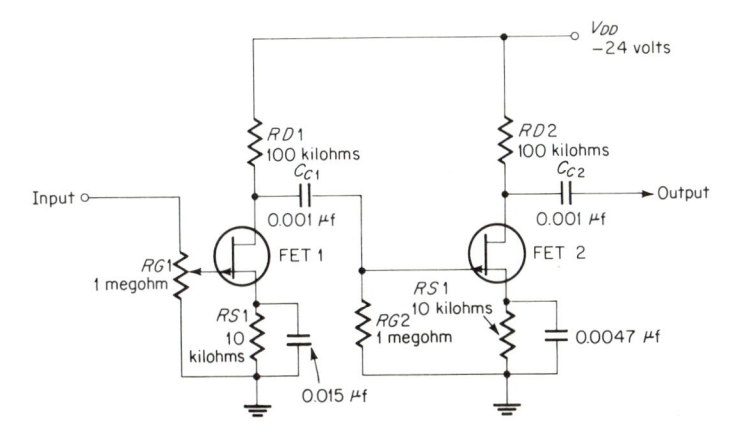

Fig. 16-50. Typical *p*-channel FET amplifier circuit.

450 PULSE AND SWITCHING CIRCUIT ACTION

than bipolar transistors to produce the same functions. One reason for this is that the MOS units occupy less room on the integrated chip. Also, there are fewer manufacturing steps involved, and hence the cost per function is smaller. It is not expected that the MOS technology will supplant bipolar units in large, high-speed equipment since their speed of response is at present somewhat slower than the bipolar equivalents. Nevertheless, the units are being used in ever-increasing numbers where high speed is not a major requirement.

To begin to understand how the MOSFET is used in digital applications, a simple, although typical, circuit is shown in Fig. 16-51. The

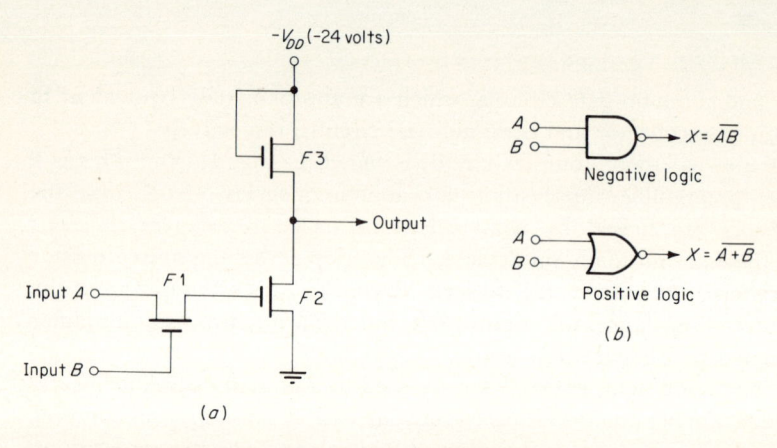

Fig. 16-51. (a) Typical circuit used in a MOSFET integrated circuit; (b) equivalent logic symbols. $F1$, $F2$, and $F3$ are p-channel, enhancement-mode MOSFET's, and the simplified schematic symbol is used to reduce drafting time on large, complex drawings.

circuit is composed of three MOSFET's, and each is a p-channel enhancement-mode transistor. In large, complex drawings the simplified symbol is used for the transistors to reduce drawing time. Once they are identified as to type, this poses no problem. The drain is supplied with -24 volts (V_{DD}), and if the proper bias is applied to the gates, the transistors will conduct. However, since this is a digital circuit, it does not use gate bias, as such, as might be expected. The signals applied to inputs A and B will determine whether or not the transistors conduct.

The overall function of such a circuit is that of a gate, and only when the input condition is correct will there be an output. Transistor $F1$ is the input gate and performs the AND/OR logic function. Transistor

*F*2 is simply an inverter. However, *F*3 here performs a function that is unlike any in typical bipolar applications. It is used in this case as a resistor, and its only function is to provide both current limitation and a voltage drop. This is, of course, what a more conventional resistor will do. By returning the gate lead to $-V_{DD}$, the transistor is *turned on* and its source-to-drain resistance is used as the collector load for *F*2. This, again, illustrates an advantage of MOSFET's used in digital applications. The resistors can be easily and simply formed at the same time as the FET's themselves.

The circuit action is actually rather simple. In terms of voltage levels, if inputs *A* and *B* are at ground (0 volts), the drain of *F*1 will also be at ground. Actually, the drain of *F*1 is at this time more or less floating since *F*1 is not conducting. At any rate, *F*2 is nonconducting since its gate is essentially at ground. If either input *A* or *B* is driven to a negative voltage, the gate of *F*2 will remain at ground and hence *F*2 will remain off. Only when inputs *A* AND *B* are at a negative voltage together will the drain of *F*1 go negative and so turn on *F*2. With *F*2 on, the drain will be driven toward ground and the output wire along with it.

In logic terms, the circuit action can be summarized as shown in the drawing of Fig. 16-52. With both inputs at a high level ($\cong$0 volts),

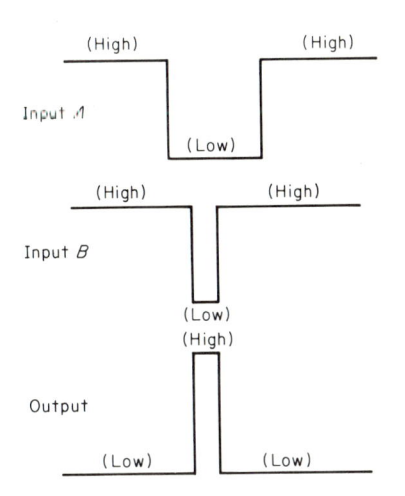

Fig. 16-52. Idealized waveforms for the MOSFET circuit in Fig. 16-51.

the output is at the low level (negative). When input *A* goes low, nothing occurs since input *B* is still high. However, when *B* goes low, at this time both *A* and *B* are low and the output goes high, as shown. Then, when *B* returns to high, the output goes back to low again.

$F1$, then, acts like a negative AND, positive OR gate and, when its output is inverted, becomes a NAND gate for negative logic and a NOR gate for positive logic, the symbols for which are shown in Fig. 16-51b.

A slightly different circuit is shown in Fig. 16-53. In this instance, if both A and B are high (ground), $F1$ and $F2$ are off and the gate of $F3$ is returned to V_{DD} through the drain resistor $F4$. Thus, $F3$ is on (gate very negative) and the output is essentially at ground. If A or B is driven negative, one of the transistors $F1$ or $F2$ is turned on and its drain falls to ground or nearly so. This removes the negative voltage from the gate of $F3$, and it turns off, allowing the drain to rise to V_{DD} through an external connection. Hence, a negative input at A OR B results in a negative at

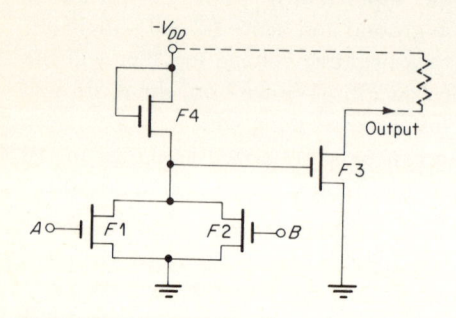

Fig. 16-53. AND/OR gate implemented by MOSFET's.

the output. The circuit is obviously working as a negative OR gate. If positive logic is being used, it functions as a positive AND gate. Hence, the circuit action in this case is relatively simple. Note that $F1$ (or $F2$) inverts the signal at its input and $F3$ reinverts it; thus the equivalent circuit action is no inversion at all.

A final example of MOSFET integrated circuitry is given in Fig. 16-54. This is a *JK* flip-flop that functions much as any similar circuit. The state of the flip-flop can be only one of two possible conditions: set or reset. If the flip-flop is set, there is a logic 1 at the Q_S output and a logic 0 at the Q_R output. If reset, a logic 1 appears at Q_R while a logic 0 appears at the Q_S output. For positive logic, ground is logic 1 and a large negative voltage (between -10 and -24 volts) is logic 0. To reset the flip-flop (assuming it is already set), both the K (clear) input and the clock pulse (*CP*) must be low or negative together. This will cause $F1$

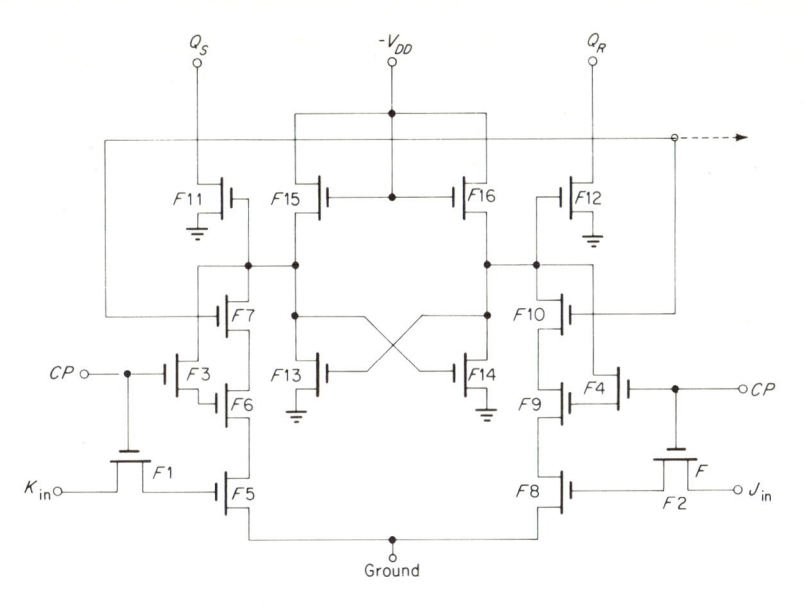

Fig. 16-54. *JK* flip-flop, MOSFET integrated circuit.

and $F3$ to conduct, which will, in turn, cause $F5$ and $F6$ to conduct. Transistors 5, 6, and 7 form a series AND gate, and hence if all three conduct, the drain of $F7$ is approximately at ground potential (if its gate is negative). This places ground at the gate of $F14$, which is therefore off. Its drain goes negative, thereby turning on $F13$, and the drain of $F13$ goes toward ground. With ground at the gate of $F11$, its drain goes toward $-V_{DD}$ (through an external return). Since a negative voltage has been defined as logic 0, the Q_S output is at logic 0. By inference, then, the Q_R output must be at logic 1, which is the reset condition.

To verify a logic 1 at the Q_R output, note that if the drain of $F14$ is negative because its gate is at ground, the gate of $F12$ is also negative. Thus, $F12$ is on and its drain must also be at ground (logic 1).

The transistors $F7$ and $F10$ are connected to an external lead, which, if it is negative, will allow the flip-flop to transfer. If this lead is at ground, the flip-flop will not respond to the normal inputs and will remain in its present state.

If the flip-flop is reset, it may be set by applying a negative voltage at the *J* input and *CP* at the same time. The corresponding action is the same, but, of course, the flip-flop will end up in the opposite state. If *both* the *J* and *K* inputs are low and then the *CP* pulse goes low, the flip-flop will toggle (reverse its state regardless of its original state).

QUESTIONS AND PROBLEMS

16-1 Briefly give your meaning of the word *monolithic*.

16-2 Briefly give your meaning of the word *epitaxial*.

16-3 Briefly describe the formation of a parasitic transistor.

16-4 Describe how a parasitic transistor can be rendered ineffectual.

16-5 Refer to Fig. 16-22a. The multiple-emitter transistor, considered alone, acts as a logical gate. If its collector is to be made the output, write the Boolean equation for logic 1 at this point for positive logic.

16-6 Refer to Question 16-5. Write the Boolean equation for the same conditions using negative logic.

16-7 Refer to Fig. 16-28. The circuit conditions are as follows: the flip-flop is reset; J is high and K is low; clear is high. Describe the circuit response to a clock pulse.

16-8 Refer to Fig. 16-28. The circuit conditions are as follows: the flip-flop is set; J is high and K is low; clear is high. Describe the circuit response to a clock pulse.

16-9 Refer to Fig. 16-37. The following inputs are standing at the inputs: present is low, clear is high; clock and D are high. Describe the condition of the flip-flop.

16-10 Refer to Fig. 16-43. Forty-three input pulses are applied to the $\overline{ADV\ A}$ input, after initially being reset. Write the equation for the logical state of the counter (reset = 00000 and LSD to the right).

APPENDIX: BOOLEAN ALGEBRA THEOREMS

THEOREM 1 $0 \cdot A = 0$

Because it is impossible for both inputs to be at logic 1 together, the output is always logic 0 and hence the gate is unnecessary.

THEOREM 2 $1 + A = 1$

Only one input is required to provide a logic 1 at the output, and therefore the output is always at logic 1 and the gate is unnecessary.

THEOREM 3 $1 \cdot A = A$

With one input leg always at logic 1:

1 If $A = 0$, $X = 0$
2 If $A = 1$, $X = 1$

Hence, the output is dependent only upon the state of A. The gate is not necessary.

THEOREM 4 $0 + A = A$

Since one input leg is always at logic 0, the output is in the same state as the A input. Again, the gate is unnecessary.

THEOREM 5 $A \cdot A = A$

Because both inputs are identical, the output follows either one and the gate is not necessary.

THEOREM 6 $A + A = A$

Because both inputs are identical, the output follows either one and the gate is not necessary.

THEOREM 7 $A \cdot \bar{A} = 0$

A logic 1 at A and $\bar{A}$ can never occur at the same time, and hence the output is always logic 0; the gate is unnecessary.

THEOREM 8 $A + \bar{A} = 1$

A logic 1 must appear at either A or $\bar{A}$, and thus the output is always at logic 1. The gate performs no logic function at all.

THEOREM 9 $AB = BA$

The order of writing does not affect the equation since the conditions must occur simultaneously.

THEOREM 10 $A + B = B + A$

The order of writing does not affect the equation since the conditions must occur simultaneously.

THEOREM 11 $ABC = (AB)C = A(BC)$

Inputs connected by the same logical sign may be grouped in any manner.

THEOREM 12 $A + B + C = A + (B + C) = (A + B) + C$

Inputs connected by the same logical sign may be grouped in any manner.

APPENDIX: BOOLEAN ALGEBRA THEOREMS 457

THEOREM 13 $\quad \overline{ABC} = \bar{A} + \bar{B} + \bar{C}$

A	B	C	X
0	0	0	1
1	0	0	1
0	1	0	1
1	1	0	1
0	0	1	1
1	0	1	1
0	1	1	1
1	1	1	0

$\overline{A \cdot B \cdot C} = \bar{A} \cdot \bar{B} \cdot \bar{C} = \bar{A} + \bar{B} + \bar{C}$

This is one of DeMorgan's theorems. The bar symbol inverts everything over which it is placed; thus

$$\overline{A \cdot B \cdot C} = \bar{A} \stackrel{-}{\cdot} \bar{B} \stackrel{-}{\cdot} \bar{C} = \bar{A} + \bar{B} + \bar{C}$$

THEOREM 14 $\quad \overline{A + B + C} = \bar{A}\bar{B}\bar{C}$

A	B	C	X
0	0	0	1
1	0	0	0
0	1	0	0
1	1	0	0
0	0	1	0
1	0	1	0
0	1	1	0
1	1	1	0

$\overline{A+B+C} = \overline{\bar{A}+\bar{B}+\bar{C}} = \bar{A} \cdot \bar{B} \cdot \bar{C} = \bar{A}\bar{B}\bar{C}$

This is the second one of DeMorgan's theorems. The bar symbol inverts everything over which it is placed; thus

$$\overline{A + B + C} = \bar{A} \mp \bar{B} \mp \bar{C} = \bar{A}\bar{B}\bar{C}$$

THEOREM 15 $\quad AB + AC = A(B + C)$

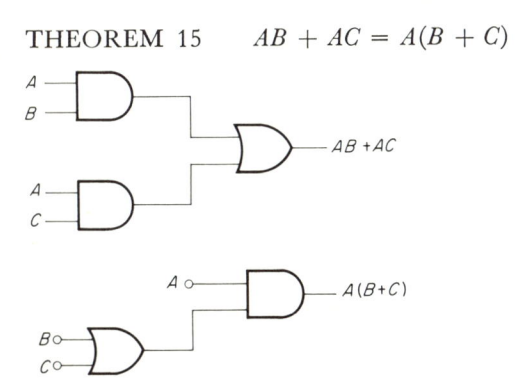

This is one of the distributive theorems. By algebraic factoring, $AB + AC = A(B + C)$; the result is a simpler although logically equivalent equation.

THEOREM 16 $\quad (A + B)(A + C) = A + BC$

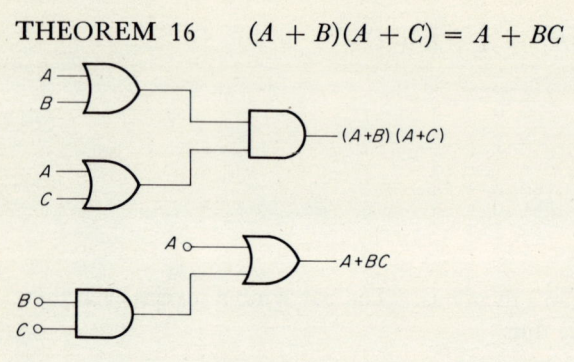

This is the other distributive theorem. It operates in reverse of Theorem 15 to provide a simpler, but equivalent, equation. The reduction is accomplished as follows:

$$(A + B)(A + C) = AA + AC + AB + BC = A + AC + AB + BC$$
$$= A + BC \quad \text{(see Theorems 19 and 20)}$$

THEOREM 17 $\quad AB + A\bar{B} = A$

Since only A has an influence on the output (B or $\bar{B} = 1$), only A is needed. This can be developed as shown:

$$AB + A\bar{B} = A(B + \bar{B}) = A \cdot 1 = A$$

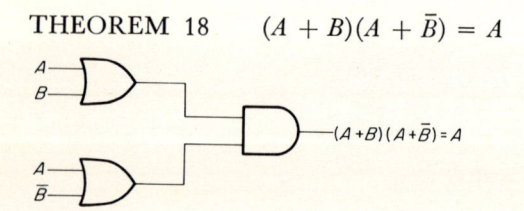

THEOREM 18 $\quad (A + B)(A + \bar{B}) = A$

Since B and $\bar{B}$ can never occur together, only A has an influence on the output:

$$(A + B)(A + \bar{B}) = AA + A\bar{B} + AB + B\bar{B}$$
$$= A\bar{B} + AB = A(B + \bar{B}) = A$$

THEOREM 19 $A + AB = A$

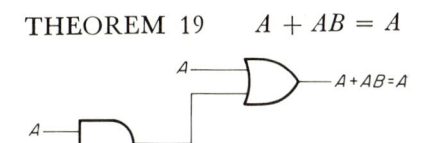

If B is to assist A in yielding an output, it must be ANDed with A. A, therefore, must be present in any case; hence B accomplishes nothing. A smaller term A, which is a part of a larger term AB, causes the larger term to be unnecessary.

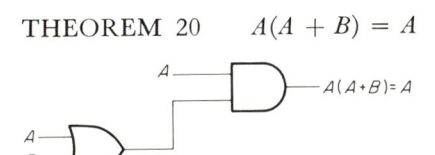

THEOREM 20 $A(A + B) = A$

Since $A(A + B) = AA + AB$, A must be present in any case. A smaller term A, which is a part of a larger term AB, causes the larger term to be unnecessary.

THEOREM 21 $A + \bar{A}B = A + B$

Since $A + \bar{A}B = (A + \bar{A})(A + B)$, the result can only be $A + B$ $(A + \bar{A} = 1)$.

THEOREM 22 $A(\bar{A} + B) = AB$

Since $A(\bar{A} + B) = A\bar{A} + AB$, the result can only be $A + B$ $(A\bar{A} = 0)$.

GLOSSARY

A_i	Circuit current gain
A_v	Circuit voltage gain
BCD	Binary-coded decimal
BV_{cbo}	Collector-base breakdown voltage (emitter open)
BV_{ebo}	Emitter-base breakdown voltage (collector open)
C_{ib}	Common-base open-circuit input capacitance
C_{ob}	Common-base open-circuit output capacitance
f_1	Low-frequency 3-db point on the response curve
f_2	High-frequency 3-db point on the response curve
$f_{\alpha co}$	Common-base cutoff frequency
h_{FE}	DC current gain (beta), base-to-collector
h_{fe}	Signal current gain, base-to-collector
I_B	DC base current
$I_{B,\text{sat}}$	Base current in saturation
I_C	DC collector current
$I_{C,\text{sat}}$	Collector current in saturation
I_{cbo}	Leakage current, transistor
I_{co}	Leakage current, diode
I_E	DC emitter current
I_D	Diode dc current
I_{th}	Thevenin-equivalent current
P_C	DC collector power dissipation
P_d	Pulse duty period
Q	Transistor
R_{BE}	Base-emitter resistance
R_{CB}	Collector-base resistance (V_{CB}/I_C)
R_{CE}	Collector-emitter resistance
$R_{CE,\text{sat}}$	Collector-emitter resistance in saturation
r_{df}	Diode dynamic forward resistance
r_e	Resistance of emitter material to a signal

R_{EE}	Ohmic, or bulk, resistance of the emitter material
R_F	Forward-diode dc resistance
r_i	Total input resistance (or impedance)
R_{ib}	Base-to-ground resistance
r_l	Total signal load
r_p	Vacuum-tube internal-plate resistance
R_{th}	Thevenin-equivalent resistance
R_Z	Zener-diode dc resistance
r_z	Zener dynamic resistance
V_B	Base-to-ground voltage
V_{BB}	Base-supply voltage
V_{BE}	Base-emitter voltage
V_C	Collector-ground voltage
V_{CB}	Collector-base voltage
V_{CC}	Collector-supply voltage
V_{CE}	Collector-emitter voltage
$V_{CE,\text{sat}}$	Collector-emitter saturation voltage
V_D	Diode dc voltage drop
V_E	Emitter-ground voltage
V_{EE}	Emitter-supply voltage
V_{th}	Thevenin-equivalent voltage
t_c	Time constant
t_d	Pulse duration time
t_f	Pulse fall time
t_{pd}	Pulse period for one complete cycle
t_r	Pulse rise time
t_s	Transistor storage time
ϵ	Epsilon (2.718)
μ	Vacuum-tube-amplification factor (also: micro-)
α	Alpha: emitter-to-collector current gain
β	Beta: base-to-collector current gain
γ	Gamma: ratio of V_{CE} to V_{CC}, or R_{CB} to $RL + R_{CB} + RE$
$+$	Logical function: OR
$\cdot$ or $\times$	Logical function: AND

BIBLIOGRAPHY

BARTEE, THOMAS C.: "Digital Computer Fundamentals," 2d ed., McGraw-Hill Book Company, New York, 1966.

BLITZER, RICHARD: "Basic Pulse Circuits," McGraw-Hill Book Company, New York, 1964.

BURROUGHS CORP.: "Digital Computer Principles," 2d ed., McGraw-Hill Book Company, New York, 1969.

COWLES, LAURENCE G.: "Transistor Circuits and Applications," Prentice-Hall, Inc., Englewood Cliffs, N.J., 1968.

GILLIE, ANGELO C.: "Binary Arithmetic and Boolean Algebra," McGraw-Hill Book Company, New York, 1965.

GILLIE, ANGELO C.: "Pulse and Logic Circuits," McGraw-Hill Book Company, New York, 1968.

HIBBERD, ROBERT G.: "Integrated Circuits," McGraw-Hill Book Company, New York, 1969.

KETCHUM, DONALD J., and E. CHARLES ALVAREZ: "Pulse and Switching Circuits," McGraw-Hill Book Company, New York, 1965.

MARCUS, MITCHELL P.: "Switching Circuits for Engineers," Prentice-Hall, Inc., Englewood Cliffs, N.J., 1962, 1967.

MILLMAN, JACOB, and HERBERT TAUB: "Pulse and Digital Circuits," McGraw-Hill Book Company, New York, 1956.

MILLMAN, JACOB, and HERBERT TAUB: "Pulse, Digital, and Switching Waveforms—Devices and Circuits," McGraw-Hill Book Company, New York, 1965.

PRESSMAN, ABRAHAM I.: "Design of Transistorized Circuits for Digital Computers," John F. Rider, Publisher, New York, 1959.

VEATCH, HENRY C.: "Transistor Circuit Action," McGraw-Hill Book Company, New York, 1968.

ANSWERS
TO ODD-NUMBERED QUESTIONS AND PROBLEMS

1-1 5 μsec
1-3 2000 Hz, or 2 kHz
1-5 10%
1-7 0.35 μsec
1-9 10 MHz

2-1 (a) 3 ohms
(b) 1 amp
(c) 3 volts
2-3 3 ohms
2-5 1.67 kilohms
2-7 4.0 volts
2-9 −2.52 volts

3-1 0.35 watt
3-3 20 ma
3-5 90 msec ($T = 5RC$)
3-7 63.2 volts
3-9 0.426 μf

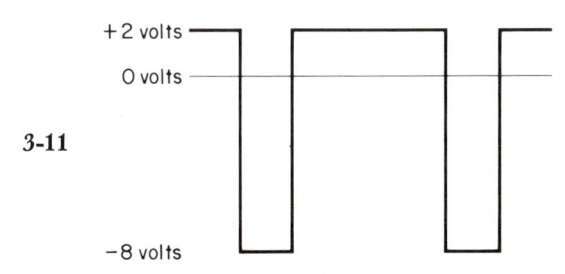

3-11

+2 volts
0 volts
−8 volts

3-13 (a) 11.11 pf
(b) 250 pf

4-1 No current

4-3 Voltage drop $A = -$, $B = +$
Induced voltage $A = +$, $B = -$
4-5 c
4-7 b
4-9 c
4-11 1 volt

5-1 True
5-3 True
5-5 False
5-7 True
5-9 False
5-11 False

6-1 $I_D = 0.012$ amp
6-3 (a) $I_D = 0.0117$ amp
(b) $R_F = 25.6$ ohms
6-5 $R_R = 10^7$ ohms
6-7 $r_{df} = 8.67$ ohms
6-9 $R_Z = 667$ ohms
6-11 $r_z = 10$ ohms

7-1 19
7-3 32.3
7-5 0.995
7-7 Low
7-9 Saturation

8-1 -6 volts
8-3 0.61 volt
8-5 6.1 ohms
8-7 CB: Current gain less than 1; voltage gain large; power gain moderate; input impedance very low; output impedance moderately high; very good at high frequencies
CE: Current gain high; voltage gain very high; power gain high; input impedance low; output impedance moderately high; poor at high frequencies
CC: Current gain high; voltage gain less than 1; power gain moderate; input impedance very high; output impedance very low; good at high frequencies
8-9 (a) 5 volts
(b) 0 ($\cong 0.1$ volt)
(c) 5 ma
(d) 0

8-11 (a) 6 volts
(b) 0
(c) −6 volts
(d) 0

9-1 50 ohms
9-3 4000 ohms
9-5 t_s (storage time, or minority-carrier storage)
9-7 $I_{C,\max} = 12$ ma
9-9 $e_{in} = -1.03$ volts
9-11 c
9-13 d
9-15 −2.18 volts
9-17 Q1 is off
9-19 The load resistors RL and RE form a voltage divider, which, when the transistor is in saturation, sets certain limits beyond which the collector and emitter cannot go.

10-1 On, or saturated
10-3 −6.3 volts
10-5 −12 volts
10-7 120 ohms
10-9 100 ohms
10-11 −2.835 volts
10-13 5.25 volts
10-15 −3.33 volts
10-17 3.61 volts

11-1 11011
11-3 1000011001
11-5 (a) 11111 (31)
(b) 1010010 (82)
11-7 (a) 00111
(b) 11100
(c) 10000

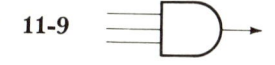

11-9

11-11

11-13

11-15

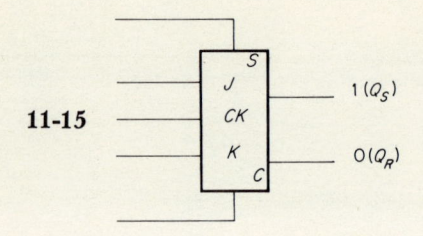

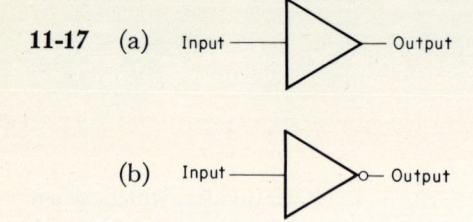

[Figure: JK flip-flop with inputs S, J, CK, K, C and outputs 1(Q_S), 0(Q_R)]

11-17 (a) Input ——▷—— Output

(b) Input ——▷∘—— Output

11-19 $AB + C + DE$
11-21 $\bar{A}\bar{B} + CD + E$
11-23 $(\bar{A} + \bar{B})(A + B)(\bar{A} + B)$
11-25 b
11-27 $BC\bar{D} + \bar{B}\bar{C}D$
11-29 $CDEF + BDEF$
  Also $DEF(B + C)$
11-31 $\bar{A}BD + \bar{A}B\bar{C} + \bar{A}\bar{C}\bar{D} = \bar{A}(BD + B\bar{C} + \bar{C}\bar{D})$
11-33 $A + \bar{B} + \bar{C} + \bar{D} + E = (A + \overline{BCD} + E)$

12-1 The reapplication of an output signal back to the input in amplified form and in phase with the original input signal.
12-3 If beta is less than approximately 33, the transistors are quiescently *in the active region* and hence oscillations must start when power is first applied.
12-5 Symmetrical
12-7 45.8 μsec
12-9 The circuit will fail to self-start since both transistors could then saturate together (loop gain less than 1). The negative limit of V_B would be excessive.
12-11 Q1 remains on and Q2 remains off.
12-13 The T input serves to provide a complementing flip-flop action.
12-15 0.833 msec (833 μsec)
12-17 Q2 is on.

ANSWERS **467**

13-1 8
13-3 128
13-5 4056
13-7 8
13-9 5
13-11 No. 6
13-13 No. 11
13-15 16
13-17 $\cong +18$ volts
13-19 6

14-1 $AB\bar{C}$
14-3 $A_R B_R C_S$
14-5
$$A,\bar{B},C,D \rightarrow \text{NAND} \rightarrow \text{INV} \rightarrow A\bar{B}CD$$
14-7
$$\bar{A},B,C,\bar{D} \rightarrow \text{NAND} \rightarrow A\bar{B}CD$$
14-9 1,000,000

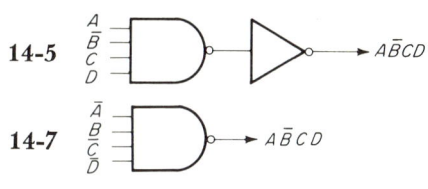

15-1 Decoupling filter
15-3 Threshold voltage
15-5 To allow the collectors to utilize a greater portion of the available 20-volt supply
15-7 The coupling resistor, by which the signal is coupled from Q2 to Q3. It also serves as the emitter resistor RE for signal conditions.
15-9 Local constant-voltage supply
15-11 Both slopes are perfectly straight (linear).
15-13 To protect $Q1$
15-15 $Q2$ is continually on, and hence no output from $Q3$ is evident.
15-17 False
15-19 False

16-1 Single-stone
16-3 Parasitic transistors are formed because the epitaxial layer is usually doped p, which together with the collector and base of the true transistor (npn), form a pnp transistor.
16-5 $\overline{A+B+C+D+E+F+G+H} = \bar{A}\bar{B}\bar{C}\bar{D}\bar{E}\bar{F}\bar{G}\bar{H}$.
16-7 The flip-flop will set.
16-9 $Q_S = 1; Q_R = 0$

INDEX

Adder:
  full, 240
  half, 238
Amplifier:
  bandpass, 9
  video, 5
  wideband, 5
AND gate, 204
Asymmetrical waveform, 273
Avalanche mode, 151

Backward diode, 404
Barkhausen criterion, 260
Barrier region, 80
Base:
  of number system, 193
  of transistor, 100
Binary arithmetic, 197
Binary code, 200
Binary-coded decimal (BCD), 200
Binary notation, 194–199
Binary point, 195
Biquinary code, 201
Boolean algebra laws:
  associative, 216
  commutative, 215
  DeMorgan's theorems, 218
  distributive, 216
  law of NOT, 217
BV_{cbo}, 108
BV_{ebo}, 108

Capacitor:
  charge equation, 35
  discharge equation, 36
Carriers:
  majority, 78
  minority, 78
Channel (FET), 392
C_{ib}, 108
Clamping diodes, 285
C_{ob}, 108
Common-load gates, 253–256
Constant-current diodes, 405
Counters:
  binary, 309
  decimal, 319
  decoding of, 312–316
  down-, 330
  hexidecimal, 317
  permuted, 317
  ring, 327
  scale-of-n, 315
  scale-of-16, 310
  up-, 328
Current-mode circuits, 174
Current-mode switching, 150

Dc restorer, 48
Decimal code, 199, 200
Decimal notation, 192–194
Decision elements (*see* AND gate; OR gate)

INDEX 469

Decoder, 340
Depletion region, 80
Differentiator, 51
Diode, 78–98
  applications, 93–98
  avalanche, 90
  characteristic curve, 88
  constant-current, 405
  equivalent circuit, 89
  junction, 79
  PIV, 89
  zener, 85
Diode gates, 243
Diode-transistor gates (DTL), 247
Duration, pulse, 2
Duration time, 2
Duty cycle (period), 2

Electric inverter, 172
Emitter follower, 145, 184
Encoder, 349
Equivalent circuit:
  Norten's, 14
  Thevenin's, 11
  vacuum-tube, 12
Excess—3 code, 202

Fall time, 2
Fan out, 167
Feedback, 260
Filter:
  high-pass, 31
  low-pass, 32
Flip-flop, 274
  basic circuit action, 274–279
  triggering, 279
Four-layer diode, 400
Fundamental frequency, 7

Gray code, 201

Harmonic frequency, 7
h_{fe}, 104, 108
h_{FE} (β), 102, 108
Hysteresis of Schmitt trigger, 299

IGFET, 393
Induced voltage, 68
Inductive load, 152
Integrated circuits, 409–453
  compatible, 411
  D-type flip-flop, 438
  DTL gates, 428
  hybrid, 411
  JK flip-flop, 432

Integrated circuits:
  LSI, 441
  monolythic, 411–425
  MOS (metal-oxide semiconductors), 445
  MSI, 441
  thin-film, 411
  TTL gates, 425
  yield, 410
Integrated squarewave, 31, 54
Internal impedance, 11
Inverter, 131, 207
Ions, 80

JFET, 393
Johnson (creeping) code, 200
Joules (watt-seconds), 25
Junction diode, 79

Karnaugh maps, 220

Leakage current:
  diode, 81
  transistor, 108, 112–113, 142–143
Limits, circuit, 122–125
Linear pulse amplifier, 352
  biased-up, 361
  differential, 357
  push-pull, 357
Load line, 117
Logic simplification, 225
Logic symbols, 202–214

Matrix:
  decoding, 342
  encoding, 349
Microelectronics, 409
Mixed transistor types, 171
Multivibrators, 259–306
  astable, 261
    nonsymmetrical, 273
    self-starting, 266
  bistable, 274
    divide-by-2, 287
    triggering, 279
  logical notation, 209
  monostable, 289
    triggering, 294
    waveforms, 291
  Schmitt trigger, 296
  Q-point, 302

NAND/NOR gates, 207, 425–431
Neon bulb ocsillator, 373
Nonsaturating circuit, 147, 150
Number codes, 199–202

Octal code, 201
Offset voltage, 3
OR gate, 206
Output resistance, emitter-follower, 189
Overtone, 7

Parasitic transistor, 421
Passivation, 410
Peak amplitude, 3
Pulse:
  definition of, 1
  droop, 39
  sag, 39
Pulse amplifiers, 352–372
Pulse amplitude, 2
Pulse duty period, 2
Pulse jitter, 5

Radix, 193
RC circuit:
  coupling, 37
  differentiating, 51
  integrating, 54
Regeneration, 260
Repetition rate, 2
Resistance, transistor: R_{BE}, 105
  R_{CB}, 104
  R_{CE}, 105
  R_{EE}, 104
Resistor-capacitor gates, 249
Resistor-transistor gates, 248
Reverse current:
  I_{cbo}, 108
  (*See also* Leakage current)
  I_{co}, 82
  I_{ebo}, 108
Rise time, 1

Saturated mode, 150
Schmitt trigger, 296
SCR, 396
SCS, 399
Shift register, 319
  dual-T input, 336

Shift register:
  left-shift, 323
  right-shift, 325
Squaring circuit (*see* **Schmitt trigger**)
Substrate (FET), 392
  IC, 412
Sync separator, 54

Td (delay time), 139
Tf (all time), 142
Theorems, circuit: **Millman's**, 20
  Norten's, 14
  superposition, 16
  Thevenin's, 11
Threshold voltage, 165
Thyristor, 399
Time constant (t_c), 28, 33, 69
Tr (rise time), 139
Transformer coupling, 70, 383
Transformers, pulse, 70
  pot-core, 71
  ribbon-core, 71
Transistor:
  alpha, 103
  beta, 102
  characteristic curves, 110–117
  equivalent circuits, 103
  power dissipation, 143
  switch, 119, 127
Transistor amplifiers:
  CB, 182
  CC, 184
  CE, 156
Triac, 401
Ts (storage time), 140
Tunnel diode, 403

Unijunction transistor, 388

Virtual ground, 362
Voltage:
  base-emitter, 108
  collector-base, 108
  collector-emitter, 108